PROGRESS IN BRAIN RESEARCH

VOLUME 41

INTEGRATIVE HYPOTHALAMIC ACTIVITY

PROGRESS IN BRAIN RESEARCH

PROGRESS IN BRAIN RESEARCH
VOLUME 41

INTEGRATIVE HYPOTHALAMIC ACTIVITY

EDITED BY

D. F. SWAAB

AND

J. P. SCHADÉ

Central Institute for Brain Research,
IJdijk 28, Amsterdam (The Netherlands)

ELSEVIER SCIENTIFIC PUBLISHING COMPANY
AMSTERDAM/OXFORD/NEW YORK
1974

ELSEVIER SCIENTIFIC PUBLISHING COMPANY
335 JAN VAN GALENSTRAAT
P.O. BOX 211, AMSTERDAM, THE NETHERLANDS

AMERICAN ELSEVIER PUBLISHING COMPANY, INC.
52 VANDERBILT AVENUE
NEW YORK, NEW YORK 10017

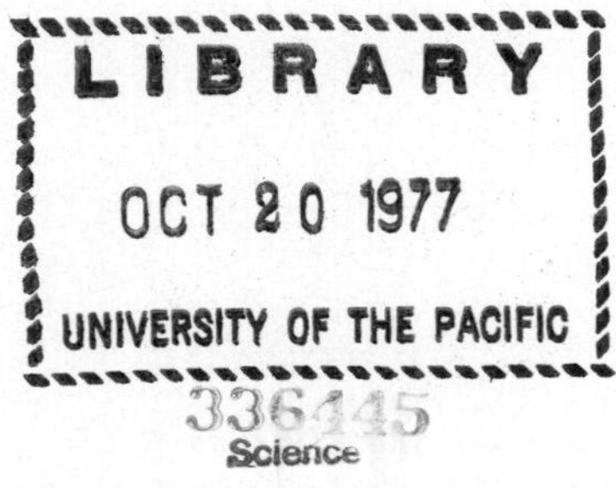

LIBRARY OF CONGRESS CARD NUMBER: 74-83317

ISBN 0-444-41239-5

WITH 179 ILLUSTRATIONS AND 29 TABLES

PRINTED IN THE NETHERLANDS

List of Contributors

E. Anderson, NASA Ames Research Center, Moffett Field, Calif. 94035, U.S.A.

J. Ariëns Kappers, Netherlands Central Institute for Brain Research, IJdijk 28, Amsterdam, The Netherlands.

C. A. Baile, Smith, Kline and French Laboratories, West Chester, Philadelphia, Pa., U.S.A.

G. J. Boer, Netherlands Central Institute for Brain Research, IJdijk 28, Amsterdam, The Netherlands.

K. Boer, Netherlands Central Institute for Brain Research, IJdijk 28, Amsterdam, The Netherlands.

B. Bohus, Rudolf Magnus Institute for Pharmacology, Medical Faculty, University of Utrecht, Vondellaan 6, Utrecht, The Netherlands.

P. F. Brain, Department of Zoology, University College of Swansea, Swansea, SA2 8PP, Wales, Great Britain.

J. A. M. De Rooij, Department of Experimental Neurology, Radboud University, Nijmegen, The Netherlands.

L. De Ruiter, Department of Zoology, The University, Groningen, The Netherlands.

R. A. C. De Vries, Netherlands Central Institute for Brain Research, IJdijk 28, Amsterdam, The Netherlands.

D. De Wied, Rudolf Magnus Institute for Pharmacology, Medical Faculty, University of Utrecht, Vondellaan 6, Utrecht, The Netherlands.

B. T. Donovan, Department of Physiology, Institute of Psychiatry, De Crespigny Park, London, SE5 8AF, Great Britain.

G. Dörner, Institute of Experimental Endocrinology, Humboldt-University, Berlin, G.F.R.

J-P. Dupouy, Laboratory of Comparative Physiology, 9 quai Saint-Bernard, Paris 5e, France.

R. G. Dyer, Department of Anatomy, University of Bristol, Bristol, BS8 1TD, Great Britain.

C. M. Evans, Department of Zoology, University College of Swansea, Swansea, SA2 8PP, Wales, Great Britain.

J. M. Forbes, Department of Animal Physiology and Nutrition, Kirkstall Laboratories, Vicarage Terrace, Leeds, LS5 3HL, Great Britain.

R. Guillemin, The Salk Institute, La Jolla, Calif. U.S.A.

W. Haymaker, NASA Ames Research Center, Moffett Field, Calif. 94035, U.S.A.

J. Herbert, Department of Anatomy, University of Cambridge, Cambridge, CB2 3DY, Great Britain.

O. R. Hommes, Department of Experimental Neurology, Radboud University, Nijmegen, The Netherlands.

W. J. Honnebier, Department of Obstetrics and Gynecology, Wilhelmina Gasthuis, Amsterdam, The Netherlands.

H. G. Jansen, Department of Anatomy, Erasmus University, Rotterdam, The Netherlands.

A. Jost, Laboratory of Comparative Physiology, 9 quai Saint-Bernard, Paris 5e, France.

U. Jürgens, Max-Planck-Institute for Psychiatry, Kraepelinstr. 2, Munich 40, G.F.R.

A. J. Kreike, Department of Anatomy, Erasmus University, Rotterdam, The Netherlands.

H. J. Lammers, Department of Anatomy, Radboud University, Nijmegen, The Netherlands.

S. Levine, Department of Psychiatry, Stanford University School of Medicine, Stanford, Calif. 94305, U.S.A.

D. W. Lincoln, Department of Anatomy, The Medical School, University Walk, Bristol, BS8 1TD, Great Britain.

A. H. M. Lohman, Department of Anatomy, Radboud University, Nijmegen, The Netherlands.

B. G. Monroe, Department of Anatomy, University of Southern California, Los Angeles, Calif. 90033, U.S.A.

W. K. Paull, Department of Anatomy, University of Southern California, Los Angeles, Calif. 90033, U.S.A.

M. J. Peddie, Department of Physiology, Institute of Psychiatry, De Crespigny Park, London, SE5 8AF, Great Britain.

C. Pilgrim, Faculty of Morphology and Anatomy, Fachbereich Biologie, Universität Regensburg, Postfach 84, Regensburg, G.F.R.

A. E. Poole, Department of Zoology, University College of Swansea, Swansea, SA2 8PP, Wales, Great Britain.

M. Rieutort, Laboratory of Comparative Physiology, 9 quai Saint-Bernard, Paris 5e, France.

D. F. SALAMAN, Department of Anatomy, The Medical School, University of Bristol, Bristol, BS8 1TD, Great Britain.

J. L. SLANGEN, Rudolf Magnus Institute for Pharmacology, University of Utrecht, Medical Faculty, Vondellaan 6, Utrecht, The Netherlands.

A. R. SMITH, Netherlands Central Institute for Brain Research, IJdijk 28, Amsterdam, The Netherlands.

J. M. STERN, Department of Psychiatry, Stanford University School of Medicine, Stanford, Calif. 94305, U.S.A.

R. STOECKART, Department of Anatomy, Erasmus University, Rotterdam, The Netherlands.

D. F. SWAAB, Netherlands Central Institute for Brain Research, IJdijk 28, Amsterdam, The Netherlands.

H. VAN DIS, Netherlands Central Institute for Brain Research, IJdijk 28, Amsterdam, The Netherlands.

N. E. VAN DE POLL, Netherlands Central Institute for Brain Research, IJdijk 28, Amsterdam, The Netherlands.

P. VAN DER SCHOOT, Faculty of Medicine, Erasmus University, Rotterdam, The Netherlands.

TJ. B. VAN WIMERSMA GREIDANUS, Rudolf Magnus Institute for Pharmacology, Medical Faculty, University of Utrecht, Vondellaan 6, Utrecht, The Netherlands.

J. G. VEENING, Department of Zoology, The University, Groningen, The Netherlands.

M. J. WAYNER, Brain Research Laboratory, Syracuse University, 601 University Avenue, Syracuse, N.Y. 13210, U.S.A.

P. R. WIEPKEMA, Department of Zoology, The University, Groningen, The Netherlands.

Preface

This volume contains the proceedings of the VIIIth International Summer School of Brain Research which has been organized by the Netherlands Central Institute for Brain Research, Amsterdam. For this conference the topic "Integrative Hypothalamic Activity" was chosen in order to stress the special properties which the hypothalamus has as one of the most important integration centres of the brain, rather than organizing lectures on different and separate subjects related to this part of the diencephalon. The interest of a number of research workers at the Institute and recent work performed by them on the hypothalamus have also been of importance in shaping the programme. This is the reason why special emphasis has been laid on the role played by the hypothalamus in sexual differentiation, in the control of parturition and in several aspects of behaviour while also the influence exerted by the pineal gland on the hypothalamo–hypophyseo–gonadal axis and on the magnocellular hypothalamic neurosecretory system is dealt with. As it was realized that structural principles lie at the base of physiological phenomena, the programme, after an introductory paper on the history of hypothalamic research, started with lectures on hypothalamic morphology.

As our knowledge of hypothalamic morphology, histo- and biochemistry and the physiological implications of the many extrinsic and intrinsic neural and hormonal hypothalamic pathways is vastly growing, it seemed right to summarize and integrate at least part of the many results obtained for an auditorium of young research workers for whom these Summer Schools are intended.

My thanks are due to all participants contributing so many valuable papers and taking such a lively part in the discussions which have been very fruitful. I am also most grateful for the invaluable help of Drs. Swaab and Schadé, the editors of this volume, in constituting the programme and in organizing this Summer School and to Miss J. Sels for her editorial assistance.

J. Ariëns Kappers
Central Institute for Brain Research
Amsterdam, The Netherlands

Contents

Breakthroughs in Hypothalamic and Pituitary Research

EVELYN ANDERSON AND WEBB HAYMAKER

NASA Ames Research Center, Moffett Field, Calif. 94035(U.S.A.)

INTRODUCTION

Sir Henry Dale (1957), in speaking of his student days, tells of learning physiology from Michael Foster's *Textbook of Physiology*, the 1891 edition. He says that the textbook was "curtly explicit in its account of the knowledge then available concerning the function of the pituitary body", for Foster wrote that "with regard to the purposes of the organ as a whole we know absolutely nothing". Dale goes on to say that this "is a statement which you may well imagine was not without an aspect of reassurance for a student with an examination in prospect". Today's students do not have that reassurance. In the past half century our knowledge, not only of the pituitary but also the hypothalamus, has undergone phenomenal growth and to attempt to grasp its significance is enough to boggle the mind of the specialist as well as the student; nor is it easy to single out the anatomists, physiologists and clinicians who over the years have been responsible for this tremendous growth and to say that these were the founders or the leaders. We must wait for Time to grade them as to the greatness of their contributions.

In accepting this assignment we were made bold by a story about Magendie which Olmsted (1938) relates in his biography of Claude Bernard regarding certain attitudes of Magendie which "forebade him to attempt to reconcile contradictory results. If he obtained one result in 1892 and another in 1893 it was all the same to him. He trusted future experiments to provide the missing explanation. Far from being ashamed of his complete lack of method in the conduct of his research, he was proud of it. He told Bernard: Everyone compares himself to something more or less majestic in his own sphere, to Archimedes, Michaelangelo, Galileo, Descartes, and so on. Louis XIV compared himself to the Sun: I am much more humble. I compare myself to a scavenger; with my hook in my hand and my pack on my back I go about the domain of science picking up what I can find" (p. 25).

We have gone through the pages of the history of the hypothalamus and the pituitary, picking up some of the great findings, which you, at this meeting, will undertake the task of integrating into a complete story. Of the many contributors, some have led the way in major breakthroughs in hypothalamic and pituitary research which have permitted others to carry through to the development of this knowledge. It is to these

References p. 52–60

Fig. 1. François Magendie (1783–1855). By courtesy of the Bibliothèque de l'Académie Nationale de Médecine cliché Assistance Publique (No. 85-1666, Médecins).

contributors whose breakthroughs we so often take for granted that we wish to pay our respects in this discussion.

THE RETE MIRABILE AND THE PITUITARY

The pituitary and its infundibulum appeared on the horizon, centuries before the hypothalamus came to be known. Galen (129(?)–199 A.D.) of Pergamon was the first to recognize their existence and to comment on their function. According to Sarton (1954), "Galen's mind was quick and tidy, but clouded by an excessive fondness for theory and classification. ... He was honest and sincere but egotistic, vain, complacent, irritable, and jealous" (p. 79).

Many of Galen's studies were based on dissection of the brains of oxen, goats, swine and monkeys (*Macaca inua* and *Macaca mulatta*), in which he described not only the pituitary and the infundibulum but also the 4 chambers (ventricles) of the brain, the fornix, the corpus callosum, the pineal, and the corpora quadrigemina. Among the chambers of the brain were the two lateral ones, which he called the "thalamus" (literally, "bridal chamber"). To Galen the chambers at the base of the brain supplied "animal spirit" to the optic nerves — hence, *thalami nervorum opticorum*,

Fig. 2. Claudius Galen (129(?)–199 A. D.). Portrait by courtesy of the Biomedical Library, Univ. of California at Los Angeles.

and its English equivalent, "optic thalamus", a term that still lingers with us (Fulton, 1943, p. 252). Galen must have seen the hypothalamus many times but found nothing remarkable to say about it.

Galen described an exceedingly prominent network of blood vessels surrounding the pituitary, which came to be called the *rete mirabile* when translated from Greek into Latin. According to Clarke and O'Malley (1968), Galen considered that "the vital spirit in the blood from the heart was changed into animal spirit in the *rete mirabile* and then, as mediator of all the brain's activities, stored in the ventricles" (p. 758). The *rete mirabile* seems to have been of central importance to Galen. He states that "the plexus that embraces the (pituitary) gland itself and extends for a great distance posteriorly is the most remarkable of bodies found in this region" (p. 758). Kuhlenbeck (1973) reminds us that such *retia mirabilia* are found in various mammals, including Edentata, Ungulata, Carnivora, Sirenia, and Cetacea, and that they are analogous to the (arterial) renal glomeruli and the (venous) portal sinusoids. O'Malley (1964) remarks that Galen's acceptance of the *rete mirabile* as a structure in the human

References p. 52–60

brain was clearly in error (p. 9). To us, on the other hand, Galen's *rete mirabile* coincides with the diminutive aggregate of vessels which Duvernoy (1972) referred to as a "surface capillary network", as a glance at the India-ink perfusion preparations published by him and by other anatomists will testify. Although in man this "rete" has been greatly reduced in magnitude, in our opinion it nonetheless qualifies in its own right as a *rete mirabile*. In short, in a sense we are in agreement with Galen on this point.

Galen's influence dominated medicine for well over a thousand years, until interrupted by Vesalius (1514–1564) with the publication of his great treatise on anatomy, *De Humani Corporis Fabrica*, in Basle in 1543. At the time, Vesalius was only 27 years of age. Singer (1946, 1957) deduces that "Vesalius was a strong, resolute man of clear, firm-knit, and unsubtle mind ...", and that he was a "noisy, bustling, exhibitionist genius".

Vesalius is said to have had the services of the artist, Jan Stefan van Kalkar, who had been a pupil of Titian. However, as is brought out by Saunders and O'Malley (1950), "There is no more contentious and difficult subject respecting the Vesalian

Fig. 3. Andreas Vesalius (1514–1564). Portrait by courtesy of the National Library of Medicine, Bethesda, Md.

problem than the question of the identity of the artist or artists responsible for the Vesalian illustrations" (p. 25). Vesalius' accurate descriptions and the artists' (including Titian's?) documentation of the dissections by over 300 engravings and woodcuts made the *Fabrica* a breakthrough of greatest importance. "With Vesalius", according to Garrison, "the anatomy of the brain became modern with one bound" (see McHenry, 1969, p. 40).

Vesalius was born in Brussels, the son of the imperial court apothecary. Attracted at an early age to the study of anatomy, he meticulously dissected all the animals he could get hold of. He went to Paris to take up the study of medicine, particularly with a view to mastering human anatomy. But he learned little during his student days. One of his professors of anatomy, Winter von Andernach, was a learned student of classical literature who was the first to translate Galen's chief anatomical treatise from Greek into Latin. Although he taught anatomy, this professor had never done a dissection. So Vesalius had to learn for himself. He again dissected animals, and from time to time obtained human bones from cemeteries or he went to places of execution in search of material. He would study bones until he could identify them with closed eyes. His zeal brought him to the attention of his professors, who asked him to undertake the job of dissection. For the first time, Sigerist (1958) relates, he stood with scalpel in hand before a human corpse in front of an audience (p. 106). The outbreak of war made him leave Paris (in 1536).

It was at Padua (starting in 1537) that Vesalius' preparation of the *Fabrica* was undertaken. He was then only 22, and had been made professor at the University. Consistent with the detestable literary manners current at that time, Vesalius mercilessly attacked Galen. In a "childishly irritable and flatulently abusive" way, he pointed to Galen's many errors, and drummed away at the fact that Galen had never dissected a human body. At first he accepted Galen's *rete mirabile* as an anatomical fact, but later denied its existence (Clarke and O'Malley, 1968, p. 768), a point on which, in our opinion, Vesalius was correct so far as gross appearances went but otherwise was clearly in error. Vesalius disagreed with Galen's view that excretory ducts ran from the lateral ventricles to the cribra ethmoidalis, but accepted the view that the brain excreted waste material *(pituita)* through the infundibulum into the pituitary, whence it somehow passed into the nasopharynx (Rolleston, 1936, p. 42). Vesalius described this pituita (as interpreted by O'Malley, 1964, p. 9) as residue left from the ultrarefinement of animal spirit (the substance responsible for sensation and motion) that had reached the brain in the form of vital spirit from the heart.

One hundred years after Vesalius — in 1664 — Thomas Willis (1621–1675) published his memorable treatise, *Cerebri Anatome*, which far surpassed anything that had been published up to that time. In addition to anatomy, this treatise dealt with functions of the nervous system. This and the six other books he wrote marked the transition between medieval and modern notions of brain structure and function.

A man of no carriage and little discourse though very congenial, Willis had the most fashionable and lucrative medical practice in London. According to Sir Michael Foster (1901), "love of truth was in him less potent than love of fame" (p. 270). Certain other critics said he was neither erudite nor original and that much of his

Fig. 4. Thomas Willis (1621–1675). Portrait by courtesy of the Biomedical Library, Univ. of California at Los Angeles.

success was attributable to the fertile mind of Richard Lower (1631–1691), who was assistant to Willis for at least 10 years. From his observations on the formation and circulation of the cerebrospinal fluid, and from a study of the mechanism of hydrocephalus, Richard Lower, in 1672, finally convinced Willis that nasal secretions do not filter down from the pituitary. In this view, Lower was in agreement with Conrad Schneider (1614–1680) of Wittenberg who had shown in 1655 that these secretions were the product of small glands in the nasal mucosa (see Mettler, 1947, p. 65).

Mention of Willis naturally leads to the arterial circle at the base of the brain. Others before Willis were aware of that circle. Among them was Gabriel Fallopius, a pupil of Vesalius, who mentioned it in 1561. Moreover, in 1632, which was 30 years prior to the publication of Willis' *Cerebri Anatome*, Guilio Casserio (1545–1605), a professor at Padua and one of Harvey's teachers, published an illustration of the base of the brain and its embracing arterial circle. Christopher Wren (1632–1723) was responsible for naming the circle after Willis. Wren, at the time a student at Oxford, was employed by Willis to prepare the anatomical plates of *Cerebri Anatome*, and in his sketch, inserted the caption, "circle of Willis". Willis had no reason to protest, for he had experimented extensively on the circle by squirting "a liquor dyed with ink" into peripheral cerebral vessels. Noting the appearance of dark spots on the cut surface of the brain, he concluded that the ink reached fine vessels in the brain substance. Willis went on to say that large vessels ("the fourfold chariot") running to the arterial

Fig. 5. Richard Lower (1631–1691). Portrait from the Wellcome Institute of the History of Medicine, London, England, by courtesy of the Trustees.

circle at the base of the brain, could, by way of their "mutual conjoinings", ... supply or fill the channels of all the rest of the cerebral vasculature, and thus prevent apoplexy. More than that, he found that some vestige of the ink reached vessels covering the base of the brain (Clarke and O'Malley, 1968, p. 778), bringing to better view the "surface capillary network", which as noted earlier, could well be a small rendition of Galen's *rete mirabile*.

INTIMATIONS OF HYPOTHALAMIC FUNCTION

Some 200 years after Willis and Lower — up to about 1875 — the hypothalamus as such was still unknown and the pituitary was considered inconsequential. Lo Monaco and van Rynberk (sic!), in two papers published in 1901, reminded their readers that the pituitary had long been regarded as an unpaired sympathetic ganglion, and that Sylvius and Magendie thought it a lymphatic gland having the function of collecting

References p. 52–60

Fig. 6. Gérard A. van Rijnberk (1875–1953). Portrait by courtesy of Professor Paul E. Voorhoeve, Amsterdam.

cerebrospinal fluid and discharging it into the circle of Willis. In their rebuttal to these views, based on extensive experimental work, Lo Monaco and van Rynberk concluded that the pituitary was an involuted organ that had "no important function, neither general nor special". This publication originated from the laboratory of no less an authority than Luigi Luciani, Professor of Physiology and Rector of the University of Rome, where van Rynberk had gone to prepare his doctor's thesis. (For a brief biography of the distinguished van Rynberk, see Duyff, 1947.)

Following suit in 1908, Schäfer and Herring referred to the anterior lobe of the pituitary as not having any physiological effect. In 1909 the pituitary was defined in Murray's *New English Dictionary on Historical Principles, Vol. 7*, as "a small bilobed body of unknown function attached to the infundibulum at the base of the brain" (see Cushing, 1932, p. 10). Reflecting the same viewpoint, Thaon in his book on the pituitary published in 1907 mentioned that the posterior lobe presents only minimal interest. But well before the early 1900s and up into the 1920s, important events were occurring in the pathological and clinical fields in the form of case reports, each giving some inkling that the pituitary was a functioning organ. It was later in this period that the significance of the hypothalamus gradually came to be realized, though confusion persisted as to where to draw the line between the function exercised by the hypothalamus, on the one hand, and the function possessed by the pituitary, on the other.

Gastric ulceration from central lesions

The earliest intimation that the base of the brain was concerned in vegetative function came from Carl von Rokitansky's observation in 1842 that infectious lesions located in this region commonly were associated with grave gastric disturbances, often in the form of perforations and, occasionally, gastric hemorrhage. Rokitansky suggested that areas of softening of the stomach occurring in association with acute affections of the brain or its membranes were probably brought about by a reflex action of the esophageal and gastric branches of the vagus.

Rokitansky's views were put to the test by Moritz Schiff 3 years later — in 1845. He found in dogs and rabbits that unilateral cerebral lesions involving the thalamus and adjacent cerebral peduncle often led, after a few days, to softening of the stomach and occasionally to perforation. Soon afterwards — in 1854 — he found that much the same effect could occur following unilateral division of the pons or even hemisection

Fig. 7. Carl Freiherr von Rokitansky (1804–1878). Portrait by courtesy of the National Library of Medicine, Bethesda, Md. (Neg. No. 61-401).

Fig. 8. Moritz Schiff (1823–1896). Portrait by courtesy of the National Library of Medicine, Bethesda, Md. (Neg. No. 68-486).

of the upper two segments of the spinal cord. Stimulation of certain structures (corpora quadrigemina, pons, and cerebral peduncle) caused gastric movements comparable to those elicited by vagus stimulation, and he found that these movements were blocked by division of the vagi. He also observed that stimulation of the splanchnics caused contraction of the vessels of the stomach. In a later summation (1867), Schiff anticipated the present-day distinction between the counterbalancing sympathetic and parasympathetic system and their role in the causation of lesions in the stomach and neighboring gut.

In the passing years more clinical cases of esophageal or gastric or duodenal softening or hemorrhage or perforation associated with central lesions got into the record: a gummatous interpeduncular tumor with softening of the right side of the pons and medulla oblongata associated with esophageal perforation (Hoffmann, 1868), a walnut-sized "sarcoma" of the meninges in the interpeduncular space associated with hyperemic softening of the stomach with numerous ecchymoses of the fundus (Arndt, 1874), and a median cerebellar tumor compressing the corpora quadrigemina and medulla oblongata associated with ecchymoses, extravasations, and hemorrhagic erosions of the stomach together with a markedly hyperemic and ecchymotic lower esophagus and duodenum (Arndt, 1888).

Fig. 9. Harvey Williams Cushing (1869–1939). Portrait by courtesy of Yale Univ. Art Gallery, gift of Mrs. Cushing; by John S. Sargent.

Nothing more of any substance on this subject appeared in the literature until Harvey Cushing's Balfour Lecture on *Peptic Ulcer and the Interbrain* delivered in 1931 and published the following year. Here he described all the kinds of cases that had been seen by Rokitansky. He stated that: "... direct stimulation of the tuber or of its descending fiber tracts, or what theoretically amounts to the same thing, a functional release of the vagus from paralysis of the antagonistic sympathetic fibers, leads to hypersecretion, hyperchlorhydria, hypermotility and hypertonicity especially marked in the pyloric segment. By the spasmodic contractions of the musculature, possibly supplemented by accompanying local spasms of the terminal blood-vessels small areas of ischemia or hemorrhagic infarction are produced, leaving the overlying mucosa exposed to the digestive effects of its own hyperacid juices" (p. 221–222). Cushing had a way of stating his ideas with ultimate clarity and logic, and with these few sentences he wiped out vagaries of the past and set the stage for the future.

Primary polydipsia of Nothnagel

Nothnagel, as early as 1881, reported on a condition which he called "primary

References p. 52–60

Fig. 10. Hermann Nothnagel (1841–1905). Portrait by courtesy of the National Library of Medicine, Bethesda, Md. (Neg. 73-207).

polydipsia". He described the case of a man who was kicked by a horse, causing him to fall backward, striking the back of his head. Within half an hour he developed a fierce thirst, drinking up to 3 liters of water and beer within a period of 3 h, and only then did he begin to urinate. The man recovered quickly without other disabilities but the thirst persisted for about 4 days. Nothnagel postulated that the site of injury responsible for the polydipsia was in the floor of the fourth ventricle. This was the first suggestion that some part of the central nervous system was sensitive to water need. Subsequent studies on the hypothalamus by Bellows and Van Wagenen (1935), Hess and Brügger (1943), Andersson (1952, 1957), Andersson and McCann (1956), Stevenson (1949), and others, have consolidated this concept.

Hypersomnia

In the field of sleep–wakefulness, chief attention turned to the reticular activating substance (or system) when its functional significance became recognized through experimentation carried out by H. W. Magoun and Giuseppe Moruzzi toward the end of the Ranson heyday at Northwestern University (in Chicago) — around 1942–1943 — and soon afterwards at the University of California at Los Angeles in association with John D. French. An account of much of this work is given in Magoun's *The Waking Brain*, published in 1958 (see also French and Magoun, 1950). Moruzzi and Magoun (1949) demonstrated the importance of the brain stem reticular substance in conscious behavior and they also found that its electrocortical accompaniment was relatively independent of specific sensory and motor pathways. In monkeys, Moruzzi and Magoun and their associates found that the reticular activating substance occupied a fairly large area of the brain stem (from the mid-pons upward through the midbrain) well into the nonspecific nuclei of the thalamus, and extending into the posterior hypothalamus.

The observations of Magoun and Moruzzi were certainly a breakthrough, but the basic idea underlying this development had been presented some 12 years earlier — in 1937 — by Wilder Penfield. Penfield's hypothesis was that "the indispensible substratum of consciousness lies outside the cerebral cortex, ... probably in the diencephalon". In a further consideration of his concept, Penfield (1952) proposed that a neuron system centrally placed in the brain stem and connected equally with the two hemispheres could be described as biencephalic, but that "centrencephalic" was perhaps more descriptive; in the "brain stem" he included the thalamus. "Let us define the centrencephalic system", Penfield (1952) stated, "as that neurone system in the higher brain stem which has ... equal functional relations with the two cerebral hemispheres". Thus, Penfield's centrencephalic system coincided in considerable degree with what Magoun later called the reticular activating substance. Subsequently, to avoid misconceptions, Penfield and Jasper (1954) referred to the structures in question as the "centrencephalic integrating system", a subject on which Jasper again expounded in his Hughlings Jackson Lecture in 1959 (Jasper, 1960).

One could of course challenge the use of the work "integrating" in the context used by Penfield. Hartwig Kuhlenbeck (1954, 1957) suggests that the grisea of the "centrencephalic system" do not integrate (in the generally accepted sense of this term) but merely provide one of the several important "activating factors" required for the occurrence of such cortico-thalamic events as are correlated with consciousness; instead, Kuhlenbeck thinks it likely that a "multifactorial" combination of input both by the specific sensory channels (optic system, cochlear pathway, and spinobulbo-thalamic channels) *and* the nonspecific (activating) reticular channels are required for the maintenance of conscious cortical (cortico-thalamic) activities in the waking state (1954, p. 128–129; 1957, p. 186–187). An interesting case in point is one described by Adolf Strümpell in 1877. The patient had lost all contact with the outer world, as manifested by anesthesias, etc., except for integrity of input from the right eye and the left ear. Whenever that eye was covered and that ear plugged, she would fall into sleep within a few minutes.

The reticular activating substance is but a small part of the reticular substance, which Jules Déjérine described in detail at the turn of the century (1901) and which Paul Yakovlev referred to in functional terms in 1949 as a portion of the "innermost system of visceration", a concept which he recently expanded (1972). Déjérine

References p. 52–60

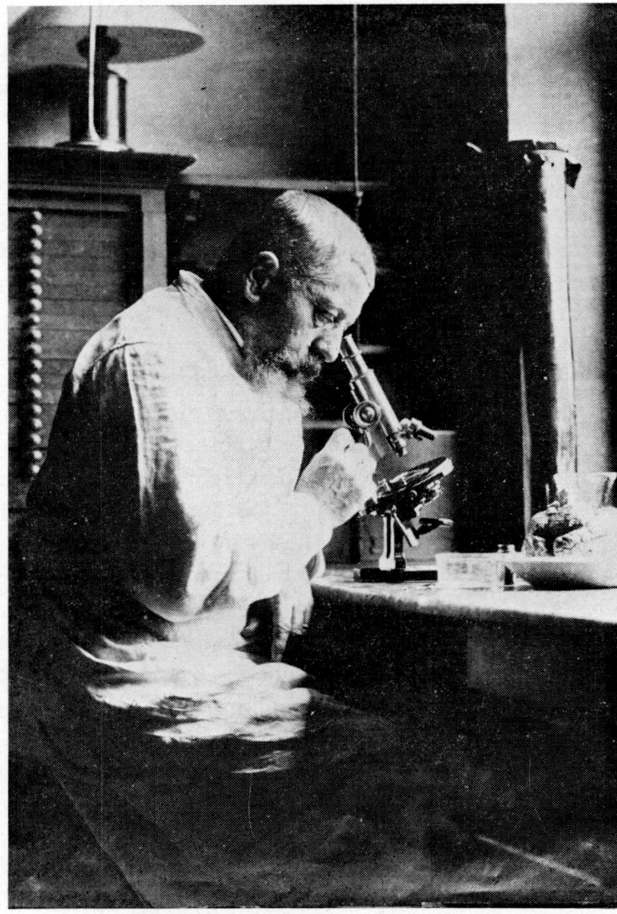

Fig. 11. Joseph Jules Déjérine (1849–1917). Portrait by courtesy of Mme. le Docteur Sorrel-Déjérine, Paris, France.

observed that the reticular substance (his "formation réticulée"), nucleated in some parts of its domain, stretches far and wide: it occupies much of the inner core of the brain stem, extends up to the subthalamic and subpallidal regions, and insinuates itself into the thalamus.

According to Ramón-Moliner and Nauta (1966), Déjérine's reticular substance is made up of small neurons with long, rectilinear and sparsely arborizing dendrites subtending large and widely overlapping dendritic fields (hence neurons of the iso-dendritic type) that mingle freely with the fascicles of transit axons. Nauta and Haymaker (1969) propose that the hypothalamus be considered a component of the reticular substance. This concept may seem unorthodox, for the neuron pattern of the hypothalamus differs in some ways from that of the brain stem. But in parts of the medial portion of the hypothalamus, more specifically in its ventromedial nucleus, "excitation can spread from a given focus in any direction and can establish an infinite number of closed self-re-exciting chains" (Szentágothai *et al.*, 1968, p. 56), as it does for the reticular substance elsewhere. The term, "reticular substance" also implies a heterogeneity of afferent connections, and the hypothalamus *does* receive heterogeneous fiber systems. Moreover, both the hypothalamus and the brain stem reticular substance contain certain highly specific neural mechanisms; the latter has the bulbar mechanism governing respiration, and the hypothalamus has highly specific mechanisms controlling vegetative and other spheres. Furthermore, there is the intriguing idea (Nauta, 1972) that the reticular substance as a whole — both that of the brain stem and the hypothalamus — has assumed the role of an internal adjustment system for the entire brain even to the point that diurnal, seasonal and other periodic fluctuations in all bodily functional realms are probably dependent on it.

The whole of the area occupied by the reticular substance seems also concerned in the maintenance of the sleep–waking cycle. However, the main thrust in the maintenance of wakefulness appears to be exercised by the posterior hypothalamus and the upper midbrain, more or less in the region occupied by the reticular activating substance. In examining the brains of victims of the epidemic of "Nona" (probably viral

Fig. 12. Ludwig Mauthner (1840–1894). Portrait by courtesy of the Biomedical Library, Univ. of California at Los Angeles.

References p. 52–60

encephalitis) occurring in the region of Vienna in the late 1880s, the ophthalmologist, Ludwig Mauthner (1890), found inflammatory lesions in the central gray *(Höhlengrau)* of the third ventricle, the region of the cerebral aqueduct, and in the floor of the fourth ventricle. He reasoned that the prolonged sleep occurring in these patients was attributable to a reduction in function of the central grey (an *Ermüdungserscheinung*), causing a break in centripetal conduction and also in centrifugal conduction from cerebral cortex. Constantin von Economo (1917a, b; 1920, 1929) was more explicit in his accounts of the small epidemic of encephalitis in Vienna in 1917 and the pandemic that shook the world in 1920, a condition for which he coined the term "Encephalitis lethargica". Like Mauthner, he regarded the prolonged somnolence together with the frequently occurring ophthalmoplegia as due to damage of the wall of the aqueduct and the wall of the caudal part of the third ventricle. Lesions located more rostrally, in the anterior part of the hypothalamus, gave rise, in his experience, to agitation and

Fig. 13. Constantin von Economo (1876–1931). Portrait by courtesy of the National Library of Medicine, Bethesda, Md. (Neg. No. 105,620).

hyperkinesias. From these observations came von Economo's turn-around *faux pas*: the caudal *Höhlengrau* was the center presiding over the mechanism of sleep, while the anterior part of the hypothalamus was the center concerned in the maintenance of waking.

A young woman in Chicago, stricken with chronic encephalitis of the von Economo type in 1935 (described by Richter and Traut in 1940), fell into an almost unbroken sleep that lasted 5 years. During this period she would awaken and be alert for 30 min to 1 h, then go back to sleep. Profound atrophy with advanced gliosis was later observed in the posterior hypothalamus (including the mammillary bodies), and the lesion extended into the subthalamus and rostral midbrain, to taper off in the tegmentum of the midbrain as a demyelinative process. All other relevant structures — such as the thalamus — were spared. This was a breakthrough of the first order in localizing the waking center; "nature's experiment" on the "waking center" had been carried out with more finesse than anything ever accomplished by physiologists in inserting needles into the brain and turning on the electricity.

That the neural sleep–waking mechanism is widespread, but concentrated in the region of the posterior hypothalamus and midbrain, has also been shown in cases of brain tumor. In this context, some of the findings of Righetti, dating back 70 years (1903), deserve mention. He tells of a boy who fell into continuous sleep, from which he could be awakened only for short periods. Eventually a tumor was found to have destroyed the optic chiasm and tuberal region and to have led to distension of the third ventricle. In an analysis of 775 patients with cerebral tumor, Righetti noted histories of pathological sleep in 115. Tumors of the thalamus and the vicinity of the third ventricle carried the highest incidence of hypersomnia. But internal hydrocephalus intervenes in the course of many growing tumors and clouds the issue; hence one can reach a conclusion only from those instances in which this objection cannot be sustained. Cox (1937), in Australia, has recorded several instances of such tumors involving the sites under discussion and causing hypersomnia, as have Fulton and Bailey (1929).

Thus, before we ever heard of an activating substance, pathologists, in their case reports, have for many years pointed to those areas in the brain which physiologists interested in the sleep–waking mechanism might have started to explore.

SEPARATION OF HYPOTHALAMIC AND PITUITARY FUNCTION

In his book on *Intracranial Tumors*, published in 1888, Byrom Bramwell wrote: "Tumors of the pituitary body are in many instances attended with an excessive development of the subcutaneous fat, and in some cases with the presence of sugar in the urine or with simple polyuria (diabetes insipidus). Whether these symptoms are due to the fact that the pituitary body itself is diseased or whether, as seems more likely, to the secondary results which tumors in this situation produce in the surrounding cerebral tissue, has not yet been decided" (p. 164–165). There were other early clinical observations relating both to the pituitary and the hypothalamus. At a

References p. 52–60

Fig. 14. Sir Byrom Bramwell (1847–1931). Portrait by courtesy of the National Library of Medicine, Bethesda, Md. (Neg. No. 73-203).

meeting of the Ophthalmological Society of the United Kingdom in 1887 there were reports, by J. B. Story of Dublin and others, of blindness, drowsiness, obesity, and irregular menstruation associated with pituitary tumors.

The power struggle between the hypothalamus and the pituitary really began with Pierre Marie's (1886) first report of two clinical cases for which he coined the term "acromegaly". This was closely followed by a similar report by Minkowski (1887) of a case of acromegaly in which he found a pituitary tumor. At this time it was generally assumed that the presence of a tumor was evidence of failure of function of the portion of the organ displaced, and thus Minkowski concluded that acromegaly was due to pituitary insufficiency. To add to the bewilderment, Babinski in 1900 and Fröhlich in 1901 reported clinical cases of pituitary tumors without acromegaly in which there was sexual infantilism and obesity.

If acromegaly were due to failure of pituitary function, then removal of the pituitary in animals should produce acromegaly. Acting on this premise, many attempts were

Fig. 15. Nicolas C. Paulesco (1874–1946). Portrait by courtesy of the New York Academy of Medicine, N. Y.

made to accomplish this very thing. It was this early interest in hypophysectomy that brought attention to the hypothalamus and helped to disentangle pituitary from hypothalamic function.

The Rumanian physiologist Paulesco was the first to devise an operative technique for hypophysectomy in dogs that was far superior to all others. This was in 1907, with a follow-up publication in 1908. His technique was a transtemporal approach which necessitated lifting the temporal lobe to gain access to the pituitary. In his series all the dogs from which the pituitary had been completely removed died within 24 h following operation; 2 of the 7 that had been partially hypophysectomized lived for 5 months; another lived for a year. From these results Paulesco concluded that the pituitary was essential to life. Cushing (1909), together with Crowe and Homans (1910), Reford (1909), and the rest of his team of surgical residents, followed up Paulesco's work by using the latter's technique in performing a large series of hypophysectomies in dogs. Their conclusions were similar to Paulesco's: the pituitary was essential to life.

It took the courage of Bernhard Aschner of Vienna to challenge Cushing's postulate that the pituitary was essential to life. Again in dogs, Aschner (1909) removed the

References p. 52–60

Fig. 16. Bernhard Aschner (1883–1941). Portrait by courtesy of the New York Academy of Medicine, N. Y.

pituitary through the roof of the mouth in order not to disturb the base of the brain. The dogs survived, but they failed to grow and did not become obese. Aschner's report seemed to irritate Cushing, for in his book, *The Pituitary Body and its Disorders* (1912), he added this footnote: "Aschner, it must be confessed, has opposed the view of essentiality of the gland to life. He is inclined to attribute the fatalities to some injury of hypothetical nerve centers of the infundibular region. As Biedl points out, the operative method Aschner employed is open to criticism. ..." (p. 12).

In the large series of dogs hypophysectomized by Cushing's team, some survived for many months, becoming "fat, loggy, sexless creatures" (Homans, cited by Fulton, 1946, p. 281). Cushing maintained that this condition was due to hypophyseal deficiency, which he called "Fröhlich's asexual adiposity". Bernhard Aschner (1912) suggested that the obesity in Cushing's dogs might be due to contusions of the base of the brain that occurred as the temporal lobe was elevated during the hypophysectomy. Aschner was supported in his proposal by Jacob Erdheim (1904, 1916), who had reported a case of adiposity and sexual dystrophy in which a suprasellar tumor deformed the interbrain. Erdheim had concluded that the adiposogenital syndrome

Fig. 17. Jacob Erdheim (1874–1937). Portrait by courtesy of the New York Academy of Medicine, N. Y.

was due to a hypothalamic and not a glandular disturbance. Aschner had effected the breakthrough in separating hypothalamic from hypophyseal function but the question was not to be settled until Camus and Roussy (1913a, b, 1914) and Bailey and Bremer (1921) had had their say.

In 1920, Percival Bailey, a house officer on Cushing's staff, was assigned the task of staining the pituitary fragments that remained after partial hypophysectomy in dogs. One day, as he was about to do a partial hypophysectomy by the transtemporal approach of Paulesco, he inadvertently severed an infundibular artery. Because of the excessive bleeding he decided not to proceed with the hypophysectomy. Instead he closed the wound and returned the dog to its cage. Next morning, on coming into the laboratory, he found a flood of urine on the floor. What amazed him was that he had produced polyuria without having touched the animal's pituitary. He spoke of this accident to Frédéric Bremer, a visiting fellow from Brussels. Bremer recalled similar experiments carried out by Jean Camus which were interrupted by World War I; Bremer, could, however, not remember the details. Bailey and Bremer decided to pursue the subject further, again in dogs. The outcome was that puncture of the

References p. 52–60

Fig. 18. Jean Camus (born 1904). Portrait by courtesy of the New York Academy of Medicine, N. Y.

postinfundibular region of the hypothalamus — without exposure of the pituitary — produced the adiposogenital syndrome and diabetes insipidus. Their publication (in 1921) contained a real blockbuster, to wit:

> "We have given proof that it is possible to provoke the cachexia 'hypophyseopriva' and the adiposogenital syndrome (together with a permanent diabetes insipidus) by puncture of the post-infundibular region of the hypothalamus without touching the pituitary. ... The question immediately arises: Cannot these symptoms nevertheless be due to the pituitary by a disturbance of its innervation? The same question was discussed in relation to diabetes insipidus. We believe that it must be answered negatively here also. ... What, then, is the function of the pituitary? It is impossible to admit that all organs with such a highly differentiated glandular structure as the pars buccalis should not have a function at some period of life or at some period in the development of the vertebrate phylum. But we must admit that we have little actual knowledge of its functional significance in the adult animal" (p. 798–799).

This manifesto on the part of Bailey and Bremer caused no small ripple. It is said that Cushing was furious and tried to stop publication of the paper (Fulton, 1954, p. 12). After a mellowing of 10 years, Cushing (1932) wrote of it as "a challenge which produced a veritable bouleversement of our cherished preconceptions" (p. 19).

Fig. 19. Percival Bailey, born 1892.

EARLIER CONCEPTS OF HYPOTHALAMIC ANATOMY

What we today call the "hypothalamus" was known for many years merely as an anatomical region lying beneath the thalamus, and given short shrift. It was another *substantia innominata*, for in drawings in various texts that part of the brain was left unlabeled. Theodor Meynert, in his chapter in Stricker's *Handbuch* published a century ago (1872), referred to the hypothalamus as *das centrale Höhlengrau des Zwischenhirnes*. According to him, this *centrale Höhlengrau* had its rostralmost portion in the tuber cinereum.

Some semblance of order began to take form in the *Habilitationsschrift* of August Forel in 1877, in which he took Meynert to task, stating that Meynert, in his discourse on the structures lying beneath the thalamus was unable to distinguish uncertain from

References p. 52–60

Fig. 20. Frédéric Bremer, born 1892.

certain and that he so interwove hypothesis with fact that a distinction between them became impossible (p. 406). In his own paper, Forel introduced the concept of a "regio subthalamica", a term he coined. In it were included Luys' body, the zona incerta (a term he introduced), and certain dorsally-located fiber tracts. Meynert's *centrales Höhlengrau*, he noted, extended forward from the regio subthalamica. Forel illustrated the hypothalamus beautifully, naming its main structures, but did not give the region a name. However, in an earlier publication (1872) Forel had regarded the regio subthalamica as extending into the area which we now know as the lateral hypothalamus, with the anterior pillar of the fornix its rostral limit.

The term "hypothalamus" was coined and introduced by Wilhelm His in 1893 in a classical paper prepared in anticipation of the pending formulation of the BNA. As based on the ontogenetic development of the human brain, His included in the hypothalamus the following: the caudal wall of the optic recess, the caudal part of the preoptic region, and the tuber cinereum and the infundibulum, all of which he included in the "pars optica hypothalami", which he regarded as belonging to the

Fig. 21. Wilhelm His (1831–1904). Portrait by courtesy of the National Library of Medicine, Bethesda, Md. (Neg. No. 72-31).

diencephalon. (He did not make clear the boundary he set for the pars mammillaris.) His took as the upper border of the hypothalamus Reichert's "sulcus Monroi" (so named by Reichert in 1859–1861), which he renamed the "hypothalamic sulcus", and he indicated that this sulcus — the sulcus limitans — continued rostralward to the optic recess. His also recognized 3 other divisions of the diencephalon: the thalamus, the epithalamus (habenular ganglion and pineal body and vicinity), and the metathalamus (the geniculate bodies). So, some 80 years ago, thanks to the meticulous observations of Wilhelm His, and despite his ambiguities as to hypothalamic boundaries, we had the beginnings of a workable scheme not only for the hypothalamus but also for other diencephalic derivatives as well.

Terminology introduced by Forel somehow became twisted when His' contribution became known. Lewellys Barker (1909), for example, referred to the corpus Luysi as the "nucleus hypothalamicus", and the regio subthalamica as the "hypothalamus" (p. 671). Subsequently, the designation "subthalamus" became variously interpreted by different authors as including miscellaneous and heterogeneous structures pertaining to ventral thalamus, dorsal thalamus, and rostral portions of the mesencephalic tegmentum (*cf.* Kuhlenbeck, 1948).

References p. 52–60

Fig. 22. Hartwig Kuhlenbeck, born 1897.

In 1910 C. Judson Herrick recognized another division of the diencephalon, the ventral thalamus, which had escaped the notice of others before him. He discovered this part of the brain in amphibians, then found it also in reptiles and mammals, and he made some desultory attempts at its homologization among these animal classes, stating, for example (in 1933), that "the ventral thalamus [in the amphibians] is probably the precursor of the mammalian subthalamus" (p. 273). Clarification of the diencephalic longitudinal zonal system — epithalamus, thalamus (or dorsal thalamus), ventral thalamus, and hypothalamus — in the entire vertebrate series was provided mainly through the comparative and embryological studies undertaken by Kuhlenbeck and his collaborators starting in 1924 (Kuhlenbeck, 1924, 1927, 1948, 1954, 1969; Kuhlenbeck and Haymaker, 1949). Thus ended the naming: the hypothalamus became merely one of the satellites of the thalamus and was not given a name of its own.

Somehow the preoptic region got into a mêlée in the early days, starting with His (1904), who, in a study of the developing human brain, classed it with the corpus striatum (p. 56). Then the region was stripped of that relationship but continued to be regarded as part of the telencephalon — more specifically, that part of the telencephalon which was thought to have remained unevaginated, that is, containing a common ventricle ("telencephalon medium" or "telencephalon impar") (Clark, 1938). This concept originated through a misunderstanding of the way in which the torus transversus, or commissural plate, develops. A telencephalon medium, or impar, is present during early ontogenetic stages of *all* vertebrates, but is located dorsal and rostral to the anterior commissure or to the primordium of the latter. In all gnathostome *Anamniota* the unpaired telencephalon persists to a greater or lesser extent at the adult stage. However, in all *Amniota* (reptiles, birds, and mammals) the telencephalon medium becomes reduced in size, occupying a negligible portion of the ventricular space dorsal to the anterior commissure and between the paired interventricular foramina (Kuhlenbeck, 1927). It likewise disappears in cyclostomes such as *Petromyzon*. Actually, the preoptic region takes origin as a rostral differentiation of the hypothalamic primordium and, accordingly, is part and parcel of the hypothalamus (Kuhlenbeck, 1954, 1969; Kuhlenbeck and Haymaker, 1949; Rose, 1942). It is hoped that this matter can, finally, be put to rest by referring to the preoptic region as the "preoptic region of the hypothalamus".

It took a while for the various components of the hypothalamus to be sorted out from the maze of hypothalamic structures and be given names. So many anatomists joined in the parcellation that it is hard to know which contributions were the real breakthroughs. As to the supraoptic nucleus, Meynert (1872) called it the "basal optic ganglion" (or *Basalganglion des Opticus*), as did also Forel (1877). Ramón y Cajal (1909–1911) referred to it by two names: the perichiasmatic and the tangential nucleus. Von Lenhossék (1887) gave it its current name, the *supraoptic nucleus*; he regarded it as the superior component of a supposed nuclear complex that included some new grisea he had discovered, the latter including the "nucleus postero-lateralis" (or nuclei tuberis laterales according to today's nomenclature) (p. 456). Later, these cell masses were brought into their proper perspective by von Kölliker (1896), who distinguished the supraoptic nucleus and lateral tuberal nuclei on the basis of cell size.

Another nucleus of interest is that lying just above the neurohypophyseal infundibulum (or medium eminence). "Arciform" and "arcuatus" are two of the aliases by which it is known. In LeGros Clark's (1938) *Opus magnum* on the hypothalamus, one looks in vain for a description or an illustration of this nucleus in the human hypothalamus, though Clark does describe and picture it in reptiles. There was so little interest in the function of this nucleus at the time that we might imagine that, like all of us, Clark and his artist (unless Clark was his own artist) had their "blind spots". Spatz and his associates, Diepen and Gaupp (1948), gave it its most appropriate name: nucleus infundibularis (or nucleus tuberis infundibularis).

In the earlier days of the exploration of hypothalamic anatomy, Malone (1910, 1914) sought to characterize hypothalamic nuclei in terms of cell type. He soon found that cell types do not respect hypothalamic boundaries. Utilizing the Nissl method, he observed that cells which he considered as of the "vegetative" type were situated not only in the hypothalamus but also in the paramedian nucleus of the thalamus and in the subthalamic nucleus. Taking up the trail, Greving (1926) added the thalamic reuniens nucleus to the structures composed of cells supposedly of this type. And

References p. 52–60

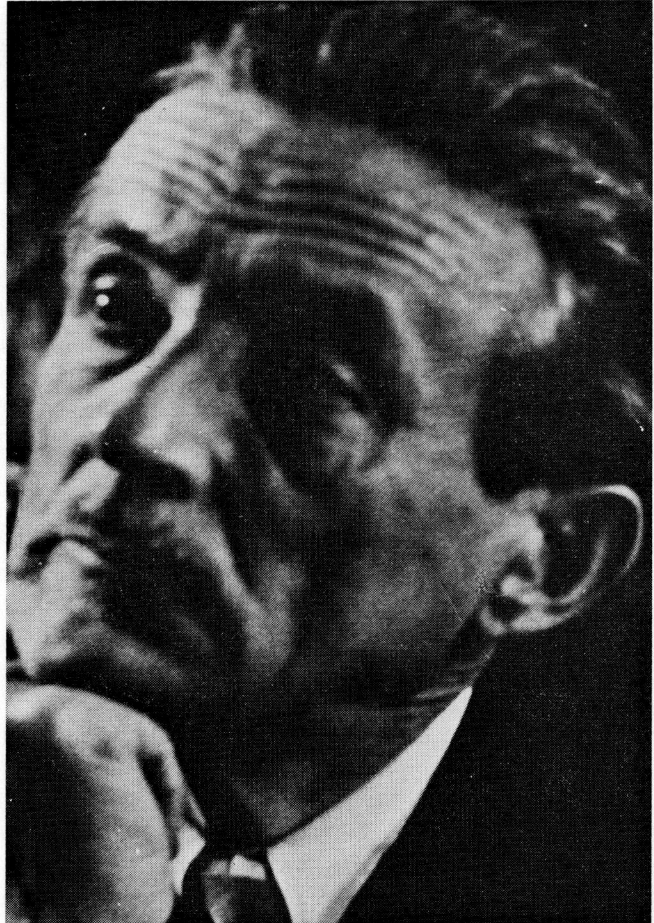

Fig. 23. Hugo Spatz (1888–1969). Portrait by courtesy of the Army Medical Library, Washington, D. C. (Neg. No. 10.5729.1).

Nicolesco and Nicolesco (1929), pointing to the resemblance of large cells in the nucleus basalis — a cell aggregate situated well in front of the hypothalamus which von Kölliker (1896) called *Meynert's Basalganglion* — to large cells in the hypothalamus, expressed the belief that this nucleus lay at the frontier between the diencephalon and the telencephalon and that it represented a vegetative portion of the diencephalon. It remains to be seen whether or not these notions have any validity, but they do give emphasis to the proposition that the hypothalamus is an "open system" so far as cell types are concerned.

Implications as to the function of individual hypothalamic cells aside, completion of the enormous task of providing a workable nomenclature of hypothalamic nuclei is to be credited more to David McK. Rioch than to any other. His chief work in this field was his description of the diencephalon of carnivora (1929a, b). His well-known

"Précis" on nomenclature, prepared in collaboration with Wislocki, O'Leary, Hinsey, and Sheehan (1940), in which terms from the past were resurrected, then buried again, still stands as the most comprehensive treatise in this field.

NEW TOOLS AND THEIR IMPACT ON HYPOTHALAMIC RESEARCH

The stereotaxic instrument

For today's students of physiology such terms as Horsley–Clarke, Fahrenheit, Celsius and galvanometer represent laboratory equipment, taken as much for granted as such items as electrical outlets, faucets, and sinks. Did you know that Gabriel Fahrenheit (1686–1736), though a native of Danzig, made his first thermometer in Amsterdam in 1720? The Horsley–Clarke stereotaxic instrument was designed and built by a surgeon

Fig. 24. Robert Henry Clarke (1850–1926). Portrait from the Wellcome Institute of the History of Medicine, London, England, by courtesy of the Trustees.

References p. 52–60

named Robert Henry Clarke — an associate of Victor Horsley in the Laboratory of Chemical Pathology, University College Hospital, London. Clarke and Horsley were studying the anatomical relations of the cortex of the cerebellum to its nuclei and peduncles and to the rest of the brain and the spinal cord. Their problem was to find a means of producing lesions in the cerebellar nuclei which would be accurate in position and limited in size and not involve other cerebellar structures. There had been attempts by others to overcome this problem. For instance, Friedrich Trendelenburg, assistant at the Physiological Institute in Freiburg i.B., had described in 1907 an instrument for making lesions deep in the brain which he called a "myelotome", but Horsley and Clarke felt that it did not satisfy the conditions they had set for themselves.

The genius behind the stereotaxic instrument belonged to Clarke, who saw the need for such an instrument in his research studies with Horsley. For a period of 10 years, from 1895 to 1905, Clarke labored over the project and two brief communications appeared on the use of the instrument, one in 1905 and one in 1906, the latter in the *British Medical Journal* under the authorship of Clarke and Horsley. Sometime around 1907 a break occurred in the longstanding and close personal friendship between Clarke and Horsley and the publication of the definitive paper on the stereotaxic apparatus in *Brain* in 1908 brought to an end their research collaboration. The authorship on this final paper read: Horsley and Clarke. Davis (1964) has discussed this rupture in friendship and suggests that "the basis of the antagonism which arose between Clarke and Horsley revolved about the fundamental difference in their temperaments and personality because Clarke was the more retiring and introverted of the two and unmistakably grew to resent Sir Victor's worldwide fame as a surgeon and scientist. ... Perhaps it was the unjust bitterness and professional jealousy, clearly recognized by his colleagues, which has relegated Clarke and the credit for his contributions to near anonymity" (p. 1337–1338).

Some interest in the instrument was aroused soon after Horsley and Clarke had published on the cerebellum (in 1908): one paper on the thalamus by Ernest Sachs appeared in 1909, and another on the corpus striatum by Kinnier Wilson in 1914. In their publications both Sachs and Wilson referred to it as the "Clarke instrument". But after these publications had appeared the interest in the instrument lagged.

Those who like to indulge in hyperbole could say that the stereotaxic instrument was a contribution of such magnitude for the neurophysiologist that it might be considered somewhat analogous to the discovery of the wheel by primitive man, for it revolutionized methodology for the study of the hypothalamus as well as the rest of the brain. World War I was undoubtedly a factor in the tardiness of investigators to make use of the Clarke instrument. Clarke himself was working on improvements of his instrument when his labors were brought to a halt by World War I. It was not until 1920 that he published the complete information on his improved instrument. This appeared in *The Johns Hopkins Hospital Reports*. In his introduction to this monograph Clarke remarked: "... the need of fresh workers in this field is urgent. Considering our ignorance of elementary anatomical and physical data, the course and connections of tracts, the functions of large centers, not to speak of smaller but important nuclei, the vast amount of indispensable work requiring no exceptional

Fig. 25. Stephen Walter Ranson (1880–1942). Portrait by courtesy of the Army Institute of Pathology, Washington, D. C. (Neg. No. 105213).

skill, the fascination of the subject, and the harvest to be gathered, the slow progress from the dearth of investigators is deplorable" (p. viii).

The first person to grasp the great significance of this new instrument as a tool for exploring the hypothalamus was Stephen Ranson. Ernest Sachs, while working with Horsley, had had Clarke's instrument maker build a copy, which he had brought with him when he returned to the U.S.A. According to Rasmussen (1947, p. 78), Ranson had his own instrument maker build a facsimile of the instrument which Sachs had brought from England, and in 1931 Ranson set about systematically to explore the entire area of the hypothalamus. There immediately followed a flood of papers from his laboratory on this subject. Within two years (1932–1934) there were 7 publications (Ingram, Hannett and Ranson, 1932; Ingram *et al.*, 1932; Magoun, Barris and Ranson, 1932; Ranson, 1934a, 1934b; Ranson and Ingram, 1932; Ranson and Magoun, 1933). One can easily imagine the excitement and activity going on in Ranson's laboratory at that time among his numerous talented co-workers (Magoun, Ingram, Fisher, Hetherington, Kabat, Clark, Harrison, Brobeck, Hare, and others). These preliminary reports were followed in the next few years by many more studies on the anatomy and physiology of the hypothalamus, culminating in an extensive review on the hypothalamus which appeared in the *Ergebnisse der Physiologie* (by Ranson and Magoun) in 1939. Its importance was that it served as the first handbook for young investigators entering the field; the second handbook was the volume entitled *The Hypothalamus* ... published by the Association for Research in Nervous and Mental Disease in 1940. The dedicatory page in this volume read: "In recognition of the distinguished contributions to knowledge of hypothalamic functions made by himself and by the

References p. 52–60

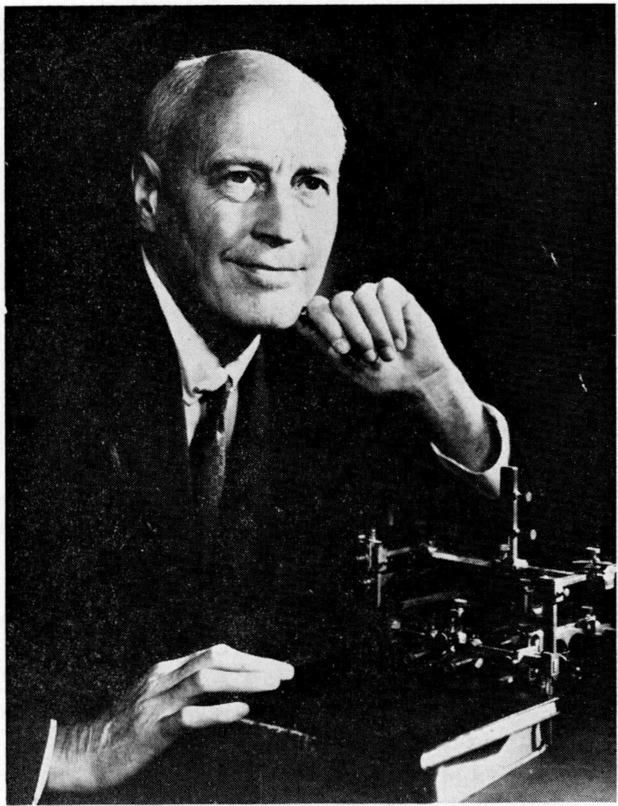

Fig. 26. Horace Winchell Magoun, born 1907. Portrait by courtesy of the Biomedical Library, Univ. of California at Los Angeles.

students he inspired, this meeting of the association is dedicated by the trustees to Stephen Walter Ranson". From 1932 to 1942, Ranson kept his research team together, until finally it was scattered by the exigencies of World War II.

One of the real breakthroughs which the Ranson group accomplished with the use of the Clarke stereotaxic instrument was to settle the vexed question of the origin of diabetes insipidus. Up until 1931, it was generally agreed that the hypothalamus controlled water metabolism but the mechanism by which diabetes insipidus was produced was still elusive. Removal of the posterior lobe alone in experimental animals did not invariably produce diabetes insipidus, nor did it occur following complete hypophysectomy. The condition was, however, easily produced by puncture of the tuber cinereum, and Pitressin controlled the polyuria that resulted, as had been shown as early as 1920 by Camus and Roussy without benefit of a stereotaxic instrument: they found that "... the opto-peduncular region alone marks the zone within which a lesion is followed by polyuria. It lies at the level of the grey substance of the *tuber cinereum* in the vicinity of the infundibulum" (p. 513).

Before the advent of the stereotaxic instrument, research in the problem of diabetes insipidus was hampered because of the difficulty in producing discrete lesions in the hypothalamus without injury to the hypophysis. With the Clarke instrument small lesions could be made in any part of the hypothalamus without disturbing the infundibulum or the hypophysis. Fisher, Ingram and Ranson (1938) were able to show that bilateral interruption of the supraoptico-hypophyseal tracts as they streamed over the optic chiasm, invariably produced diabetes insipidus in cats and in monkeys. Study of these animals revealed the supraoptic nuclei to be completely degenerated, the median eminence reduced in size, and the pars nervosa atrophic. Much of the mystery of diabetes insipidus was thus solved, and this was Fisher, Ingram and Ranson's reply to Percival Bailey, whose comment in the paper with Bremer in 1921 still staggers readers, to wit: "As to the function of the posterior lobe the experimental evidence is unequivocal. Its removal causes no symptoms. Moreover, its structure is nonglandular. Camus and Roussy were quite justified in speaking of it as an atrophied nervous lobe. ... We have, therefore, no evidence that pituitrin is anything more than a pharmacologically very interesting extract" (p. 800). This was how the situation stood some 50 years ago.

Electrical stimulation without benefit of the stereotaxic instrument

The names of Karplus and Kreidl naturally fall here, for they performed the first substantial experimental studies on the hypothalamus. Their work was begun in 1909, suffered a gap during World War I, then was recommenced and continued through 1937 (Karplus and Kreidl, 1909, 1910, 1918, 1927, 1928; Karplus 1937). True to the main title of their articles — *Gehirn und Sympathicus* — they demonstrated in cats virtually all the attributes of the hypothalamus in the sympathetic realm through stimulation of the walls of the third ventricle: pupillary dilatation, sweating, widening of the palpebral fissure, retraction of the nictitating membrane, changes in heart rate and systemic blood pressure, and inhibition of gut motility, also acceleration and increased amplitude of respiration and contraction of the urinary bladder. Thus, they had elicited practically every bodily response mediated by the sympathetic system. Completion of their painstaking work, in which they ruled out certain other parts of the cerebrum as instrumental, left the impression that the hypothalamus was simply a "sympathetic center" — in other words, that this part of the brain served only *restricted integration.*

However, in retrospect, there was something strange about the results of Karplus and Kreidl, namely the absence of the patterned somatomotor accompaniments that normally go along with a strong sympathetic discharge. The reason, according to Moruzzi (1972), was that their cats were either anesthetized or curarized, and that this was why somatomotor effects did not occur. Philip Bard (1973), in his autobiographical sketch *The Ontogenesis of One Physiologist*, points out that he had early come to the conclusion that "the hypothalamus is not an 'autonomic center', but rather a part of the brain that contains neural mechanisms requisite for complicated patterns of behavior — such as the display of emotion and defense against heat and cold — in which there are autonomic components" (p. 11). In other words, the hypothalamus

References p. 52–60

Fig. 27. Johann Paul Karplus (1866–1936). Portrait by courtesy of the Biomedical Library, Univ. of California at Los Angeles.

served *generalized integration*. Bard's conclusion was based on the results of his famous cerebral ablation experiments in cats, first published in 1928, when Karplus and Kreidl's work was still in progress. It should, however, be added, that under abnormal conditions in man, the hypothalamus can give rise to purely autonomic discharges, though when they occur they are usually of somewhat narrower scope than elicited by Karplus and Kreidl on stimulation of the hypothalamus of cats. It was for such a case, characterized by seizures largely in the autonomic realm in a patient with colloid cyst of the third ventricle, that Penfield (1929) coined the term "diencephalic autonomic epilepsy".

Walter R. Hess (1881–1973) commenced his work on subcortical electrical stimulation in 1925. Hess also used no stereotaxic instrument. Owing to the more or less standardized skulls of adult cats (cats were the only animal Hess used), fairly good localizations were obtained by inserting fine steel needles — 0.25–0.3 mm in diameter — at accurately measured placements, taking the sagittal and coronal sutures as ref-

Fig. 28. Alois Kreidl (1864–1928). Portrait by courtesy of the Biomedical Library, Univ. of California at Los Angeles.

erences, and lowering the needles, usually in two steps, to a depth which had been calculated in advance, according to the intended site. There was some scatter, but not much more than with the stereotaxic instrument. The large number of anatomically verified stimulation sites allowed the drawing of maps, in which clear correlations of stimulation sites and parameters *versus* effects were recorded in protocols and, from 1927 onward, in motion pictures. Hess used a damped, pulsating direct current, which was intended to excite the slowly responding autonomic fibers without eliciting too much response from the sensorimotor systems, as previously established in model experiments. In the stimulus interval a small current in the opposite direction obviated polarization effects.

As Hess relates in his autobiography (1963), he had prepared, while a resident in a State hospital (in about 1906), a small paper on *Viscosität des Blutes und Herzarbeit* and carried it all the way from Zürich to Bonn so that he could personally present it

References p. 52–60

Fig. 29. Philip Bard, born 1898.

to Professor Pflüger for publication in the famous *Archiv für Physiologie* (Pflügers Archiv). Pflüger wanted to know under whose direction the work had been carried out. On hearing Hess' explanation that he had done the work entirely on his own, Pflüger forthwith rejected the paper without comment. Hess recalls that this was "a rather painful experience [brought about through] the authoritarian attitude then prevalent toward youthful scientists" (p. 406). But having had ancestors belonging to a Germanic tribe in the upper Rhine valley who had withstood Roman control, he had the fortitude to "take it".

A background gained as an ophthalmologist later in his career led Hess to the investigation of mechanisms involved in the coordination of eye movements. Subsequent topics which he tackled dealt with the regulation of the circulatory system, on which he wrote a monograph in 1930, and also on the regulation of respiration, which culminated in a monograph in 1931. As he stated, he took Karplus and Kreidl's

Fig. 30. Walter R. Hess (1881–1973). Portrait by courtesy of Prof. R. Hess, Zürich, Switzerland.

"exploratory experimental investigations" into consideration. To him, their findings and the evidence he himself had accumulated on circulation and respiration pointed to a common integrating system which he suspected to be situated in the hypothalamus. This concept led him to look for experimental confirmation, which could be obtained only by subcortical stimulation. Acute experiments were usually carried out, chronic experiments only occasionally, for he found that his rigid steel electrodes too frequently caused local tissue destruction. "It was my goal", Hess (1963) stated, "to test the behavior of experimental animals (cats) subjected to electrical stimulation and diathermic lesions *without impairing the freedom of movement of the animal by immobilization or narcosis*" (p. 415) (italics are ours).

Hess was bent on an assessment of integral physiology of the nervous system, which came to fruition in the form of a masterpiece on motor function in 1947, and another, equally imposing, in 1948, on the functional organization of the vegetative nervous system. In the latter we find an in-depth discourse on the whole field of the autonomic nervous system, including concepts of organization of the sympathetic and parasympathetic systems, functional analysis of the diencephalon, sleep as a vegetative function. To him, sleep was a vegetative process by which the autonomic nervous system regulates the activity of the higher cerebral functions. In this realm the vegetative system was not the follower, but led the way, both through neural and humoral routes. One of Hess' most significant contributions lay in his demonstration that the diencephalon and upper midbrain contained highly organized systems of neurons which, upon being activated by implanted electrodes, gave rise to many kinds of behavior patterns involving the organism as a whole. His work antedated the contributions of Moruzzi and Magoun (in 1949) in this field and those by Penfield and Jasper (in 1954) on the so-called centrencephalic integrating system. Hess was not altogether in accord with the Penfield–Jasper concept that all ultimate or final sensory and motor integration of the cortex must necessarily take place in subcortical structures (*i.e.*, in the centrencephalic integrating system). More important to final integration, Hess believed, were transcortical mechanisms and the temporal patterning of impulses in sequentially interrelated systems of neurons, probably not to be located in any specific region or center. Thus, Hess left the location of the system responsible for final integration "up in the air", so to speak.

In summing up all the work he had done, Hess (1963) stated: "Only little by little and ever so slowly did the veil lift a bit here and there so that 25 years had passed before I could finally think of putting together all the many single research publications that had appeared over the years which had been concerned with individual symptoms. ... Generalization concerning symptoms, syndromes, and localizations could be supported only by such a large body of data" (p. 417). It was his discovery of the functional organization of the diencephalon and its role in the coordination of the functions of the inner organs that brought him the Nobel Prize in 1949.

THE EMERGENCE OF PRESENT CONCEPTS OF THE PATHWAYS TO THE PITUITARY

There has been no greater impact on our knowledge of pituitary function than the realization that the pituitary is innervated by two more or less separate systems: the supraoptico-hypophyseal tract, proceeding to the neurohypophysis, and the tuberoinfundibular tract, reaching the capillary plexus leading into the venous portal system that reached the anterior pituitary.

Recognition of a supraoptico-neurohypophyseal tract and its mode of functioning

As long as a century ago, many publications show their authors looking for nerve fibers in the posterior pituitary, now and then with positive results. But some said that such fibers were very few (Stendell, 1914), while others disclaimed their existence, such as von Kölliker (1896) and Thaon (1907). This was despite Ramón y Cajal's finding (in the rat), in 1894, that such a tract of fibers originated in a nucleus situated behind the optic chiasm, passed down the pituitary stalk, and ramified in the posterior lobe. Pines, in 1925 (a, b), was among those who took the lead in showing that at least some of the fibers in the neurohypophysis had their origin in the supraoptic nucleus (his "nucleus hypophyseus").

The question arose as to how the supraoptico-hypophyseal tract operated. Un-

Fig. 31. Wolfgang Bargmann, born 1906.

mistakable evidence that cells and nerve fibers — not only of the supraoptic but also of the paraventricular nucleus — exhibited secretory activity came from the elegant studies by Ernst Scharrer, first in certain fishes in 1928, then by Ernst and Berta Scharrer, mostly in invertebrates; one of their earlier summations appeared in 1940.

Here Bargmann provided the breakthrough. In Königsberg near the end of World War II, he had been carrying out studies in an effort to understand more about neurosecretion, but upon the approach of the Russian army, he thought it the better part of valor to pursue his studies elsewhere. Back in Kiel in 1948, his interests took two directions: one was alloxan diabetes, the other neurosecretion in vertebrates. The latter aroused his curiosity because of the skepticism then extant on the matter of whether nerve cells could both transmit impulses and be secretory. Seeking a new method for revealing neurosecretory material, he placed sections from a dog's brain into acid-permanganate chrome alum hematoxylin, according to Gomori's method (1939a, b), and was astonished at what he found (instead of shouting "Heureka", Bargmann exclaimed, waving the cigar which was always at hand *(magna voce)*: "Donnerwetter"!): the cells of the supraoptic and paraventricular nuclei and the fibers extending into the infundibulum and reaching the posterior lobe had selectively taken on a blue hue! His associate, Walther Hild, was at that time working on neurosecretion in the shark as the theme for his doctor's thesis, and Bargmann advised him to try the Gomori method. The account of Hild's findings soon followed Barg-

References p. 52–60

mann's first publication (1949). Bargmann's conclusion was that axons issuing from the supraoptic and paraventricular nuclei transport material from these nuclei into the neurohypophysis. He called the aggregate of fibers, the "neurosecretory pathway".

Shortly after becoming Professor of Anatomy at Kiel, Bargmann was invited to visit the United States. In his application for a visa, he gave as his occupation, *anatomist*. The immigration authorities interpreted anatomist as *an atomist* — in other words, an atomic scientist — and he was refused a visa. (The atomic bomb scientist, Klaus Fuchs, who had been a student at Kiel, was at that time under indictment for having transmitted atomic bomb secrets to the Russians.) Later in Chicago — the problem of the visa having been solved — Bargmann showed his hypothalamus preparations to Gomori. No one could have been more surprised than Gomori to see that his pancreatic island stain was also specific for the neurosecretory system. In a personal communication to one of us, Bargmann wrote: "Had I not, in the past, worked on the pancreatic islets, I would never had made much headway in the field of neurosecretion". (When, everafter, in papers dealing with hypothalamic neurosecretion "Gomori-positive" and "Gomori-negative" results were cited, Gomori would comment in conversation that he found the terms distasteful but amusing. "Right now", he once said before lunch to a fellow-Hungarian, Jacob Furth: "I feel Gomori-negative".)

The lead-up to the Bargmann discovery can be credited to Ernst Scharrer (Bargmann and Scharrer, 1951) who had earlier reached much the same conclusions as Bargmann as a result of his own morphological observations and some made by Drager (1950) on the hypophysectomized tropical indigo snake.

The initiation of a new era in neurophysiological research by Bargmann — in 1949 — came at a time when the classical studies on diabetes insipidus by the Ranson team had been completed — *circa* 1940 — and with them a full recognition of the significance of the supraoptic and paraventricular nuclei relative to that disorder. New breakthroughs followed. John Sloper (1954, 1958, 1966a, b), during the years when hardly anyone else in England was working on neurosecretion, showed that throughout the vertebrate series a cystine-rich component devoid of lipid, presumably hormone or a hormone precursor, characterized Bargmann's neurosecretory material (NSM). This disclosure disposed of the then widely held belief that NSM was a glycolipoprotein "bearer substance" which could not, by virtue of its cytochemical properties, represent the posterior pituitary principles. In a letter passed to one of us, Sloper related how, in retrospect, he had often smiled at himself with regard to this early work. He had hoped that his observations would be a step leading to the discovery of the chemical nature of antidiuretic hormone, having been oblivious of the fact that Van Dyke (Van Dyke *et al.*, 1942) had already identified cystine in his protein some 10 years earlier.

Through the use of ^{35}S-labeled cysteine, Sloper (1958) also found that the flow of neurosecretory material, about 0.2 mm/h, was in a downward direction, though, as he stated, some of the flow could also be centripetal (p. 220–223). He did not become embroiled in the argument whether bulk axoplasm flow downward occurred, as was contended in the pioneer study by Weiss and Hiscoe (1948). He championed the view,

Fig. 32. John Chaplin Sloper, born 1922.

now accepted, that the flow of NSM was at the submicroscopic level, but admitted (in a letter written in June, 1973): "I simply cannot visualise a mechanism by which hormone is transmitted". What organelles, if any, moved with the axonal flow? Palay proposed in 1955 and 1957 that NSM probably reflects the aggregation of neurosecretory vesicles 100–200 nm in diameter. In his highly definitive work in 1957, Palay showed in dehydrated rats that nerve terminals in the neurohypophysis became markedly depleted of neurosecretory vesicles and that microvesicles increased in number. The question still needs resolution whether or not NSM is in the form of such vesicles and, if so, whether the vesicles actually move from the perikaryon through the axons to be discharged into the blood stream. The transport of vesicles down the axon could well involve the participation of microtubules (Ochs, 1971) and it would now seem that the release of hormones from the nerve terminal is brought about through exocytosis.

References p. 52–60

The neurovascular link

One of the most difficult answers to the question of hypothalamic control of the anterior pituitary was: how does the tuberal region connect with the anterior pituitary? Here, the breakthrough was provided by Geoffrey Harris and his alter ego, John Green, with their concept of a neurovascular link as the morphological basis for a neurohumoral mechanism.

Harris spent his student days in just the right environment to incubate such a concept: starting his research career in the summer of 1935, he worked a few months as a junior assistant to Professor Popa of Bucharest, then a guest worker in the Anatomy School at Cambridge (Harris, 1972). Thus, Harris learned first hand about the portal circulation of the pituitary. According to Popa and his associate Fielding (1930, 1933), the blood flows from the sinusoids of the anterior pituitary to the hypothalamus through a system of portal vessels. This direction of flow was challenged by Wislocki

Fig. 33. Geoffrey Wingfield Harris (1913–1971). Portrait by courtesy of the Biomedical Library, Univ. of California at Los Angeles.

Fig. 34. John Davis Green (1917–1964). Portrait by courtesy of the Biomedical Library, Univ. of California at Los Angeles.

and King in 1936. The only way to settle the matter was to visualize these vessels directly in living (anesthetized) animals under the microscope. This was done in amphibians by Houssay, Biasotti and Sammartino (in 1935) as well as by Green (in 1947), and in rats by Green and Harris (in 1949). They saw that the direction of blood flow in the portal vessels was from the infundibulum towards the anterior pituitary. Confirmation in the dog and cat came from Török (1954); however, by opening the hypophyseal cleft from the lateral side and displacing the posterior lobe somewhat upwards, then injecting dyes into the opposite carotid, Török saw that the direction of flow in some small vessels was upwards, toward the transitional zone between the pars distalis and pars intermedia, from which the blood drained into the posterior lobe venous system. According to Szentágothai and his associates (1968), blood from some capillary loops in the upper part of the infundibulum flows upwards into the capillary network of the hypothalamus. For a current assessment of the direction of blood flow

References p. 52–60

in the pituitary one should read Duvernoy (1972), who adds to the system arteriolar branches that pass through the infundibular sulcus into the tuberal region, then break into capillaries that descend in "weeping willow" fashion to join some portions of the capillary system in the subependymal region of the infundibulum.

Among those having considerable influence on Harris were F. H. A. Marshall and E. B. Verney (1936), who at that time were studying the problem of the occurrence of ovulation in the rabbit as a result of stimulation of the central nervous system. Marshall suggested to Harris as early as 1935 that "it would be interesting to apply precisely localized electrical stimuli to different regions of the hypothalamus of estrus female rabbits to see if it were possible to obtain evidence concerning the reflex pathway normally involved in post-coital ovulation" (Harris, 1972, p. iv). Harris acted on this suggestion in 1936 and 1937 and found that a stimulus applied to the tuber cinereum *did* in fact evoke ovulation in the rabbit. In his paper reporting these findings he stated, with the brashness that characterizes the young (his own evaluation of himself), that "there is no reason to believe that the thyreotropic, adrenotropic, lactogenic, parathyrotropic and growth hormones are not similarly controlled" (p. iv). From 1935 to 1955 the work of Harris and Green (Harris, 1948a, b; Green and Harris, 1947, 1949) had a very significant impact on research on the hypothalamus, and investigators everywhere began working on the problem. With the use of the Clarke stereotaxic instrument, areas of the hypothalamus were mapped out in accordance with the pituitary hormone released on stimulation.

The mystery persisted, however, as to how the neurovascular link operated. To investigate this problem Harris and Jacobsohn (1952) initiated a study to determine whether anterior pituitary tissue vascularized by the portal system functioned differently from pituitary tissue vascularized by the systemic circulation. They found that pituitary tissue transplanted under the median eminence in hypophysectomized rats became highly vascularized and rendered the animals essentially normal as regards estrus cycles, pregnancy and lactation, while the hypophysectomized animals with pituitary tissue transplanted in the subarachnoid space under the temporal lobe showed no pituitary function even though the transplant was well vascularized.

Strong support for this concept of a neurovascular link was provided by the identification of the infundibular nucleus by Spatz (1951) and his associates, Nowakowski (1951), Christ (1951a, b), and Diepen (1953), as the main source of the tuberohypophyseal tract (or, better, the tuberoinfundibular tract), and they found that the fibers of the tract terminated along the outer edge of the infundibulum, next to the mantle capillary plexus. Szentágothai (1964) considered that the series of capillary loops in the outer portion of the infundibulum (the zona palisadica) were at best a device to enlarge the contact surface between the infundibulum and the anterior pituitary tissue. It is likely that the vascular tufts and scrawls, called *gomitoli* (Fumagalli, 1941) or *spikes* (Xuereb *et al.*, 1954), serve the same purpose. In 1964 Fuxe provided the final anatomical support by demonstrating the tuberohypophyseal tract in its entirety through use of a highly specific and sensitive cytochemical technique for dopamine. The significance of all this was that it led to the postulate that the nerve terminals release neurohumoral agents into the capillaries, through which they are carried via the portal venous system to the anterior pituitary, where they act to release pituitary hormones.

Use of the electron microscope was needed to help settle the matter. By this means a great wealth of nerve terminals was shown to impinge on the walls of the capillaries in the region of the infundibulum (median eminence), also that the capillary endothelium was of the fenestrated type such as is seen in resorptive and secretory structures elsewhere; moreover, it was found that the nerve terminals contained both electron-lucent and electron-dense structures, the former interpreted as synaptic vesicles, and the latter as neurosecretory material (Barry and Cotte, 1961, and others). Even before electron microscopic proof, it had been shown that there were chemical substances in the infundibulum capable of releasing anterior pituitary hormones. This was the work started by Guillemin (Guillemin and Rosenberg, 1955) and Saffran and Schally (Saffran, Schally and Benfey, 1955). Here again we have one of the great breakthroughs in hypothalamic research.

INTEGRATION

We should like to pursue the subject of 'Integration Hypothalamic Activity' for awhile in closing. This is an immense area, involving current concepts of hypothalamic boundaries, cellular organization within the hypothalamus, neural inputs into the hypothalamus and neural outputs, the whole subject of the limbic system, hypothalamic monoamines, cholinergic and monoaminergic fiber terminals, and feedbacks, including cellular receptivity of substances borne to the hypothalamus by the blood stream.

On the basis of ontogenetic development, the boundaries of the hypothalamus have arbitrarily been set as extending from the lamina terminalis rostrally to the mammillary–habenular border zone caudally. This has given the impression that the hypothalamus is a well-defined, sharply limited part of the encephalon. Actually the reverse is the case, for the adult hypothalamus is broadly continuous with surrounding gray matter, from which it can be delineated in sagittal sections only with difficulty. These tissue continuities also have functional associations, many of the surrounding grisea being reciprocally connected with the hypothalamus. These grisea could thus be viewed as intermediaries in the functional associations of the hypothalamus with the rest of the brain.

Grisea reciprocally connected with the hypothalamus have been under study with the use of "standard" staining techniques for more than a century. But the unravelling of the connections required a new technique, particularly with respect to the connections of nonmyelinated fibers. Walle J. H. Nauta set himself this task, first at the University of Utrecht here in Holland (until 1947) and then at the University of Zürich (until 1951). In Zürich, he finally devised a modification of the Bielschowsky method which was capable of impregnating degenerating nerve fibers, so that the axons undergoing degeneration looked something like a string of black beads against a yellow-brown background. He was urged by co-workers in Zürich to put his new

References p. 52–60

Fig. 35. Walle Jetze Harinx Nauta, born 1916.

"stain" into practice by publishing something on his findings in the rat brain. But ever so cautious, he kept watching to see how consistent his capricious silver impregnations would come out. His first publication describing the technique and some of the results he had obtained with its use appeared in 1950. It dealt with parts of the brain notoriously difficult to impregnate — the hypothalamus and neocortex. At the Walter Reed Army Institute of Research, Washington, D.C., to which he had come in 1951 at the behest of David Rioch, the "original Nauta procedure" failed to work and normal fibers were too heavily impregnated. Together with P. A. Gygax (1951, 1954) and L. F. Ryan (1952), Nauta succeeded, however, in developing methods of suppressing the staining of normal fibers, so that degenerating axoplasm came out more selectively. This became known as the "selective" or "suppressive" Nauta–Gygax method. This contribution, a *tour de force* of the first order, marked the beginning of a new chapter in the history of silver impregnation and a new era in neuroanatomical investigation. Nauta also published a critique on silver stains with Paul Glees (in 1955), whose stain of a similar nature proved of great value in the hands of certain investigators.

With the new Nauta method available, and also its subsequent modifications developed in Nauta's laboratory by R. R. Fink and L. Heimer (1967) (see also Heimer, 1967), which revealed a much greater quantity of fine degenerating terminals, and represented another technical breakthrough, Nauta and others opened up new insights into hypothalamic connections: connections with the hippocampal formation via the fornix, with the amygdala via the stria terminalis, with grisea in the region of the anterior perforated substance via small bands of fibers, with the brain stem grisea via

Fig. 36. Pierre Paul Broca (1824–1880). Portrait by courtesy of the National Medical Library, Bethesda, Md. (Armed Forces Institute of Pathology, Washington, D. C., Neg. No. 105239).

the lateral area of the hypothalamus; they even demonstrated a reversal in the direction pursued by some of the fibers of the Schütz' (1891) dorsal longitudinal fasciculus — fibers that run forward in the *centrales Höhlengrau* as far ventrally as the tuberal region (see Nauta and Haymaker, 1969, and Nauta, 1972). (In his original description, Schütz stated that all fibers in the fasciculus proceeded caudalward.) Whole new concepts were formulated from the findings of those who used this method; for example, the brains of lower animals were shown to be far less under the domination of the olfactory system than previously thought from the use of other staining methods.

But for the grand view of the setting of the hypothalamus in relation to its surroundings, we go back almost a century, to Paul Broca (1824–1880). Broca, in his 500 publications, it is said, never wrote anything mediocre, and among his best were two vast papers on the "great limbic lobe", published in 1878 and 1879. That "lobe", or "limbic convolution", surrounded the diencephalon and the cerebral peduncles. He called it "limbic" because he conceived of it as a threshold (*limbus*, border or threshold) to the newer pallium. To him the limbic lobe was a phylogenetic entity. It was composed of two arcs separated by the limbic fissure, a superior arc, which was

References p. 52–60

situated on the convex surface of the corpus callosum (the "convolution of the corpus callosum"), and an inferior arc, which comprised the "convolution of the hippocampus" — in other words, all the convolutions which we would today call the gyrus fornicatus. Moreover, Broca was convinced of the existence, at least in "osmatic" animals, of a direct connection between the olfactory bulb and the major motor pathway of the cerebral peduncles — "the middle or gray [olfactory] root". He thought that the large cells in the bulb must be motor. In following through on some of Broca's ideas on the connections of the hippocampal formation, Emil Zuckerkandl (1888), in his paper on the *Riechbündel*, and later in his paper on the *Riechstrahlung* (1904), emphasized that the *petite* septal area was the "carrefour de l'hémisphère", to which the hippocampal formation projected a major input.

In elaborating on Broca's concept of the limbic lobe Paul MacLean (1949, 1952)

Fig. 37. James Wenceslaus Papez (1883–1958). Portrait by courtesy of James P. Papez, Lancaster, Penn.

subdivided the lobe into two "rings". The *inner ring*, phylogenetically the older, included the hippocampal formation, the septal area, the parolfactory area (the carrefour de l'hémisphère), the anterior perforated substance, and the pyriform cortex and adjacent part of the amygdala. The *outer ring* included the orbitoinsulotemporal cortex rostrolaterally, the cingulate cortex superiorly, and the entorhinal cortex and subiculum inferiorly. The olfactory bulb projected its fibers into various components of this system.

This brings us to Papez' (1937) formulation of his famous concept of emotion. He stated: "The central emotive process of cortical origin may ... be conceived as being built up in the hippocampal formation and as being transferred to the mammillary body and thence through the anterior thalamic nuclei to the cortex of the gyrus cinguli. The cortex of the cingular gyrus may be looked on as the receptive region for the experiencing of emotion. ... Radiation of the emotive process from the gyrus cinguli to other regions in the cerebral cortex would add emotional coloring to psychic processes occurring elsewhere" (p. 728).

This concept was based on Papez' almost perpetual observations of the brain. Hardly anyone could have spent more time, day after day, well into the night, in viewing brain sections of a large variety of animals than Papez, with his imagination at full play. Cornelius Ariëns Kappers had an influence in the formulation of Papez' concept, for Papez alluded to Kappers' view (1928) that the gyrus hippocampi and the fascia dentata are the receptive cortex for the gyrus cinguli, while the pyramidal cells of the hippocampus form the emissive, or motor layer giving rise to the fornix, or the corticohypothalamic tract. In his paper, Papez also drew on clinical observations. For example, with reference to the hippocampal formation, he reminded his readers of the apprehension, anxiety and paroxysms of rage or terror experienced by patients with rabies, a disease in which the hippocampus is one of the most consistent sites of attack by the virus. In his laboratory at Cornell University in Ithaca, Papez once remarked (to one of us) in a wistful sort of way that somehow his concept had never caught on, for very few had requested a reprint of his article. This was in 1942. In 1949, Kuhlenbeck and Haymaker called attention to the importance of the Papez concept and interpreted the connections between gyrus cinguli, hippocampal formation, fornix, and mammillothalamic tract as a closed feedback circuit. Later (1954, 1957) Kuhlenbeck elaborated on the significance of the Papez mechanism as a circuit modulating tone, or emotion.

Paul MacLean's (1949, 1952) resynthesis of Papez' "theory of emotions" resurrected Broca's notions and breathed new life into the concept of the all-pervasive limbic system. In the 1949 paper MacLean called the limbic lobe the "visceral brain" because of the input into it of impulses from all viscera having autonomic innervation — heart, blood vessels, gastrointestinal tract, and so on (Kaada and Jasper, 1952; Kaada, Pribram and Epstein, 1949). Then in 1952 he added certain non-cortical structures (the amygdala and the fornix) to what he called the "limbic system". Of late (1972), he envisaged 3 basic brain types — reptilian, paleomammalian and mammalian — and classed the limbic system in primates as paleomammalian.

In 1958 Walle Nauta together with Kuypers added a "limbic midbrain area", which

References p. 52–60

Fig. 38. Paul Donald MacLean, born 1913.

originated in grisea of the paramedial part of the midbrain and united with many rostrally situated structures in providing reciprocal connections with the hypothalamus. Those paramedian midbrain grisea included the dorsal and ventral tegmental nuclei of von Gudden (1880) (from which von Kölliker (1896) first traced fibers into the hypothalamus), the superior central tegmental nuclei of Bekhterev (1894), and the ventral tegmental area of Tsai (1925). It was the connections of ascending fibers of Schütz's dorsal longitudinal fasciculus with the tuberal nuclei that led Nauta and Kuypers to propose that the limbic area exerts an influence on the functioning of the pituitary.

One can hardly speak of the limbic system without reference to the medial forebrain bundle. It was some 15 years after Broca's exposé — in 1893 — that Edinger, working on lower vertebrates, introduced the term "medial forebrain bundle". This was a sheaf of longitudinally conducting fibers connecting a variety of phylogenetically old (olfactory and "limbic") structures in the basal and medial walls of the hemisphere with the brain stem, a system on which Cornelius Ariëns Kappers and his associate, Hammar, later expounded (1925). The medial forebrain bundle of lower vertebrates includes the precursor of the mammalian fornix. In mammals, the portions of the medial forebrain bundle not included in the fornix system have been incorporated into the lateral forebrain bundle, that is, into the broad fiber group connecting the corpus striatum with the brain stem which, at a later phylogenetic stage, became augmented by the great afferent and efferent neocortical fiber systems of the internal capsule. What we today call the medial forebrain bundle in higher animal classes is a loose aggregate of multisynaptic fibers that occupies the lateral hypothalamus; it is the major freeway or Autobahn, with many entrances and exits, linking limbic and olfactory structures and midbrain tegmental structures with medially situated hypothalamic grisea, and so bringing all these structures into reciprocal functional relations with the hypothalamus.

The outcome of the attention given the limbic system was that, as a kind of superstructure over the hypothalamus, it was envisaged as feeding into the hypothalamus an assemblage of information which its mechanisms had sorted out and synthesized — information on the execution of complex, yet primitive, sequences of action, information on thirst and hunger (as contrasted with drinking and eating, which are not regarded as in the province of the hypothalamus), on cognition, and on the whole realm of affect and sexual function and their somatic and autonomic accompaniments (MacLean, 1949, 1972; Meissner, 1967; Pribram, 1960; Yakovlev, 1972).

This brings us to the subject of the receptivity of hypothalamic cells, about which a few final remarks seem in order. Verney, in 1947, proposed the term "osmoreceptors" to describe cells (in dogs) that are purported to sense changes in the osmotic pressure of plasma. This was in keeping with his demonstration that an increase in osmolarity of the blood perfusing the head activated an intracranial system to release larger amounts of antidiuretic hormone (ADH). Then Jewell and Verney (1957) sought by arduous and painstaking means to delimit the site of the osmoreceptors (also in dogs), but could not settle on any one of the numerous diencephalic nuclei that were suspect, including the supraoptic nucleus. Bard, Woods and Bleier, in 1966, ruled out a few of these nuclei as the sites of osmoreception (in cats), but left 7 suspect. The most interesting point made by Bard and his associates was that the hypothalamic neurons involved in the release of ADH to their primary stimulus — a change in the osmolarity of their extracellular milieu — is not materially altered by the removal of neural input. This was taken to indicate a high degree of security for the phylogenetically ancient mechanism that through the ages was developed for the production of hypertonic urine.

Then followed Stricker and Wolf's (1967) concept of volume receptors in an unidentified region of the anterior hypothalamus. In due time, the concept of glucoreceptors, thermodetectors, and others, were added. So far as the feedback of hormones to the hypothalamus is concerned, there to act on hypothalamic cells, the most work seems to have been done with estrogens. Szentágothai and his associates, Flerkó, Mess and Halász (1968), are among those who assume that specific elements situated in the region between the optic chiasm and the paraventricular nucleus are sensitive to estrogens and may be excited (or inhibited) by the estrogen level in the blood and in turn regulate (decrease or increase) the secretion of follicle-stimulating hormone. Shute and Lewis (1966) have reviewed the whole subject of cholinergic and mono-

References p. 52–60

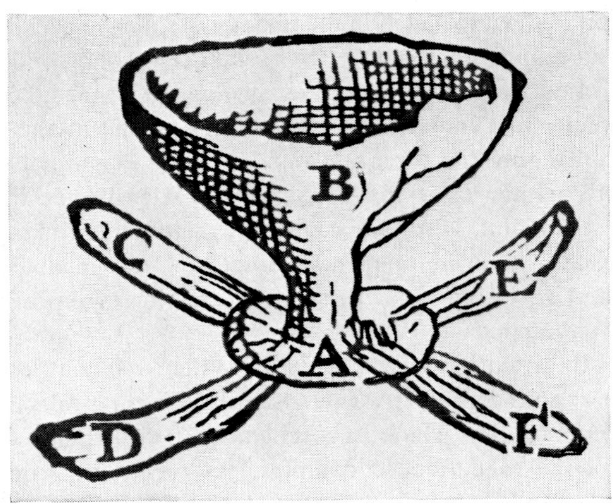

Fig. 39. Vesalius' infundibulum. Vesalius' representation of the infundibulum of the pituitary (*Ref.*, Saunders and O'Malley, 1950, p. 198; Plate 72:6 [VII: Fig. 18]). Vesalius: "In this small figure we have depicted the pelvis or cup *(cyathus)* set upright by which the pituita of the brain distills into the gland underlying it, and then we have sketched in the 4 ducts carrying the pituita down from the gland through foramina in the neighborhood of the gland. Therefore *A* indicates the gland [not shown in this figure] into which the pituita is distilled and *B*, the pelvis along which it is led. *C*, *D*, *E*, *F* are the passages provided for the easier exit of the pituita passing down there. ...".

aminergic pathways in the hypothalamus, finding that the endings of fibers in the supraoptic, paraventricular and infundibular nuclei are cholinoceptive, and that noradrenergic endings are present not only in these nuclei but also in the medial mammillary nucleus and the anterior and lateral hypothalamic areas. The full significance of these findings needs still to be determined.

These are exciting times, awaiting new methods for exploration at the molecular level. We would predict that some years hence two other people giving this opening discussion would select as their title, "Breakthroughs in the Molecular Organization of the Hypothalamus".

ACKNOWLEDGEMENT

We are most appreciative of the suggestions offered by Dr. Francis Schiller of San Francisco, in the preparation of this article.

REFERENCES

Andersson, B. (1952) Polydipsia caused by intrahypothalamic injections of hypertonic NaCl-solutions. *Experientia (Basel)*, **8**, 157–158.

Andersson, B. (1957) Polydipsia, antidiuresis and milk ejection caused by hypothalamic stimulation. In *The Neurohypophysis*, H. Heller (Ed.), Academic Press, New York, pp. 131–140.

Andersson, B. and McCann, S. M. (1956) The effect of hypothalamic lesions on the water intake of the dog. *Acta physiol. scand.*, **25**, 312–320.

ARNDT, R. (1874) Ein Tumor cerebri. *Arch. Psychiat. Nervenkr.*, **4**, 432–464.

ARNDT, R. (1888) Neubildung im Gehirn, Magenerweichung und einfaches oder rundes Magengeschwur. *Dtsch. med. Wschr.*, **14**, 83–85.

ASCHNER, B. (1909) Demonstration von Hunden nach Extirpation der Hypophyse. *Wien. klin. Wschr.*, **22**, 1730–1732.

ASCHNER, B. (1912) Über die Funktion der Hypophyse. *Pflügers Arch. ges. Physiol.*, **146**, 1–146.

BABINSKI, J. (1900) Tumeur du corps pituitaire sans acromégalie et avec arrêt de développement des organes génitaux. *Rev. neurol.*, **8**, 531–533.

BAILEY, P. AND BREMER, F. (1921) Experimental diabetes insipidus. *Arch. intern. Med.*, **28**, 773–803.

BARD, P. (1928) A diencephalic mechanism for the expression of rage with special reference to the sympathetic nervous system. *Amer. J. Physiol.*, **84**, 490–515.

BARD, P. (1973) The ontogenesis of one physiologist. *Ann. Rev. Physiol.*, **35**, 1–16.

BARD, P., WOODS, J. W. AND BLEIER, R. (1966) The locus and functional capacity of the osmoreceptors in the deafferented hypothalamus. *Trans. Ass. Amer. Phycns*, **79**, 107–121.

BARGMANN, W. (1949) Über die neurosekretorische Verknüpfung von Hypothalamus und Neurohypophyse. *Z. Zellforsch.*, **34**, 610–634.

BARGMANN, W. AND SCHARRER, E. (1951) The site of origin of the hormones of the posterior pituitary. *Amer. Sci.*, **39**, 255–259.

BARKER, L. F. (1909) *The Nervous System and its Constituent Neurones*, Stechert, New York, 671 pp.

BARRY, J. AND COTTE, G. (1961) Étude préliminaire au microscope électronique de l'éminence médiane du cobaye. *Z. Zellforsch.*, **53**, 714–724.

BEKHTEREV, V. (1894) *Die Leitungsbahnen im Gehirn und Rückenmark; ein Handbuch für das Studium des Nervensystems*, Transl. by Weinberg from the Russian, Besold, Leipzig, pp. 1877–1878.

BELLOWS, R. T. AND VAN WAGENEN, W. P. (1935) The rôle of thirst in diabetes insipidus. *Yale J. Biol. Med.*, **7**, 572–573.

BRAMWELL, B. (1888) *Intracranial Tumours*, Lippincott, Philadelphia, Pa.

BROCA, P. (1878) Le grand lobe limbique et la scissure limbique dans la série des mammifères. *Rev. d'Anthrop., 2ème Serie*, **1**, 385–498.

BROCA, P. (1879) Recherches sur les centres olfactifs. *Rev. d'Anthrop., 2ème Serie*, **2**, 385–455.

CAMUS, J. ET ROUSSY, G. (1913a) Hypophysectomie et polyurie expérimentales. *C. R. Soc. Biol. (Paris)*, **75**, 483–486.

CAMUS, J. ET ROUSSY, G. (1913b) Polyurie expérimentale par lésions de la base du cerveau; la polyurie hypophysaire. *C. R. Soc. Biol. (Paris)*, **75**, 628–633.

CAMUS, J. ET ROUSSY, G. (1914) Diabète insipide et polyurie dite hypophysaire. *Presse méd.*, **22**, 517–521.

CAMUS, J. AND ROUSSY, G. (1920) Experimental researches on the pituitary body. *Endocrinology*, **4**, 507–522.

CHRIST, J. F. (1951a) Zur Anatomie des Tuber cinereum beim erwachsenen Menschen. *Dtsch. Z. Nervenheilk.*, **165**, 340–408.

CHRIST, J. F. (1951b) Über den Nucleus infundibularis beim erwachsenen Menschen. *Acta neuroveg. (Wien)*, **3**, 267–285.

CLARK, W. E. LEG. (1938) In *The Hypothalamus. Morphological, Functional, Clinical and Surgical Aspects.* W. E. LEG. CLARK, J. BEATTIE, G. RIDDOCH AND N. M. DOTT (Eds.), Oliver and Boyd, Edinburgh, pp. 1–68.

CLARKE, E. AND O'MALLEY, C. D. (1968) *The Human Brain and Spinal Cord. A Historical Study Illustrated by Writings from Antiquity to the Twentieth Century*, Univ. California Press, Berkeley.

CLARKE, R. H. (1920) Investigation of the central nervous system; methods and instruments. *Johns Hopk. Hosp. Rep., Special Volume, Pt. I*, pp. 1–172.

CLARKE, R. H. AND HORSLEY, V. (1905) On the intrinsic fibres of the cerebellum, its nuclei and its efferent tracts. *Brain*, **28**, 13–29.

CLARKE, R. H. AND HORSLEY, V. (1906) On a method of investigating the deep ganglia and tracts of the central nervous system (cerebellum). *Brit. med. J.*, **2**, 1799–1800.

COX, L. B. (1937) Tumours of the base of the brain: their relation to pathological sleep and other changes in the conscious state. *Med. J. Austr.*, **1**, 742–759.

CROWE, S. J., CUSHING, H. AND HOMANS, J. (1910) Experimental hypophysectomy. *Bull. Johns Hopk. Hosp.*, **21**, 127–169.

CUSHING, H. (1909) The hypophysis cerebri; clinical aspects of hyperpituitarism and hypopituitarism. *J. Amer. med. Ass.*, **53**, 249–255.

CUSHING, H. (1912) *The Pituitary Body and its Disorders*, Lippincott, Philadelphia, Pa.

CUSHING, H. (1932) *Papers Relating to the Pituitary Body, Hypothalamus and Parasympathetic Nervous System*, Thomas, Springfield, Ill., pp. 175–240.

DALE, H. H. (1957) Evidence concerning the endocrine function of the neurohypophysis and its nervous control. In *The Neurohypophysis*, H. HELLER (Ed.), Academic Press, New York, pp. 1–9.

DAVIS, R. A. (1964) Victorian physician-scholar and pioneer physiologist. *Surg. Gyn. Obstet.*, **119**, 1333–1340.

DÉJÉRINE, J. (1901) *Anatomie des centres nerveux, Vol. 2*, Rueff, Paris, pp. 553–586.

DIEPEN, R. (1953) Über das Hypophysen–Hypothalamus-System bei Knochenfischen. Eine vergleichend anatomische Betrachtung. *Anat. Anz.*, **100**, 111–122.

DRAGER, G. A. (1950) Neurosecretion following hypophysectomy. *Proc. Soc. exp. Biol. (N.Y.)*, **75**, 712–713.

DUVERNOY, H. (1972) The vascular architecture of the median eminence. *Brain–Endocrine Interaction. Median Eminence: Structure and Function*, Karger, Basel, pp. 79–108.

DUYFF, J. W. (1947) In *Arch. Néerl. Physiol. de l'homme et des animaux. Livre jubilaire G. van Rynberk*, G. VAN RIJNBERK (Ed.), **28**, I–VI.

ECONOMO, C., VON (1917a) Die Encephalitis lethargica. *Wien. klin. Wschr.*, **30**, 581–585.

ECONOMO, C., VON (1917b) Neue Beiträge zur Encephalitis lethargica. *Neurol. Cbl.*, **36**, 866–868.

ECONOMO, C., VON (1920) Sleep as a problem of localization. *J. nerv. ment. Dis.*, **71**, 249–259.

ECONOMO, C., VON (1929) *Die Encephalitis lethargica, ihre Nachkrankheiten und ihre Behandlung*, Urban und Schwarzenberg, Wien.

EDINGER, L. (1893) *Vorlesungen über den Bau der nervösen Centralorgane des Menschen und der Thiers*, Vogel, Leipzig.

ERDHEIM, J. (1904) Über Hypophysenganggeschwülste und Hirncholesteatome. *Sitzber. Akad. Wiss. Wien, Pt. 3*, **113**, 537–726.

ERDHEIM, J. (1916) Nanosmia pituitaria. *Beitr. path. Anat.*, **62**, 302–377.

FINK, R. P. AND HEIMER, L. (1967) Two methods for selective silver impregnation of degenerating axons and their synaptic endings in the central nervous system. *Brain Res.*, **4**, 369–374.

FISHER, C., INGRAM, W. R. AND RANSON, S. W. (1938) *Diabetes Insipidus and the Neuro-hormonal Control of Water Balance: A Contribution to the Structure and Function of the Hypothalamico–Hypophyseal System*. Edwards Bros, Ann Arbor, Mich.

FOREL, A. (1872) Beiträge zur Kenntnis des Thalamus opticus und der ihn umgebenden Gebilde bei den Säugetieren. *Verh. Akad. Wiss. Wien, Math.-Naturw. Kl., Abt. III*, **66**, 25–58.

FOREL, A. (1877) Untersuchungen über die Haubenregion und ihre oberen Verknüpfungen im Gehirne des Menschen und einiger Saügethiere, mit Beiträgen zu den Methoden der Gehirnuntersuchung. *Arch. Phychiat. Nervenkr.*, **7**, 393–495.

FOSTER, M. (1891) *Lectures on the History of Physiology During the Sixteenth, Seventeenth and Eighteenth Centuries*, Univ. Press, Cambridge.

FRENCH, J. D. AND MAGOUN, H. W. (1952) Effects of chronic lesions in central cephalic brain stem of monkeys. *Arch. Neurol. Psychiat (Chic.)*, **68**, 591–604.

FRÖHLICH, A. (1901) Ein Fall von Tumor der Hypophysis cerebri ohne Akromegalie. *Wien. klin. Rundsch.*, **15**, 883–886, 906–908.

FULTON, J. F. (1943) *Physiology of the Nervous System, 2nd Edit.*, Oxford Univ. Press, London.

FULTON, J. F. (1946) *Harvey Cushing, a Biography*, Thomas, Springfield, Ill.

FULTON, J. F. (1954) Contemporary concepts of the hypothalamus and their origin. *Quart. Bull. Northwestern Univ. Med. School*, **28**, 10–16.

FULTON, J. F. AND BAILEY, P. (1929) Contribution to the study of tumors of the third ventricle: Their diagnosis and relation to pathological sleep. *J. nerv. ment. Dis.*, **69**, 1–25, 145–164, 261–277.

FUMAGALLI, Z. (1941) La vascolarizzazione dell'ipofisi umana. *Z. Anat. Entwicklungsgesch.*, **111**, 266–306.

FUXE, K. (1964) Cellular localization of monoamines in the median eminence and the infundibular stem of some mammals. *Z. Zellforsch.*, **61**, 710–724.

GOMORI, G. (1939a) A differential stain for cell types in the pancreatic islands. *Amer. J. Path.*, **15**, 497–499.

GOMORI, G. (1939b) Studies on the cells of the pancreatic islands. *Anat. Rec.*, **74**, 439–459.

GREEN, J. D. (1947) Vessels and nerves of amphibian hypophysis: A study of the living circulation and of the histology of the hypophyseal vessels and nerves. *Anat. Rec.*, **99**, 359–361.

GREEN, J. D. AND HARRIS, G. W. (1947) The neurovascular link between the neurohypophysis and adenohypophysis. *J. Endocrinol.*, **5**, 136–146.

GREEN, J. D. AND HARRIS, G. W. (1949) Observations of the hypophysial-portal vessels of the living rat. *J. Physiol. (Lond.)*, **108**, 359–361.

GREVING, R. (1926) Beiträge zur Anatomie der Hypophyse und ihrer Funktion: I. Eine Faserbindung zwischen Hypophyse und Zwischenhirnbasis (Tr. supraoptico-hypophyseus). *Dtsch. Z. Nervenheilk.*, **89**, 179–195.

GUDDEN, B., VON (1880) Beitrag zur Kenntnis des Corpus mammillare und der sogenannten Schenkel des Fornix. *Arch. Psychiat. Nervenkr.*, **11**, 428–452.

GUILLEMIN, R. AND ROSENBERG, B. (1955) Humoral hypothalamic control of anterior pituitary; a study with combined tissue cultures. *Endocrinology*, **57**, 599–607.

HARRIS, G. W. (1936) Induction of pseudopregnancy in rat by electrical stimulation through head. *J. Physiol. (Lond.)*, **88**, 361–367.

HARRIS, G. W. (1937) The induction of ovulation in the rabbit by electrical stimulation of the hypothalamo-hypophysial mechanism. *Proc. roy. Soc. B*, **122**, 374–394.

HARRIS, G. W. (1948a) Electrical stimulation of the hypothalamus and the mechanism of neural control of the adenohypophysis. *J. Physiol. (Lond.)*, **107**, 418–429.

HARRIS, G. W. (1948b) Neural control of the pituitary gland. *Physiol. Rev.*, **28**, 139–179.

HARRIS, G. W. (1972) Humours and hormones. The Sir Henry Dale Lecture for 1971. *Proc. Soc. Endocrinology*, **53**, ii–xxii.

HARRIS, G. W. AND JACOBSOHN, E. (1952) Functional grafts of the anterior pituitary gland. *Proc. roy. Soc. B*, **139**, 263–276.

HEIMER, L. (1967) Silver impregnation of terminal degeneration in some forebrain fiber systems: A comparative evaluation of current methods. *Brain Res.*, **5**, 86–108.

HERRICK, C. J. (1910) The morphology of the forebrain in amphibia and reptilia. *J. comp. Neurol.*, **20**, 413–546.

HERRICK, C. J. (1933) The amphibian forebrain. VI. Necturus. *J. comp. Neurol.*, **58**, 1–288.

HESS, W. R. (1930) *Die Regulierung des Blutkreislaufes*, Thieme, Leipzig.

HESS, W. R. (1931) *Die Regulierung der Atmung*, Thieme, Leipzig.

HESS, W. R. (1947) *Motorik und Zwischenhirn*, Schwabe, Basel.

HESS, W. R. (1948) *Die funktionelle Organisation des vegetativen Nervensystems*, Schwabe, Basel.

HESS, W. R. (1963) From medical practice to theoretical medicine: an autobiographic sketch. *Perspectives Biol. Med.*, **6**, 400–423.

HESS, W. R. UND BRÜGGER, M. (1943) Das subkortikale Zentrum der affektiven Abwehrreaktion. *Helv. Physiol. Acta*, **1**, 33–52.

HIS, W. (1893) Vorschläge zur Einteilung des Gehirns. *Arch. Anat. Physiol., Anat. Abt.*, 172–179.

HIS, W. (1904) *Die Entwicklung des menschlichen Gehirns während der ersten Monate*, Hirzel, Leipzig.

HOFFMANN, C. E. F. (1868) Über die Erweichung und den Durchbruch der Speiseröhre und des Magens. *Virchows Arch. path. Anat.*, **44**, 352–365.

HORSLEY, V. AND CLARKE, R. H. (1908) The structure and functions of the cerebellum examined by a new method. *Brain*, **31**, 45–124.

HOUSSAY, B. A., BIASOTTI, A. ET SAMMARTINO, R. (1935) Modifications fonctionelles de l'hypophyse après les lésions infundibulo-tubériennes chez le crapaud. *C. R. Soc. Biol. (Paris)*, **120**, 725–727.

INGRAM, W. R., HANNETT, F. I. AND RANSON, S. W. (1932) The topography of the nuclei of the diencephalon of the cat. *J. comp. Neurol.*, **55**, 333–394.

INGRAM, W. R., HANNETT, F. I., RANSON, S. W., ZEISS, F. R. AND TERWILLIGER, E. H. (1932) Results of stimulation of the tegmentum with the Horsley–Clarke stereotaxic apparatus. *Arch. Neurol. Psychiat.*, **28**, 513–541.

JASPER, H. (1960) Evolution of cerebral localization since Hughlings Jackson. *World Neurol.*, **1**, 97–126.

JEWELL, P. A. AND VERNEY, P. A. (1957) An experimental attempt to determine the site of the neurohypophyseal osmoreceptors in the dog. *Phil. Trans. B*, **240**, 197–324.

KAADA, B. R. AND JASPERS, H. (1952) Respiratory responses to stimulation of temporal pole, insula, and hippocampal and limbic gyri in man. *Arch. Neurol. Pyschiat.*, **68**, 609–619.

KAADA, B. R., PRIBRAM, K. H. AND EPSTEIN, J. A. (1949) Respiratory and vascular responses in monkeys from temporal pole, insula, orbital surface and cingulate gyrus: a preliminary report. *J. Neurophysiol.*, **12**, 347–356.

KAPPERS, C. U. A. (1928) The development of the cortex and the function of its different layers. *Acta Psychiat. Neurol.*, **3**, 115–132.

KAPPERS, C. U. A. UND HAMMAR, E. (1918) Das Zentralnervensystem des Ochsenfrosches *(Rana catesbyana)*. *Psychiat. Neurol. Bl. (Amst.)*, **22**, 368–415.

KARPLUS, J. P. (1937) Die Physiologie der vegetativen Zentren. (Auf Grund experimenteller Erfahrungen.) In *Handbuch der Neurologie, Vol. 2*, O. BUMKE, UND O. FOERSTER (Eds.), Springer, Berlin, pp. 402–475.

KARPLUS, J. P. UND KREIDL, A. (1909) Gehirn und Sympathicus. I. Mitteilung. Zwischenhirn und Halssympathicus. *Pflügers Arch. ges. Physiol.*, **129**, 138–144.

KARPLUS, J. P. UND KREIDL, A. (1910) Gehirn und Sympathicus. II. Mitteilung. Ein Sympathicuszentrum im Zwischenhirn. *Pflügers Arch. ges. Physiol.*, **135**, 401–416.

KARPLUS, J. P. UND KREIDL, A. (1918) Gehirn und Sympathicus. IV. Mitteilung. *Pflügers Arch. ges. Physiol.*, **171**, 192–200.

KARPLUS, J. P. UND KREIDL, A. (1927) Gehirn und Sympathicus. VII. Mitteilung. Über Beziehungen der Hypothalamuszentren zu Blutdruck und innerer Sekretion. *Pflügers Arch. ges. Physiol.*, **215**, 667–670.

KARPLUS, J. P. UND KREIDL, A. (1928) Gehirn und Sympathicus. VIII. Mitteilung. *Pflügers Arch. ges. Physiol.*, **219**, 613–618.

KÖLLIKER, A., VON (1896) *Handbuch der Gewebelehre des Menschen, 6th ed., Vols. 1 and 2*, Engelmann, Leipzig.

KUHLENBECK, H. (1924) Über die Homologien der Zellmassen im Hemisphärenhirn der Wirbeltiere. *Folia Anat. jap.*, **2**, 325–364.

KUHLENBECK, H. (1927) *Vorlesungen über das Zentralnervensystem der Wirbelthiere*, Fischer, Jena.

KUHLENBECK, H. (1948) The derivatives of the thalamus ventralis in the human brain and their relation to the so-called subthalamus. *Milit. Surg.*, **102**, 433–447.

KUHLENBECK, H. (1954) *The Human Diencephalon. A Summary of Development, Structure, Function, and Pathology*, Karger, Basel, pp. 21–24, 116–118, 128–129.

KUHLENBECK, H. (1957) *Brain and Consciousness*, Karger, Basel, pp. 170–177, 186–187.

KUHLENBECK, H. (1969) Ontogenetic criteria, etc. In *The Hypothalamus*, W. HAYMAKER, E. ANDERSON AND W. NAUTA (Eds.), Thomas, Springfield, Ill., pp. 13–29.

KUHLENBECK, H. (1973) *The Central Nervous System of Vertebrates. Overall Morphologic Pattern, Vol. 3, Pt. 2.*, Karger, Basel, 717 pp.

KUHLENBECK, H. AND HAYMAKER, W. (1949) The derivatives of the hypothalamus in the human brain: their relation to the extrapyramidal and autonomic systems. *Milit. Surg.*, **105**, 26–52.

LENHOSSÉK, M., VON (1887) Beobachtungen am Gehirn des Menschen. *Anat. Anz.*, 450–461.

LOMONACO, D. E VAN RYNBERK, G. (1901) Sulla funzione dell'ipofisi cerebrale. *Rend. R. Accad. dei Lincei, Ser. 5*, **10**, 172–179, 212–219, 265–270.

LOMONACO, D. E VAN RYNBERK, G. (1901) Richerche sulla funzione della ipofisi cerebrale. *Riv. Neuropat. Psichiat.*, **8**, 1–28.

MACLEAN, P. (1949) Psychosomatic disease and the "visceral brain": recent developments bearing on the Papez theory of emotion. *Psychosom. Med.*, **11**, 338–353.

MACLEAN, P. (1952) Some psychiatric implications of physiological studies on fronto-temporal portion of limbic system (visceral brain). *Electroenceph. clin. Neurophysiol.*, **4**, 407–418.

MACLEAN, P. (1972) Implications of microelectrode findings on exteroceptive inputs into the limbic cortex. In *Limbic System Mechanisms and Autonomic Function*, C. H. HOCKMAN (Ed.), Thomas, Springfield, Ill., pp. 115–136.

MAGOUN, H. W. (1958) *The Waking Brain.*, Thomas, Springfield, Ill.

MAGOUN, H. W., BARRIS, R. W. AND RANSON, S. W. (1932) Stimulation of the hypothalamus with the Horsley–Clarke instrument. *Anat. Rec.*, **52**, Suppl., p. 24.

MALONE, E. (1910) Über die Kerne des menschlichen Diencephalon. *Abh. königl. preuss. Akad. Wiss.*, Anh. Abh. **1** 1–32.

MALONE, E. F. (1914) The nuclei tuberis laterales and the so-called ganglion opticum basale. *Bull. Johns Hopk. Hosp.*, **17**, 441–480.

MARIE, P. (1886) Sur deux cas d'acromégalie, hypertrophie singulière non congénitale des extrémités, supérieures, inférieures et céphalique. *Rev. Méd.*, **6**, 297–333.

MARSHALL, F. H. A. AND VERNEY, E. B. (1936) The occurrence of ovulation and pseudopregnancy in the rabbit as a result of central nervous stimulation. *J. Physiol. (Lond.)*, **86**, 327–336.

MAUTHNER, L. (1890) Pathologie und Physiologie des Schlafes. *Wien. klin. Wschr.*, **3**, 445–446.

McHenry, L. C., Jr. (1969) *Garrison's History of Neurology, revised and enlarged.* Thomas, Springfield, Ill., pp. 17–22, 40–43, 55, 61, 78.

Meissner, W. W. (1967) Memory function in the Korsakoff syndrome. *J. nerv. ment. Dis.*, **145**, 106–122.

Mettler, C. C. (1947) *History of Medicine.* Blakiston, Philadelphia, Pa.

Meynert, T. (1872) Von Gehirne der Säugethiere. In *Handbuch der Lehre von den Geweben des Menschen und der Tieren, Vol. 2, Chap. 31,* S. Stricker (Ed.), Engelmann, Leipzig, pp. 694–696, 701, 731–733, 736.

Minkowski, O. (1887) Über einen Fall von Akromegalie. *Berl. klin. Wschr.*, **24**, 371–374.

Moruzzi, G. (1972) Foreword. In *Limbic System Mechanisms and Autonomic Function,* C. H. Hockman (Ed.), Thomas, Springfield, Ill., pp. ix–xii.

Moruzzi, G. and Magoun, H. W. (1949) Brain stem reticular formation and activation of the EEG. *Electroenceph. clin. Neurophysiol.*, **1**, 455–473.

Nauta, W. J. H. (1950) Über die sogenannte terminale Degeneration im Zentralnervensystem und ihre Darstellung durch Silberimprägnation. *Schweiz. Arch. Neurol. Psychiat.*, **66**, 353–376.

Nauta, W. J. H. (1972) The central visceromotor system: a general survey. In *Limbic System Mechanisms and Autonomic Function,* C. H. Hockman (Ed.), Thomas, Springfield, Ill., pp. 21–38.

Nauta, W. J. H. and Glees, P. (1955) A critical review of studies on axonal and terminal degeneration. *Mschr. Psychiat. Neurol.*, **129**, 74–91.

Nauta, W. J. H. and Gygax, P. A. (1951) Silver impregnation of degenerating axon terminals in the central nervous system: (1) technic. (2) chemical notes. *Stain Technol.*, **26**, 5–11.

Nauta, W. J. H. and Gygax, P. A. (1954) Silver impregnation of degenerating axons in the central nervous system: a modified technique. *Stain Technol.*, **29**, 91–93.

Nauta, W. J. H. and Haymaker, W. (1969) Hypothalamic nuclei and fiber connections. In *The Hypothalamus,* W. Haymaker, E. Anderson and W. J. H. Nauta (Eds.), Thomas, Springfield, Ill., pp. 136–309.

Nauta, W. J. H. and Kuypers, A. G. J. M. (1958) Some ascending pathways in the brain stem reticular formation. In *Reticular Formation of the Brain,* H. H. Jasper *et al.* (Eds.), Little, Brown, Boston, Mass., pp. 3–30.

Nauta, W. J. H. and Ryan, L. F. (1952) Selective silver impregnation of degenerating axons in the central nervous system. *Stain Technol.*, **27**, 175–179.

Nicolesco, I. et Nicolesco, M. (1929) Quelques données sur les centres végétatifs de la frontière diencéphalo-télencéphalique. *Rev. neurol.*, **2**, 290–317.

Nothnagel, H. (1881) Durst und Polydipsie. *Virchows Arch. path. Anat.*, **86**, 435–447.

Nowakowski, H. (1951) Infundibulum und Tuber cinereum der Katze. *Dtsch. Z. Nervenheilk.*, **165**, 261–339.

Ochs, S. (1971) The dependence of fast transport in mammalian nerve fibers on metabolism. *Acta Neuropath.*, suppl. V, 86–96.

Olmsted, J. M. D. (1938) *Claude Bernard, Physiologist,* Harper, New York, p. 25.

O'Malley, C. D. (1964) *Andreas Vesalius of Brussels,* Univ. California Press, Berkeley, pp. 1514–1564.

Palay, S. L. (1955) An electron microscopic study of the neurohypophysis in normal, hydrated, and dehydrated rats. *Anat. Rec.*, **121**, 348.

Palay, S. L. (1957) The fine structure of the neurohypophysis. In *Progress in Neurobiology, Vol. 2,* S. R. Korey and J. I. Nurnberger (Eds.), Hoeber, New York, pp. 31–49.

Papez, J. W. (1937) A proposed mechanism of emotion. *Arch. Neurol. Psychiat.*, **38**, 725–743.

Paulesco, N. C. (1907) Recherches sur la physiologie de l'hypophyse du cerveau; l'hypophysectomie et ces effects. *J. Physiol. (Paris)*, **9**, 441–456.

Paulesco, N. C. (1908) *L'hypophyse du cerveau,* Vigot Frères, Paris.

Penfield, W. (1929) Diencephalic autonomic epilepsy. *Arch. Neurol. Psychiat.*, **22**, 358–374.

Penfield, W. (1937) The cerebral cortex and consciousness. *Harvey Lect.*, **32**, 35–69.

Penfield, W. (1952) Memory mechanisms. *Arch. Neurol. Psychiat.*, **67**, 178–198.

Penfield, W. and Jasper, H. (1954) *Epilepsy and the Functional Anatomy of the Human Brain.* Little, Brown, Boston Mass., p. 198.

Pines, J. L. (1925a) Über die Innervation der Hypophysis cerebri I. *J. Psychol. Neurol.*, **32**, 80–88.

Pines, J. L. (1925b) Über die Innervation der Hypophysis cerebri II. *Z. ges. Neurol. Psychiat.*, **100**, 123–138.

Popa, G. T. and Fielding, U. (1930) A portal circulation from the pituitary to the hypothalamic region. *J. Anat. (Lond.)*, **65**, 88–91.

Popa, G. T. and Fielding, U. (1933) Hypophysio-portal vessels and their colloid accompaniment. *J. Anat. (Lond.)*, **67**, 227–232.

Pribram, K. H. (1960) A review of theory in physiological psychology. *Ann. Rev. Psychol.*, **11**, 1–40.

Ramón y Cajal, S. (1894) Algunas contribuciones al conocimiento de los ganglios del encéfale. III. Hipofisis. *Anal. Soc. espan. Hist. nat.*, **23**, 214–215.

Ramón y Cajal, S. (1909–1911) *Histologie du système nerveux de l'homme et les vertébrés, 2 vol.*, Maloine, Paris.

Ramón-Moliner, E. and Nauta, W. J. H. (1966) The isodendritic core of the brain stem. *J. comp. Neurol.*, **126**, 311–335.

Ranson, S. W. (1934a) On the use of the Horsley–Clarke stereotaxic instrument. *Psychiat. neurol. Bl. (Amst.)*, **38**, 534–543.

Ranson, S. W. (1934b) The hypothalamus: its significance for visceral innervation and emotional impression. *Trans. Coll. Phycns Philad., 4th Ser.*, **2**, 222–242.

Ranson, S. W. and Ingram, W. R. (1932) Catalepsy caused by lesions between the mammillary bodies and third nerve in the cat. *Amer. J. Physiol.*, **101**, 690–696.

Ranson, S. W. and Magoun, H. W. (1933) Respiratory and pupillary reactions induced by electrical stimulation of the hypothalamus. *Arch. Neurol. Psychiat.*, **29**, 1179–1194.

Ranson, S. W. and Magoun, H. W. (1939) The hypothalamus. *Ergebn. Physiol.*, **41**, 56–163.

Rasmussen, A. T. (1947) *Some Trends in Neuroanatomy*, Brown, Dubuque, Iowa.

Reford, L. L. and Cushing, H. C. (1909) Is the pituitary essential to the maintenance of life? *Bull. Johns Hopk. Hosp.*, **20**, 105–107.

Reichert, C. B. (1859–1861) *Der Bau des menschlichen Gehirns durch Abbildungen mit erläuteren Text, I und II Abth.*, Engelmann, Leipzig.

Richter, R. B. and Traut, E. F., (1940) Chronic encephalitis. Pathological report of a case with protracted somnolence. *Arch. Neurol. Pyschiat.*, **44**, 848–866.

Righetti, R. (1903) Contributo clinico e anatomopatologico allo studio dei gliomi cerebrali e all'anatomio delle vie ottiche centrali. *Riv. Pat. Nerv.*, **8**, 241–267. II. Esame istologica del neoplasmo. **8**, 289–312.

Rioch, D. McK. (1929a) Studies on the diencephalon of carnivora. I. The nuclear configuration of the thalamus, epithalamus, and hypothalamus of the dog and cat. *J. comp. Neurol.*, **49**, 1–119.

Rioch, D. McK. (1929b) Studies on the diencephalon of carnivora. II. Certain nuclear configurations and fiber connections of the subthalamus and midbrain of the dog and cat. *J. comp. Neurol.*, **49**, 121–153.

Rioch, D. McK., Wislocki, G. B., O'Leary, J. L., Hinsey, J. C. and Sheehan, D. (1940) A précis of preoptic, hypothalamic and hypophysial terminology with atlas. *Res. Publ. Ass. nerv. ment. Dis.*, **20**, 3–30.

Rokitansky, C., von (1842–1846) *Handbuch der pathologischen Anatomie, Vol. 3*, Braumüller und Seidel, Wien.

Rolleston, H. D. (1936) *The Endocrine Organs in the Health and Disease, With an Historical Review*, Oxford Univ. Press, London.

Rose, J. E. (1942) The ontogenetic development of the rabbit's diencephalon. *J. comp. Neurol.*, **77**, 61–129.

Sachs, E. (1909) On the structure and functional relations of the optic thalamus. *Brain*, **32**, 95–186.

Saffran, M., Schally, A. V. and Benfey, B. G. (1955) Stimulation of release of corticotrophin from adenohypophysis by a neurohypophysial factor. *Endocrinology*, **57**, 439–444.

Sarton, G. (1954) *Galen of Pergamon*, Univ. Kansas Press, Lawrence, Kans.

Saunders, J. B. deC. M. and O'Malley, C. D. (1950) *The Illustrations from the Works of Andreus Vesalius of Brussels*, World Publ., Cleveland, Ohio.

Schäfer, E. A. and Herring, P. T. (1908) The action of pituitary extracts upon the kidneys. *Phil. Trans. B*, **cxcix**, 1–29.

Scharrer, E. (1928) Die Lichtempfindlichkeit blinder Elritzen. (Untersuchungen über das Zwischenhirn der Fische, I.) *Z. vergl. Physiol.*, **7**, 1–38.

Scharrer, E. and Scharrer, B. (1940) Secretory cells within the hypothalamus. *Res. Publ. Ass. nerv. ment. Dis.*, **20**, 170–194.

Schiff, M. (1845) *De vi motoria baseos encephali inquisitiones experimentales.* Cockenheim, Frankfurt a.M.

Schiff, M. (1854) Über die Gefässnerven des Magens und die Funktion der mittleren Stränge des Rückenmarkes. *Arch. physiol. Heilk.*, **13**, 30–38.

SCHIFF, M. (1867) *Leçons sur la physiologie de la digestion, Vol. 2, Leçon 35*, Loescher, Florence and Turin, pp. 46–452.

SCHÜTZ, H. (1891) Anatomische Untersuchungen über den Faserverlauf im centralen Höhlengrau und den Nervenfaserschwund in demselben bei der progressiven Paralyse der Irren. *Arch. Psychiat. Nervenkr.*, **22**, 527–587.

SHUTE, C. C. D. AND LEWIS, P. R. (1966) Cholinergic and monoaminergic pathways in the hypothalamus. *Brit. med. Bull.*, **22**, 221–226.

SIGERIST, H. E. (1958) *The Great Doctors*, Doubleday, New York.

SINGER, C. J. (1946) Some Galenic and animal sources of Vesalius. *J. Hist. Med. allied Sci.*, **1**, 6–9.

SINGER, C. (1957) *A Short History of Anatomy and Physiology from the Greeks to Harvey*, Dover Publications, New York.

SLOPER, J. C. (1954) Histochemical observations on the neurohypophysis in dog and cat, with reference to the relationship between neurosecretory material and posterior lobe hormone. *J. Anat. (Lond.)*, **88**, 576–577.

SLOPER, J. C. (1958) Presence of a substance rich in protein-bound cystine or cysteine in the neurosecretory system of an insect. *Nature (Lond.)*, **179**, 148–149.

SLOPER, J. C. (1966a) Hypothalamic neurosecretion. The validity of the concept of neurosecretion and its physiological and pathological implications. *Brit. med. Bull.*, **22**, 209–215.

SLOPER, J. C. (1966b) The experimental and cytopathological investigation of neurosecretion in the hypothalamus and pituitary. In *The Pituitary Gland, vol. 3*, G. W. HARRIS AND B. T. DONOVAN (Eds.), Univ. California Press, Berkeley, pp. 131–239.

SPATZ, H. (1951) Neues über die Verknüpfung von Hypophyse und Hypothalamus. *Acta neuroveg.*, **3**, 5–49.

SPATZ, H., DIEPEN, R. UND GAUPP, V. (1948) Zur Anatomie des Infundibulum und des Tuber cinereum beim Kaninchen. Zur Frage der Verknüpfung von Hypophyse und Hypothalamus. *Dtsch. Z. Nervenheilk.*, **159**, 229–268.

STENDELL, W. (1914) Die Hypophysis cerebri. In *Lehrbuch der vergleichenden mikroskopischen Anatomie, pt. 8.* A. OPPEL (Ed.), Fischer, Jena, pp. 26–29.

STEVENSON, J. A. F. (1949) Effects of hypothalamic lesions on water and energy metabolisms in the rat. *Recent Progr. Hormone Res.*, **4**, 363–394.

STORY, J. B. (1887) Discussion of pituitary lesions in relation to eye symptoms, obesity, menstrual disturbances and mental manifestations. Trans. Ophthal. Soc. U. K., *Brit. med. J.*, **1**, 1334–1335.

STRICKER, E. M. AND WOLF, G. (1967) Hypovolemic thirst in comparison with thirst induced by hyperosmolarity. *Physiol. Behav.*, **2**, 33–37.

STRÜMPELL, A. (1877) Ein Beitrag zur Theorie des Schlafes. *Pflügers Arch. ges. Physiol.*, **15**, 573–574.

SZENTÁGOTHAI, J. (1964) The parvicellular neurosecretory system. In *Lectures on the Diencephalon, Progr. Brain Res.*, W. BARGMANN AND J. P. SCHADÉ (Eds.), **5**, 135–146.

SZENTÁGOTHAI, J., FLERKÓ, B., MESS, B. AND HALÁSZ, H. (1968) *Hypothalamic Control of the Anterior Pituitary. An Experimental–Morphological Study, 3rd Ed., revised and enlarged*, Kiadó, Budapest, pp. 278–287.

THAON, P. (1907) *L'hypophyse à l'état normal et dans les maladies, 2nd Ed.*, Doin, Paris, pp. 15–17.

THAON, P. (1940) *The Hypothalamus and Central Levels of Autonomic Function. Res. Publ. Ass. nerv. ment. Dis., Vol. 20*, Williams and Wilkins, Baltimore, Md.

TÖRÖK, B. (1954) Lebendbeobachtung des Hypophysenkreislaufes an Hunden. *Acta morph. Akad. Sci. Hung.*, **4**, 83–89.

TRENDELENBURG, W. (1907) Studien zur Operationstechnik am Zentralnervensystem. 1. Das Myelotom, ein Apparat zur Ausführung genau begrenzter Durchschneidungen. II. Medianspaltung des Kleinhirns am Kaninchen. *Arch. Anat. Physiol.*, 83–103.

TSIA, C. (1925) The optic tracts and centers of the opossum *Didelphis virginiana. J. comp. Neurol.*, **39**, 173–216.

VAN DYKE, H. B., CHOW, B. F., GREEP, R. C. AND ROTHEN, A. L. (1942) Isolation of protein from pars neuralis of ox pituitary with constant oxytocic, pressor and diuresis-inhibiting activities. *J. Pharmacol. exp. Ther.*, **74**, 190–209.

VERNEY, E. G. (1947) The antidiuretic hormone and the factors which determine its release. *Proc. roy. Soc. B*, **135**, 25.

WEISS, P. AND HISCOE, H. B. (1948) Experiments on the mechanism of nerve growth. *J. exp. Zool.*, **107**, 315–395.

WILSON, S. A. K. (1914) An experimental research into the anatomy and physiology of the corpus striatum. *Brain*, **36**, 427–492.

WISLOCKI, G. B. AND KING, L. S. (1936) The permeability of the hypophysis and hypothalamus to vital dyes, with a study of the vascular supply. *Amer. J. Anat.*, **58**, 421–472.

XUEREB, G. P., PRICHARD, M. M. L. AND DANIEL, P. M. (1954) The hypophysial portal system of vessels in man. *Quart. J. exp. Physiol.*, **39**, 219–230.

YAKOVLEV, P. (1949) Motility, behavior and the brain. Stereodynamic organization and neural coordinates of behavior. *J. nerv. ment. Dis.*, **107**, 313–335.

YAKOVLEV, P. (1972) A proposed definition of the limbic system. In *Limbic System Mechanisms and Autonomic Function*. C. H. Hockman (Ed.), Thomas, Springfield, Ill., pp. 241–283.

ZUCKERKANDL, E. (1888) Das Riechbundel des Ammonshornes. *Anat. Anz.*, **3**, 425–434.

ZUCKERKANDL, E. (1904) Die Riechstrahlung. *Arb. neurol. Inst. Wien*, **11**, 1–28.

Structure and Fiber Connections of the Hypothalamus in Mammals

H. J. LAMMERS AND A. H. M. LOHMAN

Department of Anatomy, Radboud University, Nijmegen (The Netherlands)

INTRODUCTION

The hypothalamus in the mammalian brain encompasses the most ventral part of the diencephalon where it forms the floor and, in parts, the walls of the third ventricle. Its upper boundary is marked by a sulcus in the ventricular wall: the ventral diencephalic or hypothalamic sulcus. This sulcus separates the hypothalamus from the dorsally located thalamus (Fig. 1).

Caudally the hypothalamus merges without any clear demarcation with the periventricular gray and the tegmentum of the mesencephalon. It is, however, customary to define the caudal boundary of the hypothalamus as represented by a plane extending

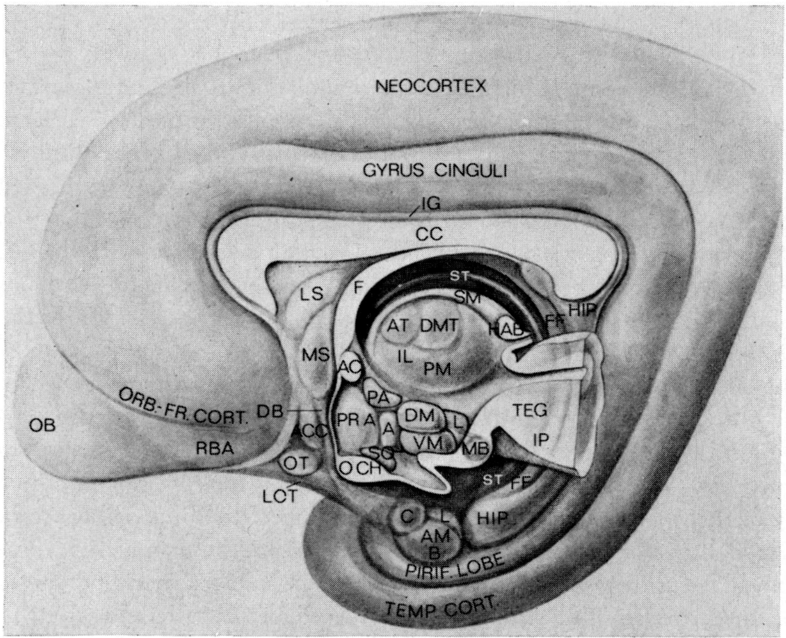

Fig. 1. Medial view of the topography of the hypothalamus and its surrounding structures in the mammalian brain. For abbreviations see p. 75.

References p. 75–78

from the caudal limit of the mammillary nuclei ventrally and the posterior commissure dorsally. Rostrally the hypothalamus is continuous with the preoptic area, which lies partly forward to and above the optic chiasma. Although it is still in dispute whether the preoptic area is of telencephalic origin or has to be considered part of the diencephalon, the similarities in histological structure and fiber connections seem to justify the concept that this area is a rostral continuation of the hypothalamus. Therefore, in the present review these two areas will be considered together.

The preoptic area at its rostral end is bounded by the anterior commissure and the lamina terminalis. Lying in front of these two structures is the septal region or septum. At their lateral side the preoptic area and the anterior part of the hypothalamus are continuous with the innominate substance (substantia innominata). This area consists of a series of cell islands, and expands laterally toward the prepyriform cortex and the amygdaloid complex. In its rostral part, the innominate substance is ventrally covered by the olfactory tubercle.

At more caudal levels the hypothalamus is limited laterally by the internal capsule and the pes pedunculi. Dorsal to the pes pedunculi it merges with the subthalamus. Ventrally the hypothalamus extends to the free, basal surface of the brain. This surface is characterized by 3 prominent structures, *viz.* the optic chiasma, the hypophyseal or infundibular stalk and the paired mammillary bodies. It is of interest to note that, although the hypophyseal stalk as well as the neural lobe of the hypophysis are of hypothalamic origin, they are never considered as parts of the hypothalamus.

STRUCTURE OF THE HYPOTHALAMUS

By means of the above-mentioned external landmarks at the ventral surface of the brain, the hypothalamus can be subdivided into an anterior part which includes the preoptic area, a middle part and a posterior part. Another subdivision is that of Crosby and Woodburne (1940) in which 3 longitudinal zones are recognized: a periventricular, a medial and a lateral zone. The periventricular zone mostly consists of small cells which, in general, are oriented along fibers parallel with the wall of the third ventricle. The medial zone is cell-rich, whereas the lateral zone contains only a small number of cells interposed between the longitudinal fiber system of the medial forebrain bundle.

On the basis of cytoarchitectonic criteria such as size and shape of cell bodies or differences in cellular density and spatial arrangement, in each of the 3 longitudinal zones a number of cell groups or *nuclei* has been described. Cell collections, that are loosely grouped, are commonly designated as *areae*. It must be noted, however, that of many of the cell groups which have been described, the boundaries are rather ill-defined, and that, therefore, each subdivision of the hypothalamic gray in separate morphological entities, including the one given in Table I, must be considered as tentative and arbitrary. For a survey of the hypothalamic nuclei in various mammals the reader is referred to Christ (1969).

TABLE I

HYPOTHALAMIC NUCLEI

Periventricular zone	
nucleus preopticus medianus	area periventricularis anterior
area preoptica periventricularis	nucleus arcuatus (infundibularis)
nucleus paraventricularis	area periventricularis posterior
nucleus suprachiasmaticus	
Medial zone	
nucleus preopticus medialis	area hypothalami posterior
nucleus hypothalami anterior	nuclei gemini (supramammillares)
nucleus dorsomedialis	nucleus premammillaris
nucleus ventromedialis	nucleus mammillaris medialis
area hypothalami dorsalis	
Lateral zone	
nucleus preopticus lateralis	nucleus perifornicalis
nucleus supraopticus	nuclei tuberis laterales
area hypothalami lateralis	nucleus mammillaris lateralis
nucleus tuberomammillaris	

It has further been revealed by Golgi studies (Szentágothai *et al.*, 1968; Millhouse, 1969) that dendrites of cells in specific nuclei may extend far beyond the cytoarchitectonic boundaries of these nuclei. Especially between the neurons of the medial and lateral zones, there exists a considerable overlap of the dendritic fields. It is tempting to suggest that this arrangement may serve as a prime morphological basis for functional interaction between neighboring hypothalamic areas.

FIBER CONNECTIONS OF THE HYPOTHALAMUS

Although it may be helpful, especially for the purpose of orientation, to study the fiber systems connecting the hypothalamus with other brain structures in normal material, determination of the precise sites of origin and termination of these longer fiber connections can be done reliably only in animal experiments. Following an experimentally induced lesion of a nerve fiber or its cell body, the fiber peripheral to the lesion undergoes an anterograde degeneration, which in essence is a breakdown of the fiber into fragments.

The first method for the study of anterograde degeneration was developed in 1885 by Marchi and Algeri, who noticed that degenerating myelin, that had been treated with a potassium dichromate solution, can be selectively stained by osmium tetroxide. Although the Marchi method has contributed considerably to our knowledge of pathways in the central nervous system, its usefulness is limited to the study of myelinated axons. Axon arborizations and synaptic end-structures, which are not covered by myelin, are beyond the reach of the Marchi technique. Fortunately, the development and gradual refinement of the reduced silver technique, which im-

References p. 75–78

pregnates degenerating axoplasm instead of degenerating myelin, has made it possible to demonstrate the course and the termination of both myelinated and unmyelinated fiber systems.

Many silver techniques were used for the study of the nervous system in the first half of this century. It remained for Glees (1946) and Nauta (1950), however, to demonstrate the extreme usefulness of silver methods for the tracing of degenerating nerve fibers in experimental material. One of the great advantages of the Nauta procedure, of which many variations are available (see Ebbesson, 1970), is that it suppresses the argyrophilia (*i.e.* affinity for silver) of normal fibers, which in turn enhances the identification of the degenerating fibers. During the last few years the value of the reduced silver technique has increased considerably because of the development of more sensitive procedures (Fink and Heimer, 1967; De Olmos, 1969; Eager, 1970; Wiitanen, 1969). These modifications are capable of demonstrating the terminal axon ramifications and boutons of most fiber pathways in the central nervous system. The electron-microscopic mapping of degenerating terminal boutons, which recently has been introduced (Heimer, 1970), not only offers the opportunity to confirm light-microscopic findings but also makes it possible to identify the anatomical relations between degenerating axons and postsynaptic neurons in a specific area of termination.

Before discussing in detail the fiber connections of the hypothalamus, it must be pointed out that most of the recent experimental studies on hypothalamic fiber anatomy have been carried out in common laboratory animals such as the cat, rat, mouse and hamster. One should, however, be aware of the possibility that there may exist considerable species differences in the origin, course and termination of analogous fiber systems. Also the composition of fiber bundles by shorter or longer neural links may vary in the various species, resulting in differences in the number of synaptic interruptions. Furthermore, it seems likely to assume that differences between lower and higher mammals in the development of specific brain areas may have their consequences on the anatomy of the hypothalamic fiber systems.

In the following account the connections of the hypothalamus will be dealt with as follows.

(1) Afferent hypothalamic connections from the basal olfactory areas and the amygdaloid complex.

(2) Afferent hypothalamic connections from the limbic forebrain area and the neocortical areas.

(3) Afferent hypothalamic connections from the mesencephalon.

(4) Retinohypothalamic connections.

(5) Thalamohypothalamic relationships.

(6) Efferent and intrinsic connections of the hypothalamus.

(1) Afferent hypothalamic connections from the basal olfactory areas and the amygdaloid complex

Ever since the first descriptions of Déjérine (1895), Herrick (1910) and Ramón y Cajal

(1911) of massive fiber systems between the rostral olfactory areas at the basal surface of the forebrain and the diencephalon, it has been assumed that there exists a close relationship between the sense of smell and the hypothalamus. The basal or secondary olfactory areas receive direct projections from the olfactory bulb and consist rostrally of the anterior olfactory nucleus, the retrobulbar area, the olfactory tubercle and the prepyriform cortex, and caudally of the periamygdaloid cortex and parts of the amygdaloid complex and entorhinal cortex (Scalia, 1968).

Very recently the projection from the rostral olfactory areas to the hypothalamus has again been confirmed in experimental material by Scott and Leonard (1971). Following lesions of the prepyriform cortex and olfactory tubercle in rats, these authors were able to trace degenerated fibers to the lateral preoptic area by the use of the Fink–Heimer technique. Part of these fibers terminate in this area and the anterior part of the lateral hypothalamic zone, whereas other fibers pass through the stria medullaris and the inferior thalamic peduncle to reach the dorsomedial nucleus of the thalamus where they terminate in its medial division. A third contingent of degenerating fibers joins the medial forebrain bundle and can be traced through the lateral hypothalamus to the nuclei gemini (supramammillares) in the posterior part of the hypothalamus just above the region of the mammillary nuclei (Fig. 2).

On the basis of a study in which light- and electron-microscopic techniques were combined, the existence of a direct, monosynaptic connection between the prepyriform cortex and the hypothalamus has, however, seriously been questioned by Heimer (1972). According to this author, the electron-microscopic mapping of anterograde degenerating terminals provides little evidence that fibers originating from rostral olfactory areas actually terminate in the preoptic and anterior hypo-

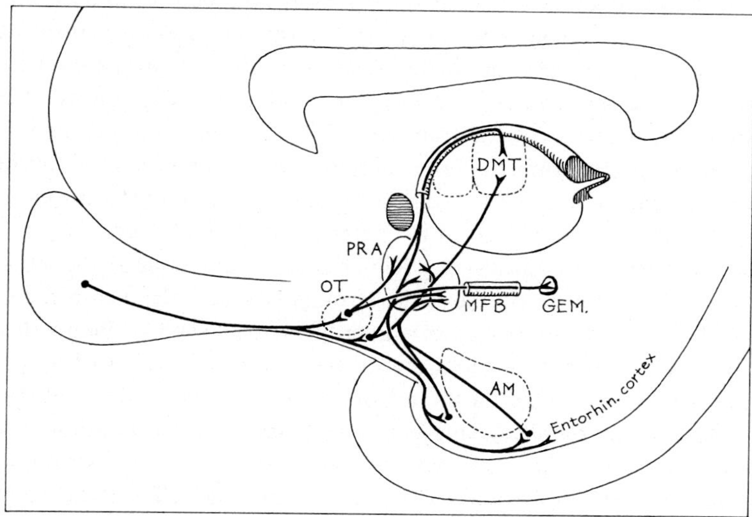

Fig. 2. Schematic drawing of the projections from the olfactory tubercle and the prepyriform and periamygdaloid cortices to the preoptic area, hypothalamus and thalamus. The secondary olfactory connections from the olfactory bulb are also illustrated. For abbreviations see p. 75.

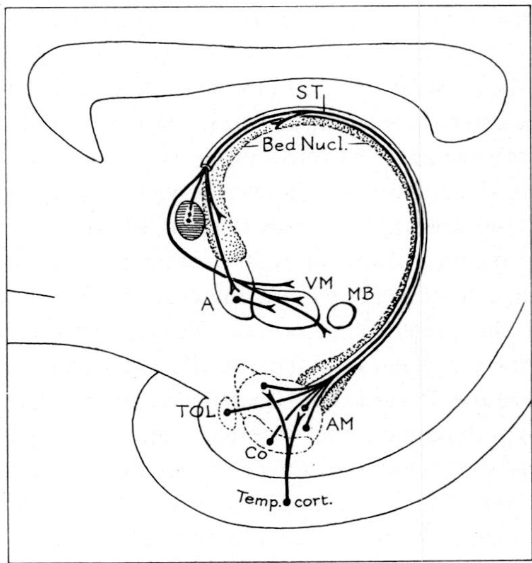

Fig. 3. Origin, course and distribution sites of the stria terminalis. Fibers from the temporal neocortex distribute to the basolateral and central amygdaloid nuclei. For abbreviations see p. 75.

thalamic areas, and this would mean that these fibers only traverse the preoptic area and terminate either in the dorsomedial thalamic nucleus or in the nuclei gemini caudally in the hypothalamus. It is of interest to note that, according to Heimer (1972), the majority of the fibers that reach the nuclei gemini seem to have their origin in the olfactory tubercle rather than in the prepyriform cortex.

The caudal basal olfactory areas, *i.e.* the periamygdaloid cortex and the amygdaloid complex, give rise to two fiber bundles: the stria terminalis and the ventral amygdalofugal pathway. After leaving the amygdaloid complex, the stria terminalis runs a curved course along the medial border of the caudate nucleus toward the anterior commissure where it divides into 3 components: (1) a commissural component which enters the anterior commissure, (2) a precommissural or supracommissural component whose fibers descend in front of the anterior commissure, and (3) a postcommissural component which descends caudal to the anterior commissure (Fig. 3).

According to the observations of Heimer and Nauta (1969), the fibers of the precommissural component of the stria terminalis reach the ventromedial hypothalamic nucleus and terminate predominantly in the "shell" around this nucleus with only a limited distribution to its central core. Additional sites of termination of the precommissural component in the hypothalamus are the nucleus tuberis lateralis and the ventral premammillary nucleus. The pattern of termination of the stria terminalis fibers around the ventromedial hypothalamic nucleus strongly suggests that the synaptic contacts between these fibers and the cells of the ventromedial nucleus are almost entirely axodendritic. It must be noted, however, that the fibers of the precommissural component of the stria terminalis which converge upon the ventromedial nucleus may also have synaptic contacts with cells of neighboring areas as, for instance, the dorsomedial hypothalamic nucleus and the lateral hypothalamic region. As is known from the Golgi studies of Millhouse (1969), dendrites of cells of these areas extend into the cell-poor zone that surrounds the ventromedial nucleus.

The site of origin of the precommissural component of the stria terminalis, namely the posterior part of the cortical amygdaloid nucleus (Leonard and Scott, 1971), is, according to the observations of Winans and Scalia (1970), also the site of termination of fibers that take their origin from the accessory olfactory bulb. The accessory bulb receives fibers from the vomeronasal organ of Jacobson which is an olfactory receptive structure in the nose. This means that there may exist a very close relationship, only interrupted by two synapses (one in the accessory olfactory bulb and one in the posterior part of the amygdaloid nucleus), between the vomeronasal epithelium and the cells of the ventromedial nucleus in the hypothalamus. It is possible that a similar relationship holds true for the projection of the main olfactory epithelium by way of the main olfactory bulb. According to the recent observations of De Olmos (1972), the precommissural fibers of the stria terminalis which reach the ventromedial hypothalamus also originate from the anterior parts of the cortical and medial amygdaloid nuclei, and these parts, as has been shown by Lohman (1963), Heimer (1968) and Scalia (1966), receive a direct projection from the main olfactory bulb. By means of his own cupric–silver method it was further found by De Olmos (1972) that the stria terminalis fibers which have their origin in the anterior parts of the cortical and medial amygdaloid nuclei distribute to the core of the ventromedial hypothalamic nucleus and not to the cell-poor zone surrounding this nucleus.

The fibers of the postcommissural component of the stria terminalis originate from the corticomedial as well as from the basolateral nuclei of the amygdaloid complex (Leonard and Scott, 1971; De Olmos, 1972). They terminate in the bed nucleus of the stria terminalis and in an area of the medial hypothalamic zone which roughly corresponds with the anterior hypothalamic nucleus. A projection from this area to the core of the ventromedial nucleus has been demonstrated by Chi (1970b). Via the postcommissural component the neocortex may have access to the medial hypothalamus, because, according to the observations of Whitlock and Nauta (1956), Druga (1969) and Lescaut (as cited by Hall, 1972), there exists convincing evidence for the existence of direct projections from the neocortex to the basolateral and central amygdaloid nuclei.

Another route, by which fibers of the olfactory cortex may reach the hypothalamus, is the ventral amygdalofugal pathway. As has been shown by Leonard and Scott (1971) in the rat, this pathway consists of short and long fibers. The short fibers distribute not further rostralward than the anterior amygdaloid area, whereas the long fibers, which take their origin from the periamygdaloid cortex, continue into the preoptic–hypothalamic region. It is not known whether these long fibers actually terminate in the lateral hypothalamic area or, as is the case with the fibers from the prepyriform cortex, are fibers of passage *en route* to the dorsomedial nucleus of the thalamus. Recent detailed experiments by De Olmos (1972) indicate that also the central amygdaloid nucleus may serve as a source from which long fibers reach the lateral hypothalamus. It is, however, also possible, as has already been pointed out by the author

References p. 75–78

himself, that the demonstration of this pathway may be attributed to a massive interruption in the rostral part of the amygdaloid complex of fibers which have their origin in the periamygdaloid cortex.

(2) Afferent hypothalamic connections from the limbic forebrain area and the neocortical areas

The limbic forebrain area is built up by two concentric ring-like structures around the hilus of the telencephalon. The inner ring is formed by the hippocampal formation which can be subdivided into pre-, supra- and retrocommissural parts. The outer ring is constituted by the fornicate gyrus which consists of the parolfactory or subcallosal area, the cingulate gyrus, the retrosplenial area and the parahippocampal gyrus.

The outer ring of the limbic forebrain area receives afferents from large portions of the neocortex (Fig. 4), and therefore, might be considered as an area of convergence of all somatic sensory modalities. As judged from its fiber projections (White, 1959), the outer ring distributes these impulses towards the inner ring so that this ring constitutes not only an histological but also a functional zone of transition between the neocortex and the hippocampal formation of the inner ring.

The efferent fiber system of the hippocampal formation is the fornix (Fig. 4). This massive fiber bundle curves underneath the corpus callosum in a forward direction towards the septal region. Here it gives off fibers to the septal nuclei and the lateral preoptic zone. The main bundle of the fornix or postcommissural fornix descends

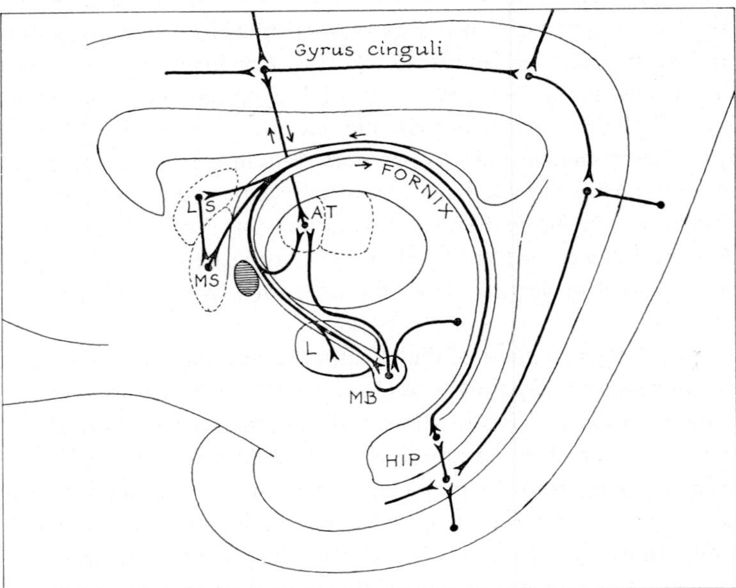

Fig. 4. Diagram illustrating the circuit of Papez. Also are indicated the projections of the fornix bundle to the septal nuclei and the lateral hypothalamic area, the intrinsic septal connections and the mammillary peduncle. The gyrus cinguli receives association fibers from large portions of the neocortex. For abbreviations see p. 75.

caudal to the anterior commissure and continues through the lateral hypothalamic zone to the mammillary nuclei.

Although the precise sites of termination of the fornix fibers in the hypothalamus have not been worked out with modern modifications of the reduced silver technique, it is quite apparent from experimental studies with the Nauta–Gygax method (Guillery, 1956; Nauta, 1956, 1958; Raisman *et al.*, 1966) that the fornix fibers distribute not only to the lateral preoptic zone and the mammillary nuclei but also to the lateral hypothalamic zone and, at least in rodents, to the anterior periventricular area and the nucleus arcuatus (tractus cortico-hypothalamicus medialis of Gurdjian, 1927). Other fibers of the fornix system reach the anterior nuclei and rostral intralaminar and paramedian nuclei of the thalamus, whereas some fibers of the postcommissural fornix bypass the mammillary nuclei and continue caudalward into the paramedian mesencephalic region (limbic midbrain area of Nauta, 1958).

The fornix system consists not only of efferent hippocampal fibers but also contains fibers that originate in the septum and are afferent to the hippocampal formation (Daitz and Powell, 1954; Raisman, 1966). As will be discussed below, the septum by way of the medial forebrain bundle also has a bi-directional relationship with the mesencephalon, and thus, obviously, constitutes an important center in the so-called limbic forebrain–limbic midbrain circuit.

Apart from its influence on the hypothalamus by way of the limbic forebrain area, the neocortex has also direct access to the hypothalamus. As reported by De Vito and Smith (1964) and Nauta (1962, 1964), the frontal lobe or prefrontal cortex of the monkey has substantial subcortical projections to the preoptic region and the hypothalamus. For further details the reader is referred to Nauta, 1971.

(3) Afferent hypothalamic connections from the mesencephalon

There are two fiber systems that connect the midbrain with the hypothalamus: a medial or periventricular system and a lateral system. Both these systems take their origin from the paramedian or limbic midbrain area which, according to Nauta (1972), is composed of the central gray substance and several tegmental cell groups including the nucleus centralis tegmenti superior of Bechterew and the ventral and dorsal tegmental nuclei of Gudden. From experimental studies it seems very unlikely that any direct projection reaches the hypothalamus from the spinal cord or the caudal part of the brain stem (Mehler *et al.*, 1960; Morest, 1967). The mesencephalon might, therefore, be considered as an intermediary interposed between the ascending impulses in the brain stem and the more rostrally situated hypothalamus.

According to Szentágothai *et al.* (1962), the ascending periventricular system has two components, a posterior and an anterior one. The posterior component distributes its fibers to the supramammillary and premammillary cell groups, the lateral hypothalamic zone and the ventromedial and dorsomedial nuclei. The anterior component continues as far forward as the anterior periventricular area and the anterior hypothalamic nucleus. In a recent study in the rat and cat with the Fink–Heimer method, Chi (1970a) was unable to demonstrate axon or terminal degeneration

References p. 75–78

in the ventromedial hypothalamic nucleus after lesions of the central gray and adjacent tegmentum of the mesencephalon.

The lateral system reaches the hypothalamus largely by way of the mammillary peduncle (Fig. 6). This bundle gives off fibers to the mammillary nuclei and ascends in the medial forebrain bundle. It apparently distributes to the lateral hypothalamic and preoptic zones and in part also reaches the medial zone of the septum (Morest, 1961; Nauta and Kuypers, 1958).

(4) Retinohypothalamic connections

There is substantial evidence from physiological studies that environmental light has an important influence upon the regulation of neuroendocrine functions (Green, 1969), and, therefore, the presence of a retinohypothalamic projection has always been under consideration. Despite intensive research in the last decades, there is as yet no general agreement whether such a pathway exists in the mammalian brain. Experimental studies utilizing the Marchi method and the reduced silver technique have largely failed to demonstrate a direct retinohypothalamic connection (see Nauta and Haymaker, 1969).

Very recently, however, a number of reports have been produced which, by the use of the newly-developed autoradiographic tracing method, have provided evidence for a localized fiber projection from the retina to the medial hypothalamic zone (Moore and Lenn, 1972; Hendrickson *et al.*, 1972; Moore, 1973). The autoradiographic method makes use of the fact that the neuronal cell bodies will incorporate tritiated amino acids such as proline and leucine into proteins which subsequently are transported along the axons to their terminals (Droz and Leblond, 1963). Application of this technique by way of intraocular injection has clearly demonstrated that in a wide range of mammals a projection exists from the retina to the suprachiasmatic nuclei lying medially in the anterior hypothalamus on either side of the floor of the third ventricle. This has been confirmed by electron-microscopic observations (Moore and Lenn, 1972; Hendrickson *et al.*, 1972).

The retinohypothalamic projection is bilateral, but the input to the contralateral side is always heavier than that to the ipsilateral nucleus. It has further been observed that most of the fibers project to the ventral half of both ipsi- and contralateral nuclei, and, according to Moore (1973), their area of termination is mostly confined to the caudal half or two-thirds of the suprachiasmatic nuclei.

(5) Thalamohypothalamic relationships

Since the experimental study of Nauta (1962) in the monkey it has commonly been believed that a direct pathway exists from the medial division of the dorsomedial nucleus of the thalamus to the hypothalamus. Because we presently know that the medial division of the dorsomedial thalamic nucleus, at least in the rat, is reached by a substantial projection from the rostral olfactory areas (see Heimer, 1972) and that this nucleus also receives fibers from the cerebellum, the spinothalamic tract and the brain

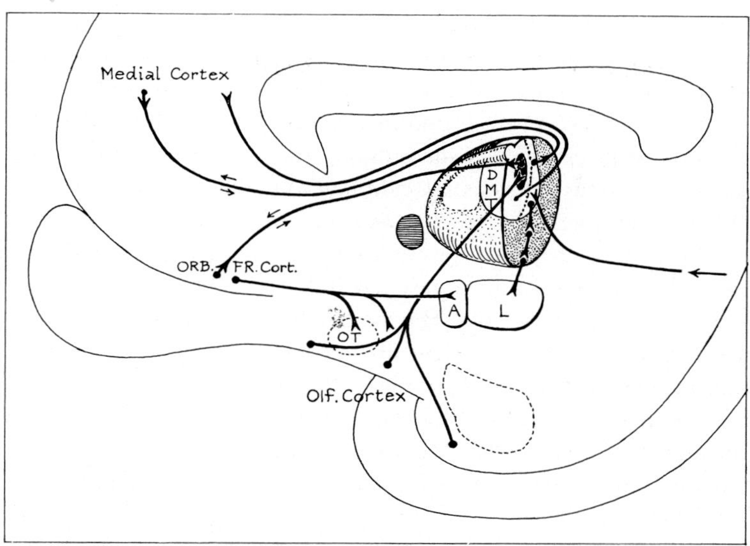

Fig. 5. Schematic representation of the afferent and efferent connections of the dorsomedial thalamic nucleus. For abbreviations see p. 75.

stem (Mehler, 1966; Chi, 1970a) (Fig. 5), the dorsomedial thalamic nucleus can thus be thought to be an important link in the conveyance of olfactory and perhaps also of somatosensory and visceral information to the hypothalamus. Recent experiments by Siegel and coworkers (1973) provide evidence that such a direct connection between the dorsomedial thalamic nucleus and the hypothalamus, as has been reported by Nauta (1964) in the monkey, may not exist in all mammals. It was found by these authors that this connection in the cat is a multineuronal pathway interrupted by at least two synapses in the midline thalamus. The final link in this pathway projects to the perifornical nucleus anteriorly in the lateral hypothalamic zone.

Another route by which the dorsomedial nucleus of the thalamus may exert influence on the hypothalamus is by way of the thalamocortical and corticohypothalamic pathways. The intermediary in this thalamo-cortico-hypothalamic circuit is in the monkey the orbital surface of the prefrontal cortex and in the rat the sulcal cortex lying along the dorsal bank of the rhinal fissure (Fig. 5). In both species these cortical areas receive projections from the medial division of the dorsomedial thalamic nucleus, and are connected by efferent fibers to the lateral preopticohypothalamic region (Freeman and Watts, 1947; Akert, 1964; Nauta, 1962; Leonard, 1969).

(6) Efferent and intrinsic connections of the hypothalamus

The best-known efferent pathways of the hypothalamus are the hypothalamohypophyseal fiber system (tractus supraoptico-paraventriculo-hypophyseus and tractus tuberoinfundibularis) and the projections from the mammillary nuclei. The supraoptico-paraventriculo-hypophyseal tract is made up by axons of the nucleus supraopticus and nucleus paraventricularis, and extends into the neurohypophysis. The fibers of this tract transport the colloid droplets containing the neurosecretory substances oxytocin and vasopressin that are formed in the supraoptic and paraventricular nuclei. The fibers of the tuberoinfundibular tract, also called tuberohypophyseal tract, arise from small nerve cells surrounding the lower part of the third ventricle with the nucleus arcuatus (nucleus infundibularis) as their main site of origin (Szentágothai, 1968). The fibers can be followed into the infundibulum of the posterior hypophyseal lobe where they have terminations on the capillaries of the hypophyseal portal system. This system forms a vascular link between the infundibulum and the anterior lobe of the hypophysis (see Haymaker, 1969, for further details).

From the mammillary nuclei efferent fibers pass caudally to the mesencephalon as the mammillotegmental tract. According to the observations of Nauta (1958), this tract terminates in the ventral and dorsal tegmental nuclei of Gudden lying in the paramedian midbrain area. Other fibers from the mammillary nuclei constitute the compact mammillothalamic tract of Vicq d'Azyr which distributes to the anterior nuclei of the thalamus. These latter nuclei project to the cingulate gyrus, and this cortical region, as has been mentioned before, is connected to the hippocampal region which, in turn, by way of the fornix is linked to the mammillary nuclei. The mammillothalamic tract thus forms part of a composite pathway, the so-called Papez circuit (Papez, 1937), which links together the hippocampal formation, the mammillary nuclei of the hypothalamus, the anterior nuclei of the thalamus, the cingulate gyrus and again the hippocampal formation.

Our present knowledge about the projections of the remaining hypothalamic nuclei is still scanty. This can be attributed to the technical difficulties in placing small, circumscript lesions in the hypothalamus, and also to the fact that the hypothalamus is traversed by numerous longitudinal fibers of extrahypothalamic origin. Lesion experiments of the hypothalamus sometimes provide more information about the termination of afferent hypothalamic fiber systems than about the efferent and intrinsic connections of the hypothalamus itself. Notwithstanding this, from the experimental studies of Guillery (1957), Nauta (1958), Wolf and Sutin (1966) and Chi and Flynn (1971) in rats and cats it seems at present well established that the medial forebrain bundle is the major efferent pathway by which the lateral preoptic–hypothalamic zone is connected with the septum and the mesencephalon. The descending fibers of the medial forebrain bundle distribute laterally to the mesencephalic tegmentum and medially to the paramedian midbrain area including the central grey substance. This latter region is also reached by fibers from periventricular and medial hypothalamic zones descending in the periventricular system. It is important to note that in none of the experimental studies cited above have hypothalamic efferent fibers been traced caudalward beyond the mesencephalon. The route over which the hypothalamus exerts its influence on the visceromotor centers of the brain stem and on the spinal cord must, therefore, be polysynaptic.

The descending medial forebrain bundle, in addition, contains fibers that arise in the septal region and terminate in the lateral preoptic–hypothalamic zone and, at least in the rat (Nauta, 1956), also in the mesencephalon (Fig. 6). This descending

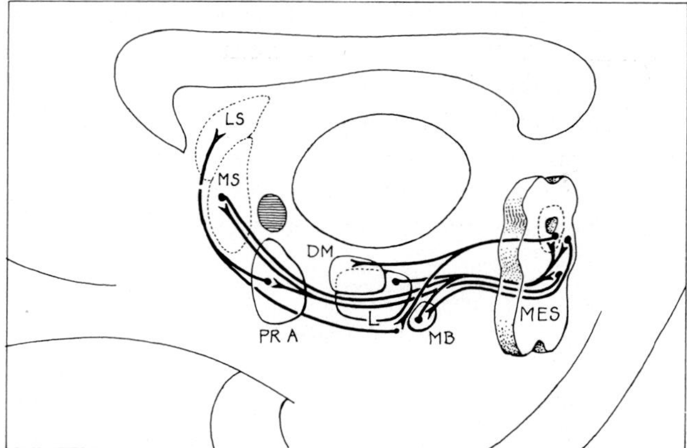

Fig. 6. Schematic drawing of the septo-hypothalamo-mesencephalic projection system. For abbreviations see p. 75.

pathway is reciprocated by ascending medial forebrain bundle fibers originating in the mesencephalon and the lateral hypothalamus and ending, respectively, in the medial and lateral septal nuclei (Guillery, 1957; Wolf and Sutin, 1966). Because the septal region also has reciprocal connections with the hippocampal formation, it thus represents a structural relay in the pathways that connect the limbic forebrain area with the limbic midbrain area and *vice-versa.*

Besides sending projections to the mesencephalon and septal region, the lateral hypothalamus is also reported to emit fibers to the amygdaloid complex. According to Nauta (1958) and Wakefield (as cited by Hall, 1972), these fibers take their origin exclusively from the lateral preoptic region. It has further been shown by Millhouse (1969) in Golgi preparations that the neurons in the lateral preoptic–hypothalamic zone give off axonal collaterals which, together with collaterals of the medial forebrain bundle fibers, enter the stria medullaris or the inferior thalamic peduncle and course in a dorsal direction toward the thalamus.

The Golgi studies of Millhouse also provide clues for an understanding of the relations between the lateral and medial hypothalamic zones. According to his observations, the fibers of the medial forebrain bundle that traverse the lateral hypothalamus make numerous synaptic contacts with the dendrites of the medial hypothalamic nuclei extending into the lateral zone. In addition, collaterals of the medial forebrain bundle fibers and from cells in the lateral hypothalamus project into the medial hypothalamus, whereas the medial zone nuclei, in turn, send their axons into the lateral zone. This apparently close relationship between the lateral and medial hypothalamic zones has recently been confirmed by experimental anatomical studies in which it was found that the lateral hypothalamus has projections to both the dorsomedial and ventromedial nuclei (Chi, 1970b; Eager *et al.*, 1971).

Much less is known about the efferent connections of the nuclei of the medial hypothalamic zone and especially of those of the ventromedial nucleus. Although the ventromedial nucleus receives a substantial projection from the amygdaloid complex, either directly (Heimer and Nauta, 1969) or indirectly, interrupted by synapses in the anterior hypothalamus (Chi, 1970b), it appears from the observations of Sutin and Eager (1969) that its neurons do not project far beyond the nuclear borders. The route by which the ventromedial nucleus may influence other hypothalamic structures must, therefore, be transsynaptic, and the experiments of Sutin and Eager indicate that these synapses take place in the area immediately surrounding the nucleus. The axons of this area project to the lateral hypothalamus, particularly to the perifornical nucleus.

CONCLUSIONS

The present paper is a survey of the structure and fiber connections of the mammalian hypothalamus. Particular attention has been paid to the anatomical techniques that presently are exploited in the experimental tracing of pathways in the central nervous system.

The hypothalamus receives its main influx from the telencephalon and the mesencephalon. Recently, by means of the autoradiographic method evidence has been provided for the existence of a direct retinohypothalamic connection.

The major fiber bundles that connect the telencephalon with the hypothalamus are the stria terminalis, the medial forebrain bundle and the fornix. The stria terminalis arises in the amygdaloid complex and terminates in the medial hypothalamic zone. Via this pathway the secondary olfactory centers of the amygdala are in direct relation with the hypothalamus.

Telencephalic sites of origin of the medial forebrain bundle, as identified in experimental material, are the anterior olfactory nucleus, the olfactory tubercle, the septal region and the frontal neocortex. Probably, further sources are the nucleus accumbens and the fundus of the striatum. These descending fibers of the medial forebrain fibers are joined by fibers of the fornix system originating from the hippocampal formation. They terminate along their course through the lateral preoptic–hypothalamic zone and extend caudally into the mesencephalon. The principal termination site in the hypothalamus of the main bundle of the fornix, which also traverses the lateral hypothalamus, is the mammillary body. It is doubtful whether the medial forebrain bundle also contains fibers that arise in the prepyriform and periamygdaloid cortices. Recent experimental observations suggest that these fibers only traverse the lateral preoptic area to terminate in the dorsomedial nucleus of the thalamus.

Besides descending fibers, the medial forebrain bundle contains ascending fibers originating in the mesencephalon and terminating partly in the lateral hypothalamus, partly in the medial zone of the septal region. Other fibers from the mesencephalon reach the hypothalamus via a medial, periventricular route.

The descending and ascending medial forebrain bundle fibers of extrahypothalamic origin are accompanied by fibers which are derived from the lateral preoptic–hypothalamic zone and distribute to the lateral septal region and the mesencephalon.

Other major efferent connections of the hypothalamus are the pathways to the hypophysis and the mammillothalamic and mammillotegmental tracts.

LIST OF ABBREVIATIONS

A	= anterior hypothalamus
AC	= anterior commissure
ACC	= nucleus accumbens
AM	= amygdaloid complex
AMB	= basal amygdaloid nucleus
AMC	= corticomedial amygdaloid nuclei
AML	= lateral amygdaloid nucleus
AT	= anterior thalamic nuclei
Bed Nucl.	= bed nucleus of stria terminalis
CC	= corpus callosum
Co	= cortical amygdaloid nucleus
DM	= dorsomedial hypothalamic nucleus
DMT	= dorsomedial thalamic nucleus
Entorhin. cortex	= entorhinal cortex
F	= fornix
FF	= fimbria fornicis
GEM	= nuclei gemini
HAB	= habenula
HIP	= hippocampus
IL	= intralaminar thalamic nuclei
IG	= indusium griseum
IP	= interpeduncular nucleus
L	= lateral hypothalamic area
LOT	= lateral olfactory tract
LS	= lateral septal nucleus
MB	= mammillary body
MES	= mesencephalon
MFB	= medial forebrain bundle
MS	= medial septal nucleus
OB	= olfactory bulb
OCH	= optic chiasma
Olf. Cortex	= olfactory cortex
Orb. Fr. Cort.	= orbitofrontal cortex
OT	= olfactory tubercle
PA	= paraventricular nucleus
PM	= paramedian thalamic nuclei
PRA	= preoptic area
Pirif. lobe	= pyriform lobe
RBA	= retrobulbar area
SM	= stria medullaris
SO	= supraoptic nucleus
ST	= stria terminalis
TEG	= tegmentum mesencephali
Temp. cort.	= temporal cortex
TOL	= nucleus of the lateral olfactory tract
VM	= ventromedial hypothalamic nucleus

REFERENCES

AKERT, K. (1964) Comparative anatomy of frontal cortex and thalamofrontal connections. In *The frontal granular Cortex and Behavior*, J. M. WARREN AND K. AKERT (Eds.), McGraw–Hill, New York, pp. 372–394.

CHI, C. C. (1970a) An experimental silver study of the ascending projections of the central grey substance and adjacent tegmentum in the rat, with observations in the cat. *J. comp. Neurol.*, **139**, 259–272.

CHI, C. C. (1970b) Afferent connections to the ventromedial nucleus of the hypothalamus in the rat. *Brain Res.*, **17**, 439–445.

CHI, C. C. AND FLYNN, J. P. (1971) Neuroanatomic projections related to biting attack elicited from hypothalamus in cats. *Brain Res.*, **35**, 49–66.

CHRIST, J. F. (1969) Derivation and boundaries of the hypothalamus, with atlas of hypothalamic grisea. In *The Hypothalamus*, W. HAYMAKER, E. ANDERSON AND W. J. H. NAUTA (Eds.), Thomas Springfield, Ill., pp. 13–60.

CROSBY, E. AND WOODBURNE, R. T. (1940) The comparative anatomy of the preoptic area and the hypothalamus. *Ann. Res. nerv. ment. Dis.*, **Proc. 20**, 52–169.

DAITZ, H. M. AND POWELL, T. P. S. (1954) Studies of the connexions of the fornix system. *J. Neurol. Neurosurg. Psychiat.*, **17**, 75–82.

DÉJÉRINE, J. J. (1895) *Anatomie des Centres Nerveux*, Rueff, Paris.

DE OLMOS, J. S. (1969) A cupric–silver method for impregnation of terminal axon degeneration and its further use in staining granular argyrophylic neurons. *Brain Behav. Evol.*, **2**, 213–237.

DE OLMOS, J. S. (1972) The amygdaloid projection field in the rat as studied with the cupric–silver method. In *The Neurobiology of the Amygdala*, B. E. ELEFTHERIOU (Ed.), Plenum Press, New York, pp. 145–204.

DE VITO, J. L. AND SMITH, O. A., JR. (1964) Subcortical projections of the prefrontal lobe of the monkey. *J. comp. Neurol.*, **123**, 413–424.

DROZ, B. AND LEBLOND, C. P. (1963) Axonal migration of proteins in the central nervous system and peripheral nerves as shown by radioautography. *J. comp. Neurol.*, **121**, 325–346.

DRUGA, R. (1969) Neocortical connections to the amygdala (An experimental study with the Nauta method). *J. Hirnforsch.*, **11**, 467–476.

EBBESSON, S. O. E. (1970) The selective silver-impregnation of degenerating axons and their synaptic endings in non-mammalian species. In *Contemporary Research Methods in Neuroanatomy*, W. J. H. NAUTA AND S. O. E. EBBESSON (Eds.), Springer, Berlin, pp. 132–161.

EAGER, R. P. (1970) Selective staining of degenerative axons in the central nervous system by a simplified silver method: spinal cord projections to external cuneate and inferior olivary nuclei in the cat. *Brain Res.*, **22**, 137–141.

EAGER, R. P., CHI, C. C. AND WOLF, G. (1971) Lateral hypothalamic projections to the hypothalamic ventromedial nucleus in the albino rat: demonstration by means of a simplified ammoniacal silver degeneration method. *Brain Res.*, **29**, 128–132.

FINK, R. P. AND HEIMER, L. (1967) Two methods for selective silver impregnation of degenerating axons and their synaptic endings in the central nervous system. *Brain Res.*, **4**, 369–374.

FREEMAN, W. AND WATTS, J. W. (1947) Retrograde degeneration of the thalamus following prefrontal lobotomy. *J. comp. Neurol.*, **86**, 65–93.

GLEES, P. (1946) Terminal degeneration within the central nervous system as studied by a new silver method. *J. Neuropathol. exp. Neurol.*, **5**, 54–59.

GREEN, J. D. (1969) Neural pathways to the hypophysis: anatomical and functional. In *The Hypothalamus*, W. HAYMAKER, E. ANDERSON AND W. J. H. NAUTA (Eds.), Thomas, Springfield, Ill., pp.276–310.

GUILLERY, R. W. (1956) Degeneration in the postcommissural fornix and the mammillary peduncle of the rat. *J. Anat. (Lond.)*, **90**, 350–370.

GUILLERY, R. W. (1957) Degeneration in the hypothalamic connexions of the albino rat. *J. Anat. (Lond.)*, **91**, 91–115.

GURDJIAN, E. S. (1927) The diencephalon of the albino rat. Studies on the brain of the rat. *J. comp. Neurol.*, **43**, 1–114.

HALL, E. (1972) Some aspects of the structural organization of the amygdala. In *The Neurobiology of the Amygdala*, B. E. ELEFTHERIOU (Ed.), Plenum Press, New York, pp. 95–121.

HAYMAKER, W. (1969) Hypothalamo-pituitary neural pathways and the circulatory system of the pituitary. In *The Hypothalamus*, W. HAYMAKER, E. ANDERSON AND W. J. H. NAUTA (Eds.), Thomas, Springfield, Ill., pp. 219–250.

HEIMER, L. (1968) Synaptic distribution of centripetal and centrifugal nerve fibers in the olfactory system of the rat. An experimental anatomical study. *J. Anat. (Lond.)*, **103**, 413–432.

HEIMER, L. (1970) Selective silver-impregnation of degenerating axoplasma. In *Contemporary Research Methods in Neuroanatomy*, W. J. H. NAUTA AND S. O. E. EBBESSON (Eds.), Springer, Berlin, pp. 106–131.

HEIMER, L. (1972) The olfactory connections of the diencephalon of the rat. An experimental light-

and electron-microscopic study with special emphasis on the problem of terminal degeneration. *Brain Behav. Evol.*, **6**, 484–523.

HEIMER, L. AND NAUTA, W. J. H. (1969) The hypothalamic distribution of the stria terminalis in the rat. *Brain Res.*, **13**, 284–297.

HENDRICKSON, A. E., WAGONER, N. AND COWAN, W. M. (1972) An autoradiographic and electron microscopic study of retino-hypothalamic connections. *Z. Zellforsch.*, **135**, 1–26.

HERRICK, C. J. (1910) The morphology of the forebrain in *Amphibia* and *Reptilia*. *J. comp. Neurol.*, **20**, 413–547.

LEONARD, C. M. (1969) The prefrontal cortex of the rat. I. Cortical connections of the mediodorsal nucleus. II. Efferent projections. *Brain Res.*, **12**, 321–343.

LEONARD, C. M. AND SCOTT, J. W. (1971) Original distribution of the amygdalofugal pathways in the rat: an experimental neuroanatomical study. *J. comp. Neurol.*, **144**, 313–330.

LOHMAN, A. H. M. (1963) The anterior olfactory lobe of the guinea pig. A descriptive and experimental anatomical study. *Acta anat. (Basel)*, **53** (Suppl. 49), 1–109.

MARCHI, V. ED ALGERI, G. (1885) Sulle degenerationi discedenti consecutive a lesioni sperimentale in diverse zone della corteccia cerebrale. *Riv. sper. Freniat.*, **11**, 492–494.

MEHLER, W. R. (1966) Further notes on the centre médian, nucleus of Luys. In *The Thalamus*, D. P. PURPURA AND M. D. YAHR (Eds.), Little, Brown, Boston, Mass., pp. 11–32.

MEHLER, W. R., FEFERMAN, M. E. AND NAUTA, W. J. H. (1960) Ascending axon degeneration following anterolateral chordotomy. An experimental study in the monkey. *Brain*, **83**, 718–750.

MILLHOUSE, O. E. (1969) A Golgi study of the descending medial forebrain bundle. *Brain Res.*, **15**, 341–363.

MOORE, R. Y. (1973) Retinohypothalamic projection in mammals: a comparative study. *Brain Res.*, **49**, 403–409.

MOORE, R. Y. AND LENN, N. J. (1972) A retinohypothalamic projection in the rat. *J. comp. Neurol.*, **146**, 1–14.

MOREST, D. K. (1961) Connexions of the dorsal tegmental nucleus in rat and rabbit. *J. Anat. (Lond.)*, **95**, 229–246.

MOREST, D. K. (1967) Experimental study of the projections of the nucleus of the tractus solitarius and the area postrema in the cat. *J. comp. Neurol.*, **130**, 277–300.

NAUTA, W. J. H. (1950) Über die sogenannte terminale Degeneration im Zentralnervensystem und ihre Darstellung durch Silberimprägnation. *Arch. Neurol. Psychiat. (Chic.)*, **66**, 353–376.

NAUTA, W. J. H. (1956) An experimental study of the fornix system in the rat. *J. comp. Neurol.*, **104**, 247–272.

NAUTA, W. J. H. (1958) Hippocampal projections and related neural pathways to the mid-brain in the cat. *Brain*, **81**, 319–340.

NAUTA, W. J. H. (1962) Neural associations of the amygdaloid complex in the monkey. *Brain*, **85**, 505–520.

NAUTA, W. J. H. (1964) Some efferent connections of the prefrontal cortex in the monkey. In *The Frontal Granular Cortex and Behavior*, J. M. WARREN AND K. AKERT (Eds.), McGraw–Hill, New York, pp. 397–407.

NAUTA, W. J. H. (1971) The problem of the frontal lobe: a reinterpretation. *J. Psychiat. Res.*, **8**, 167–187.

NAUTA, W. J. H. (1972) The central visceromotor system: a general survey. In *Limbic System Mechanics and autonomic Function*, CH. C. HOCKMAN (Ed.), Thomas, Springfield, Ill., pp. 21–38.

NAUTA, W. J. H. AND HAYMAKER, W. (1969) Hypothalamic nuclei and fiber connections. In *The Hypothalamus*, W. HAYMAKER, E. ANDERSON AND W. J. H. NAUTA (Eds.), Thomas, Springfield, Ill., 136–209.

NAUTA, W. J. H. AND KUYPERS, H. G. J. M. (1958) Some ascending pathways in the brain stem reticular formation. In *Reticular Formation of the Brain*, H. H. JASPER (Ed.), Little, Brown, Boston, Mass., pp. 3–30.

PAPEZ, J. W. (1937) A proposed mechanism of emotion. *Arch. Neurol. Psychiat. (Chic.)*, **38**, 725–743.

RAISMAN, G. (1966) The connexions of the septum. *Brain*, **89**, 317–348.

RAISMAN, G., COWAN, W. M. AND POWELL, T. P. S. (1966) An experimental analysis of the efferent projection of the hippocampus. *Brain*, **89**, 83–108.

RAMÓN Y CAJAL, S. (1911) *Histologie du Système nerveux de l'Homme et des Vertébrés, Vol. II*, Maloine, Paris.

SCALIA, F. (1966) Some olfactory pathways in the rabbit. *J. comp. Neurol.*, **126**, 285–310.

SCALIA, F. (1968) A review of recent experimental studies on the distribution of the olfactory tracts in mammals. *Brain Behav. Evol.*, **1**, 101–123.

SCOTT, J. W. AND LEONARD, C. M. (1971) The olfactory connections of the lateral hypothalamus in the rat, mouse and hamster. *J. comp. Neurol.*, **141**, 331–344.

SIEGEL, A., EDINGER, H. AND TROIANO, R. (1973) The pathway from the mediodorsal nucleus of thalamus to the hypothalamus in the cat. *Exp. Neurol.*, **38**, 202–217.

SUTIN, J. AND EAGER, R. P. (1969) Fiber degeneration following lesions in the hypothalamic ventromedial nucleus. *Ann. N.Y. Acad. Sci.*, **157**, 610–628.

SZENTÁGOTHAI, J., FLERKÓ, B., MESS, B. AND HALÁSZ, B. (1968) *Hypothalamic Control of the anterior Pituitary. An experimental-morphological Study, 3rd ed.*, Akad. Kiadó, Budapest.

WHITE, JR., L. E. (1959) Ipsilateral afferents to the hippocampal formation in the albino rat. I. Cingulum projections. *J. comp. Neurol.*, **113**, 1–42.

WHITLOCK, D. G. AND NAUTA, W. J. H. (1956) Subcortical projections from the temporal neocortex in *Macaca mulatta*. *J. comp. Neurol.*, **106**, 183–212.

WIITANEN, J. T. (1969) Selective silver impregnation of degenerating axons and axon terminals in the central nervous system of the monkey *(Macaca mulatta)*. *Brain Res.*, **14**, 546–548.

WINANS, S. S. AND SCALIA, F. (1970) Amygdaloid nucleus: new afferent input from the vomeronasal organ. *Science*, **170**, 330–332.

WOLF, G. AND SUTIN, J. (1966) Fiber degeneration after lateral hypothalamic lesions in the rat. *J. comp. Neurol.*, **127**, 137–156.

The Tubero-Infundibular Region in Man*

Structure — Monoamines — Karyometrics

J. A. M. DE ROOIJ AND O. R. HOMMES

Department of Neurology, Radboud University, Nijmegen (The Netherlands)

INTRODUCTION

The animal experiments of Szentágothai *et al.* (1968) demonstrate that the neural control of the adenohypophysis is ultimately exerted by an hypophysiotropic area in the ventral hypothalamus. This area, to a large extent occupied by the nucleus infundibularis, produces hypophysiotropic neurohormones which are transported to the gland via a neurovascular link (Green and Harris, 1947). The release of the neurohormones into the hypophyseal portal circulation is probably regulated by a monoaminergic tubero-infundibular system (Hökfelt and Fuxe, 1972). Recent studies of the human hypothalamus indicate the nucleus infundibularis as the site of the hypophysiotropic principle within the system of neural regulation of the gonads (Sheehan and Kovács, 1966; Bierich, 1971; Sheehan, 1971).

In our study of the human hypothalamus (De Rooij, 1972; Hommes, 1974) the following aspects were examined: (a) the structure of the tubero-infundibular region with special reference to the nucleus infundibularis, and the neural pathways between hypothalamus and infundibulum, including the presence of a monoaminergic tubero-infundibular system; and (b) the occurrence of morphometrically detectable changes in the nucleus infundibularis, indicative of altered cellular activity, in relation to puberty and the menopause.

MATERIAL AND METHODS

The material consists of 78 human hypothalami. For the histologic and karyometric investigations 61 hypothalami were dissected from cerebra obtained at routine autopsies performed within 24 h post mortem, and usually fixed in 4% formaldehyde. At 200 μm intervals 3 successive sections were stained respectively with hematoxylin–eosin, cresyl violet for Nissl substance and paraldehyde fuchsin for neurosecretory material. For nerve fibers, Bodian's protargol technique was used.

For the histochemical demonstration of monoamines, 17 human hypothalami were treated according to the fluorescence technique of Falck–Hillarp (Falck and Owman, 1965; Eränkö, 1967; Fuxe *et al.*, 1970). The formaldehyde-induced fluorescence,

* This investigation was supported by a grant from the Netherlands Organization for the Advancement of Pure Research (Z.W.O.).

References p. 94–96

indicating the presence of monoamines, was proven by its disappearance after reduction by sodium borohydride, and by its absence in control tissue treated identically except for the exposure to formaldehyde gas. Autofluorescent material, usually of intracellular location, was clearly distinguishable from the formaldehyde-induced fluorescence by the difference in emitted light and different morphological features.

Nuclear measurements were performed. Nuclear volume was regarded as a morphometrically detectable correlative of alterations in the functional activity of the cell (Palkovits and Fischer, 1968). In each measured cell nucleus, the maximal length (L) and the width (B), perpendicular to one another, were determined. Volume (V) was calculated according to the formula $V = \frac{1}{6}\pi LB^2$, regarding the nucleus as a spheroid. Fifty neuronal nuclei were measured in each of 3 different, topographically well-defined locations of the nucleus infundibularis and in the dorsolateral part of the nucleus supraopticus, which served as control. The mean of each sample of 50 volumes was calculated and called mean nuclear volume (MNV). Statistical tests were carried out at a level of significance of 0.05 (two-sided).

For detailed information on the technical procedures and the methodology, the reader is referred to De Rooij (1972).

RESULTS

In accordance with the subdivision of the outer surface of the tuber cinereum with reference to the sulcus infundibularis (Spatz *et al.*, 1948; Christ, 1951), its interior can be divided into the regio oralis tuberis, the regio parainfundibularis tuberis and the regio caudalis tuberis (Christ, 1951; De Rooij, 1972; Fig. 1). Two sagittal grooves in the outer surface of the regio caudalis tuberis divide this region in an eminentia postinfundibularis and two eminentiae laterales. Owing to the presence of the recessus infundibuli, the proximal part of the infundibulum (pars cava infundibuli) can be regarded as being made up of an anterior and a posterior wall and two lateral walls (Martinez, 1960).

The nucleus infundibularis, being the most mediobasal cell group of the tuber cinereum, is located directly above the insertion of the hypophysis into the hypothalamus and extends as a wedge of nerve cells into the infundibulum (Fig. 1). In an horizontal plane it is situated eccentrically around the recessus infundibuli with its principal part caudally. Anteriorly, the nucleus infundibularis occupies the mediobasal area of the regio parainfundibularis tuberis and adjacent areas of the infundibular walls. In the regio oralis tuberis only a few scattered cells of the particular cell group are found. Posteriorly, its nerve cells extend into the eminentia postinfundibularis of the regio caudalis tuberis, where they constitute in frontal sections a U-shaped cell group of high density in the floor of the third ventricle.

Nerve fibers in the tubero-infundibular region

In well made Bodian preparations many both fine and thick nerve fibers can be seen in

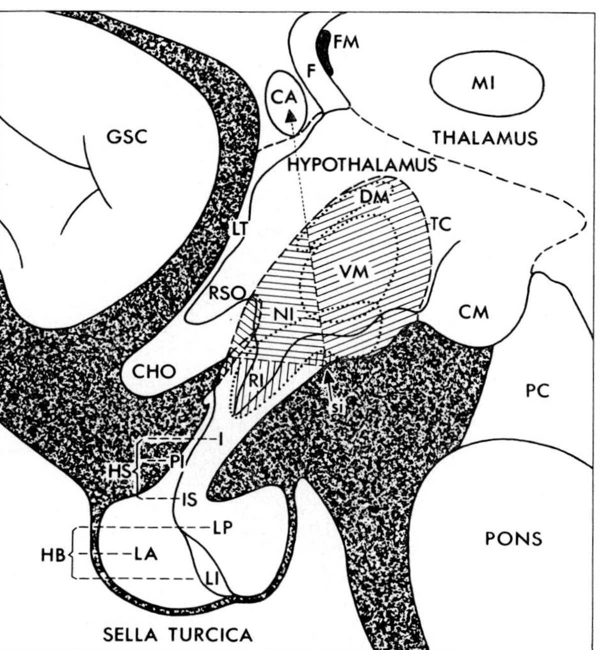

Fig. 1. ▨, regio oralis tuberis; ▤, regio parainfundibularis tuberis; ▨, regio caudalis tuberis; ||||||||, wedge of infundibular nerve cells. Schematic drawing of a sagittal section through the human hypothalamus and hypophysis near the median plane. The line through the posterior end of the sulcus infundibularis and the commissura anterior separates the regio parainfundibularis tuberis from the regio caudalis tuberis. The recessus infundibuli is surrounded by tuberal tissue consisting of the wedge of nerve cells of the nucleus infundibularis. A large part of this cell group extends into the regio caudalis tuberis. The mixed hatchings are not relevant in the present study (from De Rooij, 1972). Abbreviations: CA, commissura anterior; CHO, chiasma opticum; CM, corpus mamillare; DM, nucleus dorsomedialis; F, fornix; FM, foramen Monroi; GSC, gyrus subcallosus; HB, hypophysial body; HS, hypophysial stalk; I, infundibulum; IS, infundibular stem; LA, anterior lobe; LI, intermediate lobe; LP, posterior lobe; LT, lamina terminalis; MI, massa intermedia; NI, nucleus infundibularis; PC, pedunculus cerebri; PI, pars infundibularis adenohypophyseus; RI, recessus infundibuli; RSO, recessus supraopticus; SI, sulcus infundibularis; TC, tuber cinereum; VM, nucleus ventromedialis.

the tubero-infundibular region. The thick axons appear to belong to the tractus supraoptico-hypophyseus, which descends from the various parts of the nucleus supraopticus towards the infundibulum. In its descent, this tract covers the anterior and anterolateral surface of the rostral part of the nucleus infundibularis before entering the infundibulum (compare with the course of the neurosecretory fibers in Fig. 2). The nucleus paraventricularis also contributes to this tract as most of its axons go by way of the dorsolateral part of the nucleus supraopticus. Only some of its nerve fibers descend in the immediately periventricular area. In its course in the pars cava infundibuli, the tractus supraoptico-hypophyseus remains superficial to the wedge of infundibular nerve cells. It also remains superficial when turning inwards to pass through the zona interna at the proximal limit of the zona externa. In the zona externa no thick axons are observed.

These findings are confirmed in sections stained for neurosecretory material by

References p. 94–96

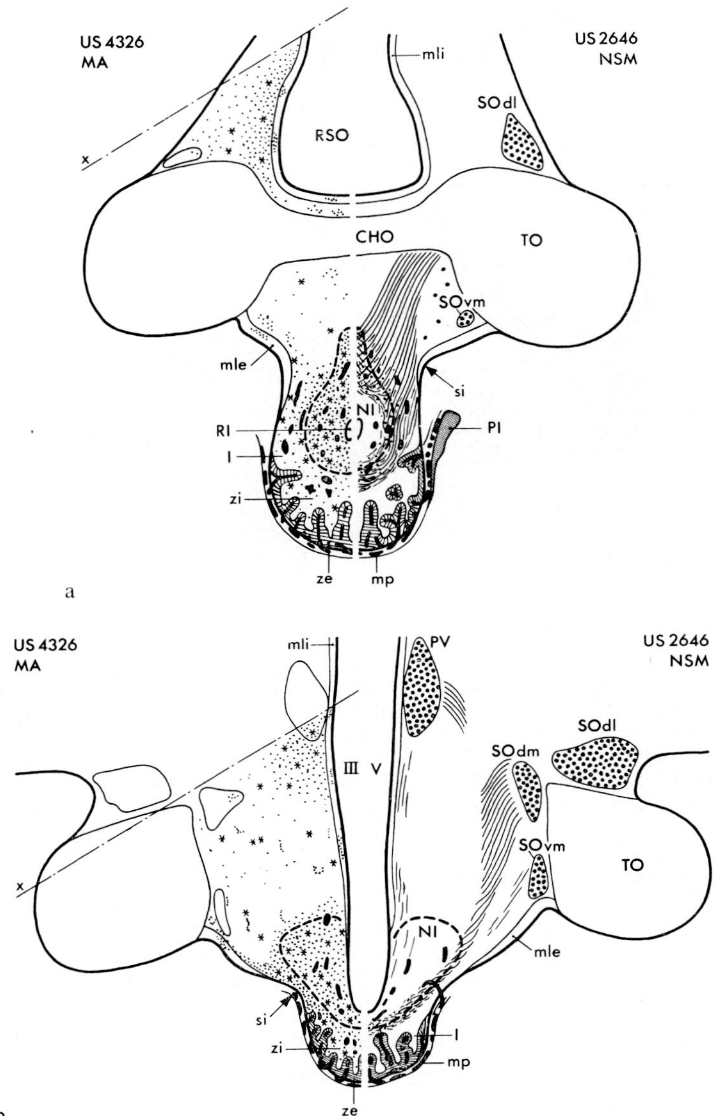

Fig. 2. Distribution of the formaldehyde-induced fluorescence of monoamines (MA) is indicated in the left half of each diagram. The course of the tractus supraoptico-hypophyseus containing the neurosecretory material (NSM) is indicated on the right. The x-line indicates the upper margin of the dissected part of the hypothalamus used for the demonstration of monoamines in this diagram. (a) Frontal section through the anterior tubero-infundibular region. (b) Frontal section through the tubero-infundibular region at the level of the posterior wall of the infundibulum. Abbreviations: CHO, chiasma opticum; I, infundibulum; NI, nucleus infundibularis; PI, pars infundibularis adenohypophyseus; PV, nucleus paraventricularis; RI, recessus infundibuli; RSO, recessus supraopticus; SOdl, SOdm, SOvm, nucleus supraopticus -pars dorsolateralis, -pars dorsomedialis, -pars ventromedialis; TO, tractus opticus; III V, third ventricle; mle, membrana limitans gliae externa; mli, membrana limitans gliae interna; mp, mantle plexus; si, sulcus infundibularis; ze, zona externa; zi, zona interna.

paraldehyde fuchsin. In the hypothalamus of some specimens the entire supraoptico-hypophyseal system appears to be loaded with this substance, both in the perikarya and the fibers. One of these cases is shown in Fig. 2. The course of the neurosecretory fibers exactly corresponds to the course of the thick axons in Bodian preparations as described above.

A large number of the many fine axons in the parainfundibular portion of the nucleus infundibularis are directed towards the entrance area of the infundibulum. In the caudal portion of the nucleus infundibularis the main direction is parallel to the U-shape of the cell group in the regio caudalis tuberis. The vertical course in the lateroposterior part of the nucleus is particularly clear.

A glial membrane-free outer zone of the regio parainfundibularis tuberis, lateral to the insertion of the posterior wall of the infundibulum (at the level of Fig. 2b), contains many delicate nerve fibers. This zone is called the "transitional outer zone" as it forms the transition between the membrana limitans gliae externa of the tuber cinereum and the zona externa of the infundibulum in the outer zone of the tubero-infundibular junction. At this level, the fine axons in this zone have the same direction as the supraoptic fibers somewhat more anteriorly but, unlike these latter, after having entered the infundibulum, they spread from both sides over the zona interna.

The vertically directed axons in the lateroposterior part of the nucleus infundibularis cannot be followed in the outer zone of the eminentia postinfundibularis. This outer zone consists of a dense layer of transversely cut fine axons and lacks a membrana limitans gliae externa. This zone is the part of the transitional outer zone which extends over the eminentia postinfundibularis. It is continuous with the fine axon-containing outer zone of the regio parainfundibularis tuberis, lateral to the insertion of the posterior infundibular wall. Caudal to the nucleus infundibularis, the dense layer of transversely cut fine axons in the eminentia postinfundibularis has disappeared. Most of the nerve fibers from the lateroposterior part of the nucleus infundibularis, and perhaps also from more dorsally situated neurons such as those of the area periventricularis posterior, thus collect on the ventral surface of the cell group. It is likely that they proceed into the outer zone of the eminentia postinfundibularis towards the anterior part and reach the outer zone of the regio parainfundibularis tuberis. From here the fine axons disperse over the zona interna on entering the infundibulum. This system of fine nerve fibers is a concrete structure that may well fit the term tractus tubero-hypophyseus.

Histochemical demonstration of monoamines in the tubero-infundibular region

Using a fluorescence microscope, monoamines appear as structures with a green, and a yellow-green to green formaldehyde-induced fluorescence. This difference in emitted light may depend on the concentration of the fluorescent substance, or may be due to the presence of different fluorophores. If the hypothalamus contained 5-hydroxytryptamine, we were unable to differentiate its fluorescence from the catecholamine fluorescence under the conditions used.

The quantity of formaldehyde-induced fluorescence (FIF) in the different hypothal-

References p. 94–96

ami varies considerably. From the series of cases not treated with drugs known to affect the monoamines, it appears that the amount of FIF is inversely related to the length of time elapsing between death and autopsy. Within the first 8 h after death much FIF can usually be demonstrated. Considerable amounts of FIF were found and distinct local differences could be established in only 5 out of 17 cases. As it was impossible to obtain the tissue immediately after death, the results are in fact a shaded impression of the real situation, if one assumes that diffusion affects the monoaminergic structures everywhere to the same extent. Besides diffusion, other limiting factors in the interpretation of the location of monoamines are the artefacts produced by autolysis and freezing, producing cracks in the tissue and affecting cell structure.

In Fig. 2 the distribution of FIF, as found in an adult human, is illustrated and is compared with the course of the neurosecretory fibers of the supraoptico-hypophyseal system. The star symbols in the diagram are large yellow-green FIF structures, representing the monoamines in nerve cell bodies and probably in a number of very large fiber varicosities. The black dots in Fig. 2 indicate the occurrence of monoamines as small FIF masses and small groups of green fluorescent dots. These mainly represent monoamine-containing nerve fibers and terminals.

In the hypothalamus schematically described in Fig. 2, the nucleus infundibularis, the nucleus dorsomedialis and the regio suprachiasmatica in particular appear to be rich in FIF. The nucleus infundibularis contains the most FIF-positive cell bodies. They are more frequent in the rostral part than in the caudal part of the nucleus. A few dots of FIF are often situated around the infundibular nerve cells. The dot-like fluorescence also appears in the intercellular substance, sometimes in a chain with the appearance of a varicose fiber. The subependymal area of the nucleus infundibularis, which lacks a glial membrane, shows much FIF, in both cells and fibers. Conversely, the membrana limitans gliae interna in other regions of the hypothalamus only occasionally contains some FIF. In the region of the nucleus infundibularis, varicose fibers and fluorescent dots are regularly seen in close contact with the ependyma. In the transitional outer zone, there are many fine FIF structures and a few varicose fibers. Laterally, this fluorescent layer, though narrowing, continues above the row of astrocyte cell bodies of the membrana limitans gliae externa. Somewhat ventral to the sulcus infundibularis, the fine dotted green FIF diminishes rather abruptly. Several large FIF structures also occur in the transitional outer zone. As this zone contains no nerve cell bodies these structures are probably large varicosities of fibers. In this zone the fluorescent dots and masses are often situated directly against the extension of the mantle plexus which covers this zone. These large masses are seen in the infundibulum as well, mainly in the zona interna and some near the blood vessels of the zona externa. In the outer zone of the eminentia postinfundibularis, FIF often occurs in close contact with an extension of the mantle plexus that covers this eminence.

As shown in Fig. 2b, the ventral part of the nucleus paraventricularis is included in this hypothalamus and has the same density of FIF as the nucleus dorsomedialis. The nucleus ventromedialis and the nuclei tuberis laterales are poor in FIF. In the anterior part of the hypothalamus, from the monoamine-fluorescent region just above the chiasma opticum towards the supraoptic crest of the lamina terminalis, the fluorescent cell bodies gradually disappear while a considerable amount of fine green FIF remains (Fig. 2a).

In two other adults a similar distribution was found, except that the regio suprachiasmatica and the nucleus dorsomedialis did not show so much FIF as in the case described in Fig. 2. In a patient treated with L-dopa, the same cell groups as mentioned above contained a considerable amount of FIF, but the transitional outer zone and infundibulum were not striking at all. In this case the autopsy delay had been very long. A 1.5-year-old child was the only individual that showed a diffuse, not dotted, greenish fluorescent layer in the zona externa. In this zone, large FIF masses were also observed in this case.

From the distribution of the FIF in the tubero-infundibular region it is obvious that there is, in man, a tubero-infundibular monoaminergic system which is located in the fine axon-containing tractus tubero-hypophyseus as found in Bodian preparations. This system also occupies the transitional outer zone of the tubero-infundibular region, located on the ventral surface of the nucleus infundibularis, proceeds in the zona interna and apparently reaches the zona externa. It enters the infundibulum at the level of its posterolateral and posterior walls, caudal to the tractus supraoptico-hypophyseus which descends more anteriorly over the anterior and anterolateral surfaces of the nucleus infundibularis to the infundibulum.

Nerve cells of the nucleus infundibularis

In the nucleus infundibularis of the newborn, 3 types of nerve cells occur, each with a preferential location within the cell group (De Rooij, 1972). A large cell type is predominantly found in the part of the nucleus infundibularis that is located in the infundibulum around the recessus infundibuli. An intermediate cell type is the principal cell in an area, dorsal and dorsolateral to the sulcus infundibularis, *i.e.* the mediobasal area of the regio parainfundibularis tuberis. The posterior part of the cell group, located in the regio caudalis tuberis, consists largely of cells of a small type. The most caudal part consists almost exclusively of these small nerve cells.

When compared with the newborn, the amount of Nissl substance in the nerve cells of the nucleus infundibularis is increased in the young adult. The nucleoli are more strongly basophilic and often larger. These changes are particularly clear in the parainfundibular part of the cell group, where the intermediate cell type was found in the newborn. These neurons have become very similar to the large cell type in size and appearance. The original large cell type hardly appears to be changed, especially regarding size. The cells in the caudal part of the nucleus infundibularis are also larger than in the newborn but the cytoplasm remains relatively small and contains less Nissl substance than the cells in the anterior part of the cell group. Though topographical differences in cell types are recognizable in the young adult, these are less obvious than in the newborn, especially with regard to the original large and intermediate types.

In the course of aging all infundibular nerve cells are enlarged considerably. In comparison with the newborn this increase in size is very apparent in the parain-

References p. 94–96

fundibular and caudal parts of the nucleus infundibularis and makes it a distinct hypothalamic cell group. Most neurons in the entire nucleus infundibularis have a comprehensive perikaryon and contain an increased amount of Nissl substance. Though this substance is still often dust-like, Nissl bodies are much more frequently seen than in young individuals. In advanced age, differences in cell type between the aforementioned locations within the infundibular nucleus are not obvious anymore.

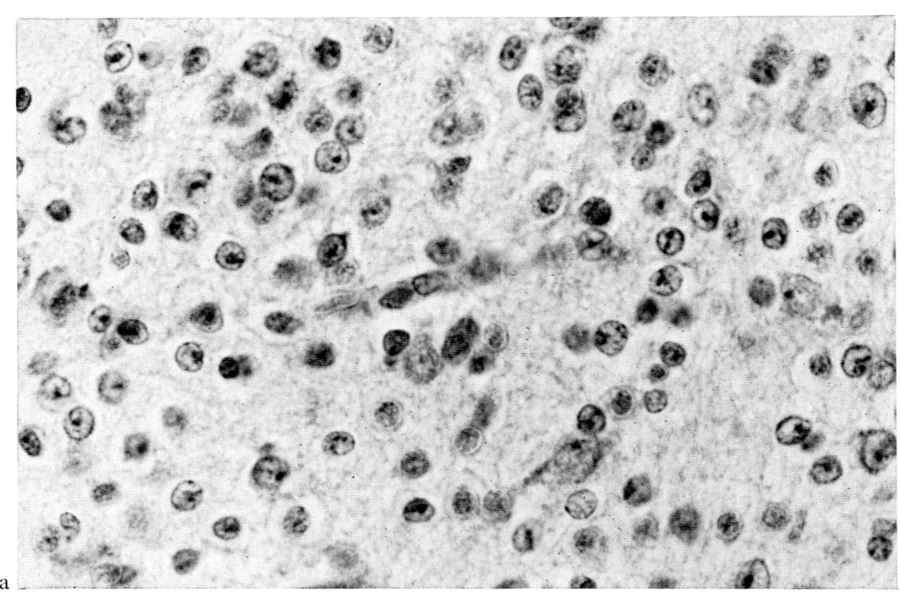

a

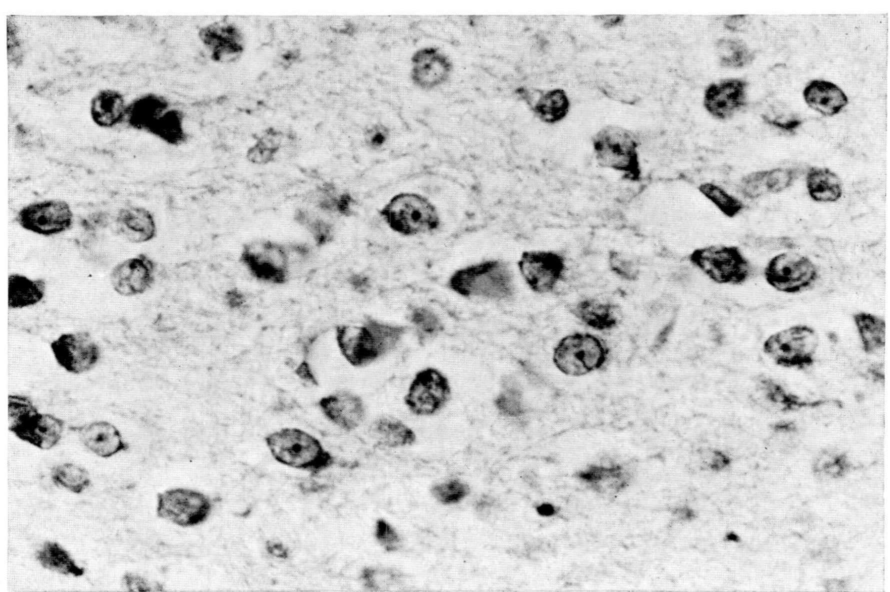

b

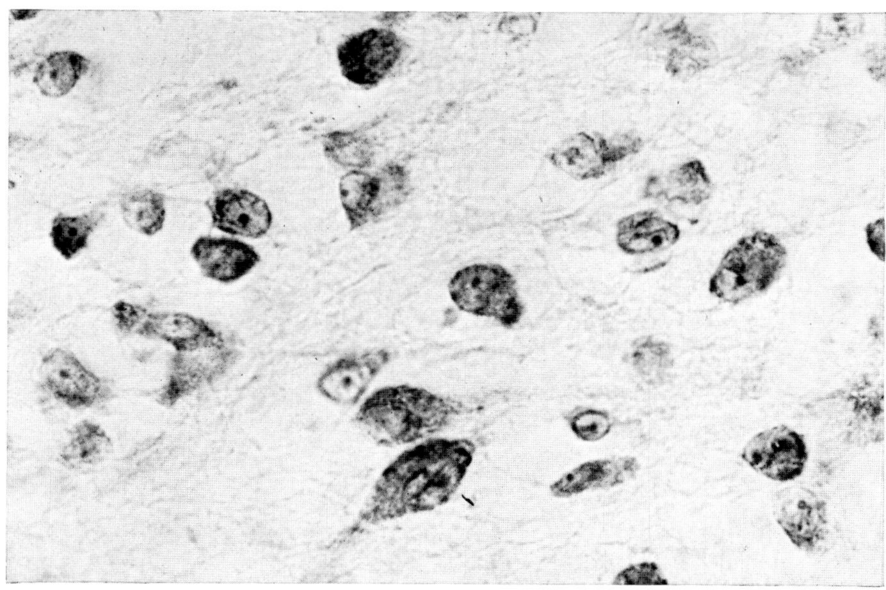

Fig. 3. Nerve cells of the lateroposterior part of the nucleus infundibularis in a newborn (a), in a young adult (b) and in an 81-year-old individual (c). In all cases the objective magnification was ×40.

In order to illustrate these cellular changes in the nucleus infundibularis, the nerve cells of the caudal part of the cell group are shown in a newborn (Fig. 3a), in a young adult (Fig. 3b) and in an old individual (Fig. 3c).

Karyometrics of the nucleus infundibularis

In view of the occurrence of 3 cell types, nuclear measurements were performed in 3 well-defined areas of the nucleus infundibularis, each of which is representative for a particular cell type. A mean nuclear volume (MNV) was calculated for: (a) the AM-part, the anteromedial part of the cell group or wedge of infundibular nerve cells located in the infundibulum; (b) the AL-part, the anterolateral part of the cell group situated in the regio parainfundibularis tuberis, dorsal and mainly lateral to the sulcus infundibularis; (c) the LP-part, the lateroposterior part of the nucleus infundibularis located in the regio caudalis tuberis, dorsomedial to the sulci bounding the eminentia postinfundibularis.

Mean nuclear volume (MNV) and age

In Fig. 4 the MNVs of the nerve cells in the different parts of the nucleus infundibularis are shown on a time scale. After 50 years of age the cell nuclei are increased in MNV, not only in the AL-part as could be expected from the findings of Hommes (1974), who measured this part of the cell group, but also in the AM- and LP-parts. The AL-part shows a considerable increase in MNV again after about 65 years of age while

References p. 94–96

such a tendency is not obvious in the AM- and LP-parts. In the young age group the MNVs of the AM-part remain rather constant. On the other hand, a distinct increase in the MNVs of the other two parts of the cell group was observed where the small and intermediate cell types were found in the newborn. This increase seems to have taken place after 10 years of age because practically all values after this age are higher than those in the range of 0–10 years. In the range of 0–10 and 11–34 years no marked

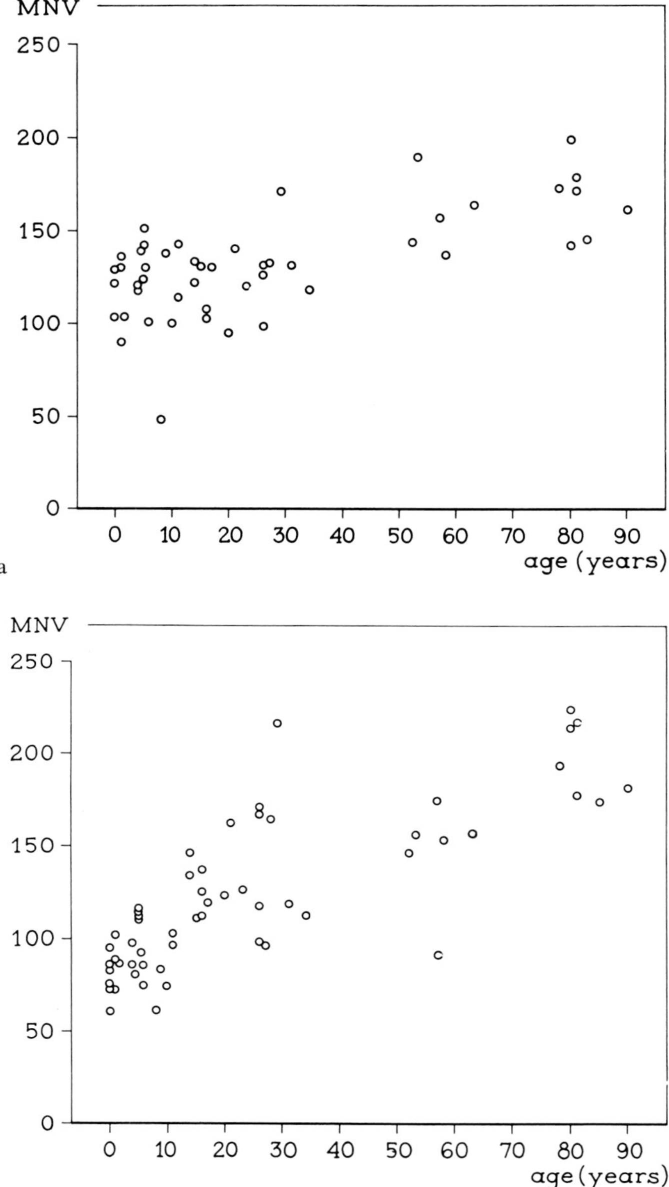

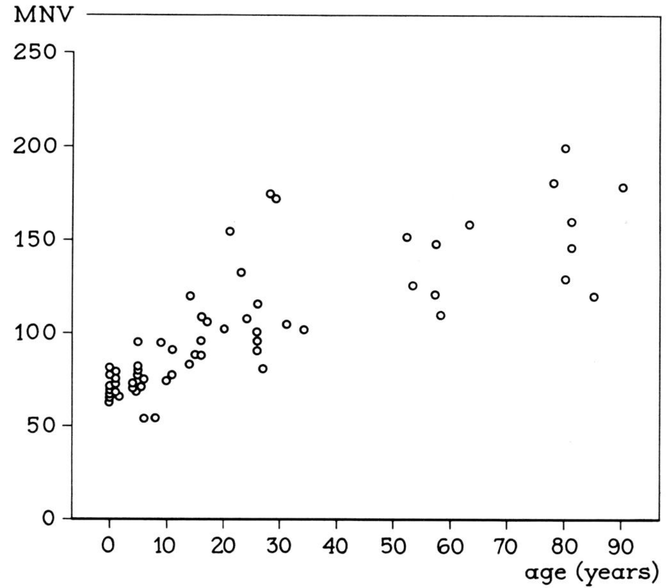

Fig. 4. Age and MNV of the AM (a)-, AL (b)- and LP (c)-parts of the nucleus infundibularis.

tendency towards increase or decrease in MNV is observed but there is a considerable variation in the latter age group.

In order to test these observations, the material was divided into different age groups, *viz.* 0–10 years, 11–34 years, and 52–90 years. The choice of these groups is based on the objective of this morphometric study and on the observations from Fig. 4. In each location we determined, firstly, whether the MNVs show a significant age-dependent correlation within the age groups, and secondly, whether there are shifts in MNV from one age group to another.

Significant rank correlations between MNV and age were not found within the age groups, except for the AL-part of the cell group in the range of 52–90 years (Spearman,

TABLE I

RESULTS OF THE WILCOXON TWO SAMPLE TEST IN COMPARING THE MNVs OF DIFFERENT AGE GROUPS FOR EACH OF THE 3 LOCATIONS IN THE NUCLEUS INFUNDIBULARIS

Age (years)	*AM* *p*	*(n/m)*	*AL* *p*	*(n/m)*	*LP* *p*	*(n/m)*
0–10/11–34	0.24	(18/18)	<0.001	(23/21)	<0.001	(25/22)
11–34/52–90	<0.001	(18/11)	<0.001	(21/13)	<0.001	(22/13)
52–63/78–90			<0.01	(6/7)		

p = p-value.
n/m = number of specimens in the age groups.

References p. 94–96

$r_s = 0.70$; $p = 0.02$). Because of this latter correlation this range was only subdivided for the AL-part, in order to compare the MNVs in the range of 52–63 years with those in the range of 78–90 years.

Table I confirms the observations mentioned above. The increase in MNV from one age group to another is significant both after about 10 and about 50 years for the AL- and LP-parts and after about 50 years for the AM-part. Within the third age group there is a significant increase in MNV of the AL-part of the nucleus infundibularis.

In the nucleus supraopticus, which was chosen as control group for the karyometric analyses, a significant rise in MNV was demonstrated between 1 and 4 years of age. After this age no further change in MNV was apparent.

MNV and body weight

The karyometric findings were also compared as a function of body weight. In the 3 parts of the nucleus infundibularis the MNV showed hardly any correlation with body weight up to about 30 kg. In the AM-part the MNV did not appear to increase before 50 kg. The MNVs of the AL- and LP-parts increase earlier, starting at a body weight of about 30 kg. The AL- and LP-parts showed an increase in MNV of almost exponential form with increasing weight. This sharp increase in MNV between 30 and 50 kg, which is represented by a shift in MNV on the age scale, must have taken place between 10 and 20 years of age.

These findings confirm the observations that the MNVs of the nerve cells of the nucleus infundibularis remain rather constant during the first 10 years of life, and even longer in the AM-part, and that they do not take part in the tendency of postnatal growth which was seen in the control group (nucleus supraopticus) and which was established for the cells of the cortex during the first 2 years after birth (up to 20 kg) by Schadé and van Groenigen (1961).

MNV and sex

Comparison of males and females within the different age groups produced no significant differences with regard to MNV of the AM-, AL- and LP-parts of the nucleus infundibularis.

MNV and location within the nucleus infundibularis

In order to analyze the relationship between the MNVs of the different parts of the nucleus infundibularis, the mean rank order for the AM-, AL- and LP-parts within each age group was calculated according to the method of m rankings of Friedman. This procedure indicated with high probability that the 3 locations in the nucleus infundibularis differ in MNV value in the age groups 0–10 years, 11–34 years and 78–90 years. The following interrelations between the MNVs of the 3 parts of the cell group, based on the mean rank order, are postulated:

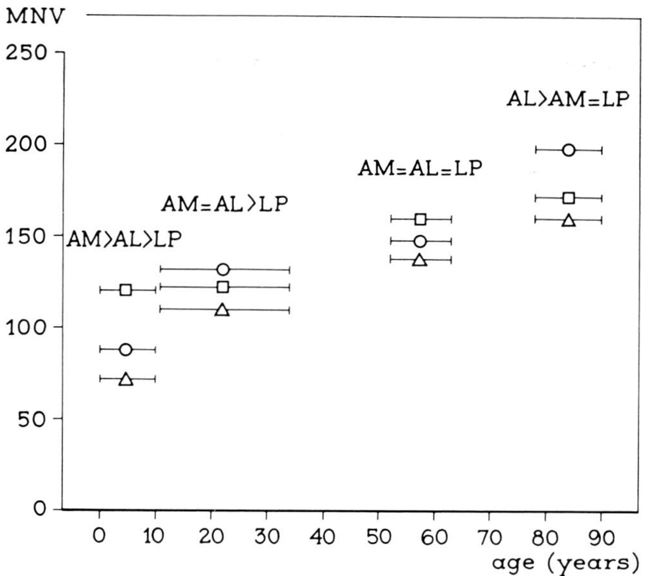

Fig. 5. Mean of MNVs within different age groups for each part of the nucleus infundibularis (□ = AM-part, ○ = AL-part and △ = LP-part). The bars indicate the range of the age groups. The level of the bar indicates the mean MNV for each age group and part of the nucleus infundibularis. The notations above the bars indicate the significant differences in MNV between the 3 parts of the cell group within the particular age groups.

0–10 years	AM > AL > LP
11–34 years	AM = AL > LP
52–63 years	AM = AL = LP
78–90 years	AL > AM = LP

The paired-off comparisons in these order relations were successively detected by application of the Wilcoxon signed rank test (= indicates no significant difference; > indicates a significant difference).

The data on the MNVs of the 3 parts of the nucleus infundibularis are assembled in Fig. 5 for comparison. The range of 52–90 years is divided into a range of 52–63 years and one of 78–90 years because of the observed increase in MNV of the AL-part.

Conclusions: The MNV of the AL- and LP-parts of the nucleus infundibularis show a rather abrupt increase after 10 years of age, while the MNV of the AM-part retains practically the same value, at least up to 34 years of age. After 50 years the MNVs of all parts of the cell group are significantly higher than those in the range of 11–34 years of age. After 50 years of age, there is still a significant increase of MNV with age for the AL-part.

DISCUSSION

Observations made in Bodian preparations provide a morphological substratum in the human brain for a neural tubero-infundibular pathway, the tractus tubero-hypophyseus. This topographically well-defined tract consists of delicate nerve fibers of which the nucleus infundibularis seems to be the main site of origin. The tract enters the infundibulum mainly through its posterior and posterolateral walls. After having entered the infundibulum the fine nerve fibers disperse over the zona interna. On the other hand, the thick axons of the tractus supraoptico-hypophyseus enter the anterior and anterolateral infundibular walls and proceed in a more tract-like manner into the zona interna. Only fibers of the tractus tubero-hypophyseus appear to reach the zona externa because only delicate nerve fibers were found in this zone.

In the course of the tractus tubero-hypophyseus much formaldehyde-induced fluorescence (FIF) was demonstrated, indicative of the presence of monoaminergic nerve fibers and terminals. The occurrence of these monoaminergic structures in the transitional outer zone, the zona interna and externa, indicates that the tractus tubero-hypophyseus comprises a monoaminergic tubero-infundibular system. A layer of diffuse FIF, which is often described in the zona externa in lower mammals, was only found in the infundibulum of a young child. Regarding its existence in man, not much can be said as its absence in other specimens may be due to diffusion. In monkeys treated similarly, but after a very short autopsy delay, the zona externa was revealed on the special vessels from the mantle plexus as a distinct FIF layer penetrating deeply into the infundibulum (unpublished data).

Regarding the occurrence of FIF cell bodies, 3 regions deserve consideration as contributing to the monoaminergic tubero-infundibular system, *viz.* the nucleus infundibularis, the nucleus dorsomedialis and the regio suprachiasmatica. Because of the distribution of FIF, the distance to the infundibulum and the course of the tractus tubero-hypophyseus, the nucleus infundibularis is a very likely source of the FIF fibers and terminals in the zones mentioned above.

The existence of such a monoaminergic tubero-infundibular system is of interest, as the effects observed in man after the administration of L-dopa on the secretion of prolactin and growth hormone (Boyd *et al.*, 1971; Eddy *et al.*, 1971; Kleinberg *et al.*, 1971) are probably mediated by the hypothalamus in view of comparable phenomena in animals (Kamberi *et al.*, 1970; Hökfelt and Fuxe, 1972; McCann *et al.*, 1972; Porter *et al.*, 1972). From animal experiments it is evident that dopamine and noradrenaline are involved in the neural regulation of the adenohypophysis. Hökfelt and Fuxe (1972) indicate that such a regulatory activity occurs in the tubero-infundibular region. The effects of L-dopa on the secretion of hypophyseal hormones are not the same in man and rat. This difference may be due to different modes of application, but the influence of monoamines on the regulation of the adenohypophysis in the rat are not necessarily similar to that in man. Moreover, the biochemical fluorimetric study of Rinne and Sonninen (1968) demonstrated the presence of much more noradrenaline than dopamine in the tubero-infundibular region in man, while in this region in several lower mammals, dopamine is the predominant catecholamine (Kobayashi *et al.*, 1970; Björklund *et al.*, 1970).

The hypophysiotropic area in the mediobasal hypothalamus is regulated by other hypothalamic and extrahypothalamic structures (De Groot, 1967; Szentágothai *et al.*,

1968; Köves and Halász, 1970) which have afferent connections with this area, mainly via the medial hypothalamus, *i.e.* the nucleus ventromedialis and the area periventricularis (Lammers, 1969). Experimental interference with monoaminergic systems is quite possible at a higher level as the parvicellular nuclei of the hypothalamus receive a considerable monoaminergic afferency.

The nucleus infundibularis of the newborn contains only a restricted number of large nerve cells when compared to that of the adult. These cells are located in the infundibulum, close to the assumed site of neurovascular transmission, *i.e.* the zona externa infundibuli. The parts of the cell group more distant to this site consist of smaller nerve cells, respectively the intermediate type situated in the area above the infundibulum, and the small type situated in the lateroposterior part of the nucleus infundibularis. In comparison with these smaller types, the large type of nerve cell appears to be a mature neuron since it possesses a considerable perikaryon, a distinct nucleolus and a fair amount of Nissl substance (De Rooij, 1972).

The intermediate and small cell types show a rather abrupt increase in MNV after 10 years of age. The shift-like changes in MNV are accompanied by an increase in cytoplasm and in the Nissl substance, indicative of an increased functional activity of the originally smaller two cell types (Palkovits and Fischer, 1968; De Rooij, 1972).

The presence of only a restricted number of mature nerve cells, assumed to be capable of specialized functional activity, may be related to the low stimulatory activity of the mediobasal tuber cinereum on the gonadotropic adenohypophyseal function prior to puberty; Bauer (1954, 1959) and Prick (1956) indicated that gonadal regulation in man takes place in the floor of the third ventricle, which is practically totally occupied by the nucleus infundibularis. That the specialized functional activity of the neurons of this cell group in man concerns the production and secretion of releasing factors, as would be expected from the research of Szentágothai *et al.* (1968), has become probable from the findings of Bierich (1971) who studied a certain type of precocious puberty due to an hamartoma of the tuber cinereum. He mentioned that such an hamartoma, which represents an increase in nerve cells in the floor of the third ventricle including those of the nucleus infundibularis (Lange-Cosack, 1951), may be accompanied by an elevation of the LH-releasing factor in the cerebrospinal fluid. This suggests that the site of production of gonadotropic releasing factors in man is situated in the floor of the third ventricle which is for the greater part occupied by the nucleus infundibularis. Topographically, this is in agreement with the hypophyseotropic area in the rat (Szentágothai *et al.*, 1968). For these reasons it is assumed that the increase in number of large nerve cells of the nucleus infundibularis after 10 years of age is a morphological sign of maturation and enhanced cellular activity related to an elevated production of gonadotropic releasing factors during puberty, in which the levels of LH and FSH rise considerably (Blizzard *et al.*, 1970). In this respect, the relatively small number of mature nerve cells in this cell group, which are located in the infundibulum, before 10 years of age, is likely to be responsible for the low gonadotropic activity of the adenohypophysis prior to puberty. A phenomenon related to the observed signs of maturation and increased cellular activity of the nucleus infundibularis in puberty might take place between about 35 and 50 years of

References p. 94–96

age. The cells of the LP-part increase in MNV and change in appearance in such a way that they cannot be distinguished anymore from those of the AM- and AL-parts, which in the meantime also increased in MNV. In view of the assumptions stated above, a cellular and nuclear hypertrophy of the nucleus infundibularis, which was demonstrated by Sheehan and Kovács (1966), Hommes (1974) as well as in this study, after about 50 years of age, may very well represent an increased production of gonadotropic releasing factors responsible for the elevated levels of gonadotropins such as observed in females during the menopause (Hart, 1971; Wide *et al.*, 1973) and in males after middle age (Ryan and Faiman, 1968; Vermeulen, personal communication; Wide *et al.*, 1973).

SUMMARY

In the human brain, a tubero-infundibular pathway called the tractus tubero-hypophyseus was established in Bodian preparations. The nucleus infundibularis seems to be the main site of origin of the fine axons of this tract which enters the infundibulum mainly through its posterior and posterolateral walls. On the other hand, the thick axons of the tractus supraoptico-hypophyseus specifically enter the anterior and anterolateral infundibular walls. The different nature of these two tracts was confirmed by the finding of neurosecretory material exclusively in the fibers of the tractus supraoptico-hypophyseus, and by the demonstration of a monoaminergic tubero-infundibular system which, topographically, corresponds exactly to the course of the tractus tubero-hypophyseus.

In the nucleus infundibularis of the newborn, 3 types of nerve cells occur, each with a preferential location within the cell group. With increasing age these cell types show a differential change in MNV, which takes place rather abruptly after 10 years of age and again in the period between about 35 and 50 years of age. In one part of the cell group, an increase in MNV is still observed after 50 years of age.

The structural plan of the tubero-infundibular region in man provides the morphological substrate for an hypophyseotropic and monoaminergic tubero-infundibular system with the nucleus infundibularis as the final neural link in the neurovascular connection with the adenohypophysis. The possibility that the described changes in cellular activity of the nucleus infundibularis are related to alterations of sexual function in puberty and the menopause is discussed.

REFERENCES

Bauer, H. G. (1954) Endocrine and other clinical manifestations of hypothalamic disease. *J. clin. Endocr.*, **14**, 13–31.

Bauer, H. G. (1959) Endocrine and metabolic conditions related to pathology in the hypothalamus: a review. *J. nerv. ment. Dis.*, **128**, 323–338.

Bierich, J. R. (1971) Frühreife. *J. Neuro-Viscer. Relat.*, **Suppl. X**, 615–626.

Björklund, A., Falck, B., Hromek, F., Owman, C. and West, K. A. (1970) Identification and terminal distribution of the tubero-hypophyseal monoamine fiber systems in the rat by means of stereotaxic and microspectrofluorimetric techniques. *Brain Res.*, **17**, 1–23.

Blizzard, R. M., Johanson, A., Guyda, H., Baghdassarian, A., Raiti, S. and Migeon, C. J. (1970) Recent developments in the study of gonadotropin secretion in adolescence. In *Adolescent Endocrinology*, F. P. Heald and W. Hung (Eds.), Butterworths, London, pp. 1–24.

Boyd, A. E., Lebovitz, H. E. and Feldman, J. M. (1971) Endocrine function and glucose metabolism in patients with Parkinson's disease and their alteration by L-dopa. *J. clin. Endocr.*, **33**, 829–837.

Christ, J. (1951) Zur Anatomie des Tuber cinereum beim erwachsenen Menschen. *Dtsch. Z. Nervenheilk.*, **165**, 340–408.

Eddy, R. L., Jones, A. L., Chakmakjian, Z. H. and Silverthorne, M. C. (1971) Effect of levodopa (L-dopa) on human hypophyseal trophic hormone release. *J. clin. Endocr.*, **33**, 709–712.

Eränkö, O. (1967) The practical histochemical demonstration of catecholamines by formaldehyde-induced fluorescence. *J. roy. micr. Soc.*, **87**, 259–276.

Falck, B. and Owman, C. (1965) A detailed methodological description of the fluorescence method for the cellular demonstration of biogenic monoamines. *Acta Univ. Lund, Sect. II*, **7**, 1–23.

Fuxe, K., Hökfelt, T., Jonsson, G. and Ungerstedt, U. (1970) Fluorescence microscopy in neuroanatomy. In *Contemporary Research in Neuroanatomy*, S. O. E. Ebbesson and W. J. H. Nauta (Eds.), Springer, New York, pp. 275–314.

Green, J. D. and Harris, G. W. (1947) The neurovascular link between the neurohypophysis and adenohypophysis. *J. Endocr.*, **5**, 136–146.

Groot, J. de (1967) Limbic and other neural pathways that regulate endocrine function. In *Neuroendocrinology, Vol. II*, L. Martini and W. F. Ganong (Eds.), Academic Press, New York, pp. 81–106.

Hart, P. G. (1971) Endocrinologische studies bij vrouwen die de menopauze naderen. *Ned. T. Geneesk.*, **115**, 1855.

Hökfelt, T. and Fuxe, K. (1972) On the morphology and the neuroendocrine role of the hypothalamic catecholamine neurons. In *Brain–Endocrine Interaction. Median Eminence: Structure and Function*, K. M. Knigge, D. E. Scott and A. Weindl (Eds.), Karger, Basel, pp. 181–223.

Hommes, O. R. (1974). *J. Endocr.*, To be published.

Kamberi, I. A., Mical, R. S. and Porter, J. C. (1970) Prolactin-inhibiting activity in hypophysial stalk blood and elevation by dopamine. *Experientia (Basel)*, **26**, 1150–1151.

Kleinberg, D. L., Noel, G. L. and Frantz, A. G. (1971) Chlorpromazine stimulation and L-dopa suppression of plasma prolactin in man. *J. clin. Endocr.*, **33**, 873–876.

Kobayashi, H., Matsui, T. and Ishii, S. (1970) Functional electron microscopy of the hypothalamic median eminence. In *International Review of Cytology, Vol. 29*, G. H. Bourne, J. F. Danielli and K. W. Jeon (Eds.), Academic Press, New York, pp. 281–381.

Köves, K. and Halász, B. (1970) Location of the neural structures triggering ovulation in the rat. *Neuroendocrinology*, **6**, 180–193.

Lammers, H. J. (1969) The neuronal connexions of the hypothalamic neurosecretory nuclei in mammals. *J. Neuro-Viscer. Relat.*, **Suppl. IX**, 311–328.

Lange-Cosack, H. (1951) Verschiedene Gruppen der hypothalamischen *Pubertas praecox. Dtsch. Z. Nervenheilk.*, **166**, 499–545.

Martinez, P. M. (1960) *The structure of the pituitary stalk and the innervation of the neurohypophysis in the cat.* Diss. Leiden. "Luctor et Emergo", Leiden.

McCann, S. M., Kalra, P. S., Donoso, A. O., Bishop, W., Schneider, H. P. G., Fawcett, C. P. and Krulich, L. (1972) The role of monoamines in the control of gonadotropin and prolactin secretion. In *Brain–Endocrine Interaction. Median Eminence: Structure and Function*, K. M. Knigge, D. E. Scott and A. Weindl (Eds.), Karger, Basel, pp. 224–235.

Palkovits, M. and Fischer, J. (1968) *Karyometric Investigations.* Akadémiai Kiadó, Budapest.

Porter, J. C., Kamberi, I. A. and Ondo, J. G. (1972) Role of biogenic amines and cerebrospinal fluid in the neurovascular transmittal of hypophysiotrophic substances. In *Brain–Endocrine Interaction. Median Eminence: Structure and Function*, K. M. Knigge, D. E. Scott and A. Weindl (Eds.), Karger, Basel, pp. 245–253.

Prick, J. J. G. (1956) Some functions of the hypothalamic-hypophysial system in relation to the pathology. *Folia psychiat. neerl.*, **59**, 408–455.

Rinne, U. K. and Sonninen, V. (1968) The occurrence of dopamine and noradrenaline in the tubero-hypophysial system. *Experientia (Basel)*, **24**, 177–178.

Rooij, J. A. M. de (1972) The nucleus infundibularis of the human hypothalamus. *Diss. Nijmegen.*

RYAN, R. J. AND FAIMAN, C. (1968) Radioimmunoassay of FSH and LH in human serum: The effects of age, and infusion of several polyamines in males. In *Gonadotropins*, E. ROSEMBERG (Ed.), Geron-X, Los Altos, pp. 333–337.

SCHADÉ, J. P. AND GROENIGEN, W. B. VAN (1961) Structural organization of the human cerebral cortex. I Maturation of the middle frontal gyrus. *Acta anat. (Basel)*, **47**, 74–111.

SHEEHAN, H. L. (1971) Der Hypothalamus beim Post-partum-Hypopituitarismus. *J. Neuro-Viscer. Relat.*, **Suppl. X**, 677–683.

SHEEHAN, H. L. AND KOVÁCS, K. (1966) The subventricular nucleus in the human hypothalamus *Brain*, **89**, 589–614.

SPATZ, H., DIEPEN, R. UND GAUPP, V. (1948) Zur Anatomie des Infundibulum und des Tuber cinereum beim Kaninchen. *Dtsch. Z. Nervenheilk.*, **159**, 229–268.

SZENTÁGOTHAI, J., FLERKÓ, B., MESS, B. AND HALÁSZ, B. (1968) *Hypothalamic Control of the Anterior Pituitary*, Akadémiai Kiadó, Budapest.

WIDE, L., NILLIUS, S. J., GEMZELL, C. AND ROOS, P. (1973) Radioimmunosorbent assay of follicle-stimulating hormone and luteinizing hormone in serum and urine from men and women. *Acta Endocr.*, **Suppl. 174**.

DISCUSSION

SMITH: I am quite interested in these fluorescence studies. Do you have any idea what kind of monoamine causes the yellow paraformaldehyde-induced fluorescence and how specific this fluorescence is?

DE ROOIJ: Since we did not measure the wavelength of the emission light of the fluorophore, we do not know what particular compound we are dealing with. As specificity test, we studied material that has not been treated with paraformaldehyde. From this procedure it appeared that autofluorescence was present in the untreated material. The material is probably lipofuchsin. There is however a clear difference between this autofluorescence and the yellow-green fluorescence in the paraformaldehyde-treated hypothalamic tissue. We applied also the sodium-borohydride test to reduce the fluorescence.

SWAAB: I should like to make a short comment on your nuclear volume measurements. We have been comparing literature data on nuclear measurements with other parameters for cellular activity in the hypothalamus, in the magnocellular as well as in the parvocellular area. It appeared that an increased cellular activity with an increased nuclear volume may go hand in hand with a decreased nuclear volume, or with no change at all. The same holds true for a decrease in cellular activity. I think therefore, that you have to be very careful to draw any conclusion about activation or inhibition of a hypothalamic area from the data you offered to us. A much better parameter for cellular activity seems to be the measurement of nucleolar volumes.

DE ROOIJ: We did not measure nucleoli, but we paid attention to the histological picture of the hypothalamic cells, *i.e.* to the size and the basophilic staining of the nucleolus and the amount of Nissl substance in the cell. The histological picture is changing parallel to the changes in mean nuclear volume. It may be difficult indeed to decide whether a change in mean nuclear volume means enhanced or decreased cellular activity. This difficult problem has been reviewed by Palkovitz and Fischer in 1968.

VAN DER SCHOOT: For the reliability of your material it seems important to have some knowledge about the duration between the moment of death and the collection of the brain tissue.

DE ROOIJ: We tested the correlation between the mean nuclear volume and the time that elapsed between death and fixation, or the length of fixation. We could however not find any significant relationship between these data. Differences in this section of the procedure will therefore not influence the results.

Histochemical Differentiation of Hypothalamic Areas

CHRISTOPH PILGRIM

Department of Biology, University of Regensburg, Regensburg (G.F.R.)

INTRODUCTION

The following contribution will only partly be concerned with ontogenetic aspects, in contrast to the impression given by the title. The main objective lies in the results of the differentiation process rather than in the process itself. In the first place, the article should be regarded as an attempt to define some of the areas of the hypothalamus on the basis of characteristic patterns of histochemical reactions. The areas chosen comprise those elements that have been central in the interest of neuroendocrinologists and histochemists during the past two decades.

Only the last part of the survey will deal with the schedule of ontogenetic development of histochemical reactions in the regions which will be discussed in the earlier sections. Most of the work reported here applies to the mammalian hypothalamus. Exceptions will be mentioned if necessary.

HISTOCHEMISTRY OF THE HYPOTHALAMO-NEUROHYPOPHYSEAL SYSTEM (HNS)

(a) The neurosecretory material (NSM)

It is known that the HNS contains specific neurosecretory material which can be identified by light and electron microscopy. The chemical composition of this so-called Gomori-positive material has been a subject of interest of histochemists for many years. The staining methods commonly used to investigate the HNS, *i.e.* chromalum hematoxylin, aldehyde fuchsin as well as their more recent, considerably improved variations (Bock, 1966; Brinkmann and Bock, 1970), are not histochemical techniques in the strict sense of the term. The reaction mechanism is not completely clear and — even if observations are restricted to the HNS — not specific. For example, lysosomes which are numerous in the HNS (see below) may be stained as well (Pilgrim, 1970b; Rodríguez, 1970).

A better insight in the chemical composition of the Gomori-positive NSM has been obtained by Sloper and coworkers who were able to show a selective coloration of the material with histochemical tests for protein-bound disulphide groups (see Sloper, 1966, for review). It could then be concluded that one was dealing with a protein

References p. 106–110

containing a relatively high number of cysteine residues. This conclusion was substantiated by Sterba (1964) by an analysis of the reaction of NSM with pseudo-isocyanin. The reaction results in a fluorescent compound which is due to a certain arrangement of oxidized SH-groups along the molecule. Since the NSM is composed of the hormones and their carrier proteins (the neurophysins) it has been debated which of the two components is stained by the histochemical tests. The hormones themselves which are known to be nonapeptides containing two cysteine residues (Boissonas and Guttman, 1968) are likely candidates, and it has indeed been shown that they can be responsible for a positive reaction with pseudo-isocyanin (Gutierrez and Sloper, 1969). Other investigations, however, indicate that it is rather the carrier proteins which react with Gomori stains (Hild and Zetler, 1953; Hadler *et al.*, 1968) or bind to histochemical reagents (Sterba, 1964; Bock and Schlüter, 1971a, 1971b). It seems quite possible that under normal circumstances both components contribute to the histochemical pictures of the HNS.

On the basis of the relatively high cysteine content it was attempted to label the HNS with [^{35}S]cysteine. A high isotope uptake in the cells of the magnocellular nuclei was recorded (Sloper, 1958). Later on, it was shown by several investigators that this technique can be usefully employed to demonstrate axonal transport of the NSM (Sloper *et al.*, 1960; Ficq and Flament-Durand, 1963; Talanti *et al.*, 1972; Norström, 1972). It is debatable, however, whether [^{35}S]cyst(e)ine-labeling of the HNS is really as selective as it was hoped to be when these studies were initiated. It has been mentioned, for example, that nerve cells in other parts of the central nervous system may also exhibit uptake of label after injection of [^{35}S]cyst(e)ine which is comparable to that of the HNS (Sloper *et al.*, 1960; Ford *et al.*, 1961). One has to bear in mind that incorporation experiments with labeled amino acids, in contrast to histochemical reactions, do in fact measure the turnover instead of the amount of a protein. Theoretically, it is possible that a high cysteine content of a certain protein fraction does not markedly influence the overall cysteine incorporation if there are many other protein fractions with relatively low cysteine content but high turnover rates (Pilgrim, 1969).

Recently we have been able to demonstrate autoradiographically a considerable uptake of [^{3}H]fucose in the perikarya of the supraoptic nucleus (Pilgrim and Wagner, in preparation). This observation suggests synthesis of glycoproteins which, moreover, seem to be subjected to axonal transport towards the pars nervosa. It has been assumed by Schiebler (1952) on the basis of a positive PAS reaction that the NSM itself might represent a glycoprotein. This conclusion was contradicted by other investigators (*cf.* Sloper, 1966). Now that the chemical compositions of the hormones as well as of the neurophysins are known (Boissonas and Guttman, 1968; Rauch *et al.*, 1969; Uttenthal and Hope, 1970), it is quite clear that the secretory product itself is no glycoprotein. It remains to be seen what other substances besides the hormones and neurophysins are produced in the neurosecretory cell. It should be recalled that a larger precursor molecule has been postulated by Sachs *et al.*, from which the hormones and neurophysins are split off after synthesis in the perikaryon (Sachs, 1969; Sachs *et al.*, 1969).

(b) Protein turnover and RNA content

Aside from the open question as to the degree of specificity of the [^{35}S]cysteine label (*cf.* above), the experiments with radioactive amino acids have demonstrated that the cells of the magnocellular nuclei are characterized by an exceptionally high rate of protein synthesis. An additional increase can be observed under osmotic stimuli, due to an increase in volume of the whole cell rather than to an increased rate of synthesis per unit volume (Pilgrim, 1969). In this respect, the neurosecretory cell seems to conform to a general rule which determines rates of protein synthesis in other cells as well (Schultze, 1968). That neurosecretory cells behave no differently from other protein synthesizing cells can also be seen from microchemical and cytophotometric measurements of the RNA content. An increase in nucleolar and cytoplasmic RNA is found under functional stimulation (Edström and Eichner, 1958, 1960; Ifft *et al.*, 1964; Ifft and Berkowitz, 1965).

(c) Hydrolytic enzymes

Protein secreting cells are further characterized by a highly developed Golgi apparatus. It is in accordance with this rule that the perikarya of the supraoptic and paraventricular nuclei exhibit an intensive reaction for thiamine pyrophosphatase (Osinchak, 1964; Pilgrim, 1967a). The pattern of the TPPase reaction shows marked variations under osmotic stress (Pilgrim, 1967b, 1969). The total distribution of the enzyme is increased (Jongkind and Swaab, 1967). The functional meaning of the localization of TPPase in the Golgi apparatus is still obscure. There is evidence that TPP plays a role in the process of excitation of the nerve cell membrane (v. Muralt, 1958). More specifically, it has been suggested that dephosphorylation of TPP and TP itself may be responsible for changes in Na^+ permeability in the course of the neuronal membrane changes (Itokawa and Cooper, 1970).

On the other hand, it has been speculated by Novikoff *et al.* (1962) that TPP in its role as a coenzyme in the citric acid cycle may influence the funneling of acetyl-CoA into the synthesis of fatty acids. This would coincide with the idea that one of the functions of the Golgi apparatus is to serve as a turnover site for cellular membranes. There is evidence, at the submicroscopic level, that the process of formation of NSM in the neurosecretory cell is associated with a considerable turnover and reorganization of cellular membranes (*cf.* Pilgrim, 1969, for discussion).

This last point leads to a discussion of lysosomal enzymes and of the role of lysosomes in the neurosecretory cell. The lysosomal localization of several hydrolytic enzymes was demonstrated and alterations under functional stimulus were recorded (Pilgrim, 1967b, 1969). The stimulus was found to lead to qualitatively and quantitatively different responses of the various enzymes which resulted in a definite change of the enzyme pattern of the magnocellular nuclei. The functional meaning of this observation is very difficult to interpret. It emphasizes, however, the role of the lysosomes in the metabolism of the neurosecretory cell and supports our ultrastructural studies on the origin and fate of lysosomes (Pilgrim, 1969, 1970a, b). These studies

References p. 106–110

suggest that lysosomes are responsible for the breakdown of cellular membranes and other constituents which are subjected to a continuous turnover. The lysosomes have to be regarded as catabolic cell organelles which are as necessary in maintaining the cell's metabolic equilibrium as *e.g.* the endoplasmic reticulum.

It is interesting to note that histochemical and ultrastructural signs of lysosomal activity can also be found in the distal parts of the HNS. They have been described in neurosecretory axons and axon endings as well as in pituicytes (Whitaker *et al.*, 1970; Whitaker and LaBella, 1972). There are several possible explanations for the function of lysosomal enzymes at these sites. Probably, a degradation or rearrangement of membrane constituents takes place also at the axon endings in connection with exocytosis of elementary granules (Douglas and Nagasawa, 1971; Douglas *et al.*, 1971; Nagasawa *et al.*, 1971). According to these authors, lysosomal activity may be necessary to take care of the "recaptured" membrane which is represented by part of the so-called "synaptic" vesicles. If it is true that the NSM is being split off from a larger precursor molecule before it is released (*cf.* above), this task may also be performed by lysosomal enzymes. As to the function of lysosomes in pituicytes, it has been discussed that these cells take part in the release of the NSM from the axon endings (Livingston, 1971). This is perhaps performed by degrading parts of the elementary granule. Furthermore, pituicytes have even been shown to take part in the removal of whole axons (Zambrano and DeRobertis, 1968; Olivieri-Sangiacomo, 1972).

(d) Oxidizing enzymes

By means of histochemical methods it should also be possible to obtain some information about the energy metabolism of the HNS cells. For that purpose a number of oxidizing enzymes were investigated with histo- and microchemical methods: α-glycerophosphate dehydrogenase, lactate dehydrogenase, succinate dehydrogenase, NADH- and NADPH-diaphorase and cytochrome oxidase are present in the HNS (see Arvy, 1962; Iijima *et al.*, 1967; Pilgrim, 1967a). None of these histochemical tests exhibits, however, any specificity for the HNS. By contrast, a remarkably high activity for glucose-6-phosphate dehydrogenase is found in the magnocellular centers (Pilgrim, 1967a). This is also true for the pars nervosa. Under osmotic stress, a marked increase of activity of this enzyme takes place, whereas activities of LDH and SDH are not significantly altered (Ifft *et al.*, 1964; Pilgrim, 1967b, 1969; Jongkind, 1967). These observations suggest that energy gain through the pentose phosphate shunt is an important metabolic pathway. It is interesting that this is true for other endocrine tissues as well (Field *et al.*, 1960). These latter authors have emphasized the importance of the pentose cycle in generating NADPH which in turn is required for the synthesis of hormones.

(e) Acetylcholinesterase (AChE) and catecholamines (CA)

A positive histochemical reaction for AChE is generally taken as an indication of cholinergic transmitter functions. AChE has been demonstrated in the perikarya of the supraoptic and paraventricular nucleus as well as in the pars nervosa (Koelle and Geesey, 1961; Arvy, 1962; Kiernan, 1964; Kobayashi and Farner, 1964; Uemura, 1965). Changes in AChE activity have been reported in experimentally induced hyper- and hypofunction (Pearse, 1958; Kivalo *et al.*, 1958). The presence of AChE in the cell bodies and axons and the obvious participation of AChE in the function of the HNS could mean that either the HNS cells are innervated by cholinergic afferent fibers or that the HNS cells are cholinergic themselves. As to the latter alternative, it has been proposed that acetylcholine liberated at the terminal in response to an impulse, might in turn control the release of the endocrine secretion from the same terminal (Koelle and Geesey, 1961). The fact that elementary granules and vesicles resembling synaptic vesicles are present in the same terminal has been taken as a corroboration of this hypothesis. In the meantime, it has become more likely that the so-called "synaptic" vesicles are in fact breakdown products of elementary granules (Livingston, 1971) or membrane fragments which have been recaptured during the process of exocytosis (*cf.* above). At present the role of AChE in the HNS cannot be explained sufficiently. Physiological experiments also indicate that the process of neurosecretion in the HNS is in some way coupled to cholinergic mechanisms (Ginsburg, 1968).

Whereas the presence of acetylcholine can be demonstrated only by indirect histochemical methods the occurrence of monoaminergic transmitters may be shown directly by the well-known formaldehyde-induced fluorescence method. If one applies this technique to the HNS it can be seen that the HNS itself does not contain monoamines though the perikarya are innervated by a dense network of afferent fibers, containing noradrenaline (Fuxe and Hökfelt, 1969). This afferent fiber system exhibits increased activity under osmotic stress *(loc. cit.)*. Parallel variations of catecholamine fluorescence and NSM content under certain experimental conditions have also been reported (Schiebler and Meinhardt, 1969). Furthermore the occurrence of a monoaminergic input is in agreement with the histochemical detection of monoamine oxidase in the perikarya of the HNS (Pilgrim, 1967a).

On the whole, it seems safe to assume that the histochemical investigations cited above favor the view that synthesis and release of NSM in the HNS are controlled by cholinergic as well as by monoaminergic mechanisms. This idea is also supported by physiological experiments, examining the action of acetylcholine and adrenaline on the release of the neurohormones (*cf.* Ginsburg, 1968).

HISTOCHEMISTRY OF THE TUBERO-INFUNDIBULAR SYSTEM (TIS)

The significance of the TIS is assumed to be the production and release of neurohormones which control the function of the adenohypophysis. The neurohormones are called releasing and inhibiting factors (RF, IF), respectively. Histochemical data about this system are much less numerous than those about the HNS. This is at least partly due to the fact that the morphological delineation of the system is not easy at all. What we know is that the terminal portion of the system is located in the zona externa

References p. 106–110

of the median eminence. We also know some of the sites of origin which consist of cell bodies in the arcuate and anterior periventricular nucleus and the retrochiasmatic area (hypophyseotropic area, Halász *et al.*, 1962). Further likely candidates are the ventromedial, anterior and premammillary nuclei which, however, after Szentágothai *et al.* (1972) belong to a higher control system and are not directly involved in the discharge of neurohormones into the portal circulation.

(a) The neurosecretory material

The NSM of the TIS is generally considered to be Gomori-negative (Diepen, 1962). Apart from the fact that this is true only for mammals (Dierickx and van den Abeele, 1959; Oksche, 1967), it is in all probability not a question of principal differences between the NSM of the TIS and the HNS, but a question of concentration. At least this is the easiest way to explain the appearance of Gomori-positive material in the terminal portions of the TIS under experimental conditions which stimulate the part of the system controlling ACTH production (*cf.* Brinkmann and Bock, 1970). Under such conditions, it is possible to apply to this NSM the same histochemical tests as described above for the NSM of the HNS with identical results (Bock und aus der Mühlen, 1968). This supports the idea that the stainable NSM in the zona externa of the median eminence is equivalent to a corticotropin-releasing factor and, probably, to a carrier protein which is chemically similar to the neurophysins (Brinkmann and Bock, 1970).

(b) Acetylcholinesterase and catecholamines

A marked reaction for AChE has been described in the terminal portions of the TIS (zona externa) as well as in the perikarya of the arcuate nucleus while in other parts of the parvicellular hypothalamus there is only weak activity or no activity at all (Kobayashi and Farner, 1964; Uemura, 1965; Hyyppä, 1969a). The significance of the presence of AChE in these neurons is similarly obscure as in the HNS (Kobayashi *et al.*, 1970).

More instructive results were obtained by the histochemical investigation of monoamines (Fuxe and Hökfelt, 1969, 1970; Lichtensteiger, 1970; Hökfelt and Fuxe, 1972). They have contributed much to the understanding of the organization of the parvicellular hypothalamus. Unlike the situation in the HNS, the formaldehyde-induced fluorescence which is due to dopamine is seen in the TIS itself. Normally, it is again restricted to the terminals in the zona externa. By so-called monoamine loading experiments the fibers can be traced to cell bodies in the arcuate and anterior periventricular nucleus. These neurons are likely to participate in the regulation of gonadotropin secretion from the anterior pituitary. Since the dopamine containing neurons constitute only part of the TIS, it is thought that they are not directly concerned with the production of RF but control the release of the neurohormones at the terminals of the RF neurons by axo-axonal synapses. The experiments carried out by Fuxe and Hökfelt (1969, 1970) are consistent with the assumption that increased activity of the dopamine neurons inhibits the release of LH- and FSH-releasing factor and stimulates the release of prolactin inhibiting factor. Physiological experiments support the role of dopamine in the control of gonadotropic hormone secretion but suggest a somewhat different mechanism in that dopamine seems to stimulate LHRF and FSHRF (Schneider and McCann, 1970). Also the question whether RF and dopamine might not be present together in the same neuron awaits final solution (Rodríguez, 1972).

Similar to the HNS, the TIS is innervated by nerve fibers containing noradrenaline and, furthermore, serotonin (Fuxe and Hökfelt, 1970; Hökfelt and Fuxe, 1972). The significance of monoaminergic mechanisms in the TIS is emphasized by a strong histochemical reaction for monoamine oxidase in the zona externa (Matsui and Kobayashi, 1965).

HISTOCHEMISTRY OF SPECIALIZED EPENDYMA AND GLIA

A conspicuous morphological variation of the ependyma of the third ventricle is represented by the so-called tanycyte-ependyma. This type of ependymal cell is found in the ventral portion of the third ventricle. The cells are characterized by long slender processes which end at the capillary loops of the median eminence or other blood vessels in the arcuate and ventromedial nucleus. Their intimate relationship to the TIS has given rise to speculations about a possible involvement of these cells in the control of pituitary function (Knowles, 1972). They could be regarded as special devices to convey information between the CSF and the portal circulation or the nerve cells of the above-mentioned nuclear regions. How and in which direction this information transmission is effected is not clear at all. It has been said that the tanycytes secrete into the portal circulation or into the CSF or that they transport substances (releasing factors?) from the CSF to the portal vessels or *vice versa*.

The distinct shape and localization of the tanycyte-ependyma has attracted attention of histochemists. Histochemical tests for various enzymes have shown that the tanycyte-ependyma is functionally different from the normal ciliated ependyma. Most oxidizing enzymes are less active in tanycyte- than in ciliated ependyma. Succinate dehydrogenase and cytochrome oxidase are completely lacking in the former. By contrast, NADPH-diaphorase and glucose-6-phosphate dehydrogenase activity exceeds that of the ciliated ependyma by far (Pilgrim, 1967a; Schachenmayr, 1967). The latter observation reveals an obvious parallel to the behavior of the magnocellular neurosecretory cells. Possibly in the tanycytes, too, it has to be taken as an indication of synthetic activity. This interpretation would agree with the presence of PAS-positive substances in the tanycytes which may be secretory in nature (Leveque and Hofkin, 1961). Recently, labeled fucose has been localized in the tanycytes of the ventral and lateral walls of the infundibular recess by electron microscopic autoradiography indicating synthesis of glycoproteins in these cells (Wagner and Pilgrim, in preparation).

Non-specific esterase is another enzyme that permits discrimination of the specialized ependyma (Colmant, 1967; Pilgrim, 1967a; Bock and Goslar, 1969; Luppa and Feustel,

References p. 106–110

1970). Apart from a small granular activity in the perikarya, which is resistant to E 600, most of the activity in the perikarya and processes is inhibited by E 600. Results obtained with other inhibitors are at variance (Goslar and Bock, 1971; Luppa and Feustel, 1971). Detailed testing with substrates of various chain length have indicated that the enzyme has neither a lipase nor a cathepsin-like function (Goslar and Bock, 1970). The natural substrate is not known. The enzyme is thought to participate in a possible transport function of the tanycytes (Luppa and Feustel, 1971). It is striking that the activity of the enzyme decreases after adrenalectomy (Bock and Goslar, 1969), because it happens at the same site where an accumulation of NSM is observed in this condition (zona externa of the median eminence, *cf.* above).

The histochemical investigations of the ependyma of the third ventricle have pointed out, furthermore, that there are functional differences even between tanycytes located at different positions along the ventricle wall. For example, the tanycytes occupying a more dorsal position exhibit a stronger reaction for non-specific esterase than those lining the infundibular recess (Colmant, 1967; Luppa and Feustel, 1971). Inside the infundibular recess itself, differences in substrate specificity can be detected between cells of the floor and those of the lateral walls. The dorsal cells located at the level of the ventromedial nucleus are, in addition, characterized by a marked reaction for ATPase and TPPase (Schachenmayr, 1967; Luppa and Feustel, 1971). The following conclusions seem to be justified. The tanycyte-ependyma is not only morphologically but also functionally different from the rest of the ependyma. It participates somehow in the regulation of pituitary function, probably by secreting substances (possibly glycoproteins) into the portal circulation.

Apart from the specialized ependyma, a special type of glial cells is present in the hypothalamus showing properties which are interesting from a histochemical viewpoint. They have been known as so-called Gomori-positive glial cells (Koritsanszky, 1969; Srebro, 1969). Their cytoplasmic granules which react with Gomori stains are rich in cysteine groups and contain endogenous peroxidase activity (Srebro and Cichocki, 1971; Srebro, 1972; Pilgrim and Wagner, 1974). Although the granules resemble lipofuscin, they lack typical lysosomal enzymes. Concerning the distribution of these cells, they are found in a narrow subependymal zone around the third ventricle which is broadened considerably at the level of the arcuate nucleus. The fact that they are especially numerous in an area which borders a region lacking a blood–brain-barrier (the median eminence) could well be significant in the function of these cells. Possibly, they are involved in sealing off the adjoining brain regions against toxic substances which penetrate the walls of the median eminence vessels and tend to leak into the extracellular space of the remaining hypothalamus. Such blood-borne substances may consist of peroxides, heavy metals and free radicals (Srebro, 1971).

DEVELOPMENTAL ASPECTS

This section is concerned with the ontogenetic development of histochemical reactions in the areas which have been discussed above. Nearly all of the results reported here have been obtained from experiments with rats. A number of authors have investigated the schedule of appearance of enzyme-histochemical reactions in the magnocellular centers of the HNS (Cohn and Richter, 1956; Smiechowska, 1964; Pilgrim, 1967a; Hyyppä, 1969a). Some of the enzymes investigated (oxidizing enzymes, acid phosphatase) are present already in primitive neuroblasts, a stage well before a morphological delineation of the nuclear areas is possible. This is true, however, for other neuroblasts as well, and does not allow any conclusion as to a special functional development during this period. Around day 18 of embryonic life of the rat the nuclear areas begin to condense from the remaining periventricular gray. This is the moment when tests for AChE become positive. Non-specific esterase is the last enzyme to give a positive reaction. The reaction seems to depend on the presence of Nissl bodies. So far, the tests for these two esterases seem to be the only histochemical reactions whose appearance can be correlated with a definite morphological event. There is an apparent lack of coordination which is also true for the postnatal period. Nevertheless in the newborn, where the neurosecretory cells still look like rather undifferentiated neuroblasts, Gomori-positive secretion granules can be detected. Although this appears to be a distinct developmental step it is not accompanied by a noticeable change of enzyme activities. At the end of the second postnatal week, development seems to be complete when judged by morphological criteria. Again, histochemical differentiation behaves differently. The typical enzyme pattern of the HNS nuclei, especially the characteristic distribution of glucose-6-phosphate dehydrogenase, is not reached until the end of the fourth week. If we ask for the functional meaning of these observations the only answer which can be given at present is that the important and specific stages of development of the HNS fall into the postnatal period and that the HNS is functionally not mature until the end of the fourth week (Pilgrim, 1967a). It is not possible to state when and to what extent the HNS controls development in other parts of the body.

The situation seems to be a little different and more exciting if we look at the development of the tubero-infundibular system. Hyyppä (1969a) has described an exceptionally early appearance of AChE in the periventricular matrix from which the ventromedial and arcuate nuclei are later formed. In this case, histochemical methods offer the opportunity to detect an important area of the hypothalamus prior to its morphological differentiation. A further point of interest lies in a temporary absolute and relative increase of AChE activity in these nuclei between postnatal days 1 and 10. At the same time, the dopamine fluorescence is first detected in the median eminence (Hyyppä, 1969b). These observations can be taken as an indication of relatively early functional maturity of at least part of the TIS and, possibly, its participation in the control of sexual development.

The histochemical differentiation of the tanycyte-ependyma has been investigated in the rat (Schachenmayr, 1967) and the chicken (Kabisch and Luppa, 1972). In the rat, the characteristic histochemical differences between the tanycyte and the ciliated ependyma (*cf.* above) essentially emerge only after birth. The demarcation is due to the fact that, starting in the first postnatal days, development of the tanycyte-ependyma lags behind that of the ciliated ependyma. The adult enzyme pattern, indicating

References p. 106–110

functional maturity, is reached only during the fifth week. This agrees with the late maturation of the HNS but is in contrast to the relatively early signs of specific function in the TIS. It is still questionable whether this discrepancy indicates that the TIS is able to perform at least part of its functions without interference of the tanycyte-ependyma (*cf.* above).

The demarcation of the tanycyte-ependyma in the chicken happens much earlier than in the rat. This is in agreement, however, with the earlier morphological differentiation of the ependyma in general. Similarly to the situation in the rat, the tanycyte-ependyma remains on a lower level of differentiation than the ciliated ependyma. In this respect, the tanycytes have been compared by Kabisch and Luppa (1972) to the neurosecretory cells whose double function as neurons and secretory cells can also be taken as indicative of a lower differentiation level. If one accepts that less specialization can mean more functional potentialities, then the tanycytes may well be able to perform not just one, but all of the tasks which have been ascribed to them as part of their role in controlling neuroendocrine activity (*cf.* above).

REFERENCES

ARVY, L. (1962) Histochemical demonstration of enzymatic activities in neurosecretory centres of some homoiothermic animals. In *Neurosecretion*, H. HELLER AND R. B. CLARK (Eds.), Academic Press, London, pp. 215–223.

BOCK, R. (1966) Über die Darstellbarkeit neurosekretorischer Substanz mit Chromalaun–Gallocyanin im supraoptico–hypophysären System beim Hund. *Histochemie*, **6**, 362–369.

BOCK, R. UND GOSLAR, H. G. (1969) Enzymhistochemische Untersuchungen an Infundibulum und Hypophysenhinterlappen der normalen und beidseitig adrenalektomierten Ratte. *Z. Zellforsch.*, **95**, 415–428.

BOCK, R. UND MÜHLEN, K. AUS DER (1968) Beiträge zur funktionellen Morphologie der Neurohypophyse. I. Über eine "gomoripositive" Substanz in der Zona externa infundibuli beidseitig adrenalektomierter weißer Mäuse. *Z. Zellforsch.*, **92**, 130–148.

BOCK, R. UND SCHLÜTER, G. (1971a) Fluoreszenzmikroskopischer Nachweis von Arginin im Neurosekret von Säugern. *Histochemie*, **25**, 152–162.

BOCK, R. UND SCHLÜTER, G. (1971b) Fluoreszenzmikroskopischer Nachweis von Arginin im Neurosekret des Schweines mit Phenanthrenchinon. *Z. Zellforsch.*, **122**, 456–459.

BOISONNAS, R. A. AND GUTTMAN, ST. (1968) Chemistry of the neurohypophysial hormones. In *Handbuch der experimentell Pharmakologie, Vol. XXIII*, B. BERDE (Ed.), Springer, Berlin, pp. 40–66.

BRINKMANN, H. UND BOCK, R. (1970) Quantitative Veränderungen "Gomori-positiver" Substanzen in Infundibulum und Hypophysen-hinterlappen der Ratte nach Adrenalektomie und Kochsalz- oder Durstbelastung. *J. neuro-visc. Relat.*, **32**, 48–64.

COHN, P. AND RICHTER, D. (1956) Enzymic development and maturation of the hypothalamus. *J. Neurochem.*, **1**, 166–172.

COLMANT, H. J. (1967) Über die Wandstruktur des dritten Ventrikels der Albinoratte. *Histochemie*, **11**, 40–61.

DIEPEN, R. (1962) Hypothalamus. In *Handbuch der mikroskopischen Anatomie des Menschen, Vol. IV/7*, W. BARGMANN (Ed.), Springer, Berlin.

DIERICKX, K. AND ABEELE, A. VAN DEN (1959) On the relations between the hypothalamus and the anterior pituitary in *Rana temporaria*. *Z. Zellforsch.*, **51**, 78–87.

DOUGLAS, W. W. AND NAGASAWA, J. (1971) Membrane vesiculation at sites of exocytosis in the neurohypophysis, adenohypophysis and adrenal medulla: a device for membrane conservation. *J. Physiol. (Lond.)*, **218**, 94P–95P.

DOUGLAS, W. W., NAGASAWA, J. AND SCHULZ, R. A. (1971) Coated microvesicles in neurosecretory terminals of posterior pituitary glands shed their coats to become smooth "synaptic" vesicles. *Nature (Lond.)*, **232**, 340–341.

EDSTRÖM, J. E. UND EICHNER, D. (1958) Quantitative Ribonucleinsäure-Untersuchungen an den Ganglienzellen des Nucleus supra-opticus der Albinoratte unter experimentellen Bedingungen (Kochsalzbelastung). *Z. Zellforsch.*, **48**, 187–200.

EDSTRÖM, J. E. UND EICHNER, D. (1960) Qualitative und quantitative Ribonucleinsäureuntersuchungen an den Ganglienzellen der NN. supraopticus und paraventricularis der Ratte unter normalen und experimentellen Bedingungen (Kochsalzbelastung). *Anat. Anz.*, **108**, 312–319.

FICQ, A. AND FLAMENT-DURAND, J. (1963) Autoradiography in endocrine research. In *Techniques in endocrine Research*, P. ECKSTEIN AND F. KNOWLES (Eds.), Academic Press, London, pp. 73–85.

FIELD, J. B., PASTAN, I., HERRING, B. AND JOHNSON, P. (1960) Studies of pathways of glucose metabolism of endocrine tissues. *Endocrinology*, **67**, 801–806.

FORD, D. H., HIRSCHMAN, A., RHINES, R. AND ZIMBERG, S. (1961) The rate of uptake and autoradiographic localization of S^{35} in the central nervous system, pituitary, and skeletal muscle of the normal male rat after the injection of S^{35}-labeled cystine. *Exp. Neurol.*, **4**, 444–459.

FUXE, K. AND HÖKFELT, T. (1969) Catecholamines in the hypothalamus and the pituitary gland. In *Frontiers in Neuroendocrinology*, W. F. GANONG AND L. MARTINI (Eds.), Oxford Univ. Press, New York, pp. 47–96.

FUXE, K. AND HÖKFELT, T. (1970) Participation of central monoamine neurons in the regulation of anterior pituitary function with special regard to the neuro-endocrine role of tubero-infundibular dopamine neurons. In *Aspects of Neuroendocrinology*, W. BARGMANN AND B. SCHARRER (Eds.), Springer, Berlin, pp. 192–205.

GINSBURG, M. (1968) Production of the neurohypophysial hormones. In *Handbuch der experimentell Pharmakologie, Vol. XXIII*, B. BERDE (Ed.), Springer, Berlin, pp. 286–371.

GOSLAR, H. G. UND BOCK, R. (1970) Weitere Befunde zum enzymhistochemischen Verhalten von Infundibulum und Hypophysenhinterlappen der Ratte nach beidseitiger Adrenalektomie. In *Aspects of Neuroendocrinology*, W. BARGMANN UND B. SCHARRER (Eds.), Springer, Berlin, pp. 322–323.

GOSLAR, H. G. UND BOCK, R. (1971) Histochemische Eigenschaften der unspezifischen Esterasen im Tanycytenependym des III. Ventrikels, im Subfornicalorgan und im Subcommissuralorgan der Wistarratte. *Histochemie*, **28**, 170–182.

GUTIERREZ, M. AND SLOPER, J. C. (1969) Reaction *in vitro* of synthetic oxytocin and lysine–vasopressin with the pseudoisocyanin-chloride technique used for the demonstration of neurohypophysial neurosecretory material. *Histochemie*, **17**, 73–77.

HADLER, W. A., PETELLI, A. S., DE LUCCA, O. AND ZITI, L. M. (1968) The meaning of the selective staining of neurosecretory substance by aldehyde–fuchsin and chrome–alum hematoxylin. A comparative study carried out on neurosecretory substance, cystine, cysteine and some unsaturated compounds subjected to several oxidizing agents. *Acta histochem. (Jena)*, **29**, 304–319.

HALÁSZ, B., PUPP, L. AND UHLARIK, S. (1962) Hypophysiotrophic area in the hypothalamus. *J. Endocr.*, **25**, 147–154.

HILD, W. UND ZETLER, G. (1953) Über die Funktion des Neurosekrets im Zwischenhirn-Neurohypophysensystem als Trägersubstanz für Vasopressin, Adiuretin und Oxytocin. *Z. ges. exp. Med.*, **120**, 236–243.

HÖKFELT, T. AND FUXE, K. (1972) On the morphology and the neuroendocrine role of the hypothalamic catecholamine neurons. In *Brain-Endocrine Interaction. Median Eminence: Structure and Function*, K. M. KNIGGE, D. E. SCOTT AND A. WEINDL (Eds.), Karger, Basel, pp. 181–223.

HYYPPÄ, M. (1969a) Histochemically demonstrable esterase activity in the hypothalamus of the developing rat. *Histochemie*, **20**, 29–39.

HYYPPÄ, M. (1969b) A histochemical study of the primary catecholamines in the hypothalamic neurons of the rat in relation to the ontogenetic and sexual differentiation. *Z. Zellforsch.*, **98**, 550–560.

IFFT, J. D. AND BERKOWITZ, W. (1965) A comparison of selected morphological and chemical methods for measuring neuron activity in the supraoptic nucleus of dehydrated rats. *Anat. Rec.*, **152**, 231–234.

IFFT, J. D., MCNARY, JR., W. F. AND SIMONEIT, L. (1964) Succinic dehydrogenase and RNA in the supraoptic nucleus and hypothalamic anterior area in dehydrated rats. *Proc. Soc. exp. Biol. (N.Y.)*, **117**, 170–171.

IIJIMA, K., SHANTHA, T. R. AND BOURNE, G. H. (1967) Enzyme-histochemical studies on the hypothalamus with special reference to the supraoptic and paraventricular nuclei of squirrel monkey *(Saimiri sciureus)*. *Z. Zellforsch.*, **79**, 76–91.

ITOKAWA, Y. AND COOPER, J. R. (1970) Ion movements and thiamine. II. The release of the vitamin from membrane fragments. *Biochim. biophys. Acta (Amst.)*, **196**, 274–284.

JONGKIND, J. F. (1967) The quantitative histochemistry of hypothalamus, I. Pentose shunt enzymes in the activated supraoptic nucleus of the rat. *J. Histochem. Cytochem.*, **15**, 394–398.

JONGKIND, J. F. AND SWAAB, D. F. (1967) The distribution of thiamine diphosphate-phosphohydrolase in the neurosecretory nuclei of the rat following osmotic stress. *Histochemie*, **11**, 319–324.

KABISCH, H. UND LUPPA, H. (1972) Ein Beitrag zur Entwicklung des Enzymmusters im Ependym des III. Ventrikels beim Hühnerembryo. *Gegenbaurs morph. Jb.*, **118**, 187–205.

KIERNAN, J. A. (1964) Carboxylic esterases of the hypothalamus and neurohypophysis of the hedgehog. *J. roy. micr. Soc.*, **83**, 297–306.

KIVALO, E., RINNE, U. K. AND MÄKELÄ, S. (1958) Acetylcholinesterase, acid phosphatase and succinic dehydrogenase in the hypothalamic magnocellular nuclei after chlorpromazine administration. *Experientia (Basel)*, **14**, 293–296.

KNOWLES, F. (1972) Ependyma of the third ventricle in relation to pituitary function. In *Topics in Neuroendocrinology, Progr. Brain Res., Vol. 38*, J. ARIËNS KAPPERS AND J. P. SCHADÉ (Eds.), Elsevier, Amsterdam, pp. 255–270.

KOBAYASHI, H. AND FARNER, D. S. (1964) Cholinesterases in the hypothalamo-hypophysial neurosecretory system of the white-crowned sparrow, *Zonotrichia leucophrys gambelii. Z. Zellforsch.*, **63**, 965–973.

KOBAYASHI, H., MATSUI, T. AND ISHII, S. (1970) Functional electron microscopy of the hypothalamic median eminence. *Int. Rev. Cytol.*, **29**, 281–381.

KOELLE, G. B. AND GEESEY, C. N. (1961) Localization of acetylcholinesterase in the neurohypophysis and its functional implications. *Proc. Soc. exp. Biol. (N.Y.)*, **106**, 625–628.

KORITSÁNSZKY, S. (1969) System of the Gomori-positive glial cells. In *Zirkumventrikuläre Organe und Liquor*, G. STERBA (Ed.), Gustav Fischer, Jena, pp. 201–203.

LEVEQUE, T. F. AND HOFKIN, G. A. (1961) Demonstration of an alcohol–chloroform insoluble periodic acid–Schiff reactive substance in the hypothalamus of the rat. *Z. Zellforsch.*, **53**, 185–191.

LICHTENSTEIGER, W. (1970) Katecholaminhaltige Neurone in der neuroendokrinen Steuerung. *Progr. Histochem. Cytochem.*, **1**/4, 1–92.

LIVINGSTON, A. (1971) Subcellular aspects of storage and release of neurohypophysial hormones. *J. Endocr.*, **49**, 357–372.

LUPPA, H. UND FEUSTEL, G. (1970) Zur Kennzeichnung der Esteraseaktivität in den Tanycyten des Recessus infundibuli der Ratte. *Acta histochem. (Jena)*, **35**, 198–199.

LUPPA, H. AND FEUSTEL, G. (1971) Location and characterization of hydrolytic enzymes of the third ventricle lining in the region of the recessus infundibularis of the rat. A study on the function of the ependyma. *Brain Res.*, **29**, 253–270.

MATSUI, T. AND KOBAYASHI, H. (1965) Histochemical demonstration of monoamine oxidase in the hypothalamo-hypophysial system of the tree sparrow and the rat. *Z. Zellforsch.*, **68**, 172–182.

v. MURALT, A. (1958) *Neue Ergebnisse der Nervenphysiologie*. Springer, Berlin.

NAGASAWA, J., DOUGLAS, W. W. AND SCHULZ, R. A. (1971) Micropinocytotic origin of coated and smooth microvesicles ("synaptic vesicles") in neurosecretory terminals of posterior pituitary glands demonstrated by incorporation of horseradish peroxidase. *Nature (Lond.)*, **232**, 341–342.

NORSTRÖM, A. (1972) *Axonal Transport and Turnover of Neurohypophysial Proteins in the Rat. Thesis*, Elanders Boktryckeri Aktiebolag, Göteborg.

NOVIKOFF, A. B., ESSNER, E., GOLDFISCHER, S. AND HEUS, M. (1962) Nucleosidephosphatase activities of cytomembranes. In *The Interpretation of Ultrastructure, ISCB Symposia, Vol. 1*, R. J. C. HARRIS (Ed.), Academic Press, New York, pp. 149–192.

OKSCHE, A. (1967) Eine licht- und elektronenmikroskopische Analyse des neuroendokrinen Zwischenhirn-Vorderlappen-Komplexes der Vögel. In *Neurosecretion*, F. STUTINSKY (Ed.), Springer, Berlin, pp. 77–88.

OLIVIERI-SANGIACOMO, C. (1972) On the fine structure of the perivascular cells in the neural lobe of rats. *Z. Zellforsch.*, **132**, 25–34.

OSINCHAK, J. (1964) Electron microscopic localization of acid phosphatase and thiamine pyrophosphatase activity in hypothalamic neurosecretory cells of the rat. *J. Cell Biol.*, **21**, 35–47.

PEARSE, A. G. E. (1958) Esterases of the hypothalamus and neurohypophysis and their functional significance. In *Pathophysiologica Diencephalica*, S. B. CURRI AND L. MARTINI (Eds.), Springer, Wien, pp. 329–335.

PILGRIM, CH. (1967a) Über die Entwicklung des Enzymmusters in den neurosekretorischen hypothalamischen Zentren der Ratte. *Histochemie*, **10**, 44–65.

PILGRIM, CH. (1967b) Enzymhistochemische Untersuchungen an durstaktivierten neurosekretorischen Zellen der Ratte. *Experientia (Basel)*, **23**, 943.

PILGRIM, CH. (1969) Morphologische und funktionelle Untersuchungen zur Neurosekretbildung. Enzymhistochemische, autoradiographische und elektronenmikroskopische Beobachtungen an Ratten unter osmotischer Belastung. *Ergebn. Anat. Entwickl.-Gesch.*, **41**, 1–79.

PILGRIM, CH. (1970a) Function of lysosomes in neurosecretory cells. In *Aspects of Neuroendocrinology*, W. BARGMANN AND B. SCHARRER (Eds.), Springer, Berlin, pp. 349–351.

PILGRIM, CH. (1970b) Altersbedingte Anhäufung von lysosomalen Residualkörpern in neurosekretorischen Zellfortsätzen. *Z. Zellforsch.*, **109**, 573–582.

PILGRIM, CH. AND WAGNER, H.-J. (1974) Tracer-Untersuchungen am Hypothalamus der Ratte. *Anat. Anz.*, in press.

RAUCH, R., HOLLENBERG, M. D. AND HOPE, D. B. (1969) Isolation of a third bovine neurophysin. *Biochem. J.*, **115**, 473–479.

RODRÍGUEZ, E. M. (1970) Morphological and functional relationships between the hypothalamo-neurohypophysial system and cerebrospinal fluid. In *Aspects of Neuroendocrinology*, W. BARGMANN AND B. SCHARRER (Eds.), Springer, Berlin, pp. 352–365.

RODRÍGUEZ, E. M. (1972) Comparative and functional morphology of the median eminence. In *Brain–endocrine Interaction. Median Eminence: Structure and Function*, K. M. KNIGGE, D. E. SCOTT AND A. WEINDL (Eds.), Karger, Basel, pp. 319–334.

SACHS, H. (1969) Neurosecretion. *Advanc. Enzymol.*, **32**, 327–372.

SACHS, H., FAWCETT, P., TAKABATAKE, Y. AND PORTANOVA, R. (1969) Biosynthesis and release of vasopressin and neurophysin. *Recent Progr. Horm. Res.*, **25**, 447–492.

SCHACHENMAYR, W. (1967) Über die Entwicklung von Ependym und Plexus chorioideus der Ratte. *Z. Zellforsch.*, **77**, 25–63.

SCHIEBLER, T. H. (1952) Zur Histochemie des neurosekretorischen hypothalamisch-neurohypophysären Systems (II. Teil). *Acta anat. (Basel)*, **15**, 393–416.

SCHIEBLER, T. H. UND MEINHARDT, D. W. (1969) Über die Wirkung von Antiandrogenen auf die neurosekretorischen Systeme des Hypothalamus. *Z. Zellforsch.*, **100**, 581–593.

SCHNEIDER, H. P. G. AND MCCANN, S. M. (1970) Dopaminergic pathways and gonadotropin releasing factors. In *Aspects of Neuroendocrinology*, W. BARGMANN AND B. SCHARRER (Eds.), Springer, Berlin, pp. 177–191.

SCHULTZE, B. (1968) Die Orthologie und Pathologie des Nucleinsäure- und Eiweißstoffwechsels der Zelle im Autoradiogramm. In *Handbuch der allgemeinen Pathologie, Vol. II/5*, H.-W. ALTMANN *et al.* (Eds.), Springer, Berlin, pp. 466–667.

SLOPER, J. C. (1958) The application of newer histochemical and isotope techniques for the localisation of protein bound cystine or cysteine to the study of hypothalamic neurosecretion in normal and pathological conditions. In *Zweites int. Symp. über Neurosekretion*, W. BARGMANN, B. HANSTRÖM AND E. SCHARRER (Eds.), Springer, Berlin, pp. 20–25.

SLOPER, J. C. (1966) The experimental and cytopathological investigation of neurosecretion in the hypothalamus and pituitary. In *The Pituitary Gland, Vol. III*, G. W. HARRIS AND B. T. DONOVAN (Eds.), Butterworths, London, pp. 131–239.

SLOPER, J. C., ARNOT, D. J. AND KING, B. C. (1960) Sulphur metabolism in the pituitary and hypothalamus of the rat: a study of radioisotope-uptake after the injection of ^{35}S DL-cysteine, methionine and sodium sulphate. *J. Endocr.*, **20**, 9–23.

SMIECHOWSKA, B. (1964) Histochemical studies of the hypothalamus in white rats during ontogenesis. *Folia Morph. (Warszawa)*, **23**, 213–242.

SREBRO, Z. (1969) A comparative and experimental study of the Gomori-positive glia. *Fol. biol. (Kraków)*, **17**, 177–192.

SREBRO, Z. (1971) Periventricular Gomori-positive glia in brains of X-irradiated rats. *Brain Res.*, **35**, 463–468.

SREBRO, Z. (1972) Ultrastructural localization of peroxidase activity in Gomori-positive glia. *Acta anat. (Basel)*, **83**, 388–397.

SREBRO, Z. AND CICHOCKI, T. (1971) A system of periventricular glia in brain characterized by large peroxisome-like cell organelles. *Acta histochem. (Jena)*, **41**, 108–114.

STERBA, G. (1964) Grundlagen des histochemischen u. biochemischen Nachweises von Neurosecret (= Trägerprotein der Oxytozine) mit Pseudoisozyaninen. *Acta histochem. (Jena)*, **17**, 268–292.

SZENTÁGOTHAI, J., FLERKO, B., MESS, B. AND HALÁSZ, B. (1972) *Hypothalamic Control of the Anterior Pituitary*, Akademiai Kiadó, Budapest.

TALANTI, S., ATTILA, U. AND KEKKI, M. (1972) The kinetics of ^{35}S-labelled cysteine in the hypothalamic–hypophyseal tract of the rat, studied by autoradiography. *Z. Zellforsch.*, **124**, 342–353.

UEMURA, H. (1965) Histochemical studies on the distribution of cholinesterase and alkaline phosphatase in the vertebrate neurosecretory system. *Ann. Zool. Jap.*, **38**, 79–96.

UTTENTHAL, L. O. AND HOPE, D. B. (1970) The isolation of three neurophysins from porcine posterior pituitary lobes. *Biochem. J.*, **166**, 899–909.

WHITAKER, S. AND LABELLA, F. S. (1972) Ultrastructural localization of acid phosphatase in the posterior pituitary of the dehydrated rat. *Z. Zellforsch.*, **125**, 1–15.

WHITAKER, S., LABELLA, F. S. AND SANWAL, M. (1970) Electron microscopic histochemistry of lysosomes in neurosecretory nerve ending and pituicytes of rat posterior pituitary. *Z. Zellforsch.*, **111**, 493–504.

ZAMBRANO, D. AND DE ROBERTIS, E. (1968) The ultrastructural changes in the neurohypophysis after destruction of the paraventricular nuclei in normal and castrated rats. *Z. Zellforsch.*, **88**, 496–510.

DISCUSSION

HERBERT: Do you think that axo-axonal synapses exist in the median eminence? I am not sure that they have ever been shown electron microscopically. There are also other systems elsewhere where you have side by side the endings of peptide- and catecholamine systems without any morphological evidence of synapses.

PILGRIM: I have only cited the picture as it has been drawn by Fuxe and Hökfelt. To my knowledge, nobody has shown electron microscopically any axo-axonal synaps in the median eminence. The only structures that have been shown are a kind of synaptoid contacts between axons and tanycytes.

Micro-Pinocytosis and Exocytosis in Nerve Terminals in the Median Eminence of the Rat

R. STOECKART, H. G. JANSEN AND A. J. KREIKE

Department of Anatomy, Erasmus University, Rotterdam (The Netherlands)

INTRODUCTION

The concept, put forward by Douglas and co-workers (1971), that posterior pituitary hormones and their carrier-proteins are released by exocytosis, is gaining support (Dreifuss *et al.*, 1972; Santolaya *et al.*, 1972; Matthews *et al.*, 1973). Ultrastructural evidence has been presented that the content of the granular vesicles in nerve terminals of the median eminence can also be released by exocytosis (Stoeckart *et al.*, 1972). It has been hypothesized that these granular vesicles contain releasing factors, inhibiting factors and/or neurotransmitters (Kobayashi *et al.*, 1970). As a consequence of release by exocytosis, during which the membranes of the granular vesicles fuse with the nerve membranes, membrane material has to be disposed of. According to Nagasawa *et al.* (1971), recapture of excess membrane occurs by micro-pinocytosis in the posterior pituitary. This phenomenon may result in coated vesicles which are supposed to transform into agranular "synaptic" vesicles (Nagasawa *et al.*, 1971). In this study, horseradish peroxidase (HRP), a marker substance capable of demonstrating micro-pinocytosis in a variety of cells, has been used to establish whether the agranular vesicles in the nerve terminals of the median eminence may arise by micro-pinocytosis. As the median eminence is located outside the blood–brain-barrier, HRP has been injected intravenously.

METHODS

Ten adult male rats (R-Amsterdam strain), anesthetized with Avertin, were given injections of either 50 mg HRP (Calbiochem B grade) in 0.5 ml Locke or 100 mg HRP in 1.0 ml Locke by way of the tail vein. In addition, 5 male rats castrated about 20 h before were used. In control experiments HRP was omitted. Intravascular perfusion fixation was performed 5–60 min after the administration of HRP or of the solvent, using a mixture of 1% formaldehyde and 2% glutaraldehyde in 0.1 *M* cacodylate buffer (pH 7.4), to which 2% polyvinylpyrrolidone was added. The tissue of the median eminence was processed for demonstration of HRP (Graham and Karnovsky, 1966) before post-fixation with osmium tetroxide. Part of the material was impregnated

References p. 114–115

with uranyl acetate (examples in Fig. 2). The ultra-thin sections were stained with lead citrate.

OBSERVATIONS

Within 5 min after the intravenous administration of HRP, electron-dense reaction product indicative of HRP could be observed in the perivascular space (Fig. 1, at the right). Small amounts of the reaction product occurred in the intercellular spaces around the plasma membranes. A small part of the neuronal vesicles measuring 30–65 nm in diameter also appeared to be filled with electron-dense reaction product. The proportion of small vesicles characterized by the reaction product was not markedly increased if fixation occurred either 10, 20 or 60 min after administration of HRP. In the control experiments none of the small vesicles contained reaction product. In many of the nerve profiles, characterized by the small "labeled" vesicles, granular vesicles of different diameters (70–130 nm) also occurred, but these granular vesicles rarely contained reaction product. Preliminary results suggest that in the castrated rats the incorporation of HRP into small vesicles of some nerve terminals is increased (Fig. 2).

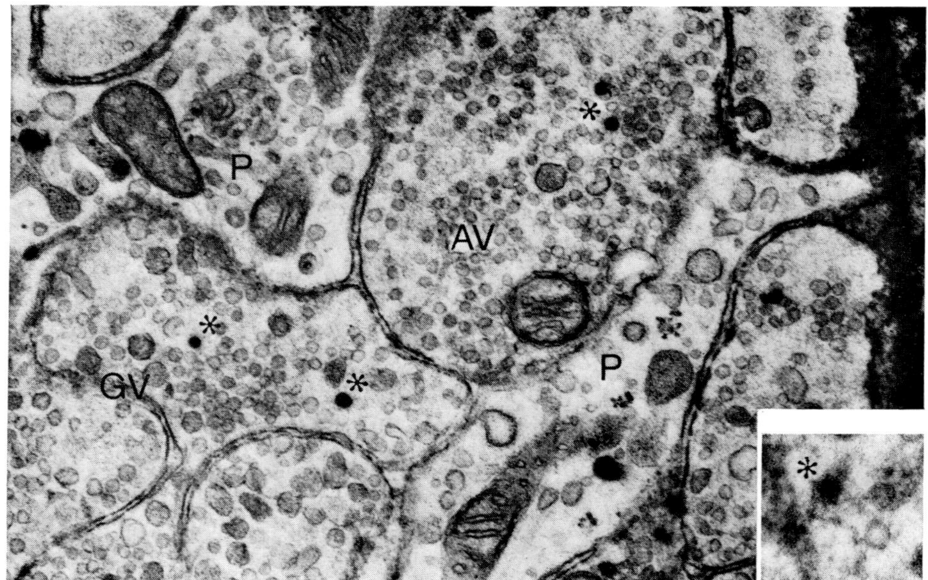

Fig. 1. Perfusion fixation 5 min after intravenous injection of 50 mg horseradish peroxidase (HRP). Nerve profiles in the external zone of the median eminence are characterized by granular vesicles (GV) of low electron-density and small agranular vesicles (AV). Some small vesicles (*) contain an electron-dense reaction product indicative of HRP. Whether the electron-dense material in the cisternae of the endoplasmic reticulum of non-nervous processes (P) represents HRP is uncertain. × 40,000. Inset: immersion fixation; coated vesicle (*) filled by HRP, close to the nerve membrane. × 70,000.

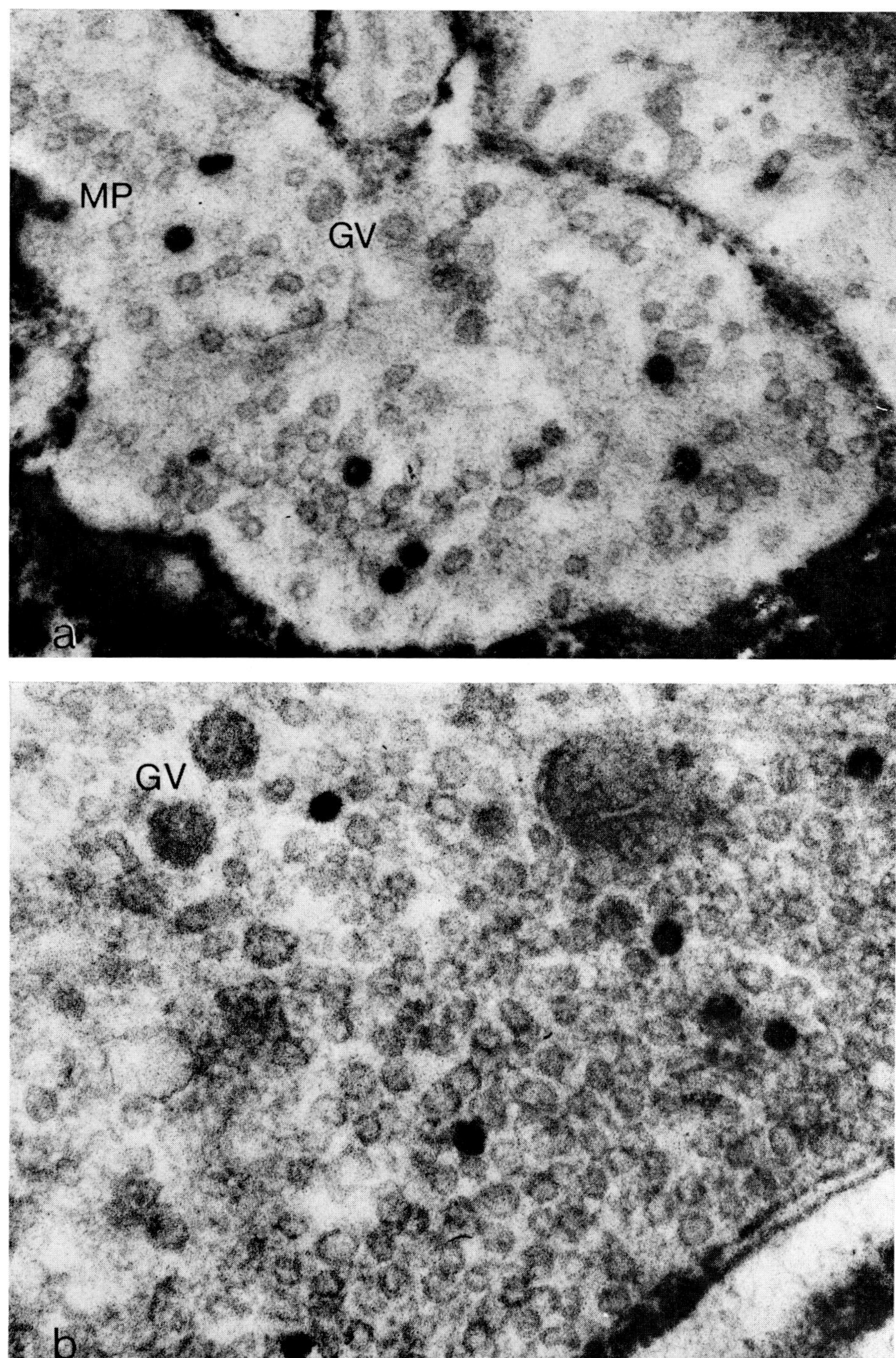

Fig. 2. Perfusion fixation 10 min after intravenous injection of 100 mg HRP; male rat, 1 day post castration. a: some vesicles of 45–55 nm contain a reaction product indicative of HRP; granular vesicles (GV) are of very low electron-density. Note micro-pinocytosis-like image (MP), containing HRP. × 75,000. b: small vesicles marked by HRP, in addition to similar but unmarked small agranular vesicles and large granular vesicles (GV). × 95,000.

References p. 114–115

DISCUSSION

The results of the present study demonstrate that at least part of the agranular "synaptic" vesicles of the median eminence arises by micro-pinocytosis. The preliminary results on incorporation of HRP in castrated rats, *i.e.* under conditions of increased release of releasing factors, suggest that a marker substance like HRP may be used to identify functional types of nerve terminals.

Although these observations do not rule out other release mechanisms, they are compatible with the exocytotic release mechanism proposed by Douglas and coworkers (1971). Moreover, evidence has been presented that the content of neuronal granular vesicles of the median eminence can be released by exocytosis (Stoeckart *et al.*, 1972). The proportion of granular vesicles observed in the process of exocytosis is small (less than 1 %). However, we calculated that at least a million granular vesicles are involved in the process of exocytosis at the moment of fixation. These granular vesicles represent the most conspicuous organelles of the preterminal and terminal nerve profiles of the median eminence which is generally accepted to be the final common pathway of the hypothalamic component of the hypothalamo–hypophyseal system. Direct evidence for the nature of the materials stored in the granular vesicles is lacking. However, studies on differential and gradient centrifugation suggest that median eminence fractions with high activity of releasing factors and biogenic amines are electron microscopically characterized by small agranular and larger granular vesicles, sometimes still contained in synaptosomes (Clementi *et al.*, 1970; Ishii, 1970; Kobayashi *et al.*, 1970; Mulder *et al.*, 1970).

SUMMARY

Incorporation of intravenously administered horseradish peroxidase into small neuronal vesicles of the rat median eminence demonstrates that at least part of these small "synaptic" vesicles of the median eminence arises by micro-pinocytosis. Preliminary studies under conditions of increased release of releasing factors suggest that horseradish peroxidase may be used to identify functionally different types of nerve terminals. The observations are compatible with an exocytotic release mechanism for neuronal granular vesicles.

REFERENCES

Clementi, F., Ceccarelli, B., Cerati, E., Demonte, M. L., Felici, M., Motta, M. and Pecile, A. (1970) Subcellular localization of neurotransmitters and releasing factors in the rat median eminence. *J. Endocr.*, **48**, 205–213.

Douglas, W. W., Nagasawa, J. and Schulz, R. (1971) Electron microscopic studies on the mechanism of secretion of posterior pituitary hormones and significance of microvesicles ('synaptic vesicles'): evidence of secretion by exocytosis and formation of microvesicles as a by-product of this process. In *Subcellular Organization and Function in Endocrine Tissues*, H. Heller and K. Lederis (Eds.), *Mem. Soc. Endocrinol., Vol. 19*, Univ. Press, Cambridge, pp. 353–378.

DREIFUSS, J. J., GRAU, J. D., LEGROS, J. J. ET NORDMANN, J. J. (1972) Sécrétion d'hormones neuro-hypophysaires par exocytose. *Schweiz. med. Wschr.*, **102**, 1275–1280.

GRAHAM, R. C. AND KARNOVSKY, M. J. (1966) The early stages of absorption of injected horseradish peroxidase in the proximal tubules of mouse kidney: ultrastructural cytochemistry by a new technique. *J. Histochem. Cytochem.*, **14**, 291–302.

ISHII, S. (1970) Isolation and identification of secretory vesicles in the axons of the equine median eminence, *Gunma Symposia on Endocrinology*, **7**, 1–11.

KOBAYASHI, H., MATSUI, T. AND ISHII, S. (1970) Functional electron microscopy of the hypothalamic median eminence. *Int. Rev. Cytol.*, **29**, 281–381.

MATTHEWS, E. K., LEGROS, J. J., GRAU, J. D., NORDMANN, J. J. AND DREIFUSS, J. J. (1973) Release of neurohypophysial hormones by exocytosis. *Nature New Biol.*, **241**, 86–88.

MULDER, A. H., GEUZE, J. J. AND DE WIED, D. (1970) Studies on the subcellular localization of corticotrophin releasing factor (CRF) and vasopressin in the median eminence of the rat. *Endocrinology*, **87**, 61–79.

NAGASAWA, J., DOUGLAS, W. W. AND SCHULZ, R. A. (1971) Micropinocytotic origin of coated and smooth microvesicles ("synaptic vesicles") in neurosecretory terminals of posterior pituitary glands demonstrated by incorporation of horseradish peroxidase. *Nature (Lond.)*, **232**, 341–342.

SANTOLAYA, R. C., BRIDGES, T. E. AND LEDERIS, K. (1972) Elementary granules, small vesicles and exocytosis in the rat neurohypophysis after acute haemorrhage. *Z. Zellforsch.*, **125**, 277–288.

STOECKART, R., JANSEN, H. G. AND KREIKE, A. J. (1972) Ultrastructural evidence for exocytosis in the median eminence of the rat. *Z. Zellforsch.*, **31**, 99–107.

DISCUSSION

MONROE: First of all, I would like to say that your pictures on peroxidase uptake are very convincing. However, there have been so many people lately showing that a variety of cells will take up peroxidase, that I wonder what your reasons are for feeling that this is quite specific for exocytosis instead of a rather general tendency of cells to take up peroxidase, just because it is there.

STOECKART: What I meant to say is that *if* exocytosis occurs, uptake of peroxidase has to occur. I certainly did not want to say that the uptake of peroxidase is proof of exocytosis. In fact, other release mechanisms cannot be ruled out as yet. However, even if micro-pinocytosis is not the consequence of release by exocytosis, it nevertheless reflects the stage of activity of the terminal.

GUILLEMIN: I personally know of no evidence whatsoever to prove that the releasing factors are really produced in neurons and not *e.g.* in tanycytes. I would like to know if the speaker or anybody in this audience is able to give me such evidence!

STOECKART: Although I am not aware of any final evidence, indirect evidence is derived from density gradient centrifugation experiments.

PILGRIM: There is some evidence in analogy to the classical hypothalamo–neurohypophyseal system. There is very good evidence for this system that the neurophysins and the posterior lobe hormones are localized in these granules. So I think it is fair that we assume in the first place that other peptides like releasing factors are synthesized in nerve cells and are present in such granules.

GUILLEMIN: I would like to say in public a few things, particularly for the young people. Because they are the ones who are once going to give us the answer to these important questions. These density gradient experiments are fine as far as the density gradients are concerned. However, I want to say that no chemistry, no simple technique of centrifugation, of separation of subcellular organs, or centrifugation of fractions of the extract of a tissue, are any better than the bio-assay that you finally use to prove your biological activity. Therefore, if the assay that is used is questionable, the conclusion as to the biochemical substrate is just as questionable.

The Hormones of the Hypothalamus

ROGER GUILLEMIN

The Salk Institute, La Jolla, Calif. 92037 (U.S.A.)

The hypothalamic hormones that we will discuss are the latest newcomers to the field of endocrinology. They represent the ultimate link between the central nervous system and the endocrine system as the two integrating orders of all homeostasis of the organism.

The hypothalamus has been known from elegant neurophysiological studies to be involved in integrating many of the visceral homeostatic mechanisms of the body such as blood pressure, body temperature, water and electrolyte metabolism, sleep patterns, thirst, hunger, to mention only a partial list of these basic functions. These multiple integrating modulating influences of the hypothalamus are effected by a complex network of nerve fibers in several relatively well-known afferent and efferent nerve pathways.

Anatomically connected to the hypothalamus from the early embryological stages, the pituitary gland or hypophysis, a structure present in all vertebrates, has long been known to be the center of all endocrine homeostasis. This it achieves by the secretion of a series of pituitary hormones, each one affecting and regulating the function of a peripheral gland such as the adrenals, the ovary, the thyroid or a peripheral non-endocrine organ, such as the kidney or the uterus.

From the embryonic development to the adult stage, the pituitary gland is composed of two distinct lobes, the anterior lobe or adenohypophysis and the posterior lobe or neurohypophysis (Fig. 1a). The posterior lobe or neurohypophysis secretes two hormones, *vasopressin* which regulates the absorption of water at the distal renal tubule, and *oxytocin* which stimulates uterine contractions and milk letdown. Vasopressin and oxytocin are two nonapeptides with a disulfide bridge, characterized and synthesized in 1952 by the pioneering work of du Vigneaud and his collaborators at Cornell University, New York.

From the days of Ramón y Cajal, it has been known that the posterior lobe of the pituitary receives a large tract of nerve fibers, coming from the neurons in two well-defined nuclei (n. paraventricularis and n. supraopticus) of the anterior hypothalamus, hence its name as neurohypophysis (Fig. 1a). The two nonapeptides, vasopressin and oxytocin, considered for many years as hormones genuinely of the posterior pituitary, are in fact manufactured in the neurons of these hypothalamic nuclei, carried within the fibers of the hypothalamic–hypophyseal tract by axoplasmic flow down to nerve endings in contact with capillary vessels in the posterior lobe of the pituitary. From

References p. 131–132

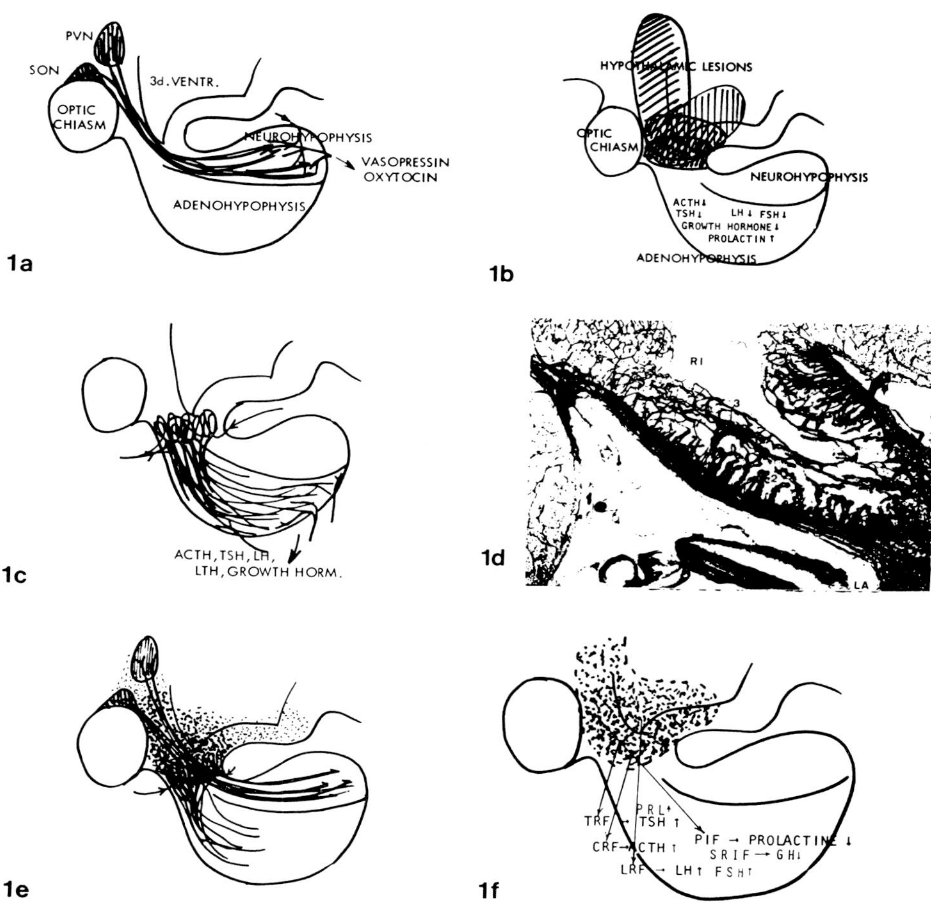

Fig. 1. a: diagrammatic representation of the pituitary gland and the innervation of the neurohypophysis by nerve fibers from the n. paraventricularis (PVN) and supra-opticus (SON). b: localized lesions in the hypothalamus produce changes in the pituitary secretion of the various adenohypophyseal hormones (increase ↑, or decrease ↓). c: diagrammatic representation of the hypothalamo–hypophyseal portal system. d: photomicrograph of the hypothalamo–hypophyseal portal system after injection with an opaque dye. e: diagrammatic representation of the hypophyseotropic area. f: changes in pituitary secretion of various adenohypophyseal hormones (increase ↑, or decrease ↓).

there, vasopressin and oxytocin are released, secreted in the general circulation upon a variety of physiological stimuli. Thus, the two peptides vasopressin and oxytocin are the first of the hypothalamic hormones to be characterized and recognized as such.

It is on this early work that the concept of *neurosecretion* was established, proposing that specialized nerve cells in the hypothalamus can make hormonally active substances. The same concept was also to serve us later, as we shall see.

Some 20 years ago, it became clear from many years of studies in experimental physiology, as well as clinical observations, that somehow, the secretory functions of the adenohypophysis were influenced by some information of hypothalamic origin. In other words, normal functions of the adenohypophysis required normal intact anatomical relationships between the pituitary gland and the hypothalamus. These could indeed be interrupted by localized hypothalamic stereotaxic lesion (Fig. 1b), transplantation of the pituitary to a site remote from the hypothalamus, etc.

After quite a few years of arguments between anatomists and physiologists it became accepted that the only possible direct connections between the hypothalamus and the adenohypophysis were not in the form of nerve fibers as in the case of the hypothalamus to neurohypophysis but were to be found in an unusually well-developed network of capillary vessels with a primary or collecting plexus spread over the floor of the third ventricle, collecting in veins along the pituitary stalk and finally distributing itself in terminal capillary throughout the parenchyma of the adenohypophysis (Fig. 1c and d).

The corollary of these conclusions was of course, that whatever information of hypothalamic origin was related to adenohypophyseal functions had to be in the form of some substance, some neurohormone of hypothalamic neural origin, carried from hypothalamus to adenohypophysis through these hypothalamic hypophyseal portal vessels. The final demonstration of the veracity of this concept was presented in 1955: when fragments of adenohypophyseal tissue were placed in culture, it was shown that they would survive and grow quite well for many weeks; however, they would stop secreting ACTH (adrenocorticotropic hormone — the only pituitary hormone we could easily measure at that time), after a few days of *in vitro* survival. We could restore their secretory ability for ACTH by introducing in the culture tubes a fragment of ventral hypothalamus in a combined culture or by adding a crude aqueous extract of the same hypothalamic tissues. The results were indeed proof that the hypothalamic tissues had contained or released some substance which was able to stimulate the secretion of a pituitary hormone. The hypothetical substances, since there were physiological reasons to suspect that there would be several, were then called *hypophyseotropic hormones* or *releasing factors* for operational facility. Indeed, stereotaxic placement of fine electrodes for electrical stimulation in various parts of the hypothalamus showed that it was possible to modulate, *i.e.* to stimulate or inhibit the secretions of practically each one of the pituitary hormones. Thus (Fig. 1e and f), a region of the hypothalamus around the floor of the third ventricle was recognized as the hypophyseotropic area, containing cells (neurons?) secreting releasing or inhibitory factors for the various pituitary hormones.

What were these substances of hypothalamic origin, which appeared to control all the secretory activity of the adenohypophysis?

It was rapidly established that the hypothalamic substances involved as the final mediators of adenohypophyseal functions were none of the classically known neurotransmitters such as epinephrine, norepinephrine, acetylcholine, etc. The hypothesis was then offered that the hypothetical hypothalamic hormones would likely be small polypeptides, like the only other substances of hypothalamic origin then known, namely vasopressin and oxytocin.

The search for the hypothalamic releasing factors was started as early as 1955 by several groups of investigators. Working with, at most, a few hundred fragments of rat, dog or sheep hypothalamic tissues, and rather simple means of purification and separation of peptides, progress at first appeared relatively rapid. Preparations of a

References p. 131–132

hypothalamic corticotropin releasing factor, or CRF, active in microgram amounts, were obtained. Concomitantly, a lively controversy of several years between proponents of vasopressin as the mediator of ACTH release *vs.* a specific CRF was eventually closed by a series of physiological studies showing no correlation between endogenous secretions of vasopressin and ACTH. After several years of work, it became apparent, however, that chemical characterization of CRF was not to be easy. This was due in part to the considerable difficulties inherent in the complicated bioassay techniques used to assess specifically CRF activity; probably even more important, it was realized that it had been a mistake to use posterior pituitary extracts as starting material, a step taken on a series of incorrect premises. It was also realized that quantities of the right starting material, *i.e.*, hypothalamic fragments much larger than had been used so far, would be necessary for eventual characterization of the still hypothetical hypothalamic releasing factors which appeared to be of high specific activity; each fragment of hypothalamus probably containing only nanogram quantities of the active substances as they were being purified.

We then started to organize for the large-scale collection of hypothalamic fragments. Over the following 3–4 years, we collected about 5 million fragments of sheep brains containing the hypothalamus. This represented the handling of more than 500 tons of sheep brains, or about 50 tons of hypothalamic fragments. The group of my former collaborator Andrew Schally, now in New Orleans, is engaged in a similar collection of about 2.5 million fragments of pig hypothalami.

Specific bioassays were devised for studying the release of several pituitary hormones, such as thyrotropin (TSH) and the gonadotropins luteinizing hormone (LH) and follicle stimulating hormone (FSH). Evidence was obtained for the existence in crude hypothalamic extracts of a TSH releasing factor, or TRF, and also of a luteinizing hormone releasing factor, or LRF. Eventually, using methodology developed over the preceding years with the use of large quantities of brain tissues, Burgus and I reported the isolation, late in 1968, of 1.0 mg of pure TRF from 300,000 sheep hypothalamic fragments. Aliquots of this material showed it to be composed *exclusively* of 3 amino acids, Glu, His, and Pro, in equimolar ratios. This information, however true and reliable in our own hands, was to be taken in the context of a report of Schally's group, concluding that porcine TRF was not a homomeric peptide, that it contained no more than 30% amino acid and that the non-peptidic moiety of its molecule was necessary for biological activity.

The preparation of ovine TRF was active in stimulating the release of TSH at 1–5 ng (1–5 $\times$ 10^{-9}g) by systemic administration to assay animals; in an *in vitro* system of pituitary incubations, it was active to release TSH at 50 pg (5 $\times$ 10^{-11}g). The biological activity was totally resistant to incubation with proteolytic enzymes and there was no evidence of a free N-terminal residue. With the few hundred micrograms remaining after ascertaining the amino acid analysis, no complete molecular structure could be ascertained by infrared spectrometry or nuclear magnetic resonance spectrometry and efforts at obtaining mass spectra on small samples were unsuccessful. However, all the spectral features observed could be explained in terms of a polypeptide structure. Since the prospect of obtaining larger quantities of the native material by extraction of more brain fragments was not particularly viewed with enthusiasm, we decided on an entirely different approach to the structure of TRF: that was to synthesize all the possible combinations of the 3 amino acids known to be constituents of the molecule of TRF, to study their biological activity. When these were obtained, they were shown to have chromatographic characteristics different from those of TRF; furthermore, they had no TRF activity at doses one million times larger than those of pure ovine TRF. Was this another dead end? Not quite. We then decided to cover the N-terminal residue of the 6 synthetic tripeptides, since we knew native TRF to be so protected. This was easily done by reacting the tripeptides with acetic anhydride. The results were unmistakable. *Only* the reaction products of the sequence Glu–His–Pro had biological activity qualitatively indistinguishable from that of native TRF. Rather than the expected acetyl-Glu–His–Pro–OH, the major reaction product was shown to be pyroGlu–His–Pro–OH, *i.e.*, a molecule with its N-terminus protected

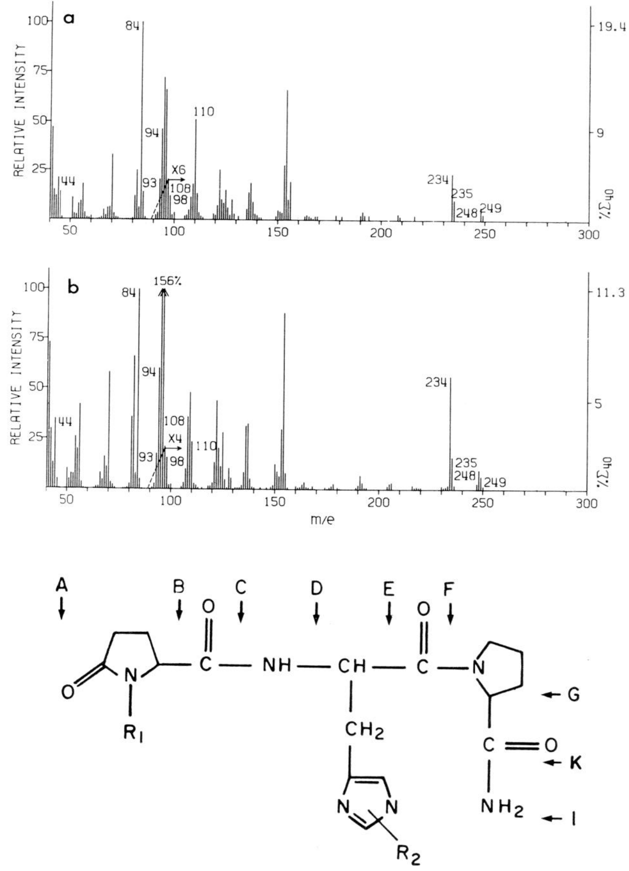

Fig. 2. Mass spectra of methylated ovine TRF (a) and (b) of synthetic pyroGlu–His–Pro–NH_2; c: shows the primary structure of TRF with indication of the fragmentation points in mass spectrometry. R_1, R_2 represent the methyl derivative prepared for mass spectrometry; in the native molecule, $R_1 = R_2 = H$.

References p. 131–132

by cyclization of Glu into pyroGlu; pGlu–His–Pro–OH was thus the first molecule of known structure to reproduce the biological activity of a hypothalamic releasing factor.

The specific activity of pGlu–His–Pro–OH was, however, about 5000 times lower than that of native TRF. It was Burgus's idea to amidate the C-terminal proline, as was known to be the case in the structure of several biologically active polypeptide hormones such as gastrin, secretin, ACTH, vasopressin, etc. This was rapidly achieved first by making the methylester of the tripeptide free acid and then preparing the primary amide by simple treatment with methanolic NH_3. pGlu–His–Pro–NH_2 was thus obtained and was shown to have quantitatively the full biological activity of native ovine TRF. We still did not know what the structure of native TRF was.

With availability of very large quantities of the synthetic peptide we were able to optimize several methods of derivatization for mass spectrometry on microgram quantities of the tripeptides; having with Horning's group modified the probe of the mass spectrometer available to us in Houston to increase its sensitivity, in September 1969 we were able to obtain by mass spectrometry the complete and unquestionable structure of native ovine TRF. It turned out to be pGlu–His–Pro–NH_2 (Fig. 2). The first of the hypothalamic releasing factors to be fully characterized had thus been synthesized before its molecular structure had been established. The same structure pGlu–His–Pro–NH_2, was subsequently demonstrated for porcine TRF by Schally's group.

Characterization of TRF was of importance as it established on definitive grounds one of the fundamental tenets of neuroendocrinology which, for 15 years, had been known to generate more than its due share of what I have called somewhere, "the prophetic literature".

Isolation and characterization of LRF, luteinizing hormone releasing factor, was to follow rapidly in 1971, the laboratory of Andrew Schally, on the basis of the elegant studies of Matsuo, first proposing the structure of *porcine* LRF as that of the decapeptide pGlu–His–Trp–Ser–Tyr–Gly–Leu–Arg–Pro–Gly–NH_2, a structure confirmed by our laboratory with *ovine* LRF. The decapeptide LRF of native or synthetic origin, stimulates not only the secretion of luteinizing hormone LH but also that of the other gonadotropin, FSH or follicle stimulating hormone. A series of physiological and clinical observations would be best explained by the presence in the hypothalamus of a specific FSH-releasing factor. No such substance has been shown to exist, so far, in hypothalamic extracts. The question of an FSH-releasing factor different from the decapeptide LRF which stimulates the secretion of both LH and FSH, is thus not settled.

Recently we observed in hypothalamic extracts the presence of a substance inhibiting the secretion of the pituitary growth hormone. The substance was isolated, characterized and synthesized as the tetradecapeptide H-Ala–Gly–|Cys–Lys–Asn–Phe–Phe–Trp–Lys–Thr–Phe–Thr–Ser–Cys|–OH; we call it somatotropin-release inhibiting factor or *somatostatin*.

Data from reliable physiological experimentation from various laboratories are best explained by the existence in hypothalamic tissues of a growth hormone releasing factor, a prolactin releasing factor, a corticotropin releasing factor and, probably, a prolactin release inhibiting factor. Nothing is known of the chemistry of these. The decapeptide Val–His–Leu–Ser–Ala–Glu–Glu–Lys–Glu–Ala, isolated by Schally from *porcine* hypothalamic tissues and proposed by that group as "growth hormone releasing hormone" on the basis of a highly controversial and difficult type of assay, was shown to be similar to a fragment of the beta-chain of porcine hemoglobin and is thus of doubtful physiological significance; this material is indeed inactive in several systems studying the secretion of immunoreactive growth hormone; it is not active either to stimulate somatic growth of laboratory animals.

Three peptides, Pro–Leu–Gly–NH_2, and |Cys–Tyr–Ile–Gln–Asn–Cys|–OH, both fragments of the molecule of oxytocin and the pentapeptide Pro–His–Phe–Asp–Gly–NH_2, have been reportedly isolated from porcine hypothalamic tissues and claimed to be involved in the control of the secretion of the melanophoretic hormones. These claims have not been confirmed by several other laboratories and thus remain for further cautious investigations.

The 3 above-mentioned hypothalamic hormones have been synthesized and are thus available now in unlimited quantities. All are of high specific activity, active *in vitro* at picomolar concentrations, *in vivo* at nanogram or microgram doses when injected in the peripheral circulation. There is no evidence of major species specificity in their biological activity, the molecules characterized from ovine or porcine hypothalamus being fully active in all mammalians tested so far, including man as we shall see later. Similarly the inhibitory factor somatostatin lowers plasma levels of growth hormone in the same time relationship.

The hypothalamic releasing factors, upon intravenous administration, act very rapidly, elevating plasma levels of pituitary hormones within a few minutes (Fig. 3).

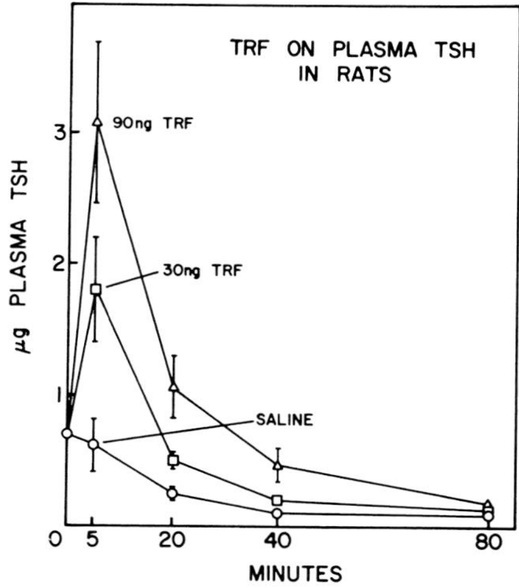

Fig. 3. Rapidity of the (pituitary) response to i.v. administration of releasing factors, in this case TRF. The figure also shows the dose-response relationship between doses of releasing factors (here TRF) and amount of pituitary hormones secreted (here TSH). On the ordinate, plasma concentrations of TSH in the rat, measured by radioimmunoassay (1 μg TSH equivalent to *ca.* 20 ng TSH USP standard).

References p. 131–132

Their biological half-life is short, of the order of a few minutes as the peptides are degraded in plasma, excreted by glomerular filtration in the kidney and simply diluted in the peripheral circulation from the injected bolus, or as the peptides reach the pituitary from the hypothalamic hypophyseal portal vessels.

The mechanism, by which the hypothalamic peptides stimulate and/or inhibit the secretion rate of anterior pituitary hormones have not, as yet, been fully elucidated. Several lines of nonexclusive evidence have been accumulated to explain the phenomena. The releasing factors increase the rate of secretion of stored hormone and no *de novo* synthesis of any protein molecule is necessary for their action (Fig. 4), as

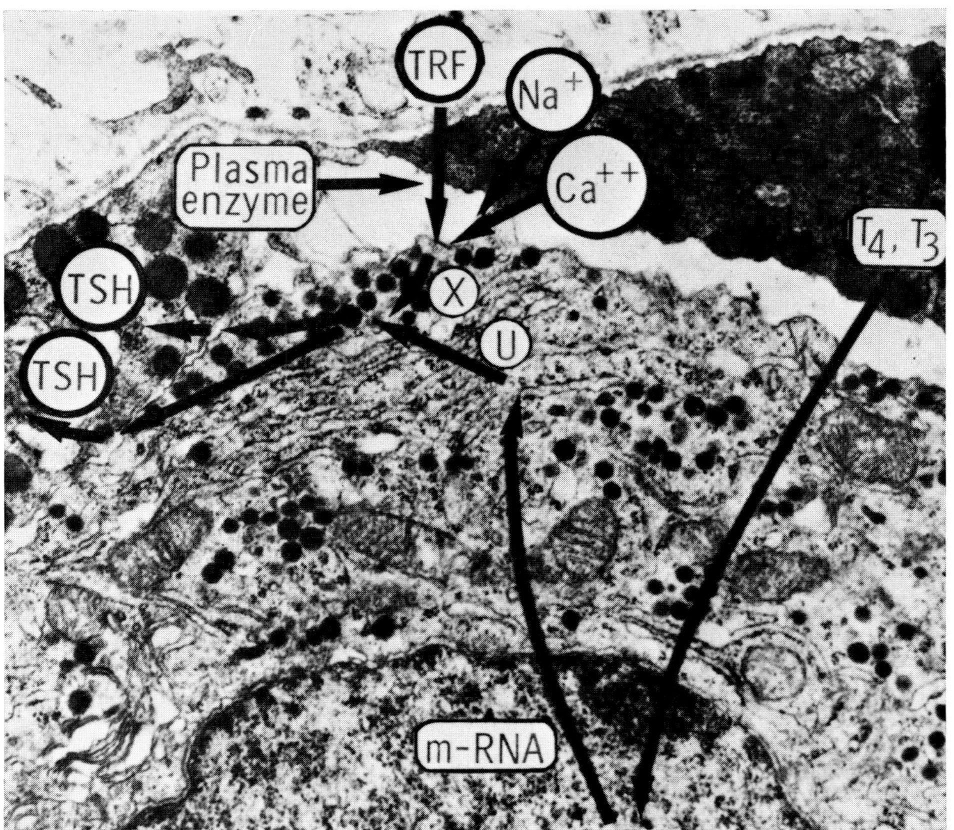

Fig. 4. Diagram of the mechanism of action of TRF and thyroid hormones (T_4, T_3) in their regulating the secretory rate of thyrotropin (TSH). TRF acts at receptor sites on the plasma membranes of thyrotropins. Ca^{2+} and Na^+ in the extracellular fluids are necessary ions for the activity of TRF. A plasma enzyme inactivates TRF. X represents a set of biochemical events following TRF reaching its pituitary receptors; these events may involve activation of the adenyl cyclase-c-AMP system; their result is release of preformed TSH in the extracellular spaces. T_4 and T_3 induce in the nucleus of the thyrotropic cell, an m-RNA which in turn is translated in a protein or polypeptide (U) of unknown nature; this substance U, somehow, interferes with the results of the events triggered by TRF at the plasma membrane, and inhibit the response to TRF. The response to TRF is rapid (sec); the inhibitory response due to T_4, T_3 requires a longer time (h).

evidenced by absence of effects on this event of the antibiotics actinomycin D, puromycin or cycloheximide.

The activity of the hypothalamic releasing factor is dependent upon a critical concentration of extracellular calcium (Ca^{2+}) ions. The action of the releasing factors can be simulated by elevated potassium (K^+) concentration in the medium of *in vitro* cultures or incubation provided sufficient Ca^{2+} is present. We proposed several years ago, that the peptides might function, as high $[K^+]$ has been suggested to, by depolarizing the pituitary cell membrane and thereby increasing the permeability of the cell to Ca^{2+}, which, in turn, would mediate hormone release. It remains to be shown, however, that individual pituitary cells depolarize (or hyperpolarize) under the influence of specific hypothalamic peptides.

In keeping with the effects of elevated K^+, there is at any rate, good evidence that TRF and LRF act at the level of the plasma membrane. Probably, our best evidence for this mechanism is the demonstration of the biological activity of TRF coupled by cyanogen bromide activation to high molecular weight Dextran with carefully controlled evidence that there is no free peptide to account for the biological activity. Furthermore, $[^3H]$TRF binds to membrane fractions or intact pituitary cells with an affinity constant *ca.* $2 \times 10^{-8} M$, a value not so different from the value observed for biological activity.

The mechanism by which hormone secretion rates may be stimulated by hypothalamic peptides may involve the activation of a system to increase the concentration of cyclic adenosine monophosphate (c-AMP). Circumstantial evidence has been accumulated to suggest that individual hypothalamic peptides stimulate the adenyl cyclase system, thereby activating a secretion process by intracellular c-AMP accumulation. As all pituitary systems studied have been comprised of heterogeneous cell types, such conclusions are at best inferential. Increasing intracellular c-AMP levels, by blocking c-AMP phosphodiesterase with theophylline, stimulation of adenyl cyclase with prostaglandins or direct addition to the cells of dibutyryl c-AMP at millimolar concentrations may indeed enhance the rate of release of GH, ACTH, TSH and LH, as shown in our own and other laboratories. In the growth hormone or ACTH systems, 10-fold rate increases occur; however dibutyryl-3′5′-c-AMP has only marginal effects on secretion rates of the two glycoproteins TSH and LH, whereas releasing factors increase hormone secretory rates $\geq 10 \times$ control rates. Prostaglandin E_1 or theophylline potentiate the effects of TRF on TSH secretion and produce as much as 30–50 × stimulation of release rates when cells are pretreated prior to TRF addition. It is thus possible that the c-AMP involvement in the secretion of the glycoproteins is involved with a process such as zymogen granule migration rates rather than in hormone release processes *per se*.

The newly characterized hypothalamic peptide somatostatin which inhibits the secretion of GH, appears to act after c-AMP, since it inhibits secretion of GH, induced by dibutyryl c-AMP.

A direct role of the hypothalamic releasing factors in affecting directly synthesis of the pituitary hormones, though claimed by several, has not been really demonstrated.

Clinicians and physiologists have known for many years that there exists some sort

References p. 131–132

of a negative feedback between the peripheral endocrine glands and the pituitary in some apparent control mechanism. This begins to be understood (see Fig. 4). The thyroid hormones, thyroxine (T_4) and triiodothyronine, block the response to TRF at the level of the pituitary cell. Experiments have shown that this block is relatively slowly established and is long lasting. The establishment of T_4 inhibition requires the induction of an intracellular protein which appears to mask the influence of TRF without modifying the basal or spontaneous TSH secretion rate. In fact, the T_4 block is prevented by DNA-directed-RNA-synthesis-inhibitors such as actinomycin D or is rapidly overcome by puromycin or cycloheximide. The data were interpreted to mean that a stable messenger RNA is made in response to T_4, and is translated into a rapidly turning-over cytoplasmic protein, a protein which in an unknown manner prevents the action of TRF.

Practically identical observations have been reported for the sex steroid feedback inhibition of LRF activity and the glucocorticoid effect on hypothalamic stimulation of ACTH release.

The rather surprising observation was made recently by Tashjian that TRF, the thyrotropin releasing factor, stimulates the secretion of prolactin by a cloned line of transformed pituitary cells. The observation has been confirmed and extended to show that TRF can stimulate the secretion of prolactin from non-transformed pituitary cells, and it also does this *in vivo* in steroid pretreated rats or in normal man, as we will see later. This raises the possibility that TRF *per se* or a closely related peptide may be the postulated PRF (Prolactin Releasing Factor) detected in hypothalamic extracts. No less surprising is the recent observation that somatostatin, the hypothalamic inhibitor of the secretion of GH, also inhibits the secretion of TSH induced by TRF while it does not inhibit the concomitant secretion of prolactin triggered by TRF. Somatostatin has no effect on the secretion of the gonadotropins LH and/or FSH either in their basal secretion or as stimulated by LRF.

Several hundred analogues of TRF and LRF have been synthesized in several laboratories throughout the world for studies aimed at correlating molecular structure and biological activity, for studies aimed at establishing a tridimensional conformation of the peptide and on that knowledge, to design possible competitive antagonists.

It would neither be possible in this article nor would it be in my competence to discuss all the structural modifications entered in the molecule of TRF and LRF as part of these programs. I will, however, summarize some of the salient conclusions that have come from these studies by our group in collaboration with the laboratory of Murray Goodman.

A possible conformation of TRF has recently been proposed by our laboratory in which the molecule would exist in a "hairpin turn" conformation stabilized by two intramolecular hydrogen bonds. Such a proposal would be in keeping with our earlier observation that placing a methyl group on the τ-nitrogen of the imidazole ring produces an analogue of TRF which is about 10 × more active than the native compound. It is also of interest that in all cases studied so far, the specific activity of a TRF analogue, when compared to TRF, has always been in the same ratio as the binding constant of that analogue when compared to that of TRF.

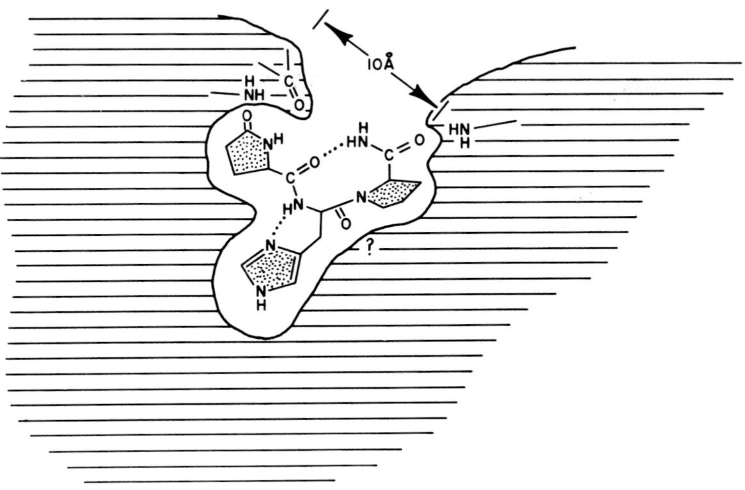

Fig. 5. A schematic representation of the TRF-receptor site showing possible sites of interaction with the molecule of TRF as deduced from conformational studies and the biological potencies of TRF analogues (after G. Grant and W. Vale. In *Current Topics in Experimental Endocrinology, Vol. 2*, JAMES (Ed.), Academic Press, New York, 1974).

The conclusions from the examination of the biological potency of TRF analogue data suggest (Fig. 5) that (1) the cis-position of the pGlu lactam function is required, but the role of the ring structure in recognition is unknown; (2) the size, position and a restricted plane of the imidazole ring is recognized by the receptor; (3) the orientation and hydrophobic ring structure of the proline moiety is required; and (4) there appears also to be a requirement for the carbonyl function of the carboxamide terminal by the receptor. The 3-dimensional model, the interpretation of conformation of analogues, the interaction of analogues with the receptor and the proposals about the topography of the TRF receptor all provide and suggest further experiments as their primary justification.

In the case of LRF, this approach has already led to a series of interesting compounds. We have already reported on several analogues of LRF, all modified in the His^2 or Trp^3 position, which behave as partial agonists or antagonists of LRF, the most active of these being des-His^2-LRF which is 5 × more potent than anything we have synthesized so far.

Secretion of the hypothalamic releasing factors represents one of the functions of some highly specialized neurons in the hypothalamus. Like any other event within the central nervous system, secretion of the releasing factors seems to be triggered by the classical neurotransmitters between neurons, *i.e.*, epinephrine, norepinephrine, dopamine, serotonin. The biosynthesis of these oligopeptides has been claimed by several groups of investigators to be non-ribosomal. This remains to be confirmed.

Thus, we see a unified concept, that of *neuroendocrinology*, describing the relationships between the central nervous system and the endocrine system, in the highest mammals. It is probably of heuristic value to remember that in the earliest stages of

References p. 131–132

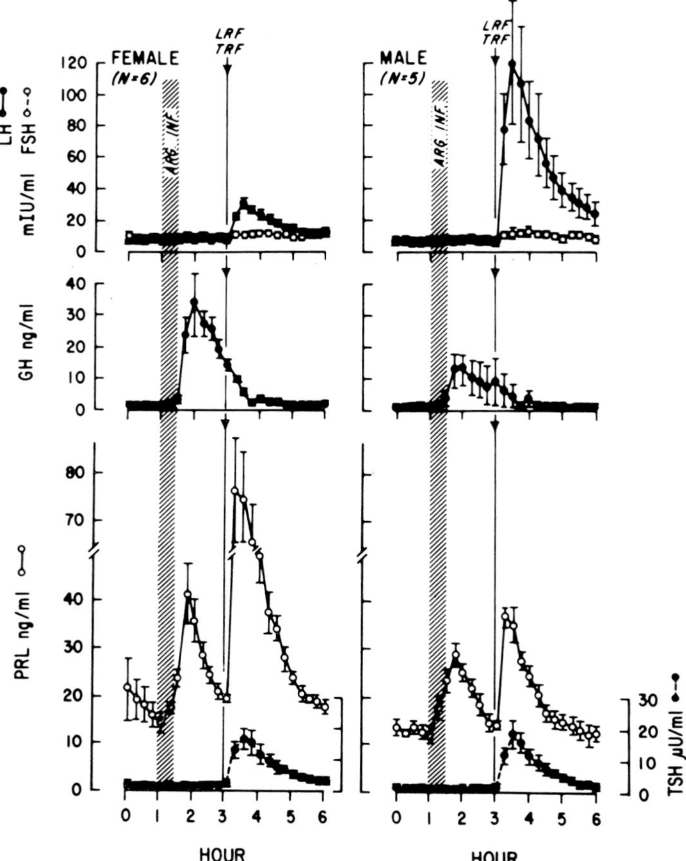

Fig. 6. Testing of the ability of the anterior pituitary to secrete GH, TSH, prolactin (PRL), LH and FSH in normal human subjects. Stimulation of the secretion of GH is achieved by i.v. administration or arginine; stimulation of the secretion of TSH and PRL, LH and FSH is produced by i.v. injection of a solution in saline of synthetic TRF (250 µg) and synthetic LRF (150 µg) — note that arginine infusion stimulates secretion of GH and PRL. All pituitary hormone plasma concentrations measured by radioimmunoassays. (From YEN AND GUILLEMIN. *Symposium on Clinical Pharmacological Methods*, New Orleans, 1973, to be published.)

the phylogeny nearly half of the cells of the primitive localized nervous systems were also endowed with neurosecretory activity.

Just as exciting as all these results in the laboratory, if not more so, are the dramatic effects observed in clinical medicine with availability of the synthetic replicates of the hypothalamic releasing factors.

Indeed, as I have mentioned earlier, the hypothalamic releasing factors characterized in ovine or porcine brains have no obvious species specificity in mammals and are highly active in humans.

For instance, it is possible with administration of a single i.v. injection containing both synthetic TRF and LRF to investigate at once, the pituitary ability to secrete LH, FSH, TSH and prolactin. Combined with infusion of arginine, this allows the

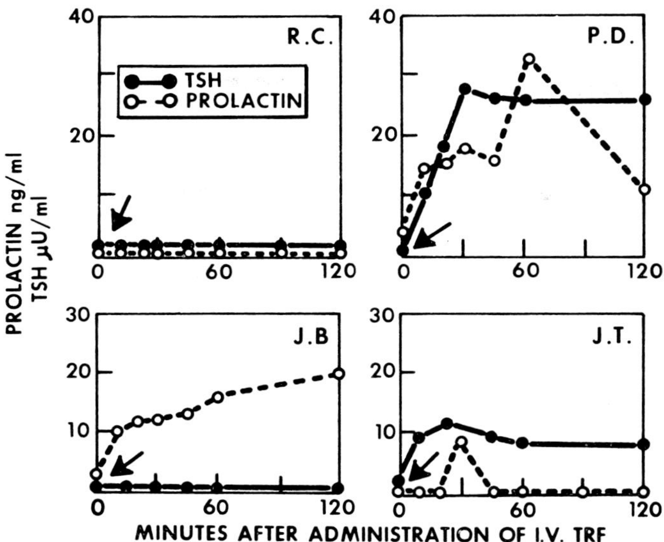

Fig. 7. The plasma TRF and plasma prolactin response to TRF is shown for 4 patients with multiple pituitary hormonal deficiencies. R.C. had craniopharyngioma removed 3 years ago and showed neither a plasma TSH or prolactin response to TRF. The other 3 patients have idiopathic hypopituitarism with multiple pituitary hormone deficiencies. J.B. showed no plasma TSH response but a normal prolactin response to TRF. J.T. had a normal plasma TSH response but a blunted plasma prolactin response to TRF. P.D. had normal plasma TSH and plasma prolactin responses to TRF. (Reproduced with the permission of KAPLAN *et al.* (1973). In *Hypothalamic Hypophysiotropic Hormones*, *Proc. Acapulco Conference*, *1972*, Excerpta Medica, Amsterdam.)

concomitant measurement of the secretion of GH (Fig. 6). A number of cases have already been reported in which abnormality of the pituitary response to such a simple secretion test has led to the early diagnosis of a pituitary tumor which could then be operated on and removed.

In children or adults with isolated or multiple pituitary deficiencies it is now apparent that some of these patients have a normal pituitary response to hypothalamic releasing factors, while others do not (Fig. 7). Thus, for some the deficiency is pituitary, for others, hypothalamic, with the possibility now available of replacement therapy by administration of the synthetic releasing factors.

Injection of LRF to normal men stimulates secretion of gonadotropins and of testosterone. The clinical implications in cases of male infertility are obvious but remain for future studies. Injection of LRF to normal women produces rapid increases in the secretion of LH and FSH, with rather striking differences in the (pituitary) response as a function of the date of the ovarian cycle (Fig. 8).

Administration of an LRF antagonist in a regimen still to be determined would thus be a powerful new means of contraception. This is indeed one of the important goals of our current efforts at denoting antagonists of LRF. On the other hand, administration of LRF to patients with hypothalamic amenorrhea who have low or normal plasma levels of gonadotropins but do not show pulsatile releases of LH, regularly leads to rapid secretions of LH and FSH usually followed by appearance of menses.

References p. 131–132

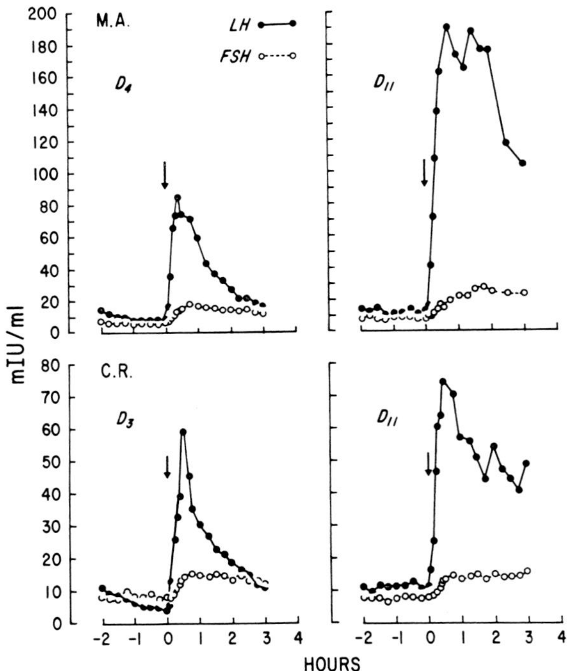

Fig. 8. Changes in gonadotropin responses to the same dose of LRF (150 μg) between the early and late follicular phases during an ovulatory cycle in two subjects. D indicates day number in menstrual cycle. (From YEN *et al.* (1973) In *Hypothalamic Hypophysiotropic Hormones*, *Proc. Acapulco Conference*, Excerpta Medica, Amsterdam.)

Thus LRF can participate in the treatment of some types of infertility as well as in a program on contraception.

Also of interest is the recent demonstration that somatostatin, the hypothalamic peptide which we have recently characterized, inhibits the secretion of GH in man (Fig. 9), in early studies in which GH secretion is stimulated by arginine infusion or L-DOPA administration. The possible significance of these observations in the management of juvenile diabetes as well as certain pituitary tumors is obvious. Somatostatin has also been shown recently to inhibit the secretion of glucagon and insulin directly at the level of the α and β cells of the endocrine pancreas (KOERKER *et al.*, *Science*, (1974), to be published).

Of extreme interest, if they are confirmed, are several recent reports claiming beneficial psychotropic effects of TRF in unipolar depression. We have early evidence that somatostatin may also have profound psychotropic effects of a different nature. Effects, if not necessarily roles, of these peptides of hypothalamic origin will thus have to be carefully studied in psychiatric medicine.

The hypothalamic hypophyseotropic hormones, as we now know them, are already opening a new chapter in medicine. In this, the physiologists and biochemists who have made this possible can take genuine pride and reward. This is indeed the best defense for basic research.

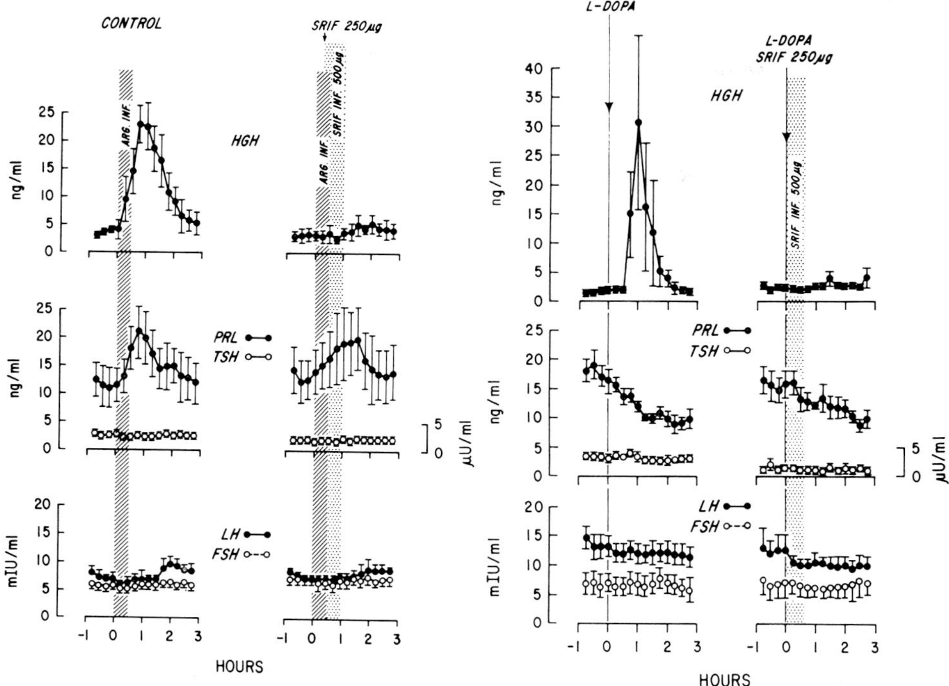

Fig. 9. Effects of the administration of synthetic somatostatin in normal human subjects. There is complete inhibition of the increase in GH secretion normally produced by infusion of arginine or oral administration of L-DOPA, where somatostatin is administered prior to or concurrently with the stimulating agent. Plasma concentrations of pituitary hormone were measured by radioimmunoassay. (From YEN *et al.* (1973). In *J. clin. Endocr.*, **37**, 632.)

ACKNOWLEDGEMENTS

Our current work is supported by the Ford Foundation, the Rockefeller Foundation and AID, Contract No. csd/2785.

REFERENCES

The following references are primarily reviews; they are offered essentially as keys to the extensive literature in neuroendocrinology.

The Hypothalamus, 1 vol. (1969) W. HAYMAKER, E. ANDERSON AND W. J. H. NAUTA (Eds.), Thomas, Springfield, Ill., 805 pp.

The Pituitary Gland, 3 vol. (1966) G. W. HARRIS AND B. DONOVAN (Eds.), Univ. California Press, Berkeley, Calif.

Frontiers in Neuroendocrinology, 1 vol. (1973) W. GANONG AND L. MARTINI (Eds.), Academic Press New York.

GUILLEMIN, R., BURGUS, R. AND VALE, W. (1971) The hypothalamic hypophysiotropic thyrotropin releasing factor (TRF). *Vitam. and Horm.*, **29**, 1–39.

SCHALLY, A. V., ARIMURA, A. AND KASTIN, A. (1973) Hypothalamic regulatory hormones. *Science*, **179**, 341–350.

BLACKWELL, R. AND GUILLEMIN, R. (1973) Hypothalamic control of adenohypophysial secretions. *Ann. Rev. Physiol.*, **35**, 357–360.

DISCUSSION

LEQUIN: I should like to ask you a question about the specificity of TRF in cases such as acromegaly. We know that TRH will give an increased release of TSH and prolactin. Lately I heard, however, that in acromegalic patients you will see a rise in growth hormone release as well. Could you please comment on that?

GUILLEMIN: This is indeed an extremely interesting observation. I can add to that, that we have many other similar observations *in vivo*, not only in acromegalic patients, but also in patients with other types of pituitary tumors such as craniopharyngiomas. We have also placed fragments of human pituitary tumors obtained by surgery, in culture, and we have confirmed *in vitro* that the effects of TRH on the pituitaries were abnormal. To me this was very interesting because I could gradually come to a working hypothesis with a rather new concept of the origin of these pituitary tumors. Acromegaly seems to be a disease at the molecular level of the pituitary receptors. If you give TRF to acromegalic patients you stimulate the secretion of growth hormone. In other words, whatever the so-called adequate stimulus reaching this abnormal pituitary is — abnormal in terms of its receptors — it will make it secrete growth hormone. So these pituitary tumors have somehow grossly modified receptors.

ISIDORIDES: I wonder whether there are different receptors for TRH that cause secretion of the different hormones.

GUILLEMIN: We have purified the receptors for TRH from normal pituitary cells as well as pituitary tumors secreting TSH and prolactin. We cannot distinguish between receptors for TRF regarding the secretion of prolactin or the secretion of TSH. This work was done, however, before we could use somatostatin. Somatostatin, we know, inhibits the secretion of TSH in response to TRF, distal to the adenyl cyclase system, but it does not block prolactin secretion in response to TRF. So, I think, we can begin now to study the steps that follow the receptor, namely the effect on the adenyl cyclase system.

HONNEBIER: By means of prostaglandins or estradiol–benzoate a rise of only LH levels can be obtained. In addition, in post-menopausal women we see an increase of FSH that is about 10-fold the normal mid-cycle peak, whereas LH remains on only a third of the mid-cyclic peak. What do you think that this tells about the existence of one or two releasing factors for LH and FSH?

GUILLEMIN: The FSH levels in post-menopausal women are of course well known, and this is one of the reasons why one compound, being the sole controller of the two gonadotropic hormones is not a satisfactory situation. Many physiologists and many physicians would like very much to see a specific FSH releasing factor. This might indeed explain the clinical and experimental circumstances where there is a dissociation between the secretion of LH and FSH.

SCHULSTER: Since we now know the structure of several of the releasing factors, would it be possible to make antibodies to these compounds, and thus localize the secretory cells, or use them as an assay system? Are you aware of any work in this direction?

GUILLEMIN: Of course this is the way to go. In fact, I know of several groups which are doing just that at the moment. Antibodies against TRF, against LRF, and against somatostatin have been obtained. However, all of these antibodies that I know are of very low binding ability. You probably could use them for immuno-precipitations, but probably not for immuno-assays. It is very difficult to make antibodies to these oligopeptides. In my own laboratory we have been totally unable, so far, to induce meaningful antibodies for TRF or LRF.

The Electrophysiology of the Hypothalamus and its Endocrinological Implications

R. G. DYER*

Department of Anatomy, University of Bristol, Bristol (Great Britain)

INTRODUCTION

"... we have to establish ... enough
of its (the single unit's) properties in
several dimensions, before we can make
reasonable statements about its defining
features, its tentative name, class or
type." T. H. BULLOCK (1966)

The original electrophysiological experiments describing the involvement of the hypothalamus in the control of the pituitary gland (*e.g.* Markee *et al.*, 1946; Harris, 1947) gave rise to the concept of neuroendocrinology and have provided the foundations of much subsequent work. Most of the early experiments involved stimulating the medial basal hypothalamus and monitoring the output from the target organs, *e.g.* the kidney or the ovary. It is now established that electrical or electrochemical stimulation of appropriate sites in the hypothalamus releases from the pituitary gland:

(a) oxytocin and vasopressin (Andersson, 1951; Andersson and McCann, 1955; Cross, 1958; Aulsebrook and Holland, 1969; Bisset *et al.*, 1971);

(b) gonadotropins (Markee *et al.*, 1946; Harris, 1948; Clemens *et al.*, 1971b; Cramer and Barraclough, 1971; Kalra *et al.*, 1971);

(c) growth hormone (Frohmann *et al.*, 1968; Martin, 1972);

(d) adrenocorticotropic hormone (D'Angelo *et al.*, 1964);

(e) thyrotropin (Reichlin, 1966; Averill and Salaman, 1967; Martin and Reichlin, 1970).

To some extent these experiments are based on the assumption that the stimulation mimics the endogenous changes in neural activity that precede pituitary activation. The possible relationship between the electrical activity of the hypothalamus and the secretion of pituitary hormones has recently been the subject of several excellent reviews (Beyer and Sawyer, 1969; Sawyer, 1970a, b; Cross, 1973). The modulation of this electrical activity by the below-mentioned substances has also been fully described:

(a) estrogen (Lincoln, 1967; Yagi, 1970; Cross and Dyer, 1972);

* Present address: ARC Institute of Animal Physiology, Babraham, Cambridge, Great Britain.

References p. 142–147

(b) progesterone (Barraclough and Cross, 1963; Cross and Silver, 1965; Komisaruk *et al.*, 1967; Endroczi, 1969; Lincoln, 1969);

(c) androgen (Pfaff and Gregory, 1971);

(d) follicle stimulating hormone (Sawyer, 1970a);

(e) luteinizing hormone (Kawakami and Sawyer, 1959; Ramirez *et al.*, 1967; Kawakami and Saito, 1967; Terasawa *et al.*, 1969; Dufy *et al.*, 1973);

(f) adrenocorticotropic hormone (Sawyer *et al.*, 1968; Steiner *et al.*, 1969; van Delft and Kitay, 1972);

(g) cortisol (Feldman and Sarne, 1970; Phillips and Dafny, 1971);

(h) prolactin (Clemens *et al.*, 1971a);

(i) oxytocin (Kawakami and Saito, 1967, 1969; Cross and Dyer, 1969; Dyball and Dyer, 1971; Moss *et al.*, 1972a);

(j) vasopressin (Capon, 1960; Beyer *et al.*, 1967; Moss *et al.*, 1972a).

It is not the purpose of this paper to present an overview of these data, the reader requiring such information should synthesize for himself from the references mentioned above and read the interesting article by Komisaruk (1971); rather it is my intention to try and pinpoint some of the technical problems, to give an example of what electrophysiology can contribute to our knowledge of neuroendocrine mechanisms and outline a strategy for further experiments in this field.

SOME TECHNICAL PROBLEMS

Preparation

The ideal electrophysiological experiment requires that the animal is unstressed, freely moving and able to perform life functions in a normal manner. Like most ideals this is difficult to obtain; although it is possible to both stimulate (Harris, 1947) and record (Hellon and White, 1966; Pfaff *et al.*, 1970) from the hypothalamus of unrestrained animals. However, the very mobility of the animal makes it difficult to obtain stable recordings for a worthwhile length of time. To overcome this problem it is necessary to immobilize the experimental animal within the rigid support of a stereotaxic apparatus, usually under anesthesia although it is also possible to use this equipment to record from conscious animals (Vincent *et al.*, 1972) and unanesthetized diencephalic island preparations (Cross and Kitay, 1967).

Very often, urethane (ethyl carbamate) has been the anesthetic of choice and this drug appears to have little direct effect upon the electrical activity of the hypothalamus (Cross and Dyer, 1971a). Furthermore, in the rat urethane anesthesia does not eliminate the milk-ejection reflex (Lincoln *et al.*, 1973) and does not always block ovulation when administered before the critical period on the afternoon of proestrus (Blake and Sawyer, 1972; Lincoln and Kelly, 1972; Dyer *et al.*, 1972). Thus, although urethane may diminish neuroendocrine reflexes, it does not always abolish them completely, unlike for example Nembutal® which will block ovulation if injected immediately prior to the critical period (Everett, 1956). However, urethane in common with ether and barbiturates (Ginsburg and Brown, 1956) greatly increased the release of oxytocin and vasopressin from the neurohypophysis and sharply elevated the systemic concentration of both ACTH and adrenaline.

Thus, in the interpretation of electrophysiological experiments some account must be given to the effect of the anesthetic used and the experimenter must ensure that the neuroendocrine reflex under investigation still occurs in the laboratory situation. For example, to relate hypothalamic unit activity directly with ovulation it is necessary to show that LHRF is secreted concurrently with any electrical changes (Blake and Sawyer, 1972).

Interpretation

The electrical changes that are indispensable for the transmission of information in the central nervous system can be induced by the application of appropriate stimuli to particular neural networks and may be recorded with either macro- or microelectrodes. (For details read Frank and Becker, 1964; Delgado, 1964.) Most of the recording methods require that the electrode be placed adjacent to a single neuron or a population of nerve cells. A number of techniques are available for recording from pools of neurons but these are only useful if the population is homogeneous in function and behavior. Most neural pools in the hypothalamus do not satisfy these criteria and adjacent cells may have different and even opposing roles, only in the paraventricular and supraoptic nuclei are similar neurons clustered together. However, even in these nuclei about a quarter of the cells do not project to the neurohypophysis (Sundsten *et al.*, 1970; Dyball and Pountney, 1973) and of the remainder it is not always possible to distinguish between neurons secreting oxytocin and vasopressin.

To fully interpret data obtained by electrophysiological techniques it is necessary to have a knowledge of the source of the signal recorded. Some of the early workers in the field of neuroendocrinology tried to relate hypothalamic EEG recordings with secretion of, for example, antidiuretic hormone (Nakayama, 1955) and ovulatory hormones (Kawakami and Sawyer, 1959; Porter *et al.*, 1957). More recently many laboratories have used the technique of multi-unit recording to investigate similar problems. However, now that the hypothalamus has been investigated with microelectrodes, records of EEG and multi-unit activity provide little new data. The electrical activity so monitored is derived from perikarya, axons and synapses (Klee *et al.*, 1965), and possibly also extracerebral sources (Cobb and Sears, 1956). For example, the increased multi-unit activity observed in the arcuate nucleus following electrical stimulation in the preoptic area (Terasawa and Sawyer, 1969) or in the prosubiculum of the hippocampus (Gallo *et al.*, 1971) may be due to an increased rate of discharge of neurons with perikarya in the arcuate nucleus or, and probably as likely, to increased activity in axons passing through the region. Gross recording techniques will never differentiate between these possibilities. In addition, quantification of multi-unit activity is very difficult since the signal is scarcely distinguishable from the intrinsic electrical noise within the system. It is unlikely that the multi-unit recording technique, at present much favored by neuroendocrinologists, will yield fundamental

References p. 142–147

insights into our understanding of the hypothalamic control of the pituitary gland. Indeed, I think it is fair to ask how much less might have been discovered by Hubel and Wiesel (1962, 1965, 1968) of coding in the visual cortex, or Eccles and his colleagues (Eccles *et al.*, 1967) of the relationship between climbing fibers, mossy fibers and Purkinje cells in the cerebellar cortex, if they had used multi-unit or EEG recording as their tool for the investigation of these problems. The hypothalamus is surely at least as complex as these structures and the best approach in any analysis of the relationship between neural activity and endocrine state is that of single-unit recording. The nature of the signal obtained from single-units is well understood. In many cases the waveform shows a notch due to the partial separation of the initial-segment and soma-dendritic spikes (Fatt, 1957; Novin *et al.*, 1970; Koizumi and Yamashita, 1972; Dyer, 1973). This inflexion is final evidence that the source of the signal is the perikaryon of the neuron and thus provides anatomical as well as electrophysiological data. Single-unit studies have already made a valuable contribution to neuroendocrine research and will undoubtedly have a prominent role in the future. However, although the single-unit approach has the potential to produce good dividends, the yield in the past has been restricted by our inability to separate recorded units into homogeneous groups. Thus to record many single units and then treat them as one population (for example, Cross and Silver, 1965; Lincoln and Cross, 1967; Findlay and Hayward, 1969; Kawakami and Saito, 1969; Haller and Barraclough, 1970; Cross and Dyer, 1971b; Dyer *et al.*, 1972), whilst attempting to relate neural changes with the secretion and feedback effects of pituitary and target organ hormones, is little better than monitoring multi-unit activity and is probably more time consuming. It is necessary to place single units in categories that depend, for example, on the afferent and efferent connections of the neuron and the pharmacology of its synapses. Only a few years ago, these requirements would have seemed nearly impossible to obtain but, as is described below, the future is really quite hopeful.

THE "OXYTOCIN CELL"

Harris (1947, 1948) was amongst the first to demonstrate that both oxytocin and vasopressin are released when electrical stimuli are applied to either the neurohypophysis or the hypothalamo–hypophyseal tract. The necessary parameters for the electrical stimulation of these fine, unmyelinated axons were described much later (Harris *et al.*, 1969) and it is now known that, in the rabbit, a frequency of sine-wave or biphasic stimulation in excess of 30 Hz must be applied to the pituitary stalk to ensure the secretion of oxytocin and, probably, vasopressin. The authors also showed that with a pulse duration of 1 msec, which is about the length of an action potential, the critical frequency rose to 50 Hz. This type of stimulation has been applied by Lincoln (1971) to the infundibulum and median eminence to induce labor in the conscious prepartum rabbit. Similar thresholds also apply to the rat (Dreifuss and Ruf, 1971).

These data from stimulation experiments do not necessarily describe how the neurosecretory cells of the hypothalamo–neurohypophyseal system behave during the endogenously triggered secretion of hormones. To answer this question it is necessary to record the activity of neurosecretory cells before, during and after the known discharge of either ADH or oxytocin. The first report (Cross and Green, 1959) of unit activity recorded from paraventricular and supraoptic neurons was based on histological reconstructions of electrode penetrations. The authors could not be certain that their neurons ended in the pituitary gland. However, by stimulating these axon terminations in the neurohypophysis and recording antidromically evoked action potentials in the paraventricular nucleus, Yagi *et al.* (1966) were able to monitor the activity of units which almost certainly contained ADH or oxytocin. It is now mandatory to use this technique when analyzing the activity of antidiuretic and oxytocic neurons and numerous papers have been published in this field (Dyball and Koizumi, 1969; Kelly and Dreifuss, 1970; Novin *et al.*, 1970; Yamashita *et al.*, 1970; Barker *et al.*, 1971a, b, c; Dyball, 1971; Dyball and Dyer, 1971; Moss *et al.*, 1971, 1972a, b; Wakerley and Lincoln, 1971; Dreifuss and Kelly, 1972a, b; Koizumi and Yamashita, 1972; Lincoln and Wakerley, 1972; Negoro and Holland, 1972; Vincent *et al.*, 1972a, b; Dyball and Pountney, 1973; Wakerley and Lincoln, 1973a, b).

With the feasibility of recording from the perikarya of known neurohypophyseal neurons established, it remained to have an on-line and continuous assay for either of the related hormones. The most operational assay is that for oxytocin, in the barbiturate anesthetized lactating rat (Bisset *et al.*, 1967), since physiological quantities of hormone cause a substantial increase in intramammary pressure which can readily be monitored with a pen recorder. Unfortunately, it was thought that anesthesia blocked the reflex release of this hormone during suckling (Yokoyama and Ota, 1965) and attempts were made to record by radiotelemetry from units in the paraventricular nucleus of freely moving rabbits whilst they suckled their young (Lincoln, unpublished observations). These difficult experiments produced few results and were terminated by the recent discovery (Lincoln *et al.*, 1973) that the anesthetized rat continues to release oxytocin in response to the suckling stimulus and even in the headholder of a stereotaxic apparatus with a stimulating electrode resting beneath the neural lobe. It was now relatively easy to record from neurosecretory cells during the discharge of oxytocin from their endings and very dramatic results were obtained. Some 15 sec before an increase in intramammary pressure heralded the arrival of oxytocin at the gland approximately half of the antidromically identified cells in both the paraventricular (Wakerley and Lincoln, 1973a) and supraoptic (Wakerley and Lincoln, 1973b) nuclei increased their firing rate from under 5 to over 50 spikes/sec (recall the stimulation experiments described above) in a brief burst of activity not exceeding 3 sec. This characteristic behavior of some units was repeated over many milk ejections (Fig. 1) and, since antidiuretic hormone is not released with oxytocin (Wakerley *et al.*, 1973), these experiments provide a technique for recording from identified "oxytocin nerve cells". The observation that supraoptic neurons are as responsive as those in the paraventricular nuclei is surprising but it is probably related to the recent report that this part of the hypothalamus contains more oxytocin than the paraventricular region (Dyer *et al.*, 1973).

References p. 142–147

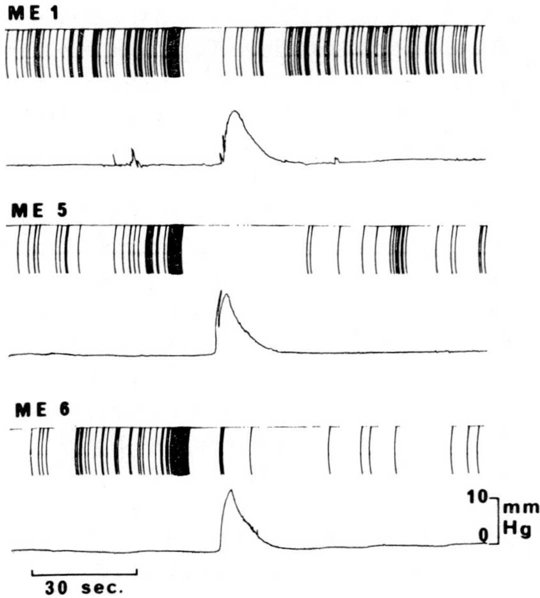

Fig. 1. Polygraph records of (a) spontaneous action potentials recorded from a neuron in the supraoptic nucleus of the rat identified by antidromic activation from the neurohypophysis, and (b) intramammary pressure recorded from a cannulated pelvic gland, both monitored through a number of milk ejections (ME) induced reflexly by suckling. Note the characteristic and repeatable increase in spike activity immediately prior to the oxytocin dependent rise in intramammary pressure. (Unpublished record of Wakerley and Lincoln.)

STRATEGY

There can be no doubt that our knowledge of the relationship between the systemic concentrations of oxytocin and the electrical activity of the neuron producing and releasing the hormone is greater than that for any other similar neuroendocrine system in the mammalian hypothalamus (for a recent appraisal read Cross, 1974) and it is salutary to draw certain conclusions from these data. In particular, it is reasonable to suggest that the experiments were only successful because: (1) the anatomical location of the neurons within the hypothalamus was well established; (2) an on-line assay was available for continuously measuring the secretion of the neurons; (3) it was possible to record the electrical activity of known neurosecretory cells.

Thus an orderly and logical sequence of experiments led to the present happy position. Today, none of the 3 criteria outlined above can be fully met for neurons containing particular releasing factors and it is foolish to attempt for CRF, TRF, PIF, LRF-FSHRF, GRF or similar cells what has been achieved for the "oxytocin cells", and what has partly been achieved for the antidiuretic neuron (Dyball, 1971; Vincent *et al.*, 1972a, b; Dyball and Pountney, 1973), until the necessary groundwork has been more fully covered. This is far from a philosophy of despair but while on-line assays for releasing factors are awaited, the electrophysiologist should devote his attentions to defining and understanding the neural connections under consideration. A start has already been made. Thus, for example, it is now possible to antidromically identify neurons in the medial basal hypothalamus with axons terminating in the median eminence (Makara *et al.*, 1972; Sawaki and Yagi, 1973) and it is probable that many of these units are releasing factor cells.

However, one should not be seduced by the attractive model system demonstrated by the "oxytocin cell", for this is a final common path neuron and its behavior may be dissimilar from the interneurons which constitute most of the electrical activity usually recorded from the hypothalamus. Thus, for example, the interneurons projecting to the paraventricular and supraoptic nuclei probably do not show the burst of activity preceding milk-ejection characteristic of the "oxytocin cell" (Fig. 1). Their changes in spike activity may be more subtle, difficult to detect and related to more than one output system. Indeed, even the "oxytocin cell" probably does not show a dramatic burst of electrical activity preceding the tonic release of its hormone.

My particular interests are centered on the preoptic and anterior hypothalamic area (PO/AH) and the electrical activity of units in this trigger zone for the pre-ovulatory secretion of luteinizing hormone in the rat. Experiments showed (Cross and Dyer, 1970, 1971b; Dyer *et al.*, 1972) that single units recorded from PO/AH on the day of proestrus were significantly faster firing than on the other days of the estrus cycle. Unfortunately, (but in view of what has been said above, not surprisingly) this observation could not be directly related to a single neuroendocrine effector system. However, the neurons controlling LH secretion are likely to proceed to the arcuate region of the hypothalamus and are thus probably dissimilar to most PO/AH cells involved in other vegetative functions. By extending the technique of antidromic stimulation to investigate this question it has proved possible to identify neurons (Type A cells) in PO/AH, whose axons terminate in the ventromedial/arcuate region (VMH/ARC) of the hypothalamus (Dyer and Cross, 1972; Dyer, 1973) and thus record from single units which may well be involved in the regulation of anterior pituitary secretions. These cells, which constitute about 41 % of the population, were very much slower firing than the adjacent units and, since about 20 % of them fired no orthodromic spikes during the entire recording session, many of the neurons in this category have probably been overlooked in previous experiments (for example, Lincoln, 1967; Moss and Law, 1971; Dyer *et al.*, 1972). Of the cells which could not be antidromically activated the spontaneous discharge rate of about 54 % was increased and/or diminished, presumably by a synaptic pathway, following stimulation of the medial basal hypothalamus (type B cells). The remaining 46 % (approx.) of the units were not affected by the stimulus (type C cells, Fig. 2). The cells which were antidromically identified did not show the proestrus increase in firing rate and were thus further distinguished from adjacent neurons which clearly demonstrated this characteristic change (Fig. 3) (Dyer, 1973).

The sensitivity of the different types of PO/AH neuron to possible neurotransmitters is now under investigation. So far 29 units have been tested with iontophoretically

References p. 142–147

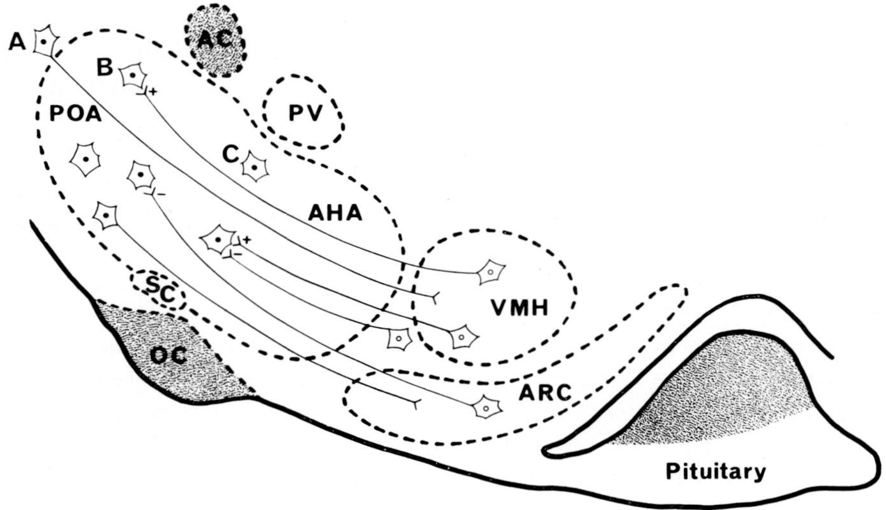

Fig. 2. Diagram of a parasagittal section through the rat hypothalamus showing the different categories of unit in the preoptic and anterior hypothalamic areas distinguished by electrical stimulation of the ventromedial/arcuate region. The type A cells are antidromically activated and have axons terminating in the zone of stimulation. Type B cells are excited (+) and/or inhibited (−) by the stimulus whilst the spontaneous discharge of type C cells is unchanged. (Details in text and Dyer, 1973.) Abbreviations; AC, anterior commissure; AHA, anterior hypothalamic area; ARC, arcuate nucleus; OC, optic chiasma; POA, preoptic area; PV, paraventricular nucleus; SC, suprachiasmatic nucleus; VMH, ventromedial nucleus.

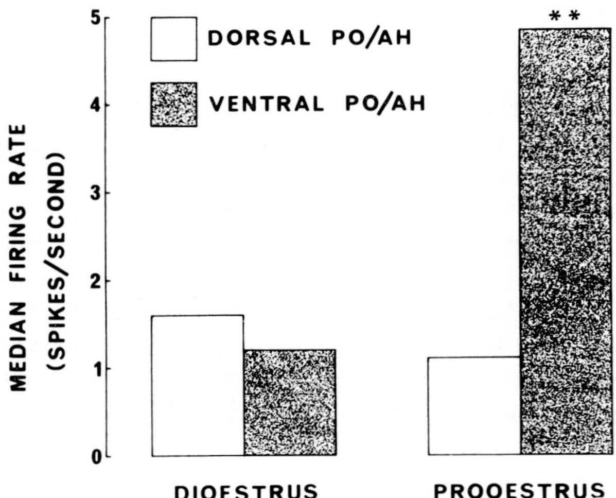

Fig. 3. Histograms of median firing rates of type B and C cells recorded from the preoptic/anterior hypothalamic area (PO/AH) of rats on the afternoons of proestrus and diestrus. Note the increased activity in the ventral PO/AH at proestrus (**). A comparable change is not observed in the slower firing type A cells (Dyer, 1973).

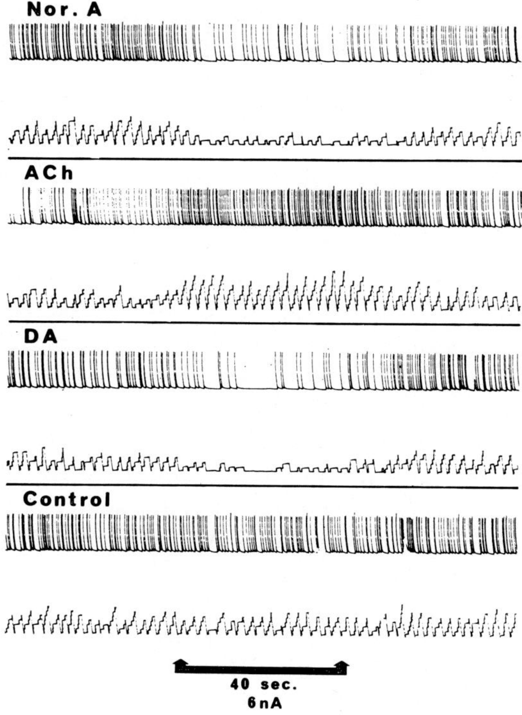

Fig. 4. Polygraph record showing the effect of a 40 sec microiontophoretic application of noradrenaline (Nor.A), acetylcholine (ACh) and dopamine (DA) on the spontaneous activity of a type B cell recorded from the preoptic area of a female rat. Sodium was applied to control for current effects. (Unpublished record of Dyer, Dyball and Drewett.)

applied acetylcholine, noradrenaline and dopamine (Fig. 4). These preliminary studies (Dyer *et al.*, unpublished observations) have already shown that the type A cells are less frequently influenced by these drugs than cell types B and C. Thus, of 25 tests with type A cells, each repeated at least twice and with stringent current controls, no change in the spontaneous activity of the unit could be detected on 12 occasions. With nearly twice as many tests (47) on cell types B and C only 7 drug applications failed to evoke a change in unit firing rate.

This electrophysiological dissection of the rostral hypothalamus has already provided the following new information.

(1) That approximately 41 % of the neurons recorded in the preoptic/anterior hypothalamic area have axons terminating in the ventromedial/arcuate region (type A cells) and that a further 32 % of the population are influenced synaptically by inputs from the medial basal hypothalamus (type B cells).

(2) That the type A cells have a relatively low spontaneous discharge rate, do not respond readily to possible neurotransmitters applied iontophoretically and do not show cyclicity during the estrus cycle.

Thus it is now possible to distinguish between units in the rostral hypothalamus on

References p. 142–147

the basis of some of their afferent and efferent connections and their sensitivity to various transmitter agents. The data obtained is a first step in an attempt to separate neurons on the basis of function and to know which units to record from when investigating particular neuroendocrine mechanisms.

CONCLUSIONS AND SUMMARY

The reader of this chapter may be disappointed by the paucity of the data connecting the electrophysiology of the hypothalamus with known endocrine conditions. A great deal could be written, for example, on the relationship between the sleep–waking state, as assessed from EEG recordings, and the secretion of hormones from the pituitary gland (for review see Findlay, 1972). However, in my view the majority of such experiments demonstrate parallelism and do not attempt to show *how* the electrophysiological and endocrine observations are connected. To achieve this basic information it is necessary to: (1) understand fully the source of the signal recorded. This demands the single-unit approach; (2) have knowledge of the function(s) of the neuron recorded; (3) have details of its afferent and efferent connections; (4) monitor changes in output alongside changes in spike activity.

These requirements will not easily be obtained but they represent to the electrophysiologist the minimum requirements needed before neural activity can be related to endocrinology and as such they represent an ideal strategy for the future.

ACKNOWLEDGEMENTS

Although I must take full responsibility for any inaccuracies in this paper, the philosophy outlined above has evolved over several years during discussions with Professor B. A. Cross and Drs. R. E. J. Dyball, D. W. Lincoln and J. B. Wakerley. I am very thankful to them for their help.

I am also very grateful to Mrs. Marilyn Jenkins for her excellent assistance in the laboratory and with the manuscript and to Drs. R. E. J. Dyball and R. F. Drewett for permission to mention our unpublished experiments and to Drs. D. W. Lincoln and J. B. Wakerley for providing Fig. 1.

Much of the original work described in this paper was supported by grants from the Medical Research Council and the Population Council.

REFERENCES

ANDERSSON, B. (1951) The effect and localisation of electrical stimulation of certain parts of the brain stem in sheep and goats. *Acta physiol. scand.*, **23**, 8–23.

ANDERSSON, B. AND MCCANN, S. M. (1955) Drinking, antidiuresis and milk-ejection from electrical stimulation within the hypothalamus of the goat. *Acta physiol. scand.*, **35**, 191–201.

AULSEBROOK, L. H. AND HOLLAND, R. C. (1969) Central regulation of oxytocin release with and without vasopressin release. *Amer. J. Physiol.*, **216**, 818–829.

AVERILL, R. L. W. AND SALAMAN, D. F. (1967) Elevation of plasma thyrotropin (TSH) during electrical stimulation in the rabbit hypothalamus. *Endocrinology*, **81**, 173–178.

BARKER, J. L., CRAYTON, J. W. AND NICOLL, R. A. (1971a) Noradrenaline and acetylcholine responses of supraoptic neurosecretory cells. *J. Physiol. (Lond.)*, **218**, 19–32.

BARKER, J. L., CRAYTON, J. W. AND NICOLL, R. A. (1971b) Supraoptic neurosecretory cells: autonomic modulation. *Science*, **171**, 206–207.

BARKER, J. L., CRAYTON, J. W. AND NICOLL, R. A. (1971c) Supraoptic neurosecretory cells: adrenergic and cholinergic sensitivity. *Science*, **171**, 208–210.

BARRACLOUGH, C. A. AND CROSS, B. A. (1963) Unit activity in the hypothalamus of the cyclic female rat: effect of genital stimuli and progesterone. *J. Endocr.*, **26**, 339–359.

BEYER, C., RAMIREZ, V. D., WHITMOYER, D. I. AND SAWYER, C. H. (1967) Effects of hormones on the electrical activity of the brain in the rat and rabbit. *Exp. Neurol.*, **18**, 313–326.

BEYER, C. AND SAWYER, C. H. (1969) Hypothalamic unit activity related to control of the pituitary gland. In *Frontiers in Neuroendocrinology*, W. F. GANONG AND L. MARTINI (Eds.), Oxford Univ. Press, New York, pp. 255–287.

BISSET, G. W., CLARK, B. J. AND ERRINGTON, M. L. (1971) The hypothalamic neurosecretory pathways for the release of oxytocin and vasopressin in the cat. *J. Physiol. (Lond.)*, **217**, 111–131.

BISSET, G. W., CLARK, B. J., HALDAR, J., HARRIS, M. C., LEWIS, G. P. AND ROCHA E SILVA, M., JR. (1967) The assay of milk-ejecting activity in the lactating rat. *Brit. J. Pharmacol.*, **31**, 537–549.

BLAKE, C. A. AND SAWYER, C. H. (1972) Effects of vaginal stimulation on hypothalamic multiple-unit activity and pituitary LH release in the rat. *Neuroendocrinology*, **10**, 358–370.

BULLOCK, T. H. (1966) Strategies for blind physiologists with elephantine problems. In *Nervous and Hormonal Mechanisms of Integration, S.E.B. Symposia XX*, Cambridge Univ. Press, Cambridge, pp. 1–10.

CAPON, A. (1960) Analyse de l'effect d'éveil excercé par l'adrénaline et d'autres amines sympathicomiméthiques sur l'électrocorticogramme du lapin non narcotisé. *Arch. Int. Pharmacodyn.*, **127**, 141–162.

CLEMENS, J. A., GALLO, R. V., WHITMOYER, D. I. AND SAWYER, C. H. (1971a) Prolactin responsive neurons in the rabbit hypothalamus. *Brain Res.*, **25**, 371–379.

CLEMENS, J. A., SHAAR, C. J., KLEBER, J. W. AND TANDY, W. A. (1971b) Areas of the brain stimulatory to LH and FSH secretion. *Endocrinology*, **88**, 180–184.

COBB, W. AND SEARS, T. A. (1956) The superficial spread of cerebral potential fields. Some evidence provided by hemispherectomy. *Electroenceph. clin. Neurophysiol.*, **8**, 717–718.

CRAMER, O. M. AND BARRACLOUGH, C. A. (1971) Effect of electrical stimulation of the preoptic area on plasma LH concentrations in proestrous rats. *Endocrinology*, **88**, 1175–1183.

CROSS, B. A. (1958) The motility and reactivity of the oestrogenised rabbit uterus *in vivo:* with comparative observations on milk ejection. *J. Endocr.*, **16**, 237–260.

CROSS, B. A. (1973) Unit responses in the hypothalamus. In "*Frontiers in Neuroendocrinology, 1973*" W. F. GANONG AND L. MARTINI (Eds.), Oxford Univ. Press, New York, pp. 133–171.

CROSS, B. A. (1974) Functional identification of hypothalamic neurones. In *Recent Advances in Hypothalamic Function*, K. LEDERIS (Ed.), University Calgary Press, Calgary, in press.

CROSS, B. A. AND DYER, R. G. (1969) Does oxytocin influence the activity of hypothalamic neurones? *J. Physiol. (Lond.)*, **203**, 70–71 P.

CROSS, B. A. AND DYER, R. G. (1970) Characterization of unit activity in hypothalamic islands with special reference to hormone effects. In *The Hypothalamus*, L. MARTINI, M. MOTTA AND F. FRASCHINI (Eds.), Academic Press, New York, pp. 115–122.

CROSS, B. A. AND DYER, R. G. (1971a) Unit activity in rat diencephalic islands — the effect of anaesthetics. *J. Physiol. (Lond.)*, **212**, 467–481.

CROSS, B. A. AND DYER, R. G. (1971b) Cyclic changes in neurones of the anterior hypothalamus during the oestrous cycle, and the effects of anaesthesia. In *Steroid Hormones and Brain Function*, R. GORSKI AND C. H. SAWYER (Eds.), Univ. California Press, Los Angeles, Calif., pp. 95–102.

CROSS, B. A. AND DYER, R. G. (1972) Ovarian modulation of unit activity in the anterior hypothalamus of the cyclic rat. *J. Physiol. (Lond.)*, **222**, 25 P.

CROSS, B. A. AND GREEN, J. D. (1959) Activity of single neurones in the hypothalamus: effect of osmotic and other stimuli. *J. Physiol. (Lond.)*, **148**, 554–569.

CROSS, B. A. AND KITAY, J. I. (1967) Unit activity in diencephalic islands. *Exp. Neurol.*, **19**, 316–330.

CROSS, B. A. AND SILVER, I. A. (1965) Effect of luteal hormone on the behaviour of hypothalamic neurones in pseudopregnant rats. *J. Endocr.*, **31**, 251–263.

D'ANGELO, S. A., SNYDER, J. AND GRODIN, J. M. (1964) Electrical stimulation of the hypothalamus: simultaneous effects on the pituitary — adrenal and thyroid systems in the rat. *Endocrinology*, **75**, 417–427.

VAN DELFT, A. M. L. AND KITAY, J. I. (1972) Effect of ACTH on single unit activity in the diencephalon of intact and hypophysectomised rats. *Neuroendocrinology*, **9**, 188–196.

DELGADO, J. M. R. (1964) Electrodes for extracellular recording and stimulation. In *Physical Techniques in Biological Research, Vol. 5A*, W. L. NASTUK (Ed.), Academic Press, New York, pp. 88–143.

DREIFUSS, J. J. AND KELLY, J. S. (1972a) Recurrent inhibition of antidromically identified rat supraoptic neurones. *J. Physiol. (Lond.)*, **220**, 87–103.

DREIFUSS, J. J. AND KELLY, J. S. (1972b) The activity of identified supraoptic neurones and their response to acetylcholine applied by iontophoresis. *J. Physiol (Lond.)*, **220**, 105–118.

DREIFUSS, J. J. AND RUF, K. B. (1971) A transpharyngeal approach to the rat hypothalamus. In *Experiments in Physiology and Biochemistry, 5*, G. A. KERKUT (Ed.), Academic Press, London, pp. 213–228.

DUFY, B., VINCENT, J. D., BENSCH, C. AND FAURE, J. M. A. (1973) Effects of vaginal stimulation and luteinizing hormone on hypothalamic single units in the freely moving rabbit. *Neuroendocrinology*, **11**, 119–129.

DYBALL, R. E. J. (1971) Oxytocin and ADH secretion in relation to electrical activity in antidromically identified supraoptic and paraventricular units. *J. Physiol. (Lond.)*, **214**, 245–256.

DYBALL, R. E. J. AND DYER, R. G. (1971) Plasma oxytocin concentration and paraventricular neurone activity in rats with diencephalic islands and intact brains. *J. Physiol. (Lond.)*, **216**, 227–235.

DYBALL, R. E. J. AND KOIZUMI, K. (1969) Electrical activity in the supraoptic and paraventricular nuclei associated with neurohypophysial hormone release. *J. Physiol. (Lond.)*, **201**, 711–722.

DYBALL, R. E. J. AND POUNTNEY, P. S. (1973) Discharge patterns of supraoptic and paraventricular neurones in rats given a 2% NaCl Solution instead of drinking water. *J. Endocr.*, **56**, 91–98.

DYER, R. G. (1973) An electrophysiological dissection of the hypothalamic regions which regulate the pre-ovulatory secretion of luteinising hormone in the rat. *J. Physiol. (Lond.)*, **234**, 421–442.

DYER, R. G. AND CROSS, B. A. (1972) Antidromic identification of units in the preoptic and anterior hypothalamic areas projecting directly to the ventromedial and arcuate nuclei. *Brain Res.*, **43**, 254–258.

DYER, R. G., DYBALL, R. E. J. AND MORRIS, J. (1973) The effect of hypothalamic deafferentation upon the ultrastructure and hormone content of the paraventricular nucleus. *J. Endocr.*, **57**, 509–516.

DYER, R. G., PRITCHETT, C. J. AND CROSS, B. A. (1972) Unit activity in the diencephalon of female rats during the oestrous cycle. *J. Endocr.*, **53**, 151–160.

ECCLES, J. C., ITO, M. AND SZENTÁGOTHAI, J. (1967) *The Cerebellum as a Neuronal Machine*, Springer, Berlin.

ENDROCZI, E. (1969) Electrophysiological studies on the mechanism of progesterone action in the rat. *Acta physiol. Acad. Sci. hung.*, **36**, 83–93.

EVERETT, J. W. (1956) The time of release of ovulating hormone from the rat hypophysis. *Endocrinology*, **59**, 580–585.

FATT, P. (1957) Electrical potentials occurring around a neurone during its antidromic activation. *J. Neurophysiol.*, **20**, 27–60.

FELDMAN, S. AND SARNE, Y. (1970) Effect of cortisol on single cell activity in hypothalamic islands. *Brain Res.*, **23**, 67–75.

FINDLAY, A. L. R. (1972) Hypothalamic inputs: methods and five examples. In *Topics in Neuroendocrinology, Progr. Brain Res., Vol. 38*, J. ARIËNS KAPPERS AND J. P. SCHADÉ (Eds.), Elsevier, Amsterdam, pp. 163–191.

FINDLAY, A. L. R. AND HAYWARD, J. N. (1969) Spontaneous activity of single neurones in the hypothalamus of rabbits during sleep and waking. *J. Physiol. (Lond.)*, **201**, 237–258.

FRANK, K. AND BECKER, M. C. (1964) Microelectrodes for recording and stimulation. In *Physical Techniques in Biological Research, 5A*, W. L. NASTUK (Ed.), Academic Press, New York, pp. 22–87.

FROHMANN, L. A., BERNARDIS, L. L. AND KANT, K. J. (1968) Hypothalamic stimulation of growth hormone secretion. *Science*, **162**, 580–582.

GALLO, R. V., JOHNSON, J. H., GOLDMAN, B. D., WHITMOYER, D. I. AND SAWYER, C. H. (1971) Effects of electrochemical stimulation of the ventral hippocampus on hypothalamic electrical activity and pituitary gonadotropin secretion in female rats. *Endocrinology*, **89**, 704–713.

GINSBURG, M. AND BROWN, L. M. (1956) Effect of anaesthetics and haemorrhage on release of neurohypophysial ADH. *Brit. J. Pharmacol.*, **11**, 236–244.

HALLER, E. W. AND BARRACLOUGH, C. A. (1970) Alterations in unit activity of hypothalamic ventromedial nuclei by stimuli which affect gonadotropic hormone secretion. *Exp. Neurol.*, **29**, 111–120.

HARRIS, G. W. (1947) The innervation and actions of the neurohypophysis: an investigation using the method of remote control stimulation. *Phil. Trans. B*, **232**, 385–441.

HARRIS, G. W. (1948) Electrical stimulation of the hypothalamus and the mechanisms of the neural control of the adenohypophysis. *J. Physiol. (Lond.)*, **107**, 418–429.

HARRIS, G. W., MANABE, Y. AND RUF, K. B. (1969) A study of the parameters of electrical stimulation of unmyelinated fibres in the pituitary stalk. *J. Physiol. (Lond.)*, **203**, 67–81.

HELLON, R. F. AND WHITE, J. G. (1966) Correlation of unit activity and local temperature in the hypothalamus of conscious rabbits. *J. Physiol. (Lond.)*, **183**, 9.

HUBEL, D. H. AND WIESEL, T. N. (1962) Receptive fields, binocular interaction and functional architecture in the cats visual cortex. *J. Physiol. (Lond.)*, **160**, 106–154.

HUBEL, D. H. AND WIESEL, T. N. (1965) Receptive fields, binocular interaction and functional architecture in two non-striate visual areas (18 and 19) of the cat. *J. Neurophysiol.*, **28**, 229–289.

HUBEL, D. H. AND WIESEL, T. N. (1968) Receptive fields and functional architecture of monkey striate cortex. *J. Physiol. (Lond.)*, **195**, 215–243.

KALRA, S. P., AJIKA, K., KRULICH, L., FAWCETT, C. P., QUIJADA, M. AND MCCANN, S. M. (1971) Effects of hypothalamic and preoptic electrochemical stimulation on gonadotropin and prolactin release in proestrous rats. *Endocrinology*, **88**, 1150–1158.

KAWAKAMI, M. AND SAITO, H. (1967) Unit activity in the hypothalamus of the cat: effect of genital stimuli, luteinizing hormone and oxytocin. *Jap. J. Physiol.*, **17**, 466–486.

KAWAKAMI, M. AND SAITO, H. (1969) The analysis of interspike interval fluctuation of hypothalamic unit activity in response to luteinizing hormone and oxytocin. *Jap. J. Physiol.*, **19**, 243–259.

KAWAKAMI, M. AND SAWYER, C. H. (1959) Neuroendocrine correlates of changes in brain activity thresholds by sex steroids and pituitary hormones. *Endocrinology*, **65**, 652–658.

KELLY, J. S. AND DREIFUSS, J. J. (1970) Antidromic inhibition of identified rat supraoptic neurones. *Brain Res.*, **22**, 406–409.

KLEE, M. R., OFFENLOCH, K. AND TIGGES, J. (1965) Cross-correlation analysis of electroencephalographic potentials and slow membrane transients. *Science*, **147**, 519–521.

KOIZUMI, K. AND YAMASHITA, H. (1972) Studies of antidromically identified neurosecretory cells of the hypothalamus by intracellular and extracellular recordings. *J. Physiol. (Lond.)*, **221**, 683–705.

KOMISARUK, B. R. (1971) Strategies in neuroendocrine neurophysiology. *Amer. Zool.*, **11**, 741–754.

KOMISARUK, B. R., MCDONALD, P. G., WHITMOYER, D. I., AND SAWYER, C. H. (1967) Effects of progesterone and sensory stimulation on EEG and neuronal activity in the rat. *Exp. Neurol.*, **19**, 494–507.

LINCOLN, D. W. (1967) Unit activity in the hypothalamus, septum and preoptic area of the rat: characteristics of spontaneous activity and the effect of oestrogen. *J. Endocr.*, **37**, 177–189.

LINCOLN, D. W. (1969) Effects of progesterone on the electrical activity of the forebrain. *J. Endocr.*, **45**, 585–596.

LINCOLN, D. W. (1971) Labour in the rabbit: effect of electrical stimulation applied to the infundibulum and median eminence. *J. Endocr.*, **50**, 607–618.

LINCOLN, D. W. AND CROSS, B. A. (1967) Effect of oestrogen on the responsiveness of neurones in the hypothalamus, septum and preoptic area of rats with light-induced persistent oestrus. *J. Endocr.*, **37**, 191–203.

LINCOLN, D. W., HILL, A. AND WAKERLEY, J. B. (1973) The milk-ejection reflex of the rat: an intermittent function not abolished by surgical levels of anaesthesia. *J. Endocr.*, **57**, 459–476.

LINCOLN, D. W. AND KELLY, W. A. (1972) The influence of urethane on ovulation in the rat. *Endocrinology*, **90**, 1594–1599.

LINCOLN, D. W. AND WAKERLEY, J. B. (1972) Accelerated discharge of paraventricular neurosecretory cells correlated with reflex release of oxytocin during suckling. *J. Physiol. (Lond.)*, **222**, 23–24.

MAKARA, G. B., HARRIS, M. C. AND SPYER, K. M. (1972) Identification and distribution of tuberoinfundibular neurones. *Brain Res.*, **40**, 283–290.

MARKEE, J. E., SAWYER, C. H. AND HOLLINSHEAD, W. H. (1946) Activation of the anterior hypophysis by electrical stimulation in the rabbit. *Endocrinology*, **38**, 345–357.

MARTIN, J. B. AND REICHLIN, S. (1970) Thyrotropin secretion in rats after hypothalamic electrical stimulation or synthetic TSH — releasing factor. *Science*, **168**, 1366–1368.

MARTIN, J. B. (1972) Plasma growth hormone (GH) response to hypothalamic or extrahypothalamic electrical stimulation. *Endocrinology*, **91**, 107–115.

MOSS, R. L., DYBALL, R. E. J. AND CROSS, B. A. (1971) Responses of antidromically identified supraoptic and paraventricular units to acetylcholine, noradrenaline and glutamate applied iontophoretically. *Brain Res.*, **35**, 573–575.

MOSS, R. L., DYBALL, R. E. J. AND CROSS, B. A. (1972a) Excitation of antidromically identified neurosecretory cells of the paraventricular nucleus by oxytocin applied iontophoretically. *Exp. Neurol.*, **34**, 95–102.

MOSS, R. L. AND LAW, O. T. (1971) The estrous cycle: its influence on single unit activity in the forebrain. *Brain Res.*, **30**, 435–438.

MOSS, R. L., URBAN, I. AND CROSS, B. A. (1972b) Microelectrophoresis of cholinergic and aminergic drugs on paraventricular neurons. *Amer. J. Physiol.*, **223**, 310–318.

NAKAYAMA, T. (1955) Hypothalamic electrical activities produced by factors causing discharge of pituitary hormones. *Jap. J. Physiol.*, **5**, 311–316.

NEGORO, H. AND HOLLAND, R. C. (1972) Inhibition of unit activity in the hypothalamic paraventricular nucleus following antidromic activation. *Brain Res.*, **42**, 385–402.

NOVIN, D., SUNDSTEN, J. W. AND CROSS, B. A. (1970) Some properties of antidromically activated units in the paraventricular nucleus of the hypothalamus. *Exp. Neurol.*, **26**, 330–341.

PFAFF, D. N. AND GREGORY, E. (1971) Correlation between pre-optic area unit activity and the cortical electroencephalogram: difference between normal and castrated male rats. *Electroenceph. clin. Neurophysiol.*, **31**, 223–230.

PFAFF, D. W., GREGORY, E. AND SILVA, M. T. A. (1970) Testosterone and corticosterone effects on single unit activity in the rat brain. In *Influence of Hormones on the Nervous System*, D. H. FORD (Ed.), Karger, Basel, pp. 269–281.

PHILLIPS, M. I. AND DAFNY, N. (1971) Effect of cortisol on unit activity in freely moving rats. *Brain Res.*, **25**, 651–655.

PORTER, R. W., CAVANAUGH, E. B., CRITCHLOW, B. V. AND SAWYER, C. H. (1957) Localised changes in electrical activity of the hypothalamus in estrous cats following vaginal stimulation. *Amer. J. Physiol.*, **189**, 145–151.

RAMIREZ, V. D., KOMISARUK, B. R., WHITMOYER, D. I. AND SAWYER, C. H. (1967) Effect of hormones and vaginal stimulation on the EEG and hypothalamic units in the rat. *Amer. J. Physiol.*, **212**, 1376–1384.

REICHLIN, S. (1966) Control of thyrotropic hormone secretion. In *Neuroendocrinology, 1*, L. MARTINI AND W. F. GANONG (Eds.), Academic Press, New York, pp. 445–536.

SAWAKI, Y. AND YAGI, K. (1973) Electrophysiological identification of cell bodies of the tuberoinfundibular neurones in the rat. *J. Physiol. (Lond.)*, **230**, 75–85.

SAWYER, C. H. (1970a) Some endocrine applications of electrophysiology. In *The Hypothalamus*, L. MARTINI, M. MOTTA AND F. FRASCHINI (Eds.), Academic Press, New York, pp. 83–101.

SAWYER, C. H. (1970b) Electrophysiological correlates of release of pituitary ovulating hormones. *Fed. Proc.*, **29**, 1895–1899.

SAWYER, C. H., KAWAKAMI, M., MEYERSON, B., WHITMOYER, D. I. AND LILLEY, J. (1968) Effects of ACTH, dexamethasone and asphyxia on electrical activity of the rat hypothalamus. *Brain Res.*, **10**, 213–226.

STEINER, F. A., RUF, K. AND AKERT, K. (1969) Steroid-sensitive neurons in rat brain; anatomical localization and responses to neurohumors and ACTH. *Brain Res.*, **12**, 74–85.

SUNDSTEN, J. W., NOVIN, D. AND CROSS, B. A. (1970) Identification and distribution of paraventricular units excited by stimulation of the neural lobe of the hypophysis. *Exp. Neurol.*, **26**, 316–329.

TERASAWA, E. AND SAWYER, C. H. (1969) Changes in electrical activity in the rat hypothalamus related to electrochemical stimulation of adenohypophyseal function. *Endocrinology*, **85**, 143–149.

TERASAWA, E., WHITMOYER, D. I. AND SAWYER, C. H. (1969) Effects of luteinising hormone on multiple-unit activity in the rat hypothalamus. *Amer. J. Physiol.*, **217**, 1119–1126.

VINCENT, J. D., ARNAULD, E., BIOULAC, B. (1972a) Activity of osmosensitive single cells in the hypothalamus of the behaving monkey during drinking. *Brain Res.*, **44**, 371–384.

VINCENT, J. D., ARNAULD, E. AND NICOLESCU-CATARGI, A. (1972b) Osmoreceptors and neurosecretory cells in the supraoptic complex of the unanaesthetised monkey. *Brain Res.*, **45**, 278–281.

WAKERLEY, J. B., DYBALL, R. E. J. AND LINCOLN, D. W. (1973) Milk ejection in the rat: the result of a selective release of oxytocin. *J. Endocr.*, **57**, 557–558.

Wakerley, J. B. and Lincoln, D. W. (1971) Phasic discharge of antidromically identified units in the paraventricular nucleus of the hypothalamus. *Brain Res.*, **25**, 192–194.

Wakerley, J. B. and Lincoln, D. W. (1973a) The milk-ejection reflex of the rat: a 20- to 40-fold acceleration in the firing of paraventricular neurones during oxytocin release. *J. Endocr.*, **57**, 477–493.

Wakerley, J. B. and Lincoln, D. W. (1973b) Unit activity in the supraoptic nucleus during reflex milk ejection. *J. Endocr.*, **59**, xlvi–xlvii.

Yagi, K. (1970) Effects of oestrogen on the unit activity of the rat hypothalamus. *J. Physiol. soc. Jap.*, **32**, 692–693.

Yagi, K., Azuma, T. and Matsuda, K. (1966) Neurosecretory cell: capable of conducting impulse in rats. *Science*, **154**, 778–779.

Yamashita, H., Koizumi, K. and Brooks, C., McC. (1970) Electrophysiological studies of neurosecretory cells in the cat hypothalamus. *Brain Res.*, **20**, 462–466.

Yokoyama, A. and Ota, K. (1965) The effect of anaesthesia on milk yield and maintenance of lactation in the goat and rat. *J. Endocr.*, **33**, 341–351.

DISCUSSION

Van der Schoot: Since the presentation of your paper was mainly on the electrophysiology of the preoptic area, I would like to know what the basis of your choice for this particular area was.

Dyer: It is well established now that the preoptic area is indispensable for the phasic release of LH during the estrus cycle. Since it was probable that it would be easier to see some phasic events than to detect tonic events, where a change might be much smaller and less dramatic, we chose the preoptic area for our study.

Herbert: May I ask you: when you stimulate in the arcuate area, how can you be sure that you are stimulating the endings of preoptic cells? Are you sure that you do not stimulate *e.g.* cell bodies or fibers *en passage*?

Dyer: We have established that it is endings and not fibers *en passage*, by repeating these experiments after placement of the stimulating electrode in the caudal part of the arcuate nucleus, or in the posterior hypothalamus, and showing that antidromically identified cells in the preoptic area could not be obtained in this way. Therefore it was definitely the endings of the neurons that we were stimulating.

The Mammalian Pineal Gland and its Control of Hypothalamic Activity

J. ARIËNS KAPPERS, A. R. SMITH AND R. A. C. DE VRIES

Netherlands Central Institute for Brain Research, Amsterdam (The Netherlands)

INTRODUCTION

Before dealing with the subject mentioned in the title of this paper, it seems fit to give first of all a general introduction surveying the essentials of mammalian pineal structure, innervation, biochemistry and function without going too much into detail or giving an exhaustive list of references. For additional information, we refer to some surveys and books, listed at the beginning of the bibliography.

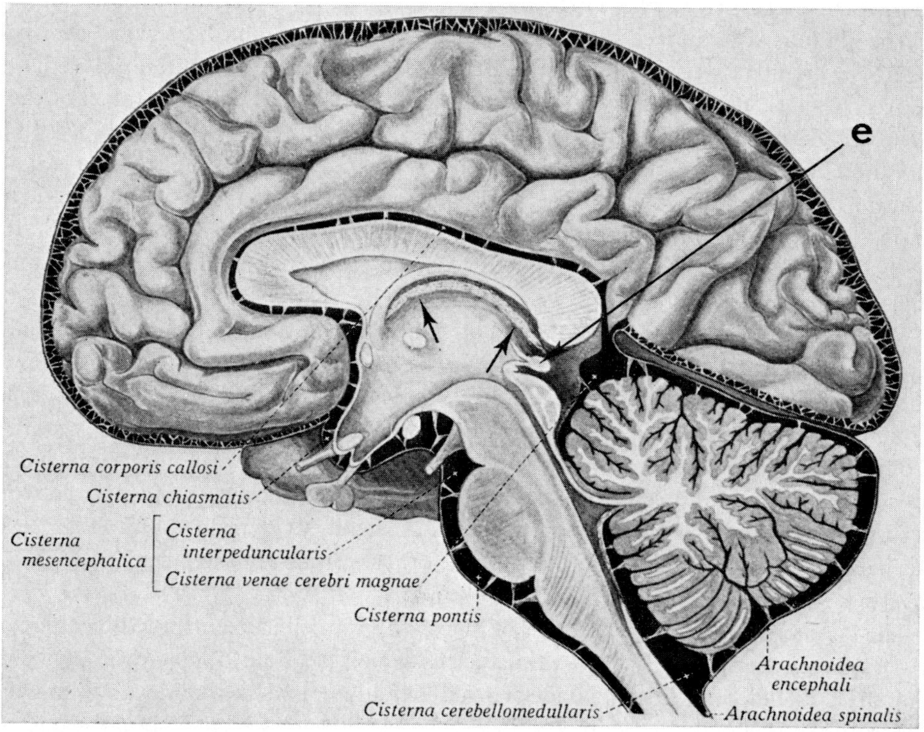

Fig. 1. Topography of the pineal gland (e) in man. The arrows indicate the position of the velum interpositum. (From M. W. WOERDEMAN (1950) *Atlas of Human Anatomy, II*, Wetensch. Uitg., Amsterdam; Plate 459, slightly altered.)

References p. 170–173

Ontogenetically originating from that part of the diencephalic roof which is situated between the habenular commissure rostrally and the posterior commissure caudally, the epiphysis cerebri or pineal gland is an epithalamic structure. In mammals, it is a solid, parenchymatous organ, except for cysts of various origin which may specially occur in the human pineal. The gland can be cone-shaped, hence the classical name "conarium" or "pineal body", as it is for instance in man, or rod-shaped as in rabbit. Its tip can extend, as it does in man, into the superior cystern, an extension of the subarachnoid space which lies dorsal to the quadrigeminal lamina (Fig. 1). A pineal recess, lined by ependyma, indicates the opening of the lumen of the originally saccular pineal "anlage" into the third ventricle. In many species, part of the dorsal surface of the organ is covered by the suprapineal recess of the third ventricle. This is a protrusion of the ventricle developing just rostral to the habenular commissure, then growing in a dorsocaudal direction. Between the ependymal wall of this recess and the pineal parenchyma at the dorsal surface of the organ, there is the pial covering of both this recess and the pineal gland, as well as a variable amount of loose pia-arachnoidal tissue forming part of the so-called velum interpositum. This velum is situated between the fornix and the roof of the third ventricle. Evidently, products produced by the pineal parenchyma cannot be released directly into the internal cerebrospinal fluid contained in the suprapineal recess, and such a release is also impossible by way of the pineal recess (Kappers, 1971a).

The gland is vascularized by branches of the posterior cerebral arteries forming an extensive capillary network in the pineal parenchyma. The topographical relation between the organ and the large intracranial veins is a very close one. In rat, the caudal tip of the gland invaginates into the floor of the sinus confluens (Kappers, 1960), while in rabbit this tip even invaginates into either the right or the left transverse sinus (Smith, 1971 and 1972; Romijn, 1973b), making pinealectomy in the rabbit practically impossible. Pineal venules in mouse (Bartheld and Moll, 1954) and in rabbit (Smith, 1971 and 1972) have been demonstrated to drain directly into the large intracranial veins.

The mammalian pineal is composed of strings of cells, often showing a pseudo-follicular arrangement. Between these strings, connective tissue strands are present containing arterioles and venules. These, as well as the parenchymal capillaries, are surrounded by perivascular spaces varying in width according to the species examined. Extensions of pericapillary spaces may penetrate deeply between the parenchymal cells to merge with intercellular spaces. In this way, an extensive canalicular network is formed in the pineal parenchyma which is especially evident in rabbit (Fig. 2), as has been shown by Romijn (1972), one of the other workers on the pineal gland in our institute.

The parenchymal cell cords consist of pineal-specific cell elements, the pinealocytes. Most of their perikarya show, *i.a.*, a nucleus which is often lobulated, many mitochondria, rough and smooth endoplasmic reticulum and a well-developed Golgi apparatus producing dense-core and lucent vesicles. These vesicles migrate along the processes of the pinealocytes finally reaching their bud-shaped terminals. Lucent vesicles may also be pinched-off from the smooth endoplasmic reticulum present in

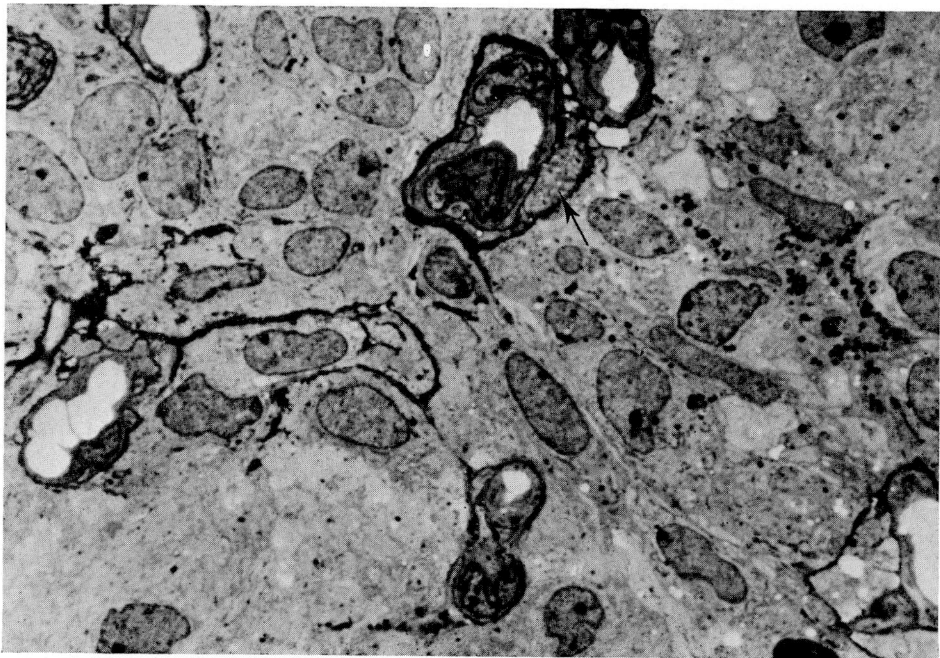

Fig. 2. Rabbit. Pineal parenchyma. Fixation and embedding as for electron micrographs, 100 nm section, silver methenamine. A number of nuclei of pinealocytes and some capillaries are seen. Pericapillary spaces (thick black lines) surround the capillaries extending deeply into the pineal parenchyma. Bundle of sympathetic nerve fibers in pericapillary space indicated by arrow. × 1570. (Photograph by H. J. Romijn.)

these terminals which, thus, would also be involved in the secretory process of the pinealocyte (Romijn, 1972, 1973c). Most of the terminals of the pinealocytic processes border on the outer or parenchymal basal membrane lining the pericapillary spaces while some may even end freely within these spaces. Depletion of the content of the vesicles present in the terminal buds of the pinealocytic processes into intercellular and pericapillary spaces has been observed electron microscopically.

In some mammals, such as rabbit, light and dark pinealocytes can be distinguished electron microscopically. They show characteristic cytological differences, a somewhat different location in the gland and may have a somewhat different function (Romijn, 1972, 1973c). Fig. 3 schematically illustrates the histology of the rabbit pineal gland which, in principle, is not very different from that of the organ in other mammalian species.

On the ground of the light and electron microscopical data mentioned, it is now generally accepted that the pinealocytes are, indeed, secretory cells. Their products are released into the capillary blood via the capillary spaces and the capillary endothelial wall which, in many species, is fenestrated, and then reach the systemic circulation directly via the venules which drain into the large intracranial veins.

As the pinealocytes are ontogenetically derived from embryonic neuroepithelium

References p. 170–173

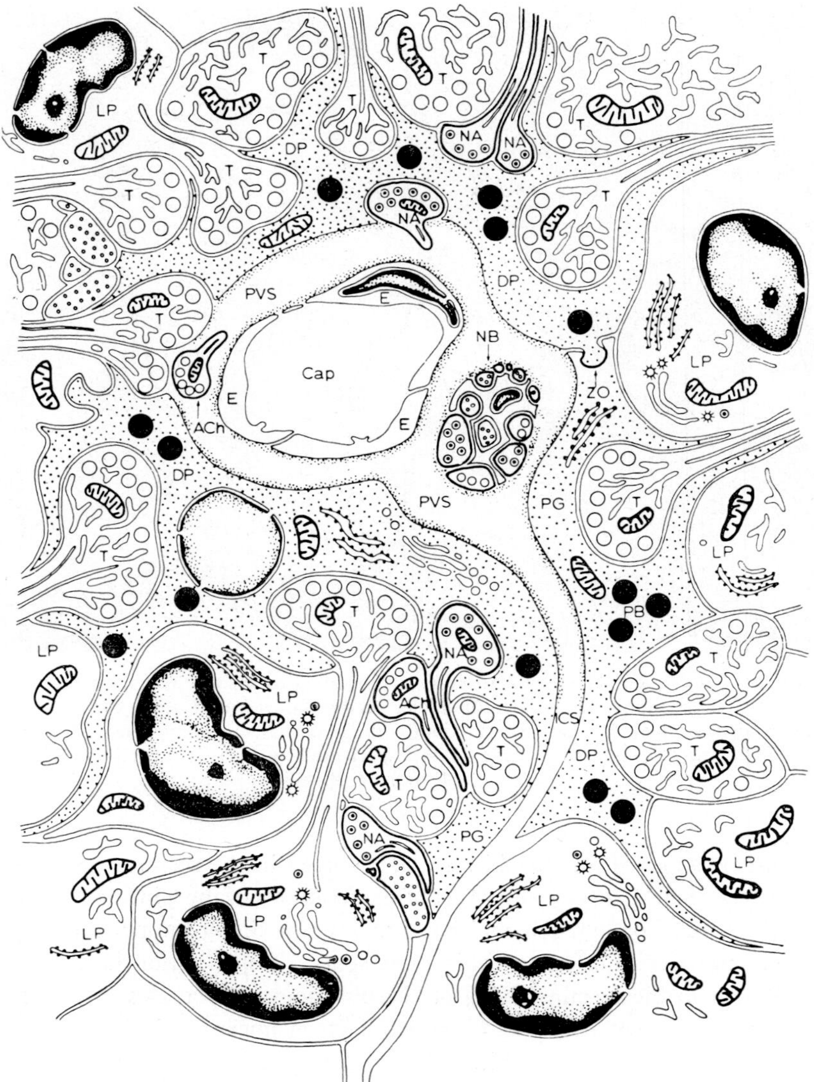

Fig. 3. Schematic representation of the medullar area of the rabbit pineal gland. ACh: cholinergic nerve ending; Cap: capillary lumen; DP: dark pinealocyte; E: endothelium; ICS: wide intercellular space; LP: light pinealocyte; NA: noradrenergic nerve ending; NB: nerve bundle, accompanied by a lemmocyte, containing noradrenergic and cholinergic axons and nerve endings; PB: pigment bodies; PG: pigment granules; PVS: perivascular space; T: club-shaped terminal of an offshoot of a light pinealocyte; ZO: zonula occludens. (Fig. 19 from H. J. Romijn, 1973c.)

and as their products are released into the blood, the mammalian pineal gland may be termed, at least on structural grounds, a neuro-endocrine organ. This does not mean that its secretory compounds are produced by nerve cells, but by cells of neural origin. By Kappers (1971b), the similarity in structure at the cell/vascular border between the pineal gland and other neuro-endocrine organs such as the posterior pituitary and the median eminence has been pointed out.

Phylogenetically, the mammalian pinealocyte is derived from neurosensory photoreceptor cells, present in the pineal organ of submammalian vertebrates. Although the mammalian pineal is no longer directly photosensitive, its function is still regulated by photic stimuli as will be mentioned later.

Besides pinealocytes, a varying number of fibrous astrocytes do occur in the gland. By some authors they have been somewhat confusingly termed "interstitial cells". Nerve cells may also be present. Their nature will be dealt with below. Furthermore, cells of mesodermal origin such as fibrocytes, mast cells and plasma cells have been demonstrated.

INNERVATION

The mammalian pineal gland is exclusively innervated by the autonomic nervous system. Sympathetic postganglionic nerve fibers, originating in the superior cervical ganglia, enter the organ along the perivascular spaces of blood vessels penetrating the gland all along its surface, and via two bilaterally symmetrical nervi conarii, coursing in the tentorium cerebelli (Kappers, 1960, also for refs., 1965; Romijn, 1972, 1973b). These nerves branch extensively in the pineal parenchyma. Nerve fibers entering the gland along the perivascular spaces may end in these spaces, but some penetrate into the pineal parenchyma while also some parenchymal nerve fibers penetrate into the perivascular spaces.

Electron microscopically, the endings of these fibers show the characteristics of sympathetic nerve terminals as do the varicosities, present along the preterminal part of the fibers. Although, earlier, some rare synaptic junctions between terminals of postganglionic sympathetic fibers and pinealocytes have been described (Kappers, 1969, 1971b, also for refs.) these do not really occur. The synapses observed should be interpreted differently as will be mentioned below. As is now known, the very extensive sympathetic innervation of the pineal is of paramount importance for the function of the organ.

Quite recently it has been ascertained that, next to a sympathetic, a parasympathetic pineal innervation is present although this has not yet been examined in many mammalian species in detail (Romijn, 1972, 1973a, 1973b, also for refs.). In the pineal of some species, the presence of a varying number of nerve cells has already been demonstrated while in other species nerve cells, which may have a similar function, are situated at the caudal tip of the organ forming the small ganglion of Pastori (1928, 1930). Besides these pineal intramural and juxtamural nerve cells, cells of a similar nature are probably situated at an even larger distance from the gland. In the rabbit pineal, synaptic junctions of cholinergic nerve terminals with the perikarya and dendrites of intramural pineal neurons have now been demonstrated by light as well as by electron microscopy (Romijn, 1973a, b and in preparation). Most probably, these are the synapses interpreted by earlier authors to be synapses between sympathetic fibers

References p. 170–173

and pinealocytic processes. The axons of the intramural neurons distribute throughout the rabbit pineal parenchyma in which a great number of cholinergic nerve terminals have been electron microscopically observed. Like the noradrenergic terminals, they do not synapse with the pinealocytes.

Most probably, the pineal intramural neurons are postganglionic parasympathetic elements which are innervated by preganglionic parasympathetic fibers. The localization of the cells of origin of the latter are, as yet, unknown. It has been suggested (Kenny, 1961) that the preganglionic parasympathetic fibers for the pineal would run in the greater petrosal nerve.

Although the pineal gland is part of the brain, it does not appear to be connected either by afferent or efferent nerve fibers with centers in the brain proper. In several mammalian species, some fibers deriving from the posterior and/or the habenular commissure have been observed to enter the pineal stalk or the most rostral part of the gland. Most of these fibers are, however, aberrant commissural fibers which make hairpin loops leaving the gland again to enter the contralateral side of the commissure from which they are derived (Kappers, 1960, 1965; Kenny, 1965; Romijn, 1972, 1973b).

BIOCHEMISTRY

For surveys of the biochemistry of the mammalian pineal organ we may refer to Arvy (1963) and Kappers (1969). For this introduction only a few data are selected which seem to be of importance for understanding pineal function. In the organ, a number of biogenic amines have been demonstrated. Of the catecholamines, histamine, noradrenaline and dopamine are present. The pineal content of histamine varies widely (Giarman and Day, 1958). This is related to the pineal mast cell population, the compound being localized primarily in these elements (Machado *et al.*, 1965). As noradrenaline is exclusively localized in the postganglionic sympathetic fibers, the pineal content of this compound is considerable due to the extensive sympathetic innervation of the organ. Pineal dopamine is not primarily located in the sympathetic nerve fibers. In the rat, a rich extraneuronal pool of this compound has been observed in pinealocytes in which it is probably stored (Pellegrino de Iraldi and Zieher, 1966; Zieher and Pellegrino de Iraldi, 1966).

In the mammalian pineal organ, indoleamines are abundant. 5-Hydroxy- as well as 5-methoxyindoles have been demonstrated. The pineal of most mammals, including man, contains a large amount of 5-hydroxytryptamine or serotonin which is, indeed, larger than in any other neural structure (Giarman and Day, 1958; Miline *et al.*, 1959; Giarman *et al.*, 1960). This compound which is not autofluorescent, but which fluoresces yellow using the Falck–Hillarp technique for histochemical fluorescence, is present in pinealocytes as well as, in some mammals, in the intrapineal part of the sympathetic nerve fibers (see Kappers, 1969, for literature). Serotonin which is produced in the pineal from tryptophan, is a precursor substance of N-acetyl-5-methoxytryptamine. Lerner, who first isolated and identified this latter substance in bovine pineals (Lerner *et al.*, 1958, 1959; Lerner and Case, 1960), termed it "melatonin" because it aggregates the pigment granules in the melanophores of amphibian skin. Melatonin has long been considered the pineal hormone "par excellence" because of its endocrinological effects which will be dealt with later and the more so because its synthesis was thought to be pineal-specific. In 1971, however, Cardinali and Rosner demonstrated the presence of both the rate-limiting enzyme of melatonin synthesis, hydroxyindole-O-methyl transferase (HIOMT), and melatonin itself, in the rat retina.

Other indole derivatives isolated and identified in the mammalian pineal are 5-methoxyindole-3-acetic acid and 5-hydroxyindole-3-acetic acid. Probably, both compounds are products of serotonin metabolism by monoamine oxidase (see Quay, 1965a, for pineal indole derivatives and their metabolic pathways). Of the other indole derivatives isolated and identified in the mammalian pineal gland, only 5-hydroxytryptophol and 5-methoxytryptophol (Delvigs *et al.*, 1965) will be mentioned here because they have also been shown to be of endocrinological consequence (see later). To our knowledge, these compounds have, so far, not been identified in other organs or tissues. Being derivatives of serotonin, it seems, however, questionable whether they are pineal-specific because serotonin is known to occur in many other tissue elements than just pinealocytes.

As has been demonstrated in rat, the pineal protein content is very high in comparison with other parts of the brain. In their electrophoretic study, Pun and Lombrozo (1964) found 4 proteins which they considered pineal-specific, but they did not identify them. As will be mentioned later, one of the authors of this paper (Smith, 1972) did identify one of these proteins while he was also able to demonstrate its presence in two hypothalamic nuclei. The striking activity of aminopeptidase observed in pineal parenchyma (Niemi and Ikonen, 1960, in rat; Bayerova and Bayer, 1967, in the human), and the fact that, biochemically, pineal aminopeptidase activity is very similar to that in the hypophysis and the cortex cerebri (Jouan and Rocaboy, 1966), led to the postulation that a very active synthesis of proteins occurs in the mammalian pineal gland.

It is of interest that not only pineal weight, the number of mitoses in the organ, and the lipid-, glycogen- and RNA-content show a circadian rhythm (Quay, 1961, 1962, 1963), but also a number of pineal biogenic amines and enzymes as far as, respectively, their concentration and activity are concerned. Table I summarizes a number of literature data on the circadian rhythm of pineal compounds under normal and experimental conditions. Many experiments point to a light-independent, endogenous regulation of pineal serotonin levels. The day/night rhythm of most other substances mentioned in Table I is probably exogenously regulated and greatly dependent on the pineal noradrenaline rhythm, which, as pineal noradrenaline is exclusively present in the sympathetic nerve fibers, links these rhythms with the sympathetic innervation of the organ.

From Table I it is quite obvious that light and darkness are among the factors which influence the pineal content of most compounds as does the innervation of the gland which is apparent when performing superior ganglionectomy or deafferentiation of the superior cervical ganglia which send postganglionic sympathetic fibers to the gland. Experimental neuroanatomical investigations (Moore *et al.*, 1968) have shown that the multineuronal pathway along which light or darkness exogenously regulate

References p. 170–173

TABLE I

DAY/NIGHT RHYTHM AND CONTENT OR ACTIVITY OF SOME SUBSTANCES OCCURRING IN THE EPIPHYSIS CEREBRI OF THE ADULT RAT UNDER NORMAL AND EXPERIMENTAL CONDITIONS ACCORDING TO DATA FROM VARIOUS AUTHORS

Substance	*Daytime*	*Night*	*Ganglion-ectomy*	*Decentr. ganglia*	*Contin. light*	*Contin. darkness*	*Enucleation of eyes*
5-hydroxytryptophan decarboxylase	activity high	activity low	rhythm vanishes activity average	not investigated	rhythm vanishes activity high	rhythm vanishes activity low	rhythm vanishes activity low
5-hydroxytryptamine = serotonin (5-HT)	content high	content low	rhythm vanishes content average	rhythm vanishes content average	rhythm vanishes content high	rhythm remains normal	rhythm remains
N-acetyl-transferase	activity low	activity high	rhythm vanishes	rhythm vanishes	rhythm vanishes activity low	rhythm remains	rhythm remains
N-acetyl-serotonin	content low	content high	not investigated	not investigated	rhythm vanishes	rhythm vanishes	not invest.
Hydroxyindole-O-methyl-transferase	activity low	activity high	rhythm vanishes activity average-low	rhythm vanishes activity average-low	rhythm vanishes activity low	rhythm vanishes activity rel. high	rhythm vanishes activity average-high
N-acetyl-5-methoxy-tryptamine (melatonin)	content low	content high	rhythm vanishes	rhythm vanishes	rhythm vanishes content rel. low	rhythm irreg.* content rel. high	not investig.*
Noradrenaline	content low	content high	rhythm vanishes content low	rhythm vanishes content low	rhythm vanishes content low	rhythm vanishes content high	rhythm vanishes content high

* According to other authors, the day/night rhythm bound to locomotor activity is maintained.
(From Kappers, 1973)

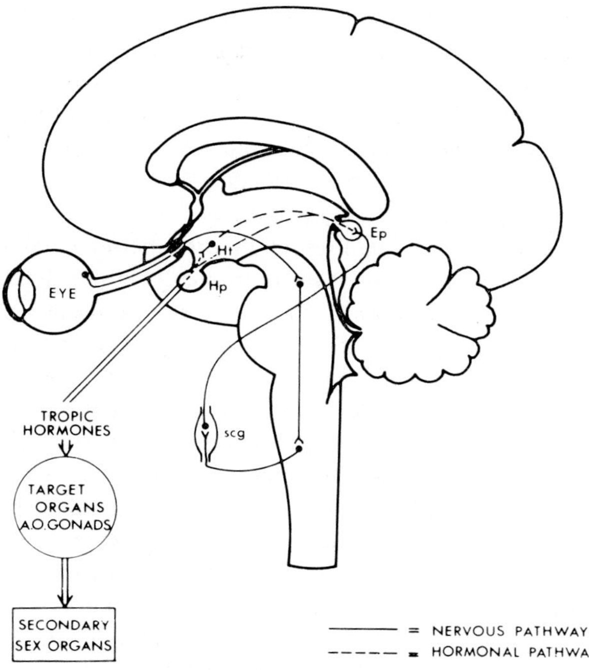

Fig. 4. Schematic diagram of the neural pathway leading from the retina of the lateral eye to the pineal organ (see text) and possible pathways by which the pineal organ may exert its functional influence on the pars distalis of the hypophysis and, via this structure, on the organs of reproduction. Ep: epiphysis; Hp: hypophysis; Ht: hypothalamus; scg: superior cervical ganglion. (From Kappers, 1969.)

the circadian rhythm of most compounds involved in the synthesis of melatonin, and also the day/night rhythm of noradrenaline, is constituted of the following links: the retinae — the optic nerves — the crossed inferior accessory optic tracts joining the medial forebrain bundle in the lateral hypothalamus and terminating in the rostral part of the mesencephalic tegmentum — a still ill-defined tegmentospinal system — the rostral part of the intermediolateral nuclei in the upper thoracic segments of the spinal cord — preganglionic nerve fibers ending in the superior cervical ganglia — and, finally, the postganglionic sympathetic fibers terminating in the pineal gland (Fig. 4). So far, the role of the pineal parasympathetic innervation is not exactly known. It has been suggested by Romijn (1972) that the endogenously regulated circadian rhythm of serotonin would possibly depend on it.

Concerning the function of the gland, we will have to restrict ourselves in this paper to its antigonadotropic activity. It has long been known that light and darkness, which, as has been mentioned, influence the pineal content of several compounds, also exert an effect on the development and function of the primary and secondary organs of reproduction. Lengthening of the daily photoperiod either under natural seasonal circumstances or under experimental conditions stimulates the functional development of the sex organs. Permanent illumination may even cause permanent estrus. Concomitantly, the pinealocytes and their nucleoli decrease in size while the pineal content of, for instance, serotonin is low. The effect of excessive illumination on the pineal is realized only if the neural pathway along which photic stimuli can reach the organ indirectly is intact. Evidently, pineal function is generally inhibited in animals, exposed to large quantities of light which influence the gland indirectly via the multineuronal pathway mentioned, while under this condition, the development and function of the reproductive system are stimulated.

On the other hand, lengthening of daily periods of darkness or permanent darkness show the opposite effect. Sexual maturation in prepubertal animals is postponed while, in mature animals, the function of the reproductive system is inhibited. Female rats, when kept in permanent darkness, may even show a more or less permanent anestrus. Concomitantly, the weight of the pineal gland and the pinealocytes increases in size while the same holds for their lipid and RNA contents. The synthesis of melatonin is activated. Removal of the eyes shows the same effect as does permanent darkness. Apparently, pineal function is enhanced in animals exposed to excessive quantities of darkness, either under natural or under experimental conditions, while the development and function of the reproductive system is inhibited. From the above, it follows that there is a direct relationship between the quantity of light to which an animal is exposed, the function of the pineal gland and the function of the reproductive organs, at least in those mammalian species investigated so far.

It has, furthermore, often been demonstrated that, in general, pinealectomy and bilateral superior ganglionectomy counteract the inhibitory effect of long periods of darkness or of blinding on the reproductive organs. Similarly, these operations counteract the stimulatory effect of additional light on the reproductive system. Pinealectomy, when performed in normal animals reared under normal conditions of environmental lighting, causes an increase in weight of the gonads, uterus, seminal vesicles and prostate as well as precocious puberty in prepubertal animals. The effects of pinealectomy can be abolished by administration of pineal extracts or of certain pineal compounds. On the other hand, chronic administration of pineal extracts to normal animals causes effects reversed to those seen after pinealectomy: the development and function of the reproductive organs are inhibited.

The question concerning the exact nature of the pineal product or products exerting an antigonadotropic effect and that about the way this effect is exerted can best be combined. Firstly, melatonin has been considered the pineal compound which would be exclusively responsible for the antigonadotropic effect of the gland. Numerous experiments with this substance have been performed. Daily administration to maturing female rats was shown to induce a decrease in weight of the ovaries and even ovarian atrophy, decrease in weight of the uterus and of the hypophysis. A significant retardation and decrease in incidence of estrus have also been observed by some authors. Furthermore, melatonin was demonstrated to block the increase in estrus frequency following pinealectomy.

It should, however, been mentioned that the results of these or similar experiments have not always been consistent or confirmed by other authors. This may be due to several factors, one of them being the time of administration in relation to the estrus cycle. For a better and somewhat more exact understanding of the effect of melatonin, the finding was important that administration of this compound causes an atrophy of the seminal vesicles and of the prostate, but not of the testes. It was also histologically demonstrated that the ovaries of female rats, injected with melatonin, did not show any corpora lutea although the many follicles present were of large size. The interstitial tissue proved to be histologically inactive, also pointing to an anti-LH effect of melatonin. In the anterior pituitaries, FSH-producing cells were quite numerous and active (Moszkowska, 1967).

The above experiments suggest that melatonin specifically reduces either hypophyseal LH synthesis, LH depletion or both, inhibiting, in consequence, the secretion of estrogen but not the synthesis and depletion of FSH. This means that melatonin cannot be responsible for all of the antigonadotropic effects of the pineal since pinealectomy is known to cause not only an increase of hypophyseal release, but also of hypophyseal synthesis of both gonadotropic hormones, LH as well as FSH (Martini *et al.*, 1968).

Regarding the question how melatonin exactly works, the following is of interest. Fraschini *et al.* (1968) found that pinealectomy causes testicular hypertrophy, increase in weight of the ventral prostate and of the seminal vesicles and increase of the pituitary LH content, suggesting that pinealectomy stimulates synthesis and depletion of pituitary LH. Implantation of pineal fragments or of melatonin in the median eminence or the mesencephalic reticular substance of castrated male rats significantly reduced pituitary LH stores as well as plasma LH levels. On the other hand, no effect on the pituitary LH content was observed when melatonin was implanted into the pituitary or the cerebral cortex. According to these authors, this suggests that pineal indole compounds, at least melatonin, modify pituitary function by acting on the median eminence and on the reticular substance. Furthermore, systemic administration of melatonin results in a significant atrophy of the prostates and the seminal vesicles while the weights of the pituitary gland and of the testes are unmodified, suggesting that melatonin inhibits LH secretion, but not that of FSH (Martini *et al.*, 1968). Furthermore, it was found (Clementi *et al.*, 1966, 1969) that pinealectomy in castrated rats did not change the ultrastructure of the castration cells while implants of melatonin or of pineal tissue into the median eminence or the midbrain reticular formation of castrated animals practically restored the morphological aspects of the gonadotropic cells, reducing the volume of the cisternae of the endoplasmic reticulum and pituitary LH storage as well. All data available suggest that melatonin inhibits pituitary LH secretion and that this effect of melatonin is exerted via structures in the brain, more especially the hypothalamus, controlling the function of the pituitary gland.

It should, however, be mentioned that, *in vitro*, Moszkowska (1967) did not observe any effect of melatonin on pituitary gonadotropin levels when incubating anterior pituitaries in the presence of both melatonin and hypothalamus, while Ellis (1969) found that melatonin, added to testis tissue cultures, caused a strong inhibition of testosterone production. This latter finding suggests that, at least in *in vitro* conditions, melatonin could exert its anti-LH effect on the gonads directly. It had also earlier been demonstrated (Wurtman, 1964) that administered melatonin concentrates in the gonads. From the foregoing it appears, at least, that the antigonadotropic activity

References p. 170–173

of melatonin is restricted to an anti-LH effect and that, *in vivo*, the hypothalamus is involved in its working mechanism.

As the next candidates for antigonadotropic pineal factors, 5-methoxytryptophol and 5-hydroxytryptophol, should be mentioned, the first compound exerting an even stronger inhibitory effect on the pituitary–gonadal axis than melatonin does in decreasing the frequency of estrus and reducing ovarian weight in maturing female rats (McIsaac *et al.*, 1964). The substance also inhibits copulation-induced release of gonadotropin in adult rabbits (Farrell *et al.*, 1966). Implantation of 5-methoxytryptophol in either the median eminence or the mesencephalic reticular formation of male castrated rats did not cause significant changes in the LH content of the hypophysis (Fraschini *et al.*, 1968, 1969) while, in contrast, hypophyseal FSH content was reduced (Clementi *et al.*, 1966). It, therefore, appears that this pineal indole derivative also exerts its antigonadotropic effect via the hypothalamus, and, evidently, the reticular substance, inhibiting FSH, but not LH secretion in the hypophysis.

5-Hydroxytryptophol has, so far, not been found to exert an antigonadotropic effect in endocrinological experiments. However, when implanted into the median eminence, the compound has been demonstrated to inhibit hypophyseal LH content, just as melatonin does (Motta *et al.*, 1967; Fraschini *et al.*, 1971). The fact that 5-hydroxytryptophol shows an effect only when implanted into the brain and not after injection in the systemic circulation has been tentatively explained by suggesting that this substance would be unable to pass the blood–brain barrier (Martini *et al.*, 1968).

As has been already mentioned, the mammalian pineal gland is rich in 5-hydroxytryptamine or serotonin which is synthesized in the pinealocytes. It has been suggested that serotonin would specially reduce LH production in the hypophysis, probably inhibiting in the hypothalamus the production and/or the release of LH-RF. This matter will be discussed in connection with the findings of Smith which will be reported later.

Leaving pineal indole derivatives as possible candidates for antigonadotropic agents produced in the gland, we now turn to pineal peptides which have already been shown to be synthesized in the organ while some have been considered to be pineal-specific (Pun and Lombrozo, 1964; see also Smith, 1972). Among others, Milcu *et al.* (1963) isolated a polypeptide from bovine pineal glands which also showed, next to pressor and oxytocic activity, an antigonadotropic effect. These authors concluded that this polypeptide would be arginine vasotocin while Pavel and Petrescu (1966) demonstrated that the gonadotropic activity exerted by pregnant mare serum on mouse uteri could, indeed, be inhibited by both purified pineal arginine vasotocin and by the same compound, synthetically produced. Following Pavel (1965), lysine vasotocin, isolated from the pineal, would also be a pineal peptide having antigonadotropic activity. Cheesman (1970) and Cheesman and Fariss (1970) isolated from bovine pineals a cyclic polypeptide which they identified as 8-arginine vasotocin and which showed a similar activity. These authors, however, attributed the antigonadotropic activity to an oxytocic effect on the immature mouse uterus, stimulated by human chorionic gonadotropin (HCG). It should also be mentioned that, according to Moszkowska and Ebels (1968), the antigonadotropic effect of arginine vasotocin, likewise observed by these authors, is not due to its inhibition of the synthesis or release of hypophyseal gonadotropic hormones, but to the direct effect of the substance on either the gonads or the gonadotropic hormones. Thieblot (1965) and Thieblot and Blaise (1963) were also convinced that the antigonadotropic agent of the pineal gland is a polypeptide. In pineal extracts, Thieblot and Menigot (1971) isolated a fraction of low molecular weight containing 5 peptides and showing antigonadotropic activity. Whether this effect is restricted to only one of these peptides and, if so, to which one, they were not able to establish as yet. From extracts of acetone-desiccated powder of sheep pineals subjected to gel filtration on Sephadex G-25 Ebels *et al.* (1965) obtained two fractions showing biological activity. One of them, fraction F_3, decreased hypophyseal FSH secretion *in vitro*. This fraction, which was further purified (Ebels, 1967), was shown to be a peptide having the relatively low molecular weight of about 700 (Moszkowska *et al.*, 1971). It appeared that, *in vitro*, the activity of this fraction differed from that of synthetic arginine vasotocin (Moszkowska and Ebels, 1970).

Recently, Benson *et al.* (1971, 1972a, b) found that melatonin-free extracts of bovine and rat pineal extracts demonstrate an antigonadotropic effect inhibiting compensatory ovarian growth in young adult mice after unilateral ovariectomy and reducing both ovarian and uterine weight in intact adult mice. Purification of the effective fraction which was shown to have a molecular weight of 500–1000, and further ultrafiltration and gel filtration yielded several peaks, one of which was ninhydrin-positive and contained the pineal antigonadotropic compound. Even in partially purified form, this was estimated to be 60–70 times as active as melatonin. Trypsin inactivated the substance. This same fraction isolated from an aqueous extract of fresh human pineal glands on Sephadex G-25 proved to possess a similar antigonadotropic activity when tested in unilaterally ovariectomized mice, while an extract of human cerebrum, treated in the same way, did not show this effect (Matthews *et al.*, 1971).

From the above mentioned data, it appears that at least some of the pineal peptides isolated should be seriously considered as candidates for pineal antigonadotropins, arginine vasotocin probably excepted. It is, however, not yet exactly known whether they are pineal-specific while the same is true regarding the mechanism by which their effect is produced.

After this survey of some data on the mammalian pineal gland indicating a number of, as yet, unsolved problems we will now deal with some recent results obtained by workers at the institute, starting with the effect of the pineal gland on the parvocellular hypothalamic neurosecretory system.

In specific areas of the rabbit pineal, Smith (1972 and Smith *et al.*, 1972a, b) observed the presence of yellow autofluorescent and of yellow non-autofluorescent pinealocytes, the fluorescence of the latter only occurring after formaldehyde treatment of the freeze-dried cryostat sections and being evidently due to the presence of serotonin. Under conditions known to influence serotonin levels in the pineal organ, such as administration of *p*-chlorophenylalanine (*p*CPA), which inhibits the activity of tryptophan hydroxylase involved in the synthesis of serotonin (Koe and Weissman, 1966), light and darkness during the normal circadian rhythm of illumination, and

References p. 170–173

permanent darkness, Smith studied the distribution of autofluorescent cells and of non-autofluorescent serotonergic cells. It appeared that the number of yellow autofluorescent pinealocytes increased after acute and chronic treatment with *p*CPA as well during the period of light under normal lighting conditions. This increase was demonstrated to occur at the cost of the number of non-autofluorescent cells, originally present. By qualitative and quantitative evaluation it could be proved that the same type of cell, the pinealocyte, is involved in the storage of the non-autofluorescent compound serotonin and of the autofluorescent substance, both being stored in granules measuring 1–2 μm.

As could be expected on the basis of earlier investigations, the circadian rhythm of non-autofluorescent serotonergic pinealocytes persisted during permanent darkness

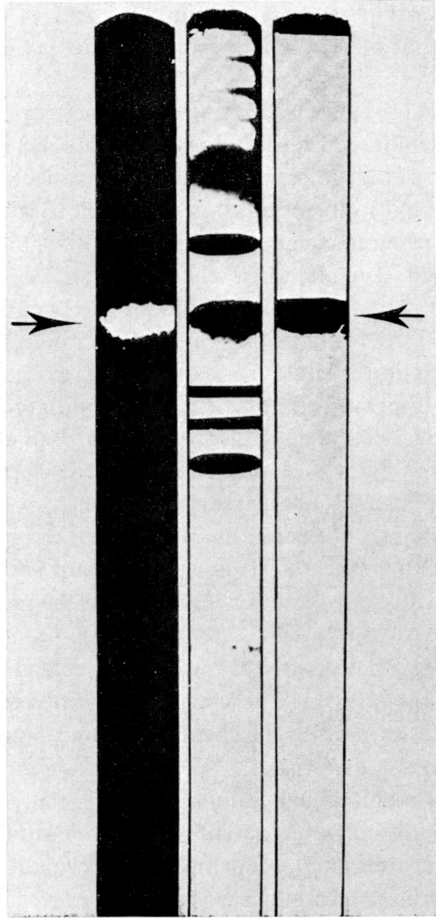

Fig. 5. Electrophoretic pattern of pineal proteins. In the middle the normal pattern of pineal proteins is shown as stained with Coomassi blue (*cf.* Smith, 1972). The gel at the right indicates the disk of the autofluorescent substance stained with the Koshland method which is specific for proteins containing relatively much tryptophan. The gel at the left demonstrates the autofluorescent disk as observed under the fluorescence microscope.

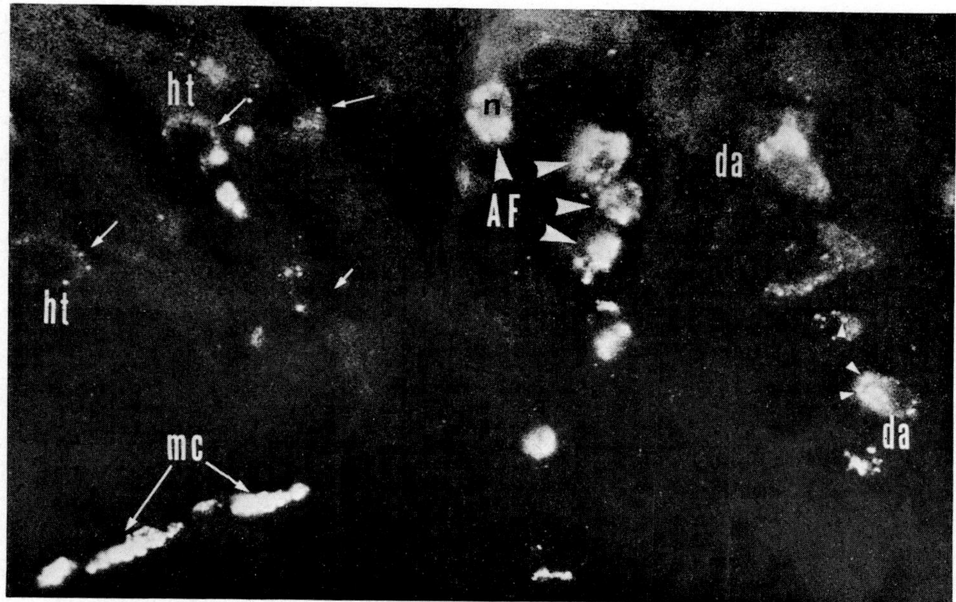

Fig. 6. Transverse section of rat hypothalamus (composed photograph). Autofluorescent and yellow formaldehyde-induced fluorescent cells, containing possibly serotonin, present in the arcuate nucleus of the rat hypothalamus. The autofluorescent cells can be demonstrated in unfixed, unthawed, unstained and only frozen-dried sections. AF: autofluorescent cells; da: dopaminergic cells; ht: serotonergic cells; mc: mast cells; n: nucleus. Small white arrows indicate intense yellow autofluorescent granules in the cytoplasm of the serotonergic neurons.

indicating that this rhythm is not controlled by exogenous environmental lighting, but endogenously. In contrast, however, no such fluctuations could be observed in the number of autofluorescent cells under conditions of permanent darkness, this number staying fixed at the high night levels as compared to control animals subjected to normal diurnal lighting conditions.

In order to correlate the fluorescence microscopical data with serotonin and tryptophan levels under the same experimental conditions mentioned, changes in serotonin and free tryptophan levels were studied using microchemical determination methods. It appeared that the levels of both compounds decreased during acute and chronic *p*CPA treatment as well as during the night under normal lighting conditions. During permanent darkness, on the other hand, the circadian rhythm of both pineal serotonin and tryptophan content persisted proving the endogenous, light-independent control of both rhythms.

In order to elucidate the nature of the autofluorescent compound, Smith, using specific staining methods for proteins containing much tryptophan as well as the microdisk electrophoresis technique, was able to prove that this compound is, indeed, a protein containing relatively much tryptophan. In the polyacrylamide gels, one disk only stained with the tryptophan-protein staining technique while the same band appeared to show yellow autofluorescence exclusively (Fig. 5). Under the experimental

References p. 170–173

conditions mentioned above, the same changes in the levels of this pineal protein occurred as were already demonstrated by fluorescence microscopy and the microchemical determination methods.

In protein precipitates of rabbit cerebral cortex, medulla oblongata, hypophysis, and pons no such a fluorescence could be demonstrated, the specific disk here being quite absent. Furthermore, it appeared that the autofluorescent protein fraction was identical with one of the 4 protein fractions found in the rat pineal by Pun and Lombrozo (1964) who thought this to be pineal-specific in comparison to proteins present in the anterior hypothalamus, putamen and several cortical areas.

In a more recent investigation, Smith was able to demonstrate the presence of the same protein, as verified with the methods used before, in neurons of the arcuate nucleus and of the basolateral part of the ventromedial nucleus of the rat hypothalamus, but in no other hypothalamic nuclei. It, therefore, appeared that this protein which is very rich in tryptophan is not pineal-specific as it also occurs in hypothalamic nuclei which are part of the hypophyseotropic hypothalamic area and, according to Mess *et al.* (1970) are especially involved in the production and release of LH-RF. Besides yellow autofluorescent cells, yellow non-autofluorescent serotonergic neurons proved to be present also in the arcuate and ventromedial hypothalamic nuclei of rat (Fig. 6). It is out of the question that the latter cells could have been confused with dopaminergic cells, known to be present in the arcuate nucleus, as Smith could distinguish these types of cells in the very same sections. In Table II, the characteristics of the autofluorescent compound present in the pineal gland of rat and rabbit, and in the rat hypothalamic nuclei are compared.

In order to study the possible functional relationship between the pineal gland and

TABLE II

COMPARISON OF CHARACTERISTICS OF AUTOFLUORESCENCE AS OBSERVED IN, RESPECTIVELY, THE MAMMALIAN PINEAL GLAND (RAT AND RABBIT) AND THE ARCUATE AND VENTROMEDIAL HYPOTHALAMIC NUCLEI (RAT)

Characteristics	*Pineal gland*	*Hypothalamic nuclei*
Emission maximum	525–543 nm	525–543 nm
Extinction maximum	405 nm	405 nm
Quenching after some time	no	no
Location	granules (1–2 μm)	granules (1–2 μm)
Cell type in which present	pinealocytes	nerve cell perikarya
Presence in nerve fibers	never	usually not (1 $\times$ only)
Stainability with Bruemmer's method	positive	positive
Reaction on specificity tests (HCl, $NaBH_4$, NH_3)	positive	positive
Effect of administration of *p*-chlorophenylalanine	*in*crease in number of autofluorescent granules	*in*crease in number of autofluorescent granules
Effect of castration	*in*crease in number of autofluorescent granules	*de*crease in number of autofluorescent granules

the two hypothalamic nuclei mentioned, Smith quantitatively examined the change in number, if any, of yellow autofluorescent cells and of serotonergic cells in the arcuate and ventromedial hypothalamic nuclei as compared to controls under the following experimental conditions: (1) castration, (2) pinealectomy, (3) pinealectomy followed by substitution using either rat or sheep pineal extract, and (4) *p*CPA administration. Young adult male rats were exclusively used while quantification was performed by counting the cells containing the two different substances present within squares measuring 800 sq. μm in a great number of sections. The results are summarized in Table III.

It appears that, in control rats, there is no difference in quantity of autofluorescent and of non-autofluorescent cells comparing the two hypothalamic nuclei. In both nuclei, however, the content of autofluorescent cells is larger than that of serotonergic cells. After castration, the number of both cell types decreases significantly in both nuclei, but the number of autofluorescent cells drops somewhat more. In the pineal, the number of both autofluorescent and non-autofluorescent cells is increased. Pinealectomy is followed by a considerable increase in number of the autofluorescent cells in both hypothalamic nuclei while the number of serotonergic cells decreases. After pinealectomy followed by substitution using either rat or sheep pineal extract, the content of both autofluorescent and non-autofluorescent serotonergic neurons is restored to normal values in both hypothalamic nuclei. Administration of *p*CPA causes a significant increase in number of autofluorescent cells in both hypothalamic nuclei, but a decrease in number of serotonergic cells. Earlier, Smith (1972) had demonstrated that, in the rabbit pineal, the increase of autofluorescent cells occurs at the cost of the non-autofluorescent serotonergic cells. The same evidently holds for the arcuate and ventromedial nuclei.

An exact interpretation of the above findings is not easy, the more so because, so far, hypophyseal and plasma levels of the gonadotropic hormones have not been determined under the experimental conditions used. In any case, however, clear histochemical evidence is given of an influence exerted by the rat pineal gland on neurons of the arcuate and ventromedial hypothalamic nuclei, nuclei which are now thought to be involved in the production of LH-RF. Some of these neurons contain serotonin while others store a protein which is extremely rich in tryptophan. On the ground of the experiments performed and on the basis of some literature data, the hypothesis can be brought forward that the serotonergic cells in the arcuate–ventromedial area of the hypothalamus inhibit the function of the neurons, present in the same area, which produce LH-RF. The function of the serotonergic cells mentioned is regulated by the pineal gland, antigonadotropic pineal compounds enhancing the activity of the hypothalamic serotonergic neurons which inhibit the activity of the hypothalamic serotonergic neurons which inhibit the activity of the LH-RF producing neurons. An extensive paper on this subject is now in preparation.

Finally we turn to the investigations of De Vries (De Vries, 1972a, b; De Vries and Kappers, 1971) concerning the influence exerted by the rat pineal gland on the magnocellular neurosecretory hypothalamic system of which a short survey will be presented here. As was observed by two other workers at the institute, Jongkind and Swaab

References p. 170–173

TABLE III

NUMBERS OF YELLOW AUTOFLUORESCENT CELLS AND OF YELLOW NON-AUTOFLUORESCENT SEROTONERGIC CELLS PER 0.08 sq. mm, PRESENT IN THE ARCUATE (AN) AND VENTROMEDIAL (VMN) HYPOTHALAMIC NUCLEI OF YOUNG, ADULT (200–250 g) MALE WISTAR RATS UNDER DIFFERENT EXPERIMENTAL CONDITIONS

Treatment	*Autofluorescent cells*	*Non-autofluorescent cells*
Controls (n = 7)	AN: 29.25 ± 4.21* VMN: 27.25 ± 3.90	AN: 9.20 ± 1.03 VMN: 9.68 ± 0.95
Castration (n = 5)	AN: 17.50 ± 3.23 VMN: 19.43 ± 2.74	AN: 4.58 ± 0.91 VMN: 6.87 ± 1.31
Pinealectomy (n = 5)	AN: 67.74 ± 5.58 VMN: 64.65 ± 5.50	AN: 5.20 ± 1.21 VMN: 4.12 ± 0.60
Pinealectomy followed by substitution using sheep pineal protein-containing extract (n = 5)	AN: 30.32 ± 4.02 VMN: 28.34 ± 3.51	AN: 8.94 ± 1.30 VMN: 10.42 ± 1.60
Pinealectomy followed by substitution using rat pineal protein-containing extract (n = 5)	AN: 29.40 ± 3.15 VMN: 30.75 ± 3.60	AN: 9.13 ± 0.90 VMN: 10.50 ± 1.00
Administration of *p*-chlorophenylalanine (n = 5)	AN: 40.30 ± 4.96 VMN: 41.45 ± 3.92	AN: 3.40 ± 0.65 VMN: 4.21 ± 0.73

* Mean ± S.E.M.

(see De Vries, 1972a, for refs.), the distribution of the Golgi-specific enzyme thiamine diphosphate-phosphohydrolase (TPP-ase) is a sensitive marker of the neurosecretory activity of both magnocellular nuclei, the supraoptic and the paraventricular nuclei. When the level of neurosecretory activity is low, the distribution of the enzyme in both nuclei is low and *vice-versa.* By a special incubation method, lead phosphate deposits are formed in the sections which can be transformed into lead sulphide. The distribution of these lead sulphide deposits have been measured using a "Hennig" ocular.

The following results were obtained by De Vries. After pinealectomy, a highly significant decrease in TPP-ase distribution was observed in the supraoptic nucleus of prepubertal and of adult male rats. The same result was obtained in the supraoptic and in the paraventricular nucleus of adult female rats, proving that the inhibitory effect of pinealectomy on neurosecretory activity in the magnocellular hypothalamic nuclei is not sex-bound.

In female pinealectomized and ovariectomized rats and in pinealectomized rats with persistent estrus due to permanent illumination which can be expected to have high plasma levels of gonadotropins, the distribution of TPP-ase in both magnocellular nuclei was also decreased. Evidently, the levels of gonadotropins in the plasma do not run parallel with magnocellular neurosecretory activity in all circumstances as was suggested by earlier experiments of Jongkind and Swaab. It appears that this activity is influenced by other factors, *i.e.* pineal function.

De Vries, furthermore, demonstrated that the decrease in activity of the magnocellular nuclei caused by pinealectomy could be completely restored to normal by 3 successive daily injections of 10 μg of melatonin, but only partially so by administration of the same dose of 5-methoxytryptophol. Moreover, pineal tissue, if implanted close to the supraoptic nucleus, completely abolished the activity-suppressing effect of pinealectomy on this nucleus at the homolateral, but not at the heterolateral side. In contrast, cerebellar tissue implanted at the same location, did not restore the activity of the supraoptic nucleus, neither at the site of implantation nor at the contralateral side. It could also be demonstrated that the positive effect on functional restoration of the activity of the supraoptic nucleus of the pineal implant was dependent on the distance between this implant and the nucleus. The larger this distance the smaller was the restoring effect of the implant in pinealectomized rats. Pineal tissue implanted exerted its restoring influence on neurosecretory activity of the supraoptic nucleus for at least 7–10 days. This finding suggests that the implant, which was observed to be revascularized, releases its products into the hypothalamic tissue during that time although it is improbable that these pineal products are newly synthesized in the implant due to the fact that this is not normally innervated.

Summarizing this short survey of De Vries' investigations, it appears that pinealectomy in rat causes a decrease in activity of the magnocellular neurosecretory hypothalamic nuclei and that this effect can be abolished by administration of melatonin and by implantation of pineal tissue into the hypothalamus during a restricted time if the implant is situated rather close to the nucleus. Evidently, a normal function of the pineal gland is necessary for a normal activity of these magnocellular neurosecretory nuclei. Furthermore, it appears that the absence of the pineal is of more consequence for neurosecretory activity than is the level of circulating gonadotropic hormones, this activity not being simply linked with the plasma level of gonadotropins. This is evident from the fact that in female rats which had either been reared in constant light or had been subjected to ovariectomy (both conditions most probably causing high plasma gonadotropin levels and high neurosecretory activity), pinealectomy, nevertheless, was followed by a decrease of neurosecretory activity to the same level as observed in non-ovariectomized rats reared under normal circumstances of illumination. Further investigations by De Vries are now in progress.

From the foregoing it appears that the pineal gland influences the function of both the parvocellular as well as the magnocellular neurosecretory hypothalamic systems, as has been shown by histochemical techniques.

CONCLUSIONS

In concluding this short and compressed survey of the mammalian pineal gland and of the recent work which has been performed at our institute, we should like to add a few words on the essential nature of the organ in general.

It is now quite evident that the mammalian pineal body is not a phylogenetic relic or a vestigial organ of no functional consequence, but a highly active structure prob-

References p. 170–173

ably functioning not only in infantile and juvenile stages but throughout life. On morphological grounds, the mammalian pineal can be considered an endocrine organ. Its products are synthesized in secretory pineal-specific cells, the pinealocytes, and released into the blood of the pineal capillaries which drain into the large intracranial veins. It has been suggested by some authors that pineal products would possibly be released directly into the cerebrospinal fluid. This contention is, however, rather improbable on the basis of pineal morphology. Whether pineal products might enter the cerebrospinal fluid indirectly via the systemic circulation is perhaps possible but has never been proven, so far.

From a functional standpoint, however, we hesitate to consider the mammalian pineal to be an endocrine organ in the classical sense. Truly, removal of the organ often causes effects which can be abolished by administration of pineal extracts. It has, however, been shown that the antigonadotropic effect of administration of pineal compounds depends on several factors such as the age of the animal, its sex, the period of the estrus cycle, conditions of environmental lighting, etc., in other words on the physiological state of the animal at the moment of administration.

From the data now available it appears that, at the input side, pineal function depends on its afferent innervation by the autonomic nervous system, its sympathetic as well as its parasympathetic division. As yet, practically nothing exact is known about the influence exerted by the parasympathic innervation, which has recently been discovered. The important part played by the sympathetic innervation of the pineal in the function of the organ has, however, been demonstrated by many authors. It should be recalled that sympathetic pineal innervation is by postganglionic fibers originating in the superior cervical ganglia which receive preganglionic fibers originating in the rostral part of the intermediolateral nucleus in the spinal cord. It, therefore, stands to reason that a multitude of various stimuli may reach these central sympathetic cells and that these stimuli, of various origin, may influence pineal function via the pre- and post-ganglionic sympathetic fibers. So far, it has been elucidated that light stimuli, or, for that matter, much more so the absence of light stimuli, influence pineal function via the sympathetic innervation. Reiter and collaborators have shown that the same probably holds for olfactory stimuli (Reiter *et al.*, 1971). Quite apart from the neural input mentioned it would be possible that a hormonal input influences pineal function. As yet, this possibility has not been sufficiently examined.

As far as the output of the mammalian pineal gland is concerned it is now well known that many compounds are synthesized in the organ. Several of these have been regarded as pineal agents causing the characteristic pineal effects. Especially melatonin, which has long been thought to be exclusively produced in the gland and which thus has the status of a pineal hormone, has been considered the pineal agent "par excellence". As melatonin is now known also to be produced in the retina it can no longer be regarded as a pineal-specific hormone. It has, moreover, been shown that melatonin is responsible for only one of the antigonadotropic effects of the pineal — inhibiting pituitary LH production. Other pineal indole derivatives have likewise been shown to have antigonadotropic effects. Recently, polypeptides stand in the center of interest as pineal agents (*e.g.* Ebels *et al.*, 1963). So far, however, the exact chemical composition of these proteins is not exactly known while there is, as yet, no certainty that they are exclusively produced by the pineal organ. It, therefore, appears that to date there is not a single pineal compound fulfilling all of the criteria of a pineal hormone, *i.e.* a pineal substance being exclusively produced in the gland and having a specific functional influence on a specific target organ.

Concerning the target organs of the pineal agents, there is, likewise, no final certainty. In the present paper, we have summarily dealt with the antigonadotropic function of the gland exclusively. It is, however, known that the organ also contains a gonadotropic factor while, moreover, it does not only influence the hypothalamo–hypophyseo–gonadal axis. A pineal effect *i.a.* on the adrenal cortex and the thyroid and parathyroid glands has also been described (see the excellent review on endocrine pineal aspects by Reiter and Fraschini, 1969). This means that the pineal produces substances having a contrasting activity and that, moreover, there would not be one single target organ but several. How far an effect of the pineal on a single target organ could influence the function of several other organs, *i.e.* by their functional interrelationship, is a question to which more attention should be paid. At least as far its antigonadotropic influence is concerned, it appears that, under normal functional conditions, the pineal is active at the hypothalamic level rather than at the levels of the pituitary or of the gonads. The influence of the pineal on adrenal cortical and on thyroid function could also point to the hypothalamus as a more generalized target center of the pineal compounds. That, indeed, the pineal gland exerts its function on hypothalamic centers involved in the magno- and parvocellular neurosecretory hypothalamic systems has been proved by the investigations surveyed in the present paper.

Wurtman (1964), in a paper dealing with the significance of melatonin as a hormone, mentioned already that the pineal is not absolutely necessary for life and that, although there is a disturbance in the estrus cycle after pinealectomy, cycles do continue, the animals go on ovulating, copulating, showing gain in weight, and living a practically normal life-span. This has since been corroborated by other authors. Evidently, after some time a new balance is established leading to a normal functioning of the functional systems originally disturbed. It would, therefore, indeed seem that the pineal is "not a prime mover" (Wurtman, 1964) but a "regulator of regulators" (Reiter and Hester, 1966), or an important center for general homeostasis (Quay, 1956b, 1972), very probably exerting its influence primarily on that most important center of integration of the vegetative as well as of the cerebrospinal nervous system: the hypothalamus.

SUMMARY

A general introduction surveying the essentials of mammalian pineal structure, innervation, biochemistry and function is given in this article. On morphological grounds, the mammalian pineal can be considered an endocrine organ. From a functional standpoint, however, we hesitate to consider the mammalian pineal to be an endocrine organ in the classical sense, because till now, there is not a single pineal

References p. 170–173

compound fulfilling all of the criteria of a pineal hormone, *i.e.* a pineal substance being exclusively produced in the gland and having a specific functional influence on a specific target organ.

In the present paper, we have also summarily dealt with the antigonadotropic function of the gland, possibly stimulating serotonergic neurons present in the arcuate and ventromedial nuclei which may inhibit the production of LH-RF in the same nucleus. The gland also exerts an influence on the supraoptic and paraventricular hypothalamic nuclei, pinealectomy decreasing the neurosecretory activity of these nuclei. From these data, it appears that the pineal gland exerts its function on hypothalamic centers involved in the magno- and parvocellular neurosecretory hypothalamic systems.

REFERENCES

General surveys

Structure and Function of the Epiphysis Cerebri, *Progr. Brain Res.*, *Vol. 10* (1965) J. Ariëns Kappers and J. P. Schadé (Eds.).

Wurtman, R. J., Axelrod, J. and Kelly, D. E. (1968) *The Pineal*, Academic Press, New York.

The Pineal Gland. A CIBA Foundation Symposium (1971) G. E. W. Wolstenholme and J. Knight (Eds.), Churchill Livingstone, Edinburgh.

Arvy, L. (1963) *Histo-Enzymologie des Glandes Endocrines*. Gauthier-Villars, Paris.

Bartheld, F. von, and Moll, J. (1954) The vascular system of the mouse epiphysis with remarks on the comparative anatomy of the venous trunks in the epiphyseal area. *Acta anat. (Basel)*, **22**, 227–235.

Bayerova, G. und Bayer, A. (1967) Beitrag zur Fermenthistochemie der menschlichen Epiphyse. *Acta histochem. (Jena)*, **28**, 169–173.

Benson, B., Matthews, M. J. and Orts, R. J. (1972b) Presence of an antigonadotropic substance in rat pineal incubation media. *Life Sci.*, **11**, 669–677.

Benson, B., Matthews, M. J. and Rodin, A. E. (1971) A melatonin-free extract of bovine pineal with antigonadotropic activity. *Life Sci.*, **10**, 607–612.

Benson, B., Matthews, M. J. and Rodin, A. E. (1972a) Studies on a non-melatonin pineal antigonadotrophin. *Acta endocr. (Kbh.)*, **69**, 257–266.

Cardinali, D. P. and Rosner, J. M. (1971) Retinal localization of the hydroxyindole-O-methyl transferase (HIOMT) in the rat. *Endocrinology*, **89**, 301–303.

Cheesman, D. W. (1970) Structural elucidation of a gonadotropin-inhibiting substance from the bovine pineal gland. *Biochim. biophys. Acta (Amst.)*, **207**, 247–253.

Cheesman, D. W. and Fariss, B. L. (1970) Isolation and characterization of a gonadotropin-inhibiting substance from the bovine pineal gland. *Proc. Soc. exp. Biol. (N.Y.)*, **133**, 1254–1256.

Clementi, F., De Virgiliis, G. and Mess, B. (1969) Influence of pineal gland principles on gonadotrophin-producing cells of the rat anterior pituitary gland: an electron microscopic study. *J. Endocrinol.* **44**, 241–246.

Clementi, F., De Virgiliis, G., Fraschini, F. and Mess, B. (1966) Modifications of pituitary morphology following pinealectomy and the implantation of the pineal body in different areas of the brain. *Proc. VIth Internat. Congr. Electr. Microsc.*, *Kyoto*, Maruzen, Tokyo, pp. 539–540.

Delvigs, P., McIsaac, W. M. and Taborsky, R. G. (1965) The metabolism of 5-methoxytryptophol. *J. biol. Chem.*, **240**, 348–350.

De Vries, R. A. C. (1972a) *Some Aspects of the Functional Relationship of the Rat Pineal Gland to the Hypothalamic Magnocellular Neurosecretory Nuclei under Different Experimental Conditions*, Med. Thesis, Univ. of Amsterdam, Nooy, Purmerend.

De Vries, R. A. C. (1972b) Influence of pinealectomy on hypothalamic magnocellular neurosecretory activity in the female rat during normal light conditions, light-induced persistent oestrus, and after gonadectomy. *Neuroendocrinology*, **9**, 244–249.

DE VRIES, R. A. C. AND KAPPERS, J. ARIËNS (1971) Influence of the pineal gland on the neurosecretory activity of the supraoptic hypothalamic nucleus in the male rat. *Neuroendocrinology*, **8**, 359–366.

EBELS, I. (1967) Étude chimique des extraits épiphysaires fractionnés. *Biol. Méd. (Paris)*, **56**, 395–402.

EBELS, I., MOSZKOWSKA, A. ET SCEMAMA, A. (1965) Étude *in vitro* des extraits épiphysaires fractionnés. Résultats préliminaires. *C. R. Acad. Sci. (Paris)*, **260**, 5126–5129.

ELLIS, L. (1969) The direct action of melatonin and serotonin on testicular androgen production *in vitro*. *J. Reprod. Fert.*, **18**, 159.

FARRELL, G., MCISAAC, W. M. AND POWERS, D. (1966) *Program 48th Meeting Endocrine Society*, Chicago, Ill., p. 98.

FRASCHINI, F., COLLU, R. AND MARTINI, L. (1971) Mechanisms of inhibitory action of pineal principles on gonadotropin secretion. In *The Pineal Gland. A CIBA Foundation Symposium*, G. E. W. WOLSTENHOLME AND J. KNIGHT (Eds.), Churchill Livingstone, Edinburgh, pp. 259–273.

FRASCHINI, F., MESS, B. AND MARTINI, L. (1968) Pineal gland, melatonin and the control of luteinizing hormone secretion. *Endocrinology*, **82**, 919–924.

FRASCHINI, F., MESS, B., PIVA, F. AND MARTINI, L. (1969) Brain receptors sensitive to indole compounds: function in control of luteinizing hormone secretion. *Science*, **159**, 1104–1105.

GIARMAN, N. J. AND DAY, M. (1958) Presence of biogenic amines in the bovine pineal body. *Biochem. Pharmacol.*, **1**, 235.

GIARMAN, N. J., FREEDMAN, D. X. AND PICARD-AMI, L. (1960) Serotonin content of the pineal glands of man and monkey. *Nature (Lond.)*, **186**, 480–481.

JOUAN, P. ET ROCABOY, J.-C. (1966) Étude de l'activité peptidasique de la glande pinéale du porc. *C. R. Soc. Biol. (Paris)*, **160**, 859.

KAPPERS, J. ARIËNS (1960) The development, topographical relations and innervation of the epiphysis cerebri in the albino rat. *Z. Zellforsch.*, **52**, 163–215.

KAPPERS, J. ARIËNS (1965) Survey of the innervation of the epiphysis cerebri and the accessory pineal organs of vertebrates. In *Structure and Function of the Epiphysis Cerebri, Progr. Brain Res. Vol. 10*, J. ARIËNS KAPPERS AND J. P. SCHADÉ (Eds.), Elsevier, Amsterdam, pp. 87–153.

KAPPERS, J. ARIËNS (1969) The mammalian pineal organ. *J. Neurovisc. Rel.*, **Suppl. IX**, 40–184.

KAPPERS, J. ARIËNS (1971a) The pineal gland: an introduction. In *The Pineal Gland. A CIBA Foundation Symposium*, G. E. W. WOLSTENHOLME AND J. KNIGHT (Eds.), Churchill Livingstone, Edinburgh, pp. 3–34.

KAPPERS, J. ARIËNS (1971b) Innervation of the pineal organ: phylogenetic aspects and comparison of the neural control of the mammalian pineal with that of other neuroendocrine systems. In *Subcellular Organization and Function in Endocrine Tissues*, H. HELLER AND K. LEDERIS (Eds.), *Mem. Soc. Endocrinol.*, **19**, 27–47.

KAPPERS, J. ARIËNS (1973) The epiphysis cerebri. *Organorama*, **10**/3, 3–14.

KENNY, G. C. T. (1961) The "nervus conarii" of the monkey. An experimental study. *J. Neuropath. exp. Neurol.*, **20**, 563–570.

KENNY, G. C. T. (1965) The innervation of the mammalian pineal body (a comparative study). *Proc. Austral. Ass. Neurologists*, **3**, 133–141.

KOE, K. AND WEISSMAN, A. (1966) p-Chlorophenylalanine: a specific depletor of brain serotonin. *J. Pharmacol. exp. Ther.*, **154**, 499–516.

LERNER, A. B. AND CASE, J. D. (1960) Melatonin. *Fed. Proc.*, **19**, 590–592.

LERNER, A. B., CASE, J. D. AND HEINZELMAN, R. V. (1959) Structure of melatonin. *J. Amer. Chem. Soc.*, **81**, 6084.

LERNER, A. B., CASE, J. D., TAKAHASHI, Y., LEE, T. H. AND MORI, W. (1958) Isolation of melatonin, the pineal gland factor that lightens melanocytes. *J. Amer. Chem. Soc.*, **80**, 2587.

MACHADO, A. B. M., FALEIRO, L. C. M. AND DIAS DA SILVA, W. (1965) Study of mast cell and histamine contents of the pineal body. *Z. Zellforsch.*, **65**, 521–529.

MARTINI, L., FRASCHINI, F. AND MOTTA, M. (1968) Neural control of anterior pituitary functions. In *Recent Progress in Hormone Research, Vol. 24, V. Peptide Hormones*, E. D. ASTWOOD (Ed.), Acad. Press, New York, pp. 439–496.

MATTHEWS, M. J., BENSON, B. AND RODIN, A. (1971) Antigonadotropic activity in a melatonin-free extract of human pineal glands. *Life Sci.*, **10**, 1375–1379.

MESS, B., ZANISI, M. AND TIMA, L. (1970) Site of production of releasing and inhibiting factors. In *The Hypothalamus*, L. MARTINI, M. MOTTA AND F. FRASCHINI (Eds.), Academic Press, New York, pp. 259–276.

McIsaac, W. M., Taborsky, R. G. and Farrell, G. (1964) 5-Methoxytryptophol: effect on estrus and ovarian weight. *Science*, **145**, 63–64.

Milcu, S. M., Pavel, S. and Neascu, C. (1963) Biological and chromatographic characterization of a polypeptide with pressor and oxytocic activities isolated from bovine pineal gland. *Endocrinology*, **72**, 563–566.

Miline, R., Stern, P. et Huković, S. (1959) Sur la présence de la sérotonine dans la glande pinéale. *Bull. Sci.*, **4**, 75.

Moore, R. Y., Heller, A., Bhatnager, R. K., Wurtman, R. J. and Axelrod, J. (1968) Central control of the pineal gland: visual pathways. *Arch. Neurol. (Chic.)*, **18**, 208–218.

Moszkowska, A. (1967) Étude des extraits épiphysaires fractionnés–Physiologie. *Biol. Méd. (Paris)*, **56**, 403–412.

Moszkowska, A. and Ebels, I. (1968) A study of the antigonadotrophic action of synthetic arginine vasotocin. *Experientia (Basel)*, **24**, 610–611.

Moszkowska, A., Kordon, C. and Ebels, I. (1971) Biochemical fractions and mechanisms involved in the pineal modulation of pituitary gonadotropin release. In *The Pineal Gland. CIBA Foundation Symposium*. G. E. W. Wolstenholme and J. Knight (Eds.), Churchill Livingstone, Edinburgh, pp. 241–255.

Moszkowska, A. and Ebels, I. (1970) The influence of the pineal body on the gonadotropic function of the hypophysis. *J. Neuro-Visc. Rel.*, **Suppl. 10**, 160–176.

Motta, M., Fraschini, F. and Martini, L. (1967) Endocrine effects of pineal gland and of melatonin. *Proc. Soc. exp. Biol. (N.Y.)*, **126**, 431–435.

Niemi, M. and Ikonen, M. (1960) Histochemical evidence of amino peptidase activity in rat pineal gland. *Nature (Lond.)*, **185**, 928.

Pastori, G. (1928) Über Nervenfasern und Nervenzellen in der "Epiphysis cerebri". *Z. ges. Neurol. Psychiat.*, **117**, 202–211.

Pastori, G. (1930) Ein bis jetzt noch nicht beschriebenes sympathisches Ganglion und dessen Beziehungen zum Nervus conari sowie zur Vena magna Galeni. *Z. ges. Neurol. Psychiat.*, **123**, 81–90.

Pavel, S. (1965) Evidence for the presence of lysine vasotocin in the pig pineal gland. *Endocrinology*, **77**, 812–817.

Pavel, S. and Petrescu, S. (1966) Inhibition of gonadotrophin by a highly purified pineal peptide and by synthetic arginine vasotocin. *Nature (Lond.)*, **212**, 1054.

Pellegrino de Iraldi, A. and Zieher, L. M. (1966) Noradrenaline and dopamine content of normal, decentralized and denervated pineal gland in rat. *Life Sci.*, **5**, 149–154.

Pun, J. Y. and Lombrozo, L. (1964) Microelectrophoresis of brain and pineal proteins in polyacrylamide gel. *Analyt. Biochem.*, **9**, 9–20.

Quay, W. B. (1961) Reduction of mammalian pineal weight and lipid during continuous light. *Gen. comp. Endocrinol.*, **I**, 211–217.

Quay, W. B. (1962) Metabolic and cytologic evidence of pineal inhibition by continuous light. *Amer. Zool.*, **2**, 550.

Quay, W. B. (1963) Cytologic and metabolic parameters of pineal inhibition by continuous light in the rat, *Rattus norvegicus*, *Z. Zellforsch.*, **60**, 479–490.

Quay, W. B. (1965a) Indole derivatives of pineal and related neural and retinal tissues. *Pharmacol. Rev.*, **17**, 321–345.

Quay, W. B. (1965b) Experimental evidence for pineal participation in homeostasis of brain composition. In *Structure and Function of the Epiphysis Cerebri*, *Progr. Brain Res.*, *Vol. 10*, J. Ariëns Kappers and J. P. Schadé (Eds.), pp. 646–653.

Quay, W. B. (1972) Pineal homeostatic regulation of shifts in the circadian activity rhythm during maturation and aging. *Trans. N. Y. Acad. Sci. ser. 11*, **34**, 239–254.

Reiter, R. J. and Fraschini, F. (1969) Endocrine aspects of the mammalian pineal gland. *Neuroendocrinology*, **5**, 219–255.

Reiter, R. J. and Hester, R. J. (1966) Neuroendocrinological interrelationships. In *Metabolic Regulation of Physiological Activity*, B. Sacktor, R. J. Reiter, J. E. Wilson, H. J. Smith, C. G. Tiekert and R. J. Hester (Eds.), Medical Research Laboratory, Research Laboratories US Army, Edgewood Arsenal, Md., pp. 13–18.

Reiter, R. J., Sorrentino, S. and Jarrow, E. L. (1971) Central and peripheral neural pathways necessary for pineal function in the adult female rat. *Neuroendocrinology*, **8**, 321–333.

Romijn, H. J. (1972) *Structure and Innervation of the Pineal Gland of the Rabbit*, *Oryctolagus cuniculus*

(L.), *with some Functional Considerations. A Light and Electron Microscopic Investigation*, Biol. Thesis, Free Univ., Amsterdam, Nooy, Purmerend.

ROMIJN, H. J. (1973a) Parasympathetic innervation of the rabbit pineal gland. *Brain Res.*, **55**, 431–436.

ROMIJN, H. J. (1973b) Structure and innervation of the pineal gland in the rabbit, *Oryctolagus cuniculus* (L.), I. A light microscopic investigation. *Z. Zellforsch.*, **139**, 473–485.

ROMIJN, H. J. (1973c) Structure and innervation of the pineal gland of the rabbit, *Oryctolagus cuniculus* (L.), II. An electron microscopic investigation of the pinealocytes. *Z. Zellforsch.*, **141**, 545–560.

SMITH, A. R. (1971) The topographical relations of the rabbit pineal gland to the large intracranial veins. *Brain Res.*, **33**, 339–348.

SMITH, A. R. (1972) *Conditions Influencing the Serotonin and Tryptophan Metabolism in the Epiphysis Cerebri of the Rabbit; a Fluorescence Histochemical, Microchemical and Electrophoretic Study*. Med. Thesis, Univ. of Amsterdam, Nooy, Purmerend.

SMITH, A. R., JONGKIND, J. F. AND KAPPERS, J. ARIËNS (1972a) Distribution and quantification of serotonin-containing and autofluorescent cells in the rabbit pineal organ. *Gen. comp. Endocr.*, **18**, 364–371.

SMITH, A. R., KAPPERS, J. ARIËNS AND JONGKIND, J. F. (1972b) Alterations in the distribution of yellow fluorescing rabbit pinealocytes produced by p-chlorophenylalanine and different conditions of illumination. *J. Neural Transm.*, **33**, 91–111.

THIEBLOT, L. (1965) Physiology of the pineal body. In *Structure and Function of the Epiphysis Cerebri, Progr. Brain Res., Vol. 10*, J. ARIËNS KAPPERS AND J. P. SCHADÉ (Eds.), Elsevier, Amsterdam, pp. 479–488.

THIEBLOT, L. ET BLAISE, S. (1963) Influence de la glande pinéale sur les gonades. *Ann. Endocr. (Paris)*, **24**, 270–286.

THIEBLOT, L. ET MENIGOT, M. (1971) Acquisitions récentes sur le facteur antigonadotrope de la glande pinéale. *J. Neuro-Visc. Rel.*, **Suppl. 10**, 153–159.

WURTMAN, R. J. (1964) In *Some Clinical, Biochemical, and Physiological Actions of the Pineal Gland.* Combined clinical staff conference at the National Institutes of Health. *Ann. Intern. Med.*, **61**, 1153–1161.

ZIEHER, L. M. AND PELLEGRINO DE IRALDI, A. (1966) Central control of noradrenaline content in rat pineal and submaxillary glands. *Life Sci.*, **5**, 155–161.

DISCUSSION

(Because the questions were all concerned with the part of the paper referring to the work of Smith, Kappers invited him to answer them.)

VAN DER SCHOOT: From your presentation on the distribution of autofluorescent cells in the pineal during various experimental conditions, it was somewhat unexpected that you could produce a distribution that was exactly that of the control levels, by different pineal extracts. Could this point to an all-or-none effect of your extracts? My second question concerns the measurements of gonadotropic hormone levels during all the experimental circumstances that affect the hypothalamo–hypophyseal–gonadal system.

SMITH: We did not measure gonadotropic hormone levels. The dose of pineal extracts we used was the same as Moszkowska applied. It struck us, too, that sheep pineal extracts did exactly the same as rat pineal extracts. In the literature, data are presented about bovine pineal extracts injected in rats. The extracts are apparently not species-specific.

HERBERT: Do you realize that you cannot just take sheep pineal extracts, because there are considerable differences in the extracts throughout the breeding season.

EBELS: We have studied the sheep pineal gland, and last year we found an antigonadotropic activity in the peptide fraction. We could detect this activity very well in winter pineals, whereas in summer pineals it was very difficult to detect.

SMITH: We used the pineal extracts of winter pineals for our study.

PILGRIM: I would like to ask a question about the nature of the autofluorescent cells in the hypothalamus. You will of course find small autofluorescent granules in almost every nerve cell. As to the very bright autofluorescent granules you showed, I wonder whether they are not present in perivascular glial cells instead of nerve cells. In one picture, Prof. Kappers showed a very bright autofluorescent cell, that looked exactly like a perivascular glial cell. I also wonder whether in the pineal the autofluorescent cells should not be a special type of cells instead of pinealocytes, because I noticed that the localization of these cells was again perivascular.

SMITH: As the hypothalamus is concerned, we did exclude the possibility that the autofluorescent cells would be glial elements by the application of the Cajal staining technique for gliocytes. On the ground of the position and the shape of the autofluorescent cells when compared with the gliocytes it was quite evident that the cells were, indeed, neurons. In the pineal, the perivascular position of the autofluorescent cells is quite characteristic of the dark pinealocytes. In rabbit, the few astrocytes present are never situated perivascularly.

The Influence of Pituitary Peptides on Brain Centers Controlling Autonomic Responses

B. BOHUS

Rudolf Magnus Institute for Pharmacology, Utrecht (The Netherlands)

INTRODUCTION

The study of the adaptation of the organism to changes of the environment is a fascinating but complex subject. The organism's actual response and the subsequent adjustment to environmental stimuli consist of a chain of physiological and behavioral events. A common aspect of these responses is that the central nervous system plays a key role in the control of all these processes and the hypothalamus is of specific importance in the chain between higher nervous structures and the periphery. The organization of the responses, therefore, may already be integrated on the level of the central nervous system, but peripheral interactions contribute to the final response patterns.

Endocrine mechanisms represent an important integrative part in adaptive processes. The release of pituitary–adrenal system hormones or hormones of intermediate or posterior pituitary origin (MSH, vasopressin) in response to environmental changes and their effect on peripheral metabolic processes is one of the most basic phenomena in adaptation. However, the central nervous system should also be considered as a target organ of these hormones. The effect of pituitary–adrenal system hormones, MSH and vasopressin on the brain involves feedback control of the release of pituitary tropic hormones, general metabolic functions, but also a profound influence on adaptive behavior (De Wied, 1969; Bohus, 1970; 1973; Kastin *et al.*, 1973; De Wied *et al.*, 1972a; De Wied *et al.*, 1974, this volume). The behavior is but one aspect of the adaptation of the organisms. Therefore, it was deemed of interest to investigate the influence of pituitary hormones on the organization of physiological responses related to behavioral adaptation. It was observed that pituitary–adrenal hormones and vasopressin profoundly influence the extinction of a classically conditioned heart rate response. The influence of ACTH on the conditioned heart rate response is of an extra-adrenal nature, since $ACTH_{4-10}$, a fragment of the ACTH molecule practically devoid of adrenal stimulating effect, bears the same influence as the whole ACTH molecule (Bohus *et al.*, 1970, 1971; Bohus, 1973). The pattern of heart rate changes during passive avoidance behavior is also affected by peptides of pituitary origin, like vasopressin and $ACTH_{4-10}$ (Bohus *et al.*, to be published). This latter study seems to suggest that the cardiovascular effect of the peptides may be a primary

References p. 181–183

influence on the integrated control of the cardiovascular system rather than the consequence of their behavioral action. Therefore, experiments were aimed to investigate whether pituitary peptides in themselves affect the central control of the cardiovascular system.

Based upon the fact that the hypothalamus and brain stem areas are intimately involved in the central control of the cardiovascular system and in the organization of cardiovascular responses during defense reactions (*e.g.* Ingram, 1960; Uvnäs, 1960; Hilton, 1966; Zanchetti, 1970), the influence of peptides on the cardiovascular response to electrical stimulation of these areas was studied in urethane-anesthetized, hypophysectomized female rats. Hypophysectomized rats were used because the endogenous presence of pituitary peptides due to anesthesia might reset the influence of exogenously administered peptides. The removal of the pituitary results in cardiovascular changes (*e.g.* decrease in blood pressure) and a temporary impairment of centrally evoked pressor response occurs which does not last longer than 8–12 days after the operation (Bohus, in press). Accordingly, the experiments were performed 14–28 days after hypophysectomy.

MATERIAL AND METHOD

A total of 98 female albino rats of a Wistar strain, weighing 130–140 g, were used. The rats were hypophysectomized through transauricular approach and kept at a room temperature of 24 °C.

Blood pressure response to brain stem or hypothalamic stimulation was studied 14–28 days after hypophysectomy. The rats were anesthetized with urethane (1.5 g/kg i.p.); the left carotid artery was then cannulated and connected to a pressure transducer. The animal's head was fixed in a stereotaxic apparatus, and a bipolar electrode (0.2 mm diameter, insulated except for the tip) was introduced through a trephine hole. Coordinates for the brain stem reticular formation were A = 1.4; L = 1.5; H = 3.5 and for the posterior hypothalamus A = 5.0; L = 1.0; H = 2.5 according to Albe-Fessard *et al.* (1966). Centrally evoked pressor responses were obtained with rectangular, monophasic pulses of 1 msec duration, delivered from a Grass S4 stimulator through a stimulus isolation unit. The train duration was 5 sec, and the interstimulation interval 60 sec. While stimulating the brain stem reticular formation, the threshold to evoke 10 mm Hg mean blood pressure (BP) increase was determined with stimulation frequencies of 10, 20, 30, 50 and 100 Hz by stepwise increase of amplitude, each step being 0.5 V. The posterior hypothalamus was stimulated in a both amplitude and frequency dependent manner. Intensity dependent stimulation was performed with 50 Hz frequency and 1, 2, 3, 4, 5, 7 and 10 V amplitudes. Frequency dependent stimulation was carried out with 10 V intensity and 10, 20, 30, 40 and 50 Hz frequencies. The rectal temperature of the rats was continuously monitored and adjusted when decreased below 37 °C.

The experimental schedule was as follows: the centrally evoked pressor responses were determined with the stimulation protocols described above. Then 0.2 μg/rat synthetic lysine vasopressin or 2 μg/rat $ACTH_{4-10}$ (brain stem stimulations) or 0.05, 0.2 and 1.0 μg/rat lysine vasopressin or 0.05, 0.2 and 0.5 μg/rat desglycinamide-lysine vasopressin was injected intravenously. Saline-injected rats served as controls. The stimulation protocols were repeated twice, 20 and 60 min after the treatment.

At the conclusion of the experiments, the tip of the electrode was marked in the brain with a high intensity DC current. Then the rats were decapitated and the effectiveness of hypophysectomy was checked by inspection of the sellar region and determining endocrine organ weights. Observations on 4 rats were not evaluated because of the incompleteness of hypophysectomy. The brains were removed, fixed in formalin and the placement of electrodes was verified on unstained, frozen sections according to Guzman-Flores *et al.* (1958).

RESULTS

Lysine vasopressin (LVP) and $ACTH_{4-10}$, when given intravenously, increase the threshold of a centrally induced pressor response: higher amounts of current are necessary to evoke 10 mm Hg mean blood pressure (BP) increase by the stimulation of brain stem than in saline treated rats (Fig. 1). The influence of both peptides is stronger at 60 min than at 20 min after the administration and can be observed at each stimulation frequency. The increase in threshold current seems to be independent of the pressor effect of LVP: BP increase due to the amounts of peptide used does not last longer than a few minutes. $ACTH_{4-10}$, on the other hand, does not affect BP *per se*, but does increase the stimulation threshold.

Lysine vasopressin suppresses the pressor responses evoked by electrical stimulation of the posterior hypothalamus. The influence of the peptide is present whether the stimulation is intensity or frequency dependent (Fig. 2). Furthermore, the effect is dose dependent and more pronounced at 60 min than at 20 min after intravenous LVP administration. Each dose of LVP resulted in a pressor response itself, but the BP returned to basal level 12–15 min after the administration of even the highest dose: the suppressive effect of the peptide upon the hypothalamically evoked pressor response appears when the pressor response of LVP has disappeared.

One should be cautious, however, to state that the suppressive effect is independent of the cardiovascular influence of LVP because of its complex nature. Besides increasing BP, vasopressin reduces the cardiac output and the coronary blood flow (Longo *et al.*, 1964; Maxwell, 1964; Ericsson, 1971), but increases the cardiac output fraction and the blood flow of the brain (Ericsson, 1971). Therefore, the effect of LVP was compared with that of desglycinamide-lysine vasopressin (DG-LVP). This peptide was isolated from hog pituitaries and, like LVP, is capable of restoring the deficient learning behavior of hypophysectomized rats (Lande *et al.*, 1973). The behavioral activity of DG-LVP is of a similar magnitude while the pressor, antidiuretic and oxytocic activities are about a hundred times less than that of synthetic LVP (De Wied *et al.*, 1972b).

References p. 181–183

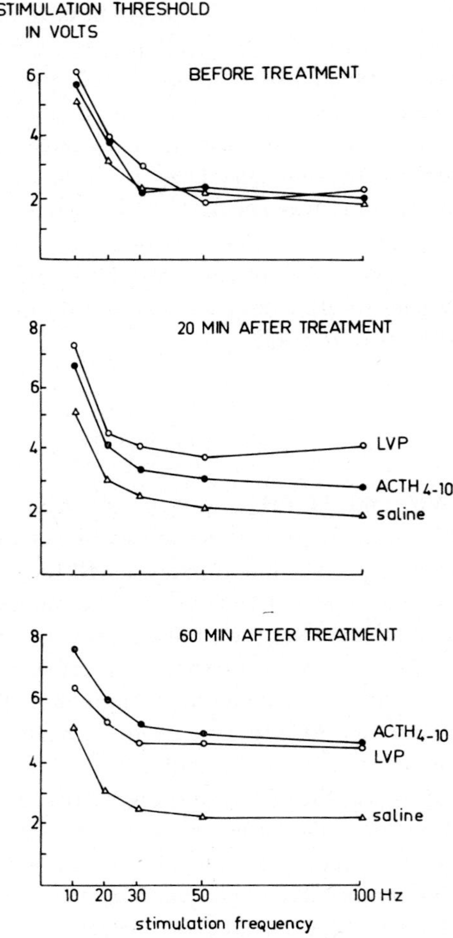

Fig. 1. Changes in the threshold of electrical stimulation of brain stem to evoke 10 mm Hg mean blood pressure increase after intravenous administration of lysine vasopressin (0.2 μg) or $ACTH_{4-10}$ (2.0 μg) in urethane-anesthetized, hypophysectomized female rats. Ten observations per each treatment group.

The influence of DG-LVP on the hypothalamically induced pressor response appeared to be comparable to the effect of LVP (Fig. 3). Both the intensity and frequency dependent BP increase were suppressed in a dose related manner. The effect of this peptide was seen only 60 min after intravenous administration. Accordingly, a vasopressin-like peptide without an obvious cardiovascular effect simulates the influence of LVP on the hypothalamically evoked pressor response.

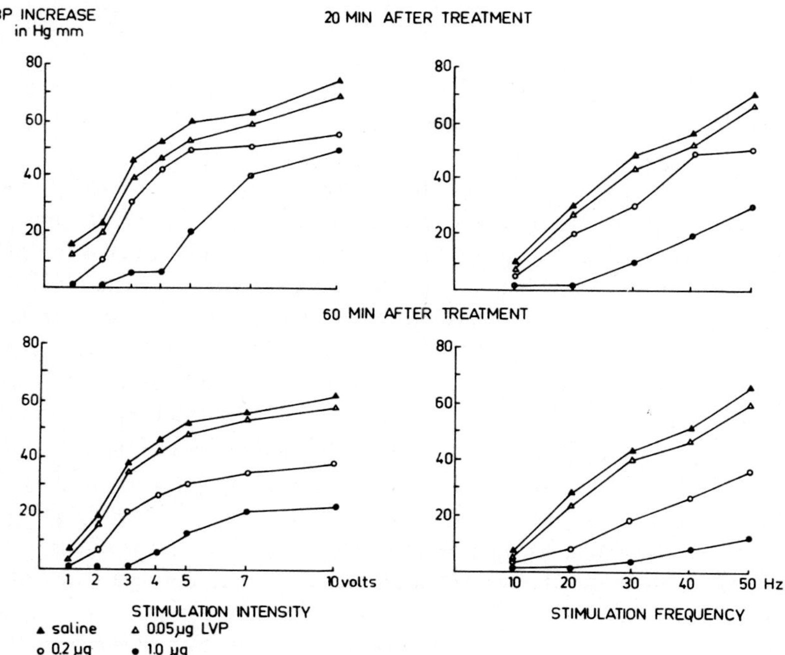

Fig. 2. The effect of lysine vasopressin on the rise in blood pressure by intensity (50 Hz, 1 msec square wave pulses) and frequency (10 V, 1 msec pulses) dependent stimulation of the posterior hypothalamus of urethane-anesthetized hypophysectomized female rats. Eight observations per each treatment group.

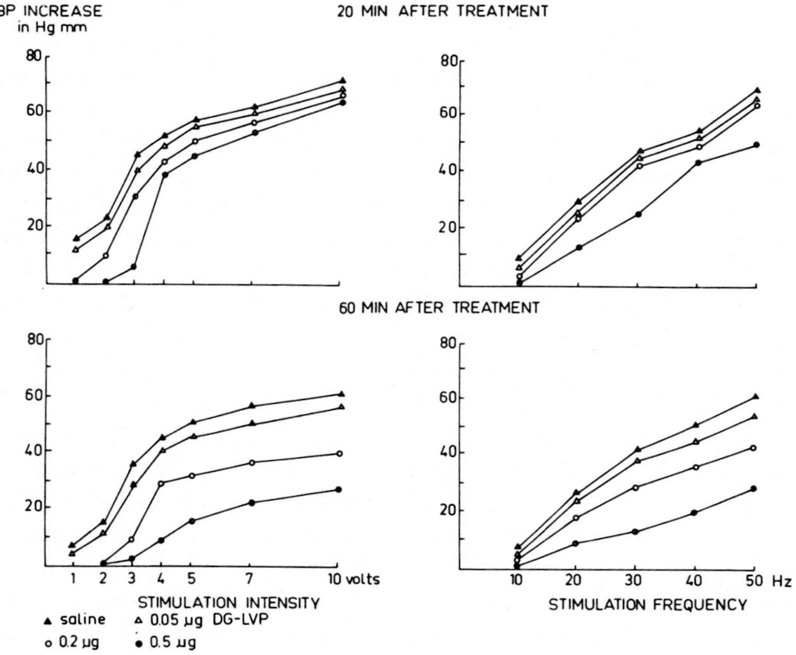

Fig. 3. The influence of desglycinamide-lysine vasopressin (DG-LVP) on the pressor response to intensity and frequency dependent stimulation of the posterior hypothalamus of urethane-anesthetized, hypophysectomized female rats. Eight observations per each treatment group.

References p. 181–183

DISCUSSION

A number of observations indicate that adrenocortical hormones potentiate the pressor effect of endogenously released or exogenously administered catecholamines or the reactivity to carotid occlusion (*e.g.* Cartoni *et al.*, 1969; Kalsner, 1969; Drew and Leach, 1971; Kadowitz and Yard, 1971; Yard and Kadowitz, 1972). Little is known about the circulatory effects of ACTH independent of the adrenal cortex. ACTH analogs, like $ACTH_{1-28}$, $ACTH_{1-24}$, and $ACTH_{5-14}$ increase adrenal and ovarian blood flow, while $ACTH_{5-14}$ is ineffective (Stark *et al.*, 1967; Stark and Varga, 1968; Varga *et al.*, 1969; Stark *et al.*, 1970). Mathieu (1972) reported that ACTH slightly elevates the blood pressure of Nembutal-anesthetized rats. Small doses of ACTH increase, while higher amounts suppress the BP response to noradrenaline. This effect is still present after adrenalectomy which indicates that its influence is extra-adrenal. α-MSH which shares the amino acid chain 1–13 of the ACTH molecule increases the ventricular contractile force and heart rate of anesthetized dogs with no change in BP (Aldinger *et al.*, 1973).

Apart from the pressor effect of vasopressin, a potentiation of the pressor effect of endogenously released or exogenously administered catecholamines has been observed (Traber *et al.*, 1967, 1968). This is also present when non-pressor amounts of vasopressin are infused and the effects last at least 2 h (Bartelstone and Nasmyth, 1965). We have also observed that both LVP and DG-LVP in doses which influence hypothalamically induced BP-rise slightly potentiate the pressor effect of noradrenaline 20 and 60 min after the administration of these peptides. The reports cited above conclude that the circulatory effects of vasopressin are localized on the blood vessels directly.

The present observations suggest that vasopressin and ACTH-like peptides influence the centrally induced pressor response. Since a number of observations suggest a peripheral potentiation of the BP increase by sympathetic stimulation during or after vasopressin infusion (Bartelstone and Nasmyth, 1965; Traber *et al.*, 1967, 1968) or ACTH administration (Mathieu, 1972), it is more probable that the suppression of centrally evoked pressor response is the result of a central action of these peptides on cardiovascular control systems. Central nervous localization of the effect of these peptides on avoidance behavior (Bohus and De Wied, 1967; Van Wimersma Greidanus and De Wied, 1971; Van Wimersma Greidanus *et al.*, 1973) lends support to this suggestion. Furthermore, Unger and Schwarzberg (1970) reported that intracisternal injection of vasopressin increases the respiration frequency and suppresses the spontaneous activity of depressor and laryngeal nerves. Depression of background multi-unit activity in the diencephalon and appearance of EEG spindles have been observed by Beyer *et al.* (1967) after the administration of vasopressin in urethane-anesthetized rats. This effect of the peptide, however, appeared to be related to the blood pressure increase and evoked by the activation of carotid baroreceptors, which in turn produce EEG synchronization. Since DG-LVP, which is practically without pressor activity, suppressed the centrally evoked pressor response as effectively as LVP and the effect of LVP occurs after the disappearance of the pressor effect of this peptide, it is unlikely that a baroreceptor reflex mechanism is responsible for the suppression of centrally evoked pressor response. Therefore, it is suggested that vasopressin and ACTH-like peptides may either suppress pressor, or activate depressor, mechanisms at a supraspinal level which results in a diminished pressor response after brain stem or hypothalamic stimulation. The precise mechanism, the site of action as well as the physiological significance of the phenomenon remain to be solved.

SUMMARY

Peptides related to ACTH or vasopressin influence adaptive behavior of rats. Autonomic correlates of classically conditioned emotional response or of passive avoidance behavior are also affected by these peptides. That is, the influence of peptides includes actions on cardiovascular adaptive responses. Brain centers controlling autonomic responses may be involved in this action.

Electrical stimulation of the brain stem or the posterior hypothalamus of urethane-anesthetized, hypophysectomized female rats produces an increase in blood pressure, the rate of which depends upon the frequency and intensity of the stimulation. Administration of peptides related to vasopressin or ACTH elevates the threshold to evoke blood pressure increase by brain stem stimulation. Pressor response to posterior hypothalamic stimulation is also suppressed by vasopressin-like peptides. These effects seem to be independent of the pressor activity of the peptides. The observations suggest that pituitary hormones may modulate the responsiveness of higher integrating autonomic responses.

REFERENCES

Albe-Fessard, D., Stutinsky, F. et Libouban, S. (1966) *Atlas Stéréotaxique du Diencéphale du Rat Blanc*, Editions du Centre National de la Recherche Scientifique, Paris.

Aldinger, E. E., Hawley, W. D., Schally, A. V. and Kastin, A. J. (1973) Cardiovascular actions of melanocyte-stimulating hormone in the dog. *J. Endocr.*, **56**, 613–614.

Bartelstone, H. J. and Nasmyth, P. A. (1965) Vasopressin potentiation of catecholamine actions in dog, rat, cat, and rat aortic strip. *Amer. J. Physiol.*, **208**, 754–762.

Beyer, C., Ramirez, V. D., Whitmoyer, D. I. and Sawyer, C. H. (1967) Effects of hormones on the electrical activity of the brain in the rat and rabbit. *Exp. Neurol.*, **18**, 313–326.

Bohus, B. and De Wied, D. (1967) Failure of α-MSH to delay extinction of conditioned avoidance behavior in rats with lesions in the parafascicular nuclei of the thalamus. *Physiol. Behav.*, **2**, 221–223.

Bohus, B. (1970) Central nervous structures and the effect of ACTH and corticosteroids on avoidance behaviour: a study with intracerebral implantation of corticosteroids in the rat. In *Pituitary, Adrenal and the Brain, Progr. Brain Res., Vol. 32*, D. de Wied and J. A. W. M. Weijnen (Eds.), Elsevier, Amsterdam, pp. 171–184.

Bohus, B., Grubits, J. and Lissák, K. (1970) Influence of cortisone on heart rate during fear extinction in the rat. *Acta physiol. Acad. Sci. hung.*, **37**, 265–272.

Bohus, B., De Wied, D. and Lissák, K. (1971) Heart rate changes during fear extinction in rats treated with pituitary peptides or corticosteroids. *Proc. Int. Union Physiol. Sci., Vol. IX*, German Physiol. Soc., Munich, p. 72.

Bohus, B. (1973) Pituitary–adrenal influences on avoidance and approach behavior of the rat. In

Drug Effects on Neuroendocrine Regulation, *Progr. Brain Res.*, *Vol. 39*, E. Zimmermann, W. H. Gispen, B. H. Marks and D. de Wied (Eds.), Elsevier, Amsterdam, pp. 407–420.

Bohus, B. (1973) Centrally evoked pressor responses in hypophysectomized rats. *J. Endocr.*, in press.

Cartoni, C., Girgis Awad, A. and Carpi, A. (1969) Effects of aldosterone, dexamethasone, and corticosterone on the cardiovascular reactivity of adrenalectomized rats. *Arch. int. Pharmacodyn.*, **182**, 98–111.

Drew, G. M. and Leach, G. D. H. (1971) Corticosteroids and their effects on cardiovascular sensitivity in the pithed adrenalectomized rat. *Arch. int. Pharmacodyn.*, **191**, 255–260.

Ericsson, B. F. (1971) Hemodynamic effects of vasopressin, *Acta chir. scand.*, Suppl. **414**, 1–29.

Guzman-Flores, C., Alcaraz, M. and Fernandez-Guardiola, A. (1958) Rapid procedure to localize electrodes in experimental neurophysiology. *Bul. Inst. Estud. med. Mex.*, **16**, 29–31.

Hilton, S. M. (1966) Hypothalamic regulation of the cardiovascular system. In *Recent Studies on the Hypothalamus*, *Brit. med. Bull.*, *Vol. 22*, K. Brown-Grant and B. A. Cross (Eds.), pp. 243–248.

Ingram, W. R. (1960) Central autonomic mechanism. In *Handbook of Physiology*, *Section 1: Neurophysiology*, *Vol. II*, J. Field, H. W. Magoun and V. E. Hall (Eds.), American Physiol. Soc., Washington, D.C., pp. 951–978.

Kadowitz, P. J. and Yard, A. C. (1971) Influence of hydrocortisone on cardiovascular responses to epinephrine. *Europ. J. Pharmacol.*, **13**, 281–286.

Kalsner, S. (1969) Mechanism of hydrocortisone potentiation of response to epinephrine and norepinephrine in rabbit aorta. *Circulat. Res.*, **24**, 383–395.

Kastin, A. J., Miller, L. M., Nockton, R., Sandman, C. A., Schally, A. V. and Stratton, L. O. (1973) Behavioral aspects of melanocyte-stimulating hormone (MSH). In *Drug Effects on Neuroendocrine Regulation*, *Progr. Brain Res.*, *Vol. 39*, E. Zimmermann, W. H. Gispen, B. H. Marks and D. de Wied (Eds.), Elsevier, Amsterdam, pp. 461–470.

Lande, S., Witter, A. and De Wied, D. (1973) Pituitary peptides. An octapeptide that stimulates conditioned avoidance acquisition in hypophysectomized rats. *J. biol. Chem.*, **246**, 2058–2062.

Longo, L. D., Morris, J. A., Smith, R. W., Beck, R. and Assali, N. S. (1964) Hemodynamic and renal effects of octapressin. *Proc. Soc. exp. Biol. (N.Y.)*, **115**, 766–770.

Mathieu, F. (1972) Effets de l'ACTH porcine sur l'hypertension artérielle provoquée par la noradrénaline chez le rat. *C.R. Soc. Biol. (Paris)*, **166**, 1394–1396.

Maxwell, G. M. (1964) The cardiovascular effects of octapressin. *Arch. int. Pharmacodyn.*, **158**, 17–23.

Stark, E., Varga, B. and Acs, Zs. (1967) An extra-adrenal effect of corticotrophin. *J. Endocr.*, **37**, 245–252.

Stark, E. and Varga, B. (1968) Effect of ACTH on target-organ blood flow; with special reference to an extra-adrenal effect. *Acta med. Acad. Sci. hung.*, **25**, 367–381.

Stark, E., Varga, B., Medzihradszky, K., Bajusz, S. and Hajtman, B. (1970) Relationships between the structure and the adrenal and extra-adrenal effects of ACTH fragments. *Acta physiol. Acad. Sci. hung.*, **38**, 193–197.

Traber, D. L., Gary, H. H. and Gardier, R. W. (1967) The involvement of the sympathetic nervous system in the pressor response to vasopressin. *Arch. int. Pharmacodyn.*, **168**, 288–295.

Traber, D. L., Wilson, R. D. and Gardier, R. W. (1968) A pressor response produced by the endogenous release of vasopressin. *Arch. int. Pharmacodyn.*, **176**, 360–366.

Unger, H. und Schwarzberg, H. (1970) Untersuchungen über Vorkommen und Bedeutung von Vasopressin und Oxytozin in Liquor cerebrospinalis und Blut für nervöse Funktionen. *Acta biol. med. germ.*, **25**, 267–280.

Uvnäs, B. (1960) Central cardiovascular control. In *Handbook of Physiology. Section 1: Neurophysiology. Vol. II*, J. Field, H. W. Magoun and V. E. Hall (Eds.), Amer. Physiol. Soc., Washington, D.C., pp. 1131–1162.

Varga, B., Stark, E., Csáki, L. and Marton, J. (1969) Effect of ACTH on gonadal blood flow in the golden hamster and the rat. *Gen. comp. Endocr.*, **13**, 468–473.

Wied, D. de (1969) Effects of peptide hormones on behavior. In *Frontiers in Neuroendocrinology*, W. F. Ganong and L. Martini (Eds.), Oxford Univ. Press, New York, pp. 97–140.

Wied, D. de, Delft, A. M. L. van, Gispen, W. H., Weijnen, J. A. W. M., and Wimersma Greidanus, Tj. B. van (1972a) The role of pituitary-adrenal system hormones in active avoidance conditioning. In *Hormones and Behavior*, S. Levine (Ed.), Academic Press, New York, pp. 135–171.

Wied, D. de, Greven, H. M., Lande, S. and Witter, A. (1972b) Dissociation of the behavioural and endocrine effects of lysine vasopressin by tryptic digestion. *Brit. J. Pharmacol.*, **45**, 118–122.

Wied, D. de, Bohus, B. and Wimersma Greidanus, Tj. B. van (1974) The hypothalamo-neurohypophyseal system and the preservation of conditioned avoidance behavior in rats. In *Integrative Hypothalamic Activity, Progr. Brain Res., Vol. 41*, D. F. Swaab and J. P. Schadé (Eds.), Elsevier, Amsterdam, pp. 417–428.

Wimersma Greidanus, Tj. B. van and Wied, D. de (1971) Effects of systemic and intracerebral administration of two opposite acting ACTH-related peptides on extinction of conditioned avoidance behavior. *Neuroendocrinology*, **7**, 291–301.

Wimersma Greidanus, Tj. B. van, Bohus, B. and Wied, D. de (1973) Effects of peptide hormones on behaviour. In *Endocrinology, Proc. 4th Intern. Congr. Endocrin., Int. Congr. Ser. no. 273*, R. Scow (Ed.), Excerpta Medica, Amsterdam, pp. 197–201.

Yard, A. C. and Kadowitz, P. J. (1972) Studies on the mechanism of hydrocortisone potentiation of vasoconstrictor responses to epinephrine in the anesthetized animal. *Europ. J. Pharmacol.*, **20**, 1–9.

Zanchetti, A. (1970) Control of the cardiovascular system. In *The Hypothalamus*, L. Martini, M. Motta and F. Fraschini (Eds.), Academic Press, New York, pp. 233–244.

Ultrastructural Changes in the Hypothalamus during Development and Hypothalamic Activity: the Median Eminence

BARBARA G. MONROE AND WILLIS K. PAULL

Department of Anatomy, University of Southern California, Los Angeles, Calif. 90033, and Department of Anatomy, University of Vermont, Burlington, Vt. 05401 (U.S.A.)

INTRODUCTION

In the intensive investigation of the hypothalamus and particularly of its role in neuroendocrine physiology, the efforts of electron microscopists over the past decade have supplied some of the essential pieces of information. Much of their attention has been focused on the median eminence, which represents the final common pathway by which hypothalamic influences are conveyed to the pituitary. The main features of the ultrastructural organization of this strategic area have been outlined and to some degree correlated with physiological and histochemical data, but many fundamental questions remain unanswered. Some of the problems presently under attack include the categorization of the kinds of axons terminating in the median eminence, the precise intracellular localization of the releasing hormones and of the several monoamines known to be present in the median eminence, the role of the ependyma, and

TABLE I

ELECTRON MICROSCOPIC STUDIES OF THE DEVELOPING MEDIAN EMINENCE (ME)

Investigator	*Tissue*	*Fetal (days)*	*Postnatal days*
		Rat	
1968 Kobayashi *et al.*	ME	20	1–30
1969 Ajika	ME (and NL)	20	1–30
1968 Daikoku *et al.*	ME (and SON)		2 h–5
1971 Daikoku *et al.*	ME–ext.	16–20	5 h–8
1971 Fink and Smith	ME (and NL)	15–20	1–10
1972 Monroe *et al.*	ME		1–6
1972 Halász *et al.*	ME–ext. (and arcuate nuc.)	16–18	
1973 (a and b) Paull	ME (and NL)	15–21	
		Mouse	
1971 Eurenius and Jarskar	ME–ext.	15–19	1–24
1973 Beauvillain	ME	12–18	1–28

References p. 207–208

the possible interplay between the axonal and ependymal components of the median eminence.

Seeking answers to such problems and hoping to provide morphological information to assist in understanding hypothalamo–hypophyseal relations in the fetus and newborn, a number of electron microscopists have recently attempted to study the development and maturation of the median eminence at the ultrastructural level. The supraoptic nucleus, arcuate nucleus and neural lobe of the pituitary were also examined in some instances. Eight of these studies were made on the rat; two on the mouse. These contributions are listed in Table I.

From these reports, and from our own observations, a general outline of events in the development of the median eminence in these animals has emerged. The following review will attempt to summarize these developmental changes, compare the early median eminence with that of adults, and assess the functional capabilities of the developing median eminence in the light of these morphological data.

RESULTS

The early cellular median eminence

The median eminence of both rats and mice begins as an essentially cellular affair devoid of nerve fibers which characterize it in the adult. In the earliest stages studied (rat, 15th fetal day; mouse, 12th fetal day) it is seen to consist of 6–10 layers of cells, organized in the manner of a stratified epithelium. The most dorsal cells line the floor of the third ventricle while the most ventral ones rest on a basal lamina facing the connective tissue where a small number of capillaries are already present. From the data currently available, all of these cells appear to belong to the family of ependymal cells. Mitotic figures are common, especially in the cells facing the ventricle (Fink and Smith, 1971; Paull, 1973a, b; and Beauvillain, 1973).

At the ultrastructural level these ependymal cells have the features typical of immaturity: a rich complement of free ribosomes, some mitochondria, and a notable scarcity of other organelles. However, the cells facing the ventricular lumen have bleb-like projections on their free surfaces and already contact their neighbors with specialized junctions (Figs. 1 and 2).

The appearance of axons

At the light microscopic level the first sign of change from this cellular state is the appearance of a narrow zone of lightly stained tissue along the ventral surface of the median eminence (Fig. 3a). Here, 5 or 6 days prior to birth in both rats and mice, the electron microscope reveals small numbers of clearly identifiable axon profiles (Eurenius and Jarskar, 1971; Fink and Smith, 1971; Beauvillain, 1973; and Paull, 1973a and b). Single axons, or more often small groups of axons, appear to have insinuated themselves between the more ventral portions of the ependymal cells. This

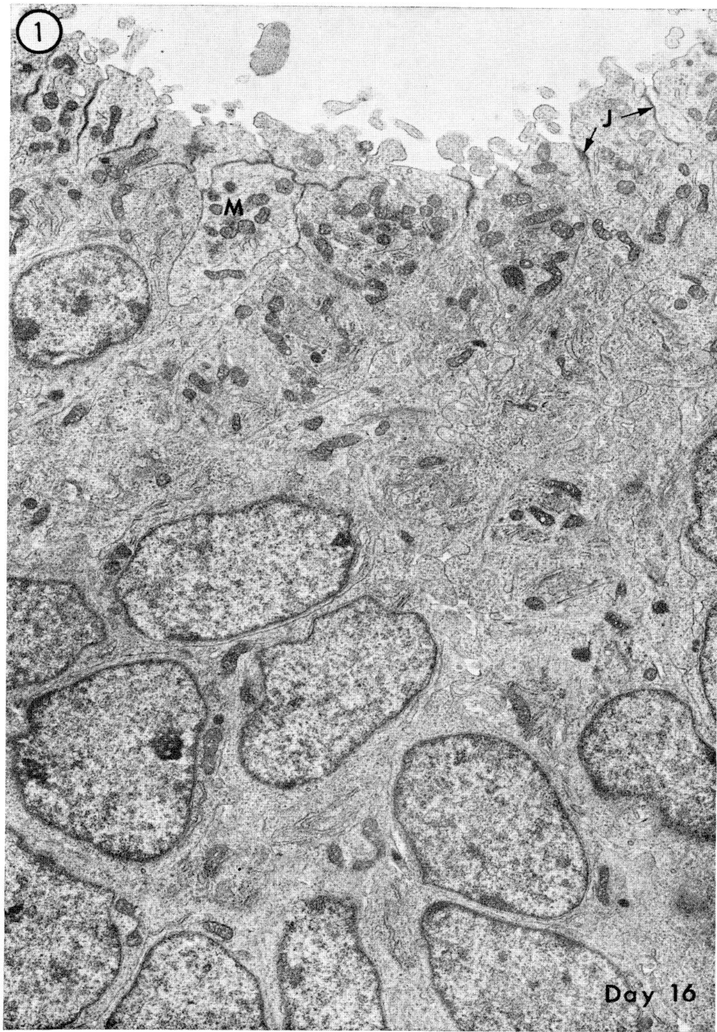

Fig. 1. Dorsal portion of the highly cellular median eminence of the rat on the 16th fetal day. The immature ependymal cells are rich in ribosomes and mitochondria (M). Other organelles are scarce. Cells facing the ventricular cavity have bulbous projections and apical junctions (J). Tissue of this and all other figures was fixed in aldehydes and post fixed with osmium tetroxide. × 6000.

results in the creation of a ventral region which consists of plump cytoplasmic processes of ependymal cells separating small groups of ingrowing axons (Fig. 4). As later events indicate, this may be regarded as the beginning of the external zone of the median eminence. However at this time, and for several days, the ependymal processes remain in contact with each other along the ventral surface, providing a continuous covering for the basal lamina and shielding the axons from it. These processes, like the rest of the ependymal cytoplasm, are characterized chiefly by their ribosomal

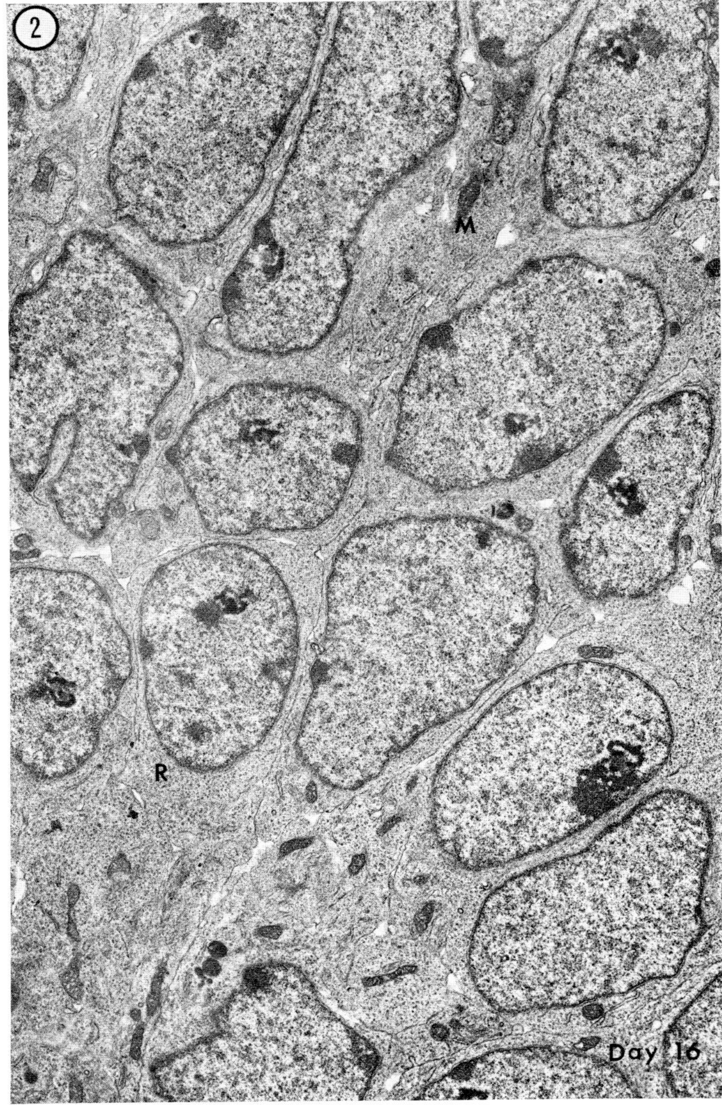

Fig. 2. Ventral portion of the median eminence of the rat on the 16th fetal day. Note the uniform character of the ependymal cells and their similarity to the dorsal ones of Fig. 1. M = mitochondria; R = ribosomes. × 6000.

content although some mitochondria and small amounts of rough endoplasmic reticulum may also be present.

According to our observations in the rat, the axonal elements present at this time (16th fetal day) are all axonal cylinders with growing tips and do not include formed end bulbs (Fig. 4). Growing tips are characterized by the fine filamentous matrix or, less commonly, groups of vesicular structures which have been described by others in

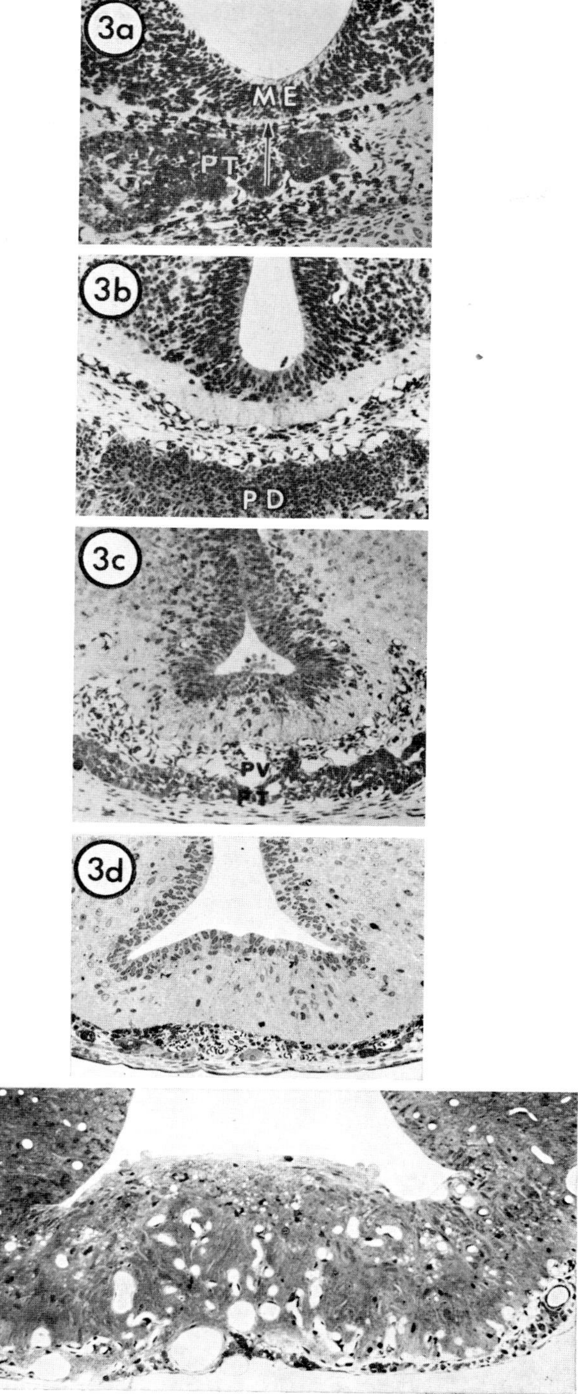

Fig. 3. Light micrographs of frontal, 1 μm sections through the median eminence of rats at levels of the widest lateral recess of the third ventricle: a, b, and c, from fetuses of 16, 18 and 21 days, respectively; d, from a 5 day neonate; e, from an adult. ME, median eminence; PD, pars distalis; PT, pars tuberalis; PV, portal vein.
Note the change in gross dimensions and in proportions of ependymal to pale-staining axonal layers ($\rightarrow$ in a). All $\times$ 100.

References p. 207–208

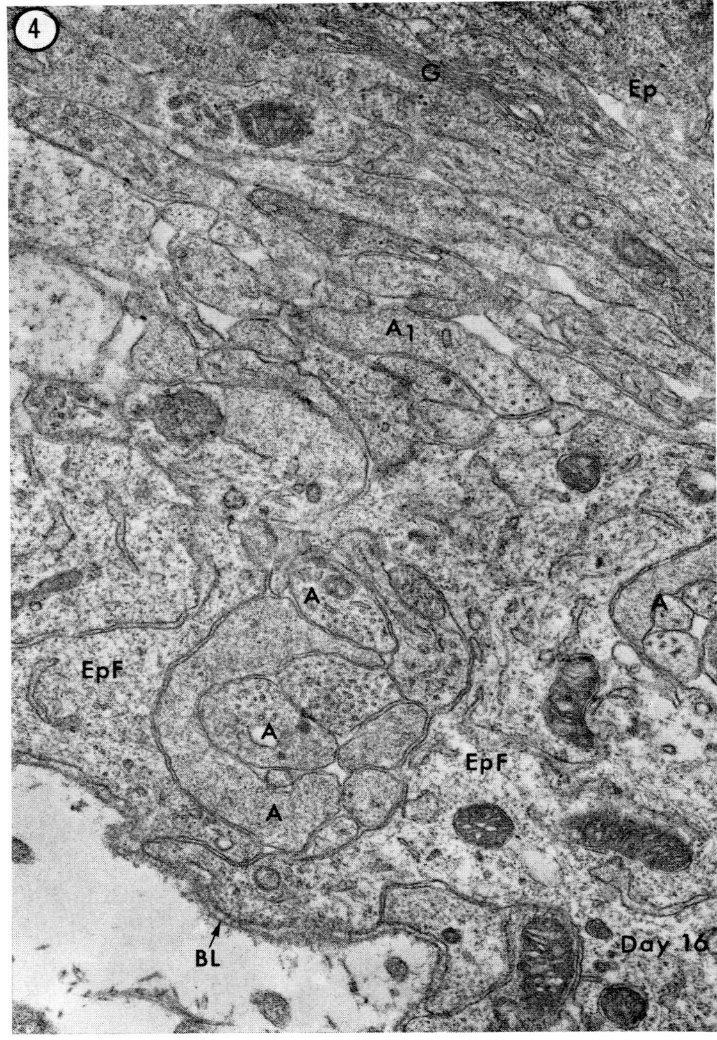

Fig. 4. Ventral surface of the median eminence of the rat on the 16th fetal day. Groups of the first axons (A) lie between plump ependymal foot processes (EpF) which abut on the basal lamina (BL). Axons contain neurotubules or a fine filamentous matrix. One (A_1) displays both. Part of an ependymal cell (Ep) with its Golgi zone (G) marks the dorsal edge of the narrow area of axon ingrowth. Ventrally the basal lamina faces the perivascular space which contains connective tissue elements. × 20,475.

growing axons (Tennyson, 1970; Yamada *et al.*, 1971; and Bunge, 1973) while more proximal regions of axons are recognizable by their content of neurotubules very like those of adult axons. We find neither agranular nor granular vesicles in the axons of this early stage, but others have reported their presence in small numbers (Eurenius and Jarskar, 1971; Fink and Smith, 1971; and Beauvillain, 1973). All agree, however, that true axon terminals have not yet developed.

Period of rapid axon ingrowth

After the initial appearance of axons, during the following two days of fetal life (rat, fetal days 17–18), some important changes in the median eminence may be seen with

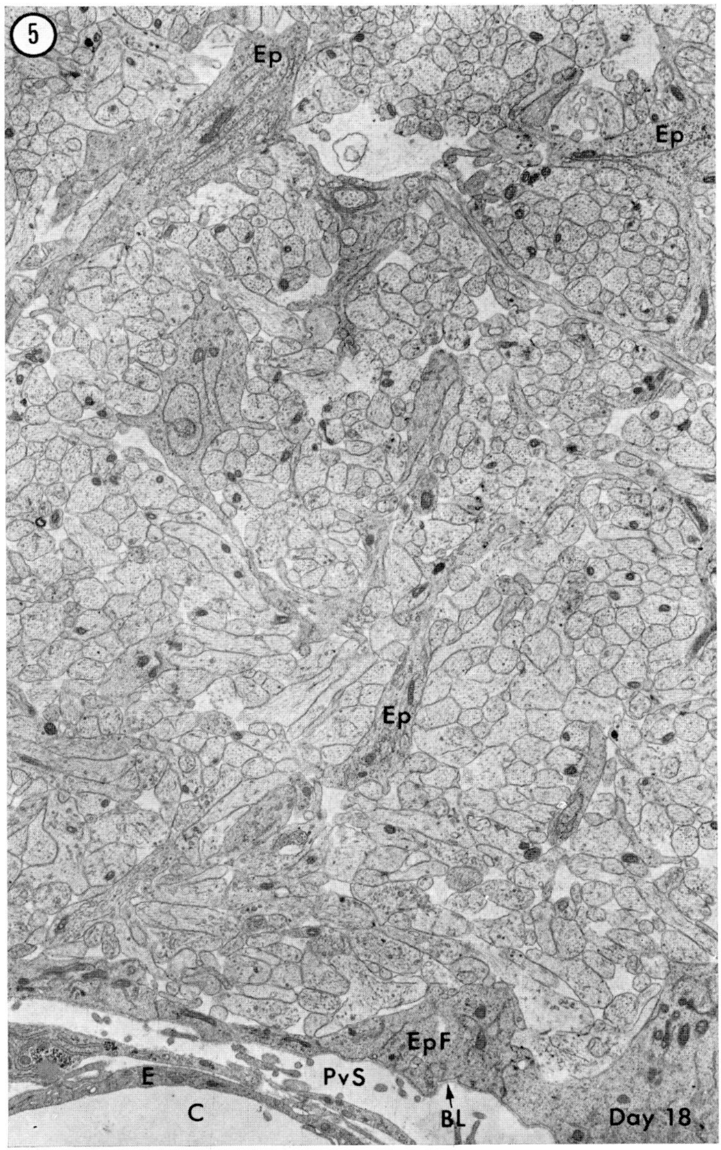

Fig. 5. A portion of the axon rich, ventral half of the median eminence of a rat on the 18th fetal day. Fascicles of many axons are separated by slender, branching ependymal processes (Ep) which end in pedestal-like feet (EpF). These feet and their flat expansions provide a continuous layer between the axons and the basal lamina (BL). C, capillary lumen; E, endothelium; PvS, perivascular space. × 6000.

References p. 207–208

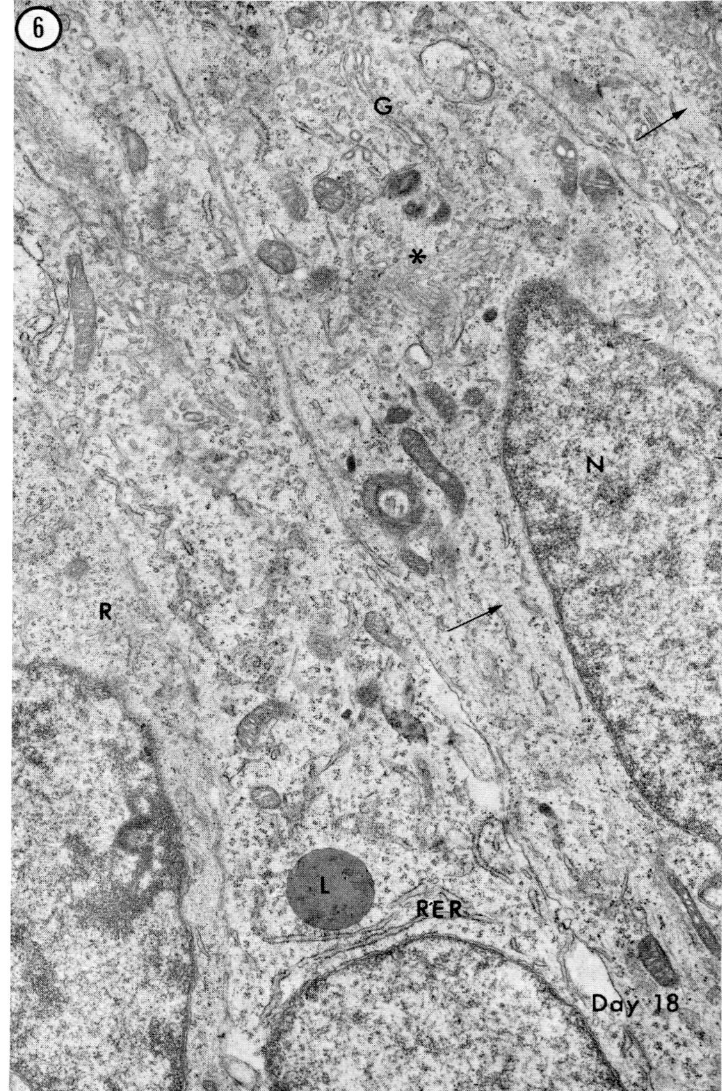

Fig. 6. Typical ependymal cells of the median eminence of the rat on the 18th fetal day. Note the increased variety of organelles: Golgi, G; lipid droplet, L; rough endoplasmic reticulum, RER; tubular and vesicular formations (*), and microfilaments (→). Some cells are still rich in ribosomes (R). Portions of 3 nuclei (N) are shown. × 12,750.

the light microscope (Fig. 3b). One of these is the development of many more primary capillaries distributed all along the ventral surface. The other is a great increase in the thickness of the ventral, pale-staining zone, so that it comes to constitute about one-half of the total depth of the median eminence. Daikoku (1971) also noted this change in the proportions of its layers.

At the ultrastructural level it is readily seen that the increased thickness of the ventral zone is due to the dramatic ingrowth of axons which has occurred in the course of two days. Fascicles of axons, now numbering in dozens, are separated by slender ependymal processes (Fig. 5). The resulting configuration suggests that the axons have crowded in between the ependymal processes, causing them to become stretched out and attenuated. Most ventrally, however, the processes expand into broader, pedestal-like feet which finally terminate in rather flat expansions spread out along the basal lamina. These expansions meet, still forming a continuous layer between the axons and the basal lamina (Fig. 5).

While the vast majority of the axon profiles contain only mitochondria and neurotubules, or the filamentous and vesicular material of growing tips, a few display, for the first time in our material, occasional dense-core vesicles and/or a few lucent vesicles resembling synaptic ones. Such a period of rapid axon ingrowth and consequent increased thickness of the ventral or neural layer has also been referred to by others for both the rat and the mouse (rat: Daikoku, 1971; Fink and Smith, 1971; mouse: Eurenius and Jarskar, 1971).

Simultaneously with the ingrowth of axons, and with great rapidity, the immature ependymal cells of the 16th fetal day elongate and acquire a wealth of new organelles in the course of two days (Fig. 6). Dorsally, where the elongated cells still lie in close array, their cytoplasm now contains well developed Golgi complexes often associated with small vesicles, sometimes coated, and with elaborate tubular formations. In addition there are moderate amounts of rough endoplasmic reticulum and occasional lipid droplets. In some cells, numerous microfilaments are seen. Ribosomes, often in polysomal formation, are still abundant but no longer dominate the scene. As is true of adult ependymal cells of the median eminence, there is considerable cytoplasmic variation from cell to cell and even from one area to another within a single cell.

The establishment of neurohemal contact

According to present concepts, the presence of axon terminals of tuberal neurons on the primary capillaries of the portal system constitutes one of the principal morphological requirements for the exercise of hypothalamic control over the pars distalis. The establishment of this neurohemal contact can therefore be regarded as an event of primary importance in the development of the median eminence.

From the studies to date there is general agreement that this occurs, at least to a limited degree, a few days before birth in both rats and mice (Kobayashi *et al.*, 1968; Ajika, 1969; Eurenius and Jarskar, 1971; Halász *et al.*, 1972; Beauvillain, 1973; and Paull, 1973a, b). At this time axon terminals, defined by their content of small, electron lucent, synaptic-like vesicles, appear among the axonal cylinders and ependymal processes and a few arrive at the basal lamina facing the primary capillaries. A very small number of dense-core (or granulated) vesicles are also seen in most of these terminals.

During the last 2 or 3 days of fetal life the further development of axon endings continues so that by birth a band of tissue about 3–6 μm in depth adjacent to the basal lamina consists of more axon endings than axonal cylinders (Eurenius and Jarskar,

References p. 207–208

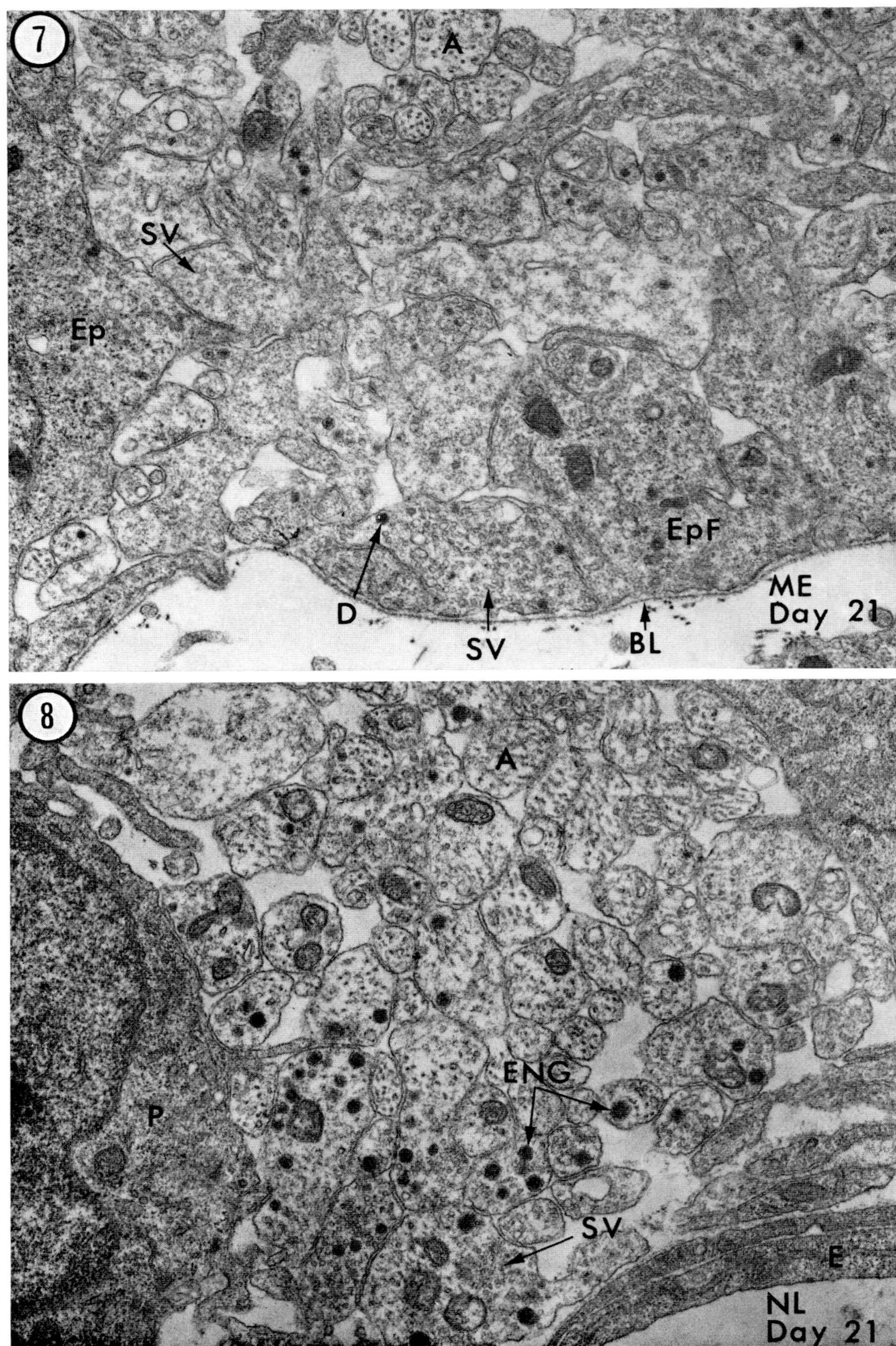
7
A
SV
Ep
EpF
D
SV
BL
ME
Day 21
8
A
P
ENG
SV
E
NL
Day 21

1971; Paull, 1973a, b). At the same time more axons insinuate themselves between the ependymal foot processes to establish endings which contact the basal lamina. Simultaneously both dense-core and lucent vesicles become somewhat numerous. With these events, this region may now be regarded as the early contact zone, or perivascular zone, of the external median eminence (Fig. 7).

In the establishment of this neurohemal contact the pattern observed is one of gradual insertion of more axon endings between the ependymal foot processes which previously had constituted a continuous layer along the basal lamina. In attempting to quantify this development for the mouse, Eurenius and Jarskar (1971) have estimated that the length of basal lamina contacted by nerve terminals as opposed to ependymal feet increases from 1/10 to 1/3 during the last 3 days of fetal life. Along the same lines, Beauvillain (1973) reported that the number of terminals per 100 sq. μm of the external zone more than doubles while the area occupied by glial elements declines during this time period.

There appears to be little doubt that these axon endings of the fetal median eminence belong to the tubero-hypophyseal system. This conclusion is based on their content of dense-core vesicles. When direct comparison is made between these endings and those being established in the neural lobes of the same fetuses (Paull, 1973b), a size differential in the granulated vesicles is apparent (compare Figs. 7 and 8). As in adults, the elementary neurosecretory vesicles present in the neural lobe are of larger average size than the dense-core vesicles of the external median eminence. Furthermore the figures compiled by Kobayashi *et al.* (1968), Ajika (1969), Daikoku *et al.* (1971), Fink and Smith (1971), and Monroe *et al.* (1972) show that the sizes of the granular vesicles in the external median eminence of fetal and neonatal rats are within the range typical of dense-core vesicles found in tubero-hypophyseal endings of adults (*i.e.* 60–120 nm).

In summary, these studies show that neurohemal contact is established before birth between axons of tuberal neurons and the early primary capillaries already present on the surface of the median eminence in rats and mice.

Following its establishment in the fetus, the contact zone undergoes further ultrastructural changes during early postnatal life. These appear to be mainly quantitative, at least in so far as they are revealed by routine electron microscopic studies (see reports listed in Table I). The arrival of more axons and the formation of additional

---

Fig. 7. Newly formed contact (perivascular) zone of the median eminence of a rat on the 21st fetal day. In the lower center an axon terminal containing synaptic-like vesicles (SV) and a few dense-core vesicles (D) lies against the basal lamina (BL). Other terminals marked by synaptic-like vesicles are also present in this zone. A, axis cylinder containing only neurotubules; Ep, ependymal process; EpF, ependymal foot. Compare with Fig. 8. × 17,000.

Fig. 8. Neural lobe of a rat on the 21st fetal day. Axon terminals of the neurohypophyseal system with synaptic-like vesicles (SV) also contain elementary neurosecretory vesicles (ENG) which are larger than the dense-core vesicles of the tubero-infundibular terminals in Fig. 7. Axis cylinders (A) contain neurotubules and resemble those in median eminence of Fig. 7. P, pituicyte; E, endothelium of a capillary. × 17,000.

References p. 207–208

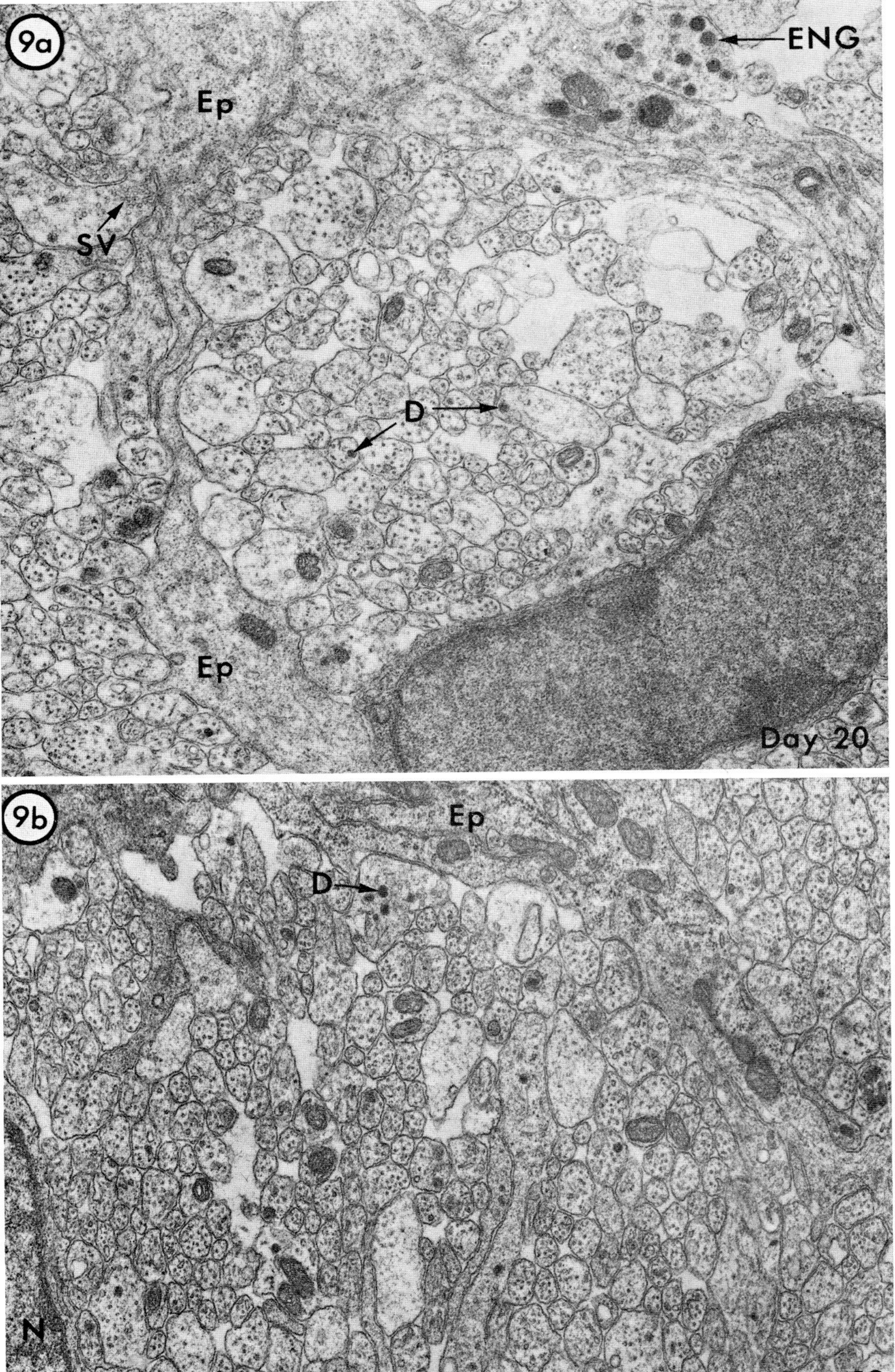
9a
ENG
Ep
SV
D
Ep
Day 20
9b
Ep
D
N
Day 19

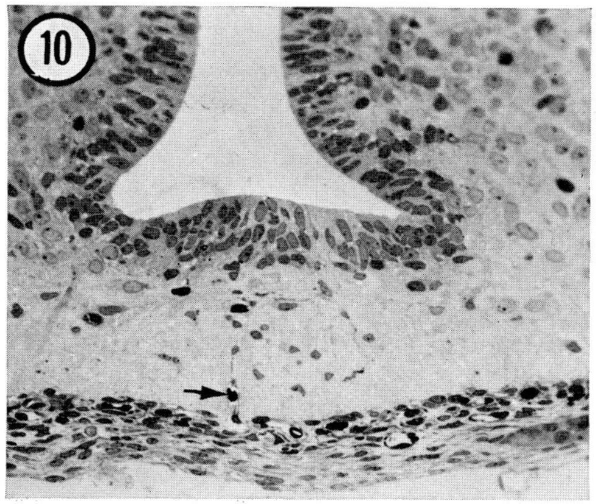

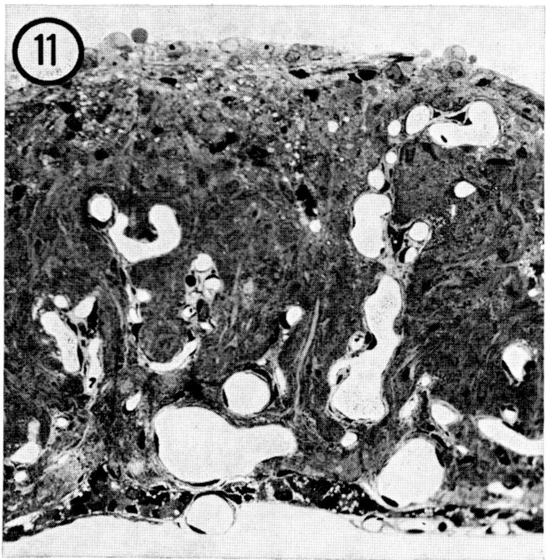

Figs. 10 and 11. Light micrographs of frontal, 1 μm sections of the median eminence. Fig. 10 from a 5 day neonatal rat shows the superficial capillary plexus along the ventral surface and a rare capillary loop containing a red blood cell (→). Loop formation has barely begun. Fig. 11 from an adult rat shows a strip of the mid-portion of the median eminence with its elaborate capillary loops. This specimen fixed by perfusion. Both × 200.

Fig. 9. Portions of the palisade zone of the median eminence of fetal rats. a: dorsal area from a 20 day fetus; b: more ventral area from a 19 day fetus. Many axons of varying diameters contain neurotubules. Note the very sparse distribution of dense-core vesicles (D). In the upper part of a, one axon contains a rare group of larger elementary neurosecretory vesicles (ENG) and may mark the edge of the internal (fiber) zone which is difficult to define in the fetus. In both micrographs some axons are closely apposed to the surfaces of ependymal processes (Ep). In the upper left of 'a' an immature synaptoid contact is suggested by the presence of synaptic-like vesicles (SV) in an axon. Both × 17,000.

terminals both contribute to the progressive thickening of this portion of the median eminence. As the capillary loops develop, still more terminals are established along them, so that in adults the perivascular zone of contact follows the ramifications of the loops deep into the substance of the median eminence (Fig. 11 and see later section on vasculature). The total area of the contact zone is thus further enlarged. At the same time the numbers of vesicles, both dense-cored and lucent, gradually increase, as evidenced by the counts made by Kobayashi *et al.* (1968) in rats up to 30 days of age. As a result the axon terminals in adults contain huge masses of the synaptic-like, lucent vesicles which always outnumber the dense-cored type, even though the latter are also more numerous than they were in early life (*cf.* Figs. 7 and 13).

In considering the relationship of this developmental morphology to the onset of hypothalamic control of the pars distalis, it is tempting to conclude that the minimal structural requirements, as understood at present, exist in late fetal life of the mouse and rat. The neurohemal contacts present at this time appear to be morphologically like those of adults. They may indeed provide sufficient basis for that hypothalamic control which is in force in the fetus (see review by Jost *et al.*, 1970, and reports in this volume). Nevertheless the quantitative differences between the contact zones of the fetus, neonate, and adults must not be overlooked. Great differences in total area, in numbers of axon terminals, and in numbers of vesicular inclusions are obvious. Beyond this, it must be recognized that the axon terminals present in the fetus and neonate may not include all the *kinds* of neurons which have an input to the median eminence in adults. For example, there is the question of whether dopaminergic components are present in the newly established contact zone. Thus, caution must be dictated in assuming that such hypothalamic control as is exercised in the fetus is necessarily like that in the adult. Furthermore, the developmental patterns in other zones of the median eminence bear importantly on functional questions. These will be discussed in the following paragraphs.

Development of the palisade zone

It has been pointed out that in the fetal rat the ventral, axon rich portion of the developing median eminence occupies about half of its depth by the 18th day while the layer of closely arranged ependymal cell bodies composes the dorsal half. With further growth, the depth of the axonal portion comes to exceed that of the ependymal

---

Fig. 12. Palisade zone of the median eminence of an adult female rat. Note the large diameter of many of these axons and the very numerous dense-core vesicles which characterize this zone in adult rats. Compare with the palisade of the late fetus in Fig. 9. Ep, ependymal process. × 17,000.

Fig. 13. Contact (perivascular) zone of the median eminence of an adult female rat. Numerous axon terminals contain huge numbers of synaptic-like vesicles (SV) and some dense-core vesicles (D). Many terminals contact the basal lamina (BL). Typical invaginations of the perivascular space and basal lamina are seen at the upper right. Compare with Fig. 7. × 17,000.

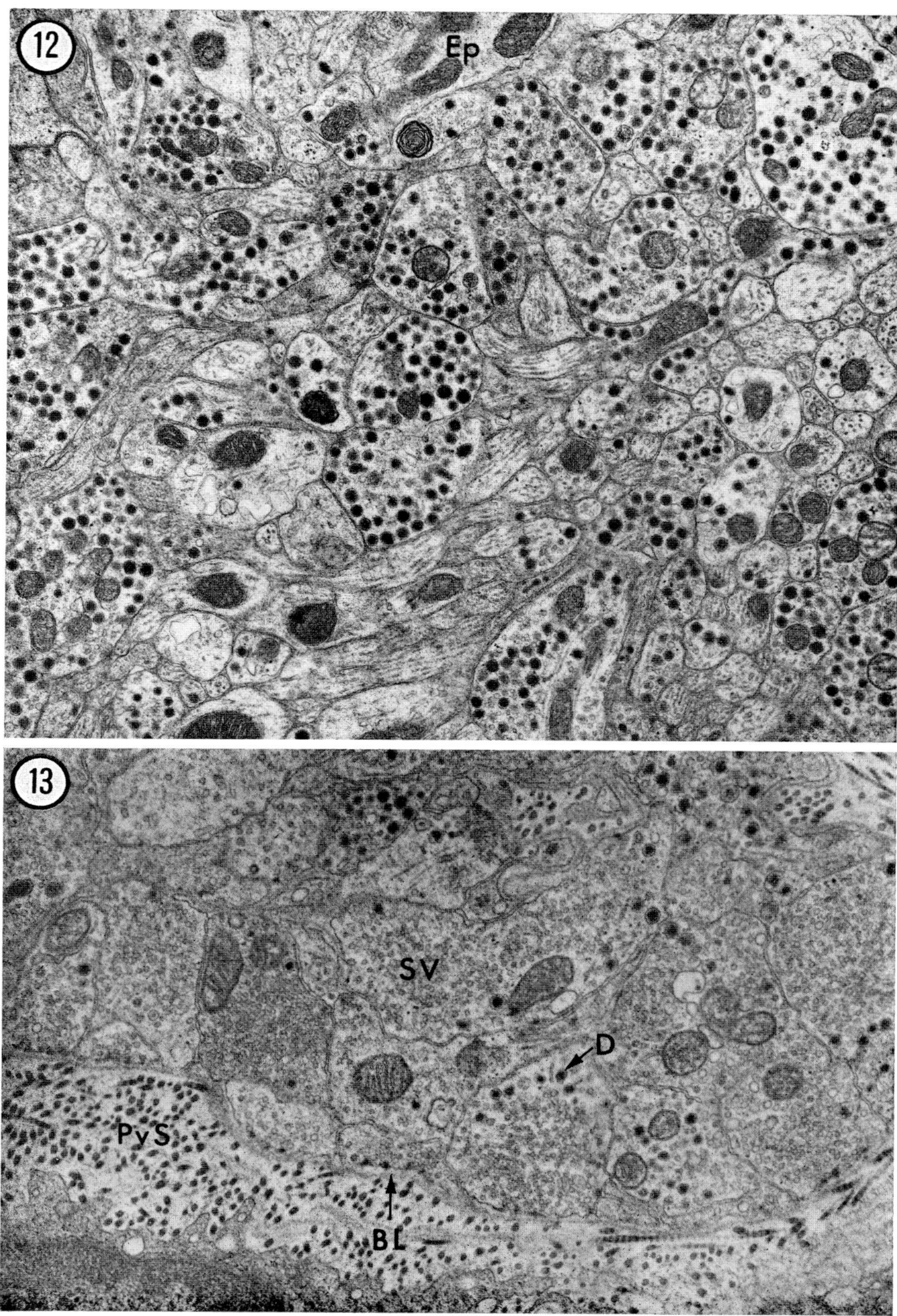

References p. 207–208

portion by the time of birth (Fig. 3b–d). However, the perivascular or contact zone with its axon endings, although established by the end of fetal life, is actually very narrow and accounts for only a small part of the total axon dominated region. The remaining portion which extends dorsally toward the most ventral ependymal cells is actually wider and may be separately designated as the *palisade* zone.

In considering the possible functional capacities of the fetal median eminence it is important to appreciate the ultrastructural organization of this palisade zone. As Monroe *et al.* (1972) have reported, this area in the neonatal rat differs markedly from the same area in the adult. This difference is even more pronounced in the fetus. In representative electron micrographs of this area from late rat fetuses (Fig. 9a and b) axonal cylinders are the dominant component. The majority are small calibered, with a few of larger diameter interspersed. Most striking is the extremely small number of dense-core vesicles displayed. This picture remains essentially very much the same during the first 5 days of postnatal life. Despite a small increase, dense-core vesicles are still only sparsely distributed by the 5th day (Monroe *et al.*, 1972). In marked contrast, comparable palisade areas of the adult median eminence (Fig. 12), although containing many axonal cylinders as slender as those of the fetus, also contain a wealth of dilated axons in which great numbers of dense-core vesicles are present. Comparison of Figs. 9a, b, and 12, taken at the same magnification, will emphasize this difference.

Immaturity of both fetal and neonatal palisades of the median eminence is also evident in the relations between ependymal elements and axons, especially in the lack of development of "synaptoid" contacts. Such contacts have been repeatedly described in the adult median eminence (see review by Knigge and Scott, 1970, and Wittkowski, 1973). They are marked by clusters of electron lucent, synaptic-like vesicles massed along the axon membrane facing the membrane of an ependymal process. Since increased densities of the membranes are neither well defined nor consistently present, these configurations are often designated as synaptoids rather than synapses. Although in the palisade of the fetal median eminence axon profiles are already seen lined up in close approximation with ependymal processes as in Fig. 9a and b, these axons rarely contain the synaptic-like vesicles and even more infrequently are such vesicles massed against the axon membrane. In the late fetus there are, at best, mere suggestions of synaptoid contacts, as in the upper left of Fig. 9a. Similarly in neonatal rats of 1–5 days Monroe *et al.* (1972) found only occasional and relatively undeveloped synaptoids and other investigators of the developing median eminence have made no mention of their presence.

By contrast, in the adult median eminence such synaptoids are so striking and so frequently encountered that their possible significance as points of communication between ependymal and axonal elements must be considered. While they occur in all zones of the median eminence, they are especially numerous in the palisade. As in Fig. 14, a single ependymal process may be plastered with repeated synaptoid contacts along its length. It is to be emphasized that the axons involved are definable as terminals, at least at this time, by their content of synaptic-like vesicles, and that they lie deep in the palisade, away from the perivascular zone of either superficial or pene-

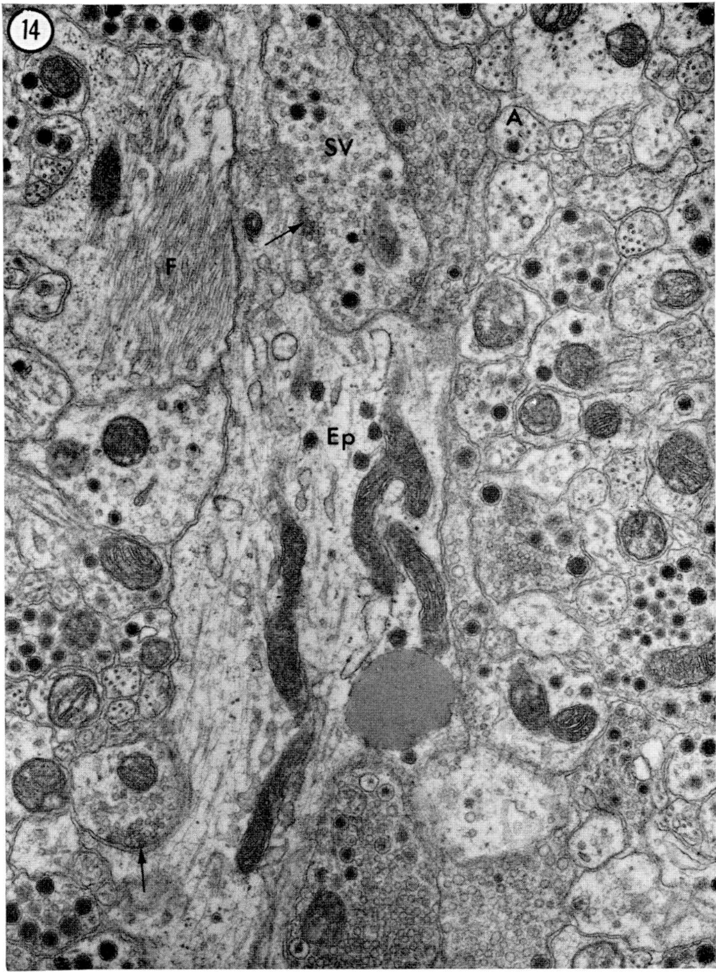

Fig. 14. Higher magnification of a part of the palisade zone of the median eminence of an adult female rat. A longitudinally sectioned ependymal process (Ep) is closely contacted by axon profiles along the greater portion of both surfaces. Synaptic-like vesicles (SV) indicate the terminal nature of many of the axons. Clusters of these vesicles are seen at synaptoid contact points (→). The ependymal process displays typically varied organelles including several dense, granule-like inclusions. In the upper left corner another ependymal profile contains more filaments (F) and ribosomes. A, preterminal axon. × 20,475.

trating capillaries. While the products of such axon endings could conceivably be delivered via the intercellular spaces and tissue fluid to the perivascular area, the morphological picture lends more weight to the possibility that these endings are significantly related to the ependyma at the synaptoid points. Whatever the answer may prove to be, it seems safe to assume that these terminals play an important role in the function of the mature median eminence. The point to be emphasized in the framework of the present topic is that their scarcity in the fetal and neonatal palisades

References p. 207–208

is suggestive of some limitation in the functional capacity of the median eminence in early life.

Unfortunately, not much information is available on the subsequent progression of changes in the palisade zone. Those investigators who have observed the median eminence in animals of intermediate ages between neonates and adults have focused their attention on the contact zone and made no distinction between events there and those in the palisade. In a series of female rats of 34–39 days of age studied in this laboratory the median eminence had a well developed palisade zone, rich in stores of dense-core vesicles, but whether these are accumulated gradually or at a certain time remains a question. If it is eventually proven that the dense-core vesicles of the median eminence comprise multiple and significant subpopulations as urged by some (Kobayashi *et al.*, 1970; Ishii, 1972), the need for further study of the maturing palisade zone will become even more imperative.

Components of the hypothalamo–neurohypophyseal system in the developing median eminence

Observations of the development of the neural lobe made in this laboratory (Paull, 1973a, b) and those of Ajika (1969) indicate that in the rat, axons invade the originally cellular neural lobe and contact its capillaries before birth. Small numbers of elementary neurosecretory granules are present. Therefore, in the late fetal and neonatal median eminence, the widened zone dominated by axons must include some components of the hypothalamo–neurohypophyseal system. Nevertheless, identification of these has presented difficulties. Despite extensive sampling of the fetal median eminence along its rostro-caudal extent, we have only rarely seen elementary neurosecretory vesicles in its axons (Fig. 9a). Furthermore, at this time all the axons and terminals in the median eminence have about the same diameter as those which have arrived in the neural lobe (*cf.* Figs. 7 and 8). Hence the criteria of content and diameter, which permit the recognition of an internal zone of neurohypophyseal axons in the adult median eminence, cannot be readily applied in the fetus or newborn. Kobayashi *et al.* (1968) also found it difficult to define an internal zone. However, Fink and Smith (1971), while noting that all the fibers in the late fetal median eminence are similar in

---

Fig. 15. Low magnification view of a population of axons lying just ventral to the ependymal lining in the adult rat median eminence. The top of a capillary loop (C) with its endothelium (E) and basal lamina (BL) occupies the lower left corner. Processes (Ep) from the overlying ependymal cells embrace the groups of axons and also contact the basal lamina. × 6000.

Fig. 16. Ependymal cells in the median eminence of an adult rat. These flattened cells (Ep) form 2–3 layers on the ventricular floor and also send long processes ventrally (*). Just beneath and between the cells are small axons and axon terminals with many synaptic-like vesicles and some dense-core vesicles. Synaptoid contacts (→) are common between these terminals and ependymal cells or processes. The more ventral part of this group of axons is shown in Fig. 15. × 17,000.

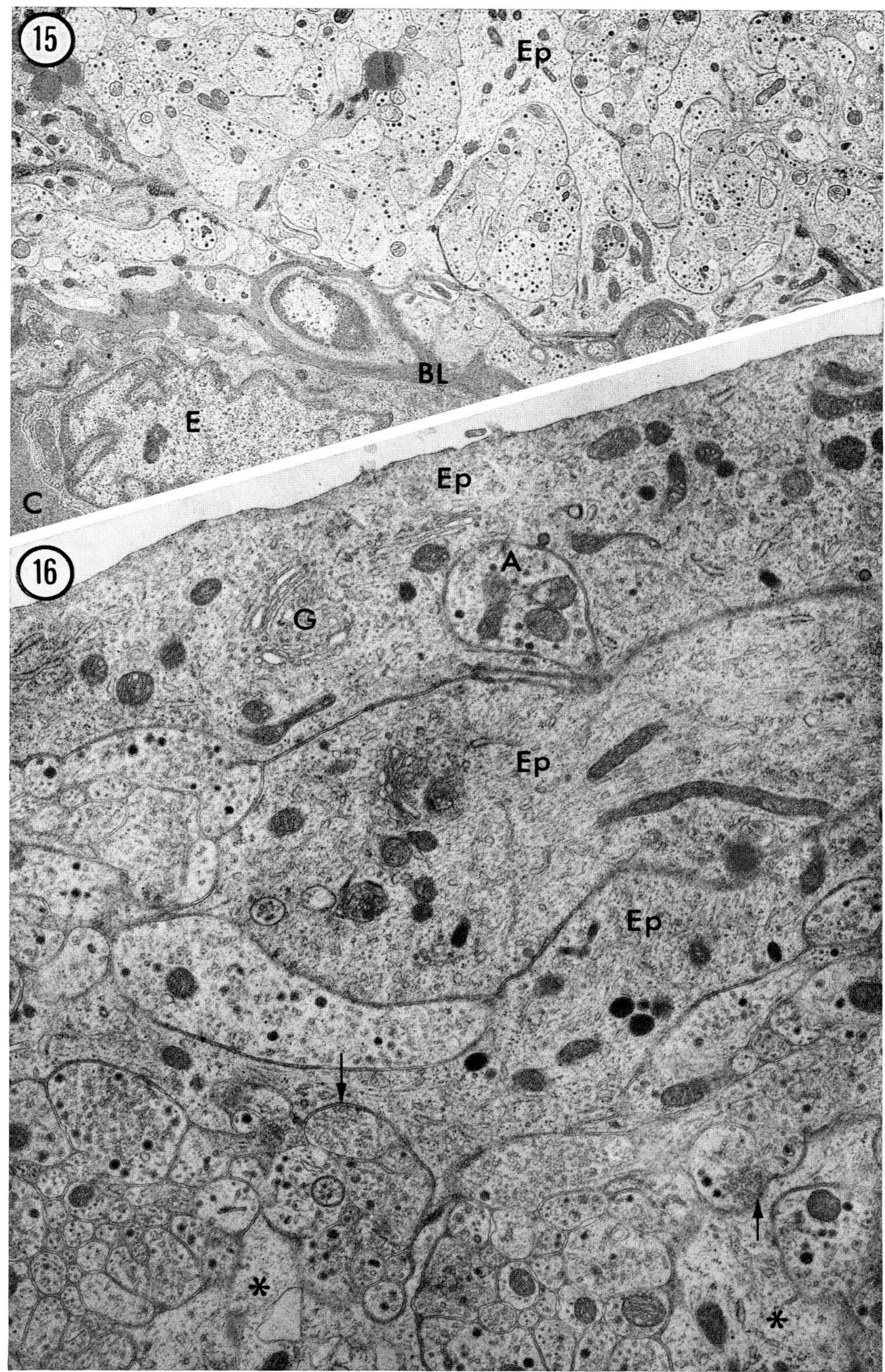

References p. 207–208

appearance (diameter?), did identify an internal zone by the presence of large neurosecretory granules in some, as had been done earlier by Daikoku *et al.* (1968) for neonates. Beauvillain (1973) also has described the internal (fiber) zone as distinct in the 1 day mouse. These discrepancies point to the need for more extended study of the developing internal zone in correlation with observations of the neural lobe.

Vasculature of the developing median eminence

Although the main thrust of this report is toward the median eminence itself, a brief account of the development of its vasculature is pertinent. Capillaries have been found to be already present along the ventral surface of the median eminence in the youngest fetuses examined (Fink and Smith, 1971; Halász *et al.*, 1972; Beauvillain, 1973; and Paull, 1973a and b). These vessels already display the essential characteristics of primary portal capillaries: *i.e.*, an attenuated, fenestrated endothelium, a basal lamina, and a surrounding perivascular space containing connective tissue elements. During late fetal and neonatal life this capillary bed is extended, fenestrations become more frequent, and there is an increasing intimacy between the capillaries and the adjacent surface of the external median eminence. Nevertheless this is essentially a superficial plexus (Eurenius and Jarskar, 1971; Fink and Smith, 1971; Beauvillain, 1973; and Paull, 1973b).

In the rat and the mouse, capillary loops only begin to penetrate the median eminence after birth, usually appearing in the latter part of the first postnatal week, and becoming more numerous during the second week (Glydon, 1957; Enemar, 1960; Campbell, 1966; Daikoku *et al.*, 1968, 1971; Eurenius and Jarskar, 1971; Fink and Smith, 1971; Monroe *et al.*, 1972; and Beauvillain, 1973). In the light of present knowledge, loops appear to be significant only in providing increased area for the establishment of additional neurohemal contacts. Their presence cannot be directly related to the onset of function but may be regarded as another quantitative difference between the adult and fetal or neonatal median eminence (Halász *et al.*, 1972; Monroe *et al.*, 1972).

In respect to the development of the portal system, it is worth noting that species differences exist. Loop development before birth has been reported for the rabbit (Campbell, 1966) and the portal system is said to be fully established during fetal life in the guinea pig (Donovan and Peddie, 1973). Whether the ultrastructural organization of the fetal median eminence is equally advanced in these animals remains to be seen.

Changes in the ependyma

As indicated early in this review, studies thus far have emphasized the initial, purely ependymal character of the fetal median eminence, the gradual separation of the ventral extensions of the ependymal cells by ingrowing axons, and the proportionate reduction in the area occupied by the ependymal layer. Little attention has been paid to changes in ependymal ultrastructure and its relations with the neural components, or to comparison with the adult ependyma. Our observations, while by no means exhaustive, provide some information which seems to bear on the question of the functional capacities of the developing median eminence.

The ependymal cells, at first immature, undergo a striking increase in numbers and kinds of organelles between the 16th and 18th days of fetal life. Indeed their organellar equipment from this time on strongly resembles that of adult eminential ependyma, at least qualitatively, if not necessarily quantitatively. Thus the well developed Golgi apparatus, the vesicular and tubular formations of smooth membrane, the filaments, microtubules, small to moderate amounts of rough endoplasmic reticulum, lipid droplets and lysosomes are characteristic of the late fetal neonatal and adult ependyma (Figs. 6, 14 and 16). In addition the peculiar pleomorphic dense inclusions, sometimes called granules, which are particularly prominent in the ependymal foot processes of adults, have also been found in those of the fetus and neonate (Monroe *et al.*, 1972; Paull, 1973b). It is these same features, plus the fact that these cells, with their processes, extend from the ventricular lumen to the perivascular spaces surrounding the primary portal capillaries, that have formed the basis for speculations by morphologists that the ependyma may serve in a secretory and/or an absorptive capacity in the functional median eminence. More substantial evidence for the latter role is found in some physiological data reviewed by Knigge and Scott (1970) and in the data assembled by Knigge and his associates on the special transport capacity of the incubated or organ-cultured median eminence (Knigge, 1973; Vaala, 1973). The fact that the ependymal cells of the developing median eminence acquire typical ultrastructural features before birth suggests that they might be capable of contributing in such a way to the function of the median eminence in very early life.

Nevertheless, this must remain a tentative conclusion for there are several other ways in which the early ependyma differs from that of adults. In the first place, as noted earlier, the synaptoid contacts between ependymal processes and axons are not fully developed in the fetus or neonate. Secondly, in the adult rat the ependymal cells covering much of the floor of the infundibular recess are quite flat as illustrated in Figs. 11 and 16 and described by Raisman and Field (1971). This is in contrast to the cuboidal and columnar ependymal cells seen in the fetus and neonate (Figs. 1 and 10). Furthermore, just beneath these flattened cells of the adult, there is a congregation of rather small axons which are unlike those of the hypothalamo–neurohypophyseal tract located more deeply in the internal (fiber) zone (Figs. 15 and 16). These axons resemble those of the contact zone in size and in their content of dense-core vesicles. Many also contain synaptic-like vesicles and appear to form synaptoid contacts with the flattened ependymal cells or their processes, or in some cases to contact the tops of the capillary loops near by. Little information is available on the nature of these subependymal axons. Possibly they are noradrenergic fibers whose existence in the internal median eminence has been indicated by histochemical fluorescence techniques (Jonsson *et al.*, 1972). The point to be emphasized here is that such fibers have not been found in the fetal or newborn median eminence and have just begun to make their appearance in the 5 day neonate. Hence, whatever the significance of the synaptoids in all layers, and of the flatter ependymal cells and related subependymal axons proves to be, their

References p. 207–208

relative paucity may well be another sign of limited function in the still developing median eminence.

CONCLUSIONS

In the mouse and rat, contacts established a day or two prior to birth between the axons of tuberal neurons and primary capillaries along the surface of the median eminence may constitute the morphological basis for the exercise of some hypothalamic control over the pars distalis in the late fetus and neonate. However, it should not be assumed that this control is entirely like that in adults. The development of a number of the other ultrastructural features of the median eminence proceeds more slowly and is not completed until some time after the 5th day of postnatal life. Therefore the very early median eminence appears to be significantly different from that of the adult. This fact suggests that its functional capacities in the fetus and neonate may be limited.

SUMMARY

Recent electron microscopic studies have revealed the general pattern of ultrastructural development of the median eminence in the rat and mouse. Beginning 5 or 6 days before birth the median eminence, initially composed of immature stratified ependymal cells, undergoes a 3 day period of rapid ingrowth by tubero-infundibular axons. Simultaneously ependymal cells mature, acquiring a wide assortment of organelles. With the subsequent development of terminals containing synaptic-like and a few dense-core vesicles, some of the axons establish typical contact with primary capillaries of the superficially located plexus 2 or 3 days before birth. These neurohemal contacts could conceivably provide the basis for some hypothalamic influences to be transmitted to the pars distalis in the fetus and neonate.

In other respects ultrastructural maturation proceeds more slowly. Away from the perivascular contact zone axons contain very few dense-core vesicles and exhibit only small numbers of immature synaptoid contacts with ependymal elements in the fetus and neonate. Later developments include accumulation of large stores of dense-core vesicles, formation of more synaptoid configurations, increases in numbers, and perhaps kinds, of axon terminals and contained vesicles, and extensive vascularization by capillary loops. All of these may be essential before mature hypothalamo–hypophyseal relationships are achieved.

ACKNOWLEDGEMENTS

The authors would like to express their thanks to Nora Tong for her technical assistance in the work done in this laboratory and to Caroline Brown for her help in the preparation of the manuscript.

This work was aided by U.S. Public Health Service Grants 5R01 NS 02321 and 5 T01-GM-00585-09.

REFERENCES

AJIKA, K. (1969) Ultrafine structure of the developing median eminence and pars nervosa of the rat. *Acta obstet. gynaec. jap.*, **16**, 143–155.

BEAUVILLAIN, J. C. (1973) Structure fine de l'éminence médiane de souris au cours de son ontogenèse. *Z. Zellforsch.*, **139**, 201–215.

BUNGE, M. B. (1973) Fine structure of nerve fibers and growth cones of isolated sympathetic neurons in culture. *J. Cell Biol.*, **56**, 713–735.

CAMPBELL, H. J. (1966) The development of the primary portal plexus in the median eminence of the rabbit. *J. Anat. (Lond.)*, **100**, 381–387.

DAIKOKU, S., KOTSU, T. AND HASHIMOTO, M. (1971) Electron microscopic observations on the development of the median eminence in perinatal rats. *Z. Anat. Entwickl.-Gesch.*, **134**, 311–327.

DAIKOKU, S., SATO, T. J. A., HASHIMOTO, T. AND MORISHITA, H. (1968) Development of the ultrastructures of the median eminence and supraoptic nuclei in rats. *J. exp. med. Tokus.*, **15**, 1–15.

DONOVAN, B. T. AND PEDDIE, M. J. (1973) The development of the hypophysial portal system in the guinea-pig. *J. Anat. (Lond.)*, **114**, 292–293.

ENEMAR, A. (1960) The structure and development of the hypophysial portal system in the laboratory mouse, with particular regard to the primary plexus. *Arkiv. Zool.*, **13**, 203–252.

EURENIUS, L. AND JARSKAR, R. (1971) Electron microscope studies on the development of the external zone of the mouse median eminence. *Z. Zellforsch.*, **122**, 488–502.

FINK, G. AND SMITH, G. C. (1971) Ultrastructural features of the developing hypothalamo-hypophysial axis in the rat. (A correlative study). *Z. Zellforsch.*, **119**, 208–226.

GLYDON, R. ST. J. (1957) The development of the blood supply of the pituitary in the albino rat with special reference to the portal vessels. *J. Anat. (Lond.)*, **91**, 237–244.

HALÁSZ, B., KOSARAS, B. AND LENGVÁRI, I. (1972) Ontogenesis of the neurovascular link between the hypothalamus and the anterior pituitary in the rat. In *Brain–Endocrine Interaction, Median Eminence: Structure and Function, Int. Symp. Munich, 1971*, K. M. KNIGGE, D. E. SCOTT AND A. WEINDL (Eds.), Karger, Basel, pp. 27–34.

ISHII, S. (1972) Classification and identification of neurosecretory granules in the median eminence. In *Brain–Endocrine Interaction, Median Eminence: Structure and Function, Int. Symp. Munich, 1971*, K. M. KNIGGE, D. E. SCOTT AND A. WEINDL (Eds.), Karger, Basel, pp. 119–141.

JONSSON, G., FUXE, K. AND HÖKFELT, T. (1972) On the catecholamine innervation of the hypothalamus, with special reference to the median-eminence, *Brain Res.*, **40**, 271–281.

JOST, A., DUPOUY, J. P. AND GELOSO-MEYER, A. (1970) Hypothalamo-hypophyseal relationships in the fetus. In *The Hypothalamus*, L. MARTINI, M. MOTTA AND F. FRASCHINI, (Eds.), Academic Press, New York, pp. 605–615.

KNIGGE, K. M. (1973) Characteristics of the *in vitro* binding of ^{3}H-thyrotropin releasing factor (TRF) by organ-cultured median-eminence. *Anat. Rec.*, **175**, 360.

KNIGGE, K. M. AND SCOTT, D. E. (1970) Structure and function of the median eminence. *Amer. J. Anat.*, **129**. 233–244.

KOBAYASHI, H., MATSUI, T. AND ISHII, S. (1970) Functional electron microscopy of the hypothalamic median eminence. *Int. Rev. Cytol.*, **29**, 281–381.

KOBAYASHI, T., KOBAYASHI, T., YAMAMOTO, K., KAIBARA, M. AND AJIKA, K. (1968) Electron microscopic observation on the hypothalamo-hypophyseal system in rats: IV Ultrafine structure of the developing median eminence. *Endocr. jap.*, **15**, 337–363.

MONROE, B. G., NEWMAN, B. L. AND SCHAPIRO, S. (1972) Ultrastructure of the median eminence of neonatal and adult rats. In *Brain–Endocrine Interaction, Median Eminence: Structure and Function, Int. Symp. Munich, 1971*, K. M. KNIGGE, D. E. SCOTT AND A. WEINDL (Eds.), Karger, Basel, pp. 7–26.

PAULL, W. K. (1973a) A light and electron microscopic study of the development of the neurohypophysis of the fetal rat. *Anat. Rec.*, **175**, 407–408.

PAULL, W. K. (1973b) *A light and electron microscopic Study of the Development of the Neurohypophysis of the fetal Rat.* Ph.D. Thesis, Univ. Southern California.

RAISMAN, G. AND FIELD, P. M. (1971) Anatomical considerations relevant to the interpretation of neuroendocrine experiments. In *Frontiers in Neuroendocrinology*, Oxford Univ. Press, New York, pp. 3–44.

TENNYSON, V. M. (1970) The fine structure of the axon and growth cone of the dorsal root neuroblast of the rabbit embryo. *J. Cell Biol.*, **44**, 62–79.

VAALA, S. S. (1973) Uptake of ^{3}H-LRF by median eminence. *Anat. Rec.*, **175**, 459–460.

WITTKOWSKI, W. (1973) Elektronenmikroskopische Untersuchungen zur funktionellen Morphologie des Tubero-hypophysären Systems der Ratte. *Z. Zellforsch.*, **139**, 101–148.

YAMADA, K. M., SPOONER, B. S. AND WESSELLS, N. K. (1971) Ultrastructure and function of growth cones and axons of cultured nerve cells. *J. Cell Biol.*, **49**, 614–635.

DISCUSSION

STOECKART: Because of the important influence of anesthetics on electron microscopical pictures, I should like to know what type of anesthesia you used for your studies.

MONROE: Many of the tissues were fixed by immersion after decapitation of the rat without any kind of anesthesia. When we did perfusions, we used Nembutal anesthesia.

PILGRIM: Are you implying that synaptoid contacts between the axons and the tanycytes are only found in the more caudal part?

MONROE: No, certainly not. I think they are much more numerous than we realized first. I have recently found a whole collection of synaptoid formations just under the flattened ependymal cells in the middle part of the median eminence. The axons that are making synaptoid contacts with the processes of these flattened ependymal cells have dense-core vesicles. They do not belong to the neurohypophyseal system.

JOST: This was a very beautiful presentation, and the pictures were extremely fine. As far as I remember, Fink and Smith (1971) described very early in development fenestrated capillaries beneath the hypothalamus, not inside the hypothalamus. Did you see such fenestrated capillaries?

MONROE: We have also found fenestrated capillaries at about the same developmental period as Fink and Smith. The fenestrations are, however, not as numerous as they are in a later stage.

GUILLEMIN: Do you think that these heavy silver stained granules, in 1 μm sections of the median eminence might belong to axons of the supraoptico-hypophyseal tract and do they contain neurophysins?

MONROE: I might add that in combination with silver staining we also tried aldehyde fuchsin staining of 1 μm sections. Actually, some of those silver granules were light pink by aldehyde fuchsin staining. I think therefore that Dr. Pilgrim might be quite right that there is a neurophysin-like substance in those granules, and that you don't see it ordinarily in the external median eminence because it is not concentrated enough. But in the 1 μm sections the external median eminence actually looks more aldehyde fuchsin-positive than we had expected.

The Ontogenetic Development of Hypothalamo-Hypophyseal Relations

ALFRED JOST, JEAN-PAUL DUPOUY AND MICHEL RIEUTORT

Laboratory of Comparative Physiology, University of Paris VI, Paris (France)

INTRODUCTION

An exhaustive discussion on the ontogenetic development of hypothalamo–hypophyseal relations raises many questions which cannot all be dealt with here, but a few should be recalled.

(1) Which are the early embryological interrelations between the neural and the epithelial primordia in differentiating the hypothalamo–hypophyseal complex as a defined structure? This embryological problem will not be considered here.

(2) Does the hypothalamus play any role in the early differentiation of specialized cell types in the pituitary, for instance corticotropic or thyrotropic cells? Some authors supposed that this might be the case, because the hypothalamic neurons seemed to mature earlier than pituitary cells (Fink and Smith, 1971); this idea was however based on a too preliminary study: the pituitary cells differentiate earlier than was assumed (unpublished data of Dupouy).

(3) Does a pituitary controlled positive endocrine feedback mechanism influence the differentiation of the mammalian hypothalamus, in a way similar to that described in *Rana pipiens* by Etkin (1963)? In this frog the thyroid hormone under pituitary control regulates the morphological and functional differentiation of the hypothalamus. Similar studies still have to be made in mammals.

(4) Does the fetal hypothalamus control the production and the release of pituitary hormones, and thus participate in the fetal endocrine correlations? This aspect will be discussed in detail in the present paper. Of course, it should first be ascertained whether the fetal pituitary produces and releases hormones, in other words, whether there is anything for the hypothalamus to control, in terms of hypophyseal functional mechanisms.

METHODS

Pituitary function

In order to explore the role of the pituitary gland in fetal endocrinology, hypophysec-

References p. 217–219

tomy of the fetus is the first obvious technique to be used. In small laboratory animals (rabbits and rats) we have long been using intra-uterine fetal decapitation as a simple procedure (Jost, 1947); it produces testicular deficiency in the rabbit fetus, thyroid and adrenocortical atrophy in both animal species (see Jost, 1966a). These symptoms can be prevented by giving the appropriate pituitary hormones to the decapitated fetus. In larger animal species and at later developmental stages, selective hypophysectomy could be done, for instance in sheep (*cf.* Liggins, 1969).

Many other indications on the function of the fetal pituitary were derived from histological, histochemical or ultrastructural studies, or from the detection of hormones either in the pituitary cells, by immunofluorescence, or in the fetal pituitary and plasma, especially with radio-immunological methods. The combined use of several techniques (surgical, morphological, biochemical) is advisable.

It is important to keep in mind that the pituitary function may be of special importance at certain developmental periods. This had been suggested a long time ago, when the effects of decapitation of rabbit fetuses on testicular functioning at different developmental periods were related to variations of the PAS stainable material present in the pituitary gland (Jost, 1953). The concept that the hypophyseal activity could pass through a phase of particular high intensity followed by a decline (Jost, 1961) has recently been substantiated. In the human fetus, Grumbach and Kaplan (1973) observed a peak in plasma FSH and plasma GH at mid-pregnancy. Fetal endocrinology necessitates therefore research at different developmental stages.

Role of the hypothalamus

Techniques for investigating the role of the hypothalamus comprise *i.a.* the experimental removal of the hypothalamus, the histological, histochemical or ultrastructural study of the hypothalamus and vascular system and the determination of hypothalamic hormones in the tissue.

(1) In order to remove the hypothalamus, rat or rabbit fetuses were surgically encephalectomized by removing the whole brain and leaving the pituitary gland *in situ* (Jost, 1966a, b; Jost *et al.*, 1966). At term the body weight of the encephalectomized fetuses was found to be reduced. The completeness of hypothalamectomy was verified in histological sections. Other authors adapted this technique and aspirated the brain tissue with a glass tube (Daikoku, 1966) or with a syringe (Mitskevich *et al.*, 1970) or electrocoagulated it (Fujita *et al.*, 1970). In the sheep fetus, Liggins (1969) could section the pituitary stalk.

(2) Many contributions were devoted either to the differentiation of the hypothalamus itself or to the vascularization of the median eminence and the development of the portal system. Only a few recent papers pertaining to the rat fetus will be quoted here. Daikoku *et al.* (1971) observed nerve fibers containing electron-lucent and dense-core vesicles in the median eminence on day 18 and thereafter, whereas Fink and Smith (1971) observed them already on day 16, in agreement with Halász *et al.* (1972) who found them in the arcuate nucleus on day 16. The monoamine-fluorescence becomes detectable only later (Hyyppä, 1969; Smith and Simpson, 1970). It has long

been known that vascular loops penetrating the median eminence and a true portal vessel system differentiate only after birth (Glydon, 1957). The capillaries of the supratuberal plexus are however connected with the hypophyseal capillaries much earlier (Halász *et al.*, 1972) and fenestrated capillaries are observed from day 15 onwards (Fink and Smith, 1971). Although these recent studies cannot prove nor disprove a physiological role of the fetal hypothalamus, they lend some morphological support to the direct experimentation which demonstrated the function of the fetal hypothalamus.

(3) The detection of releasing activity in the fetal hypothalamus depends to a great extent on the reliability and sensitivity of the assay method. Therefore, a negative result, *e.g.* no detectable TRF activity in the rat hypothalamus before postnatal day 6, as seen with an *in vivo* test (Mess, 1970) must yield precedence to a positive result, *e.g.* presence of TRF in the hypothalamus of pooled 17.5- and 18.5-day-old fetuses, and thereafter, assayed on an *in vitro* test (Conklin *et al.*, 1973; it is unfortunate that these authors pooled fetuses at different developmental stages for their study). CRF activity was also detected in the rat hypothalamus the day before birth (Hiroshige and Sato, 1971).

RESULTS

Some results concerning thyrostimulating hormone (TSH), growth hormone (GH) and adrenocorticotropic hormone (ACTH) will be summarized. Data concerning the hypothalamic control of gonadostimulating hormones will not be considered in the present review (see Jost, 1970; Nakai *et al.*, 1972).

Thyrostimulating activity

Some experiments made on rat fetuses (Jost and Geloso, 1967; Jost *et al.*, 1970) will be briefly summarized. It should first be recalled that the onset of thyroid functioning, as judged by thyroxine release in the fetal blood, takes place on day 18 (*cf.* for discussion Geloso, 1967). The morphological (Jost, 1957a and b) and functional (Geloso, 1967) differentiation of the thyroid gland depends on the fetal pituitary gland as seen in decapitated fetuses. The negative feedback mechanism of thyroxine on the pituitary thyrostimulating activity can be explored with the use of antithyroid drugs. In rat (Jost, 1957 a and b) or in rabbit (Jost, 1959) pregnant females given propyl-thio-uracil (PTU), the thyroid of control fetuses hypertrophies; this hypertrophy does not occur if thyroxine is injected into the fetus which blocks the release of TSH, or if the fetus is decapitated. The latter observation indicates that maternal TSH does not cross the placental barrier.

In order to explore the role of the hypothalamus in the hypophyseal thyrostimulating activity, we encephalectomized rat fetuses on day 18 and we studied them on day 21; other fetuses were decapitated.

The histological picture of the thyroid was normal in the encephalectomized fetuses as compared to littermate controls. The uptake of ^{131}I given to the mother was

References p. 217–219

normal 1 h after the iodide injection; it showed a statistically not significant trend to be lower 24 h after the injection.

In mother rats given PTU (50 mg orally per day), the thyroid of the encephalectomized fetuses displayed the same hyperplastic histology as in the entire fetuses, and the whole thyroid was hypertrophied. Again the uptake of ^{131}I was similar in the encephalectomized and in the entire fetuses after 1 h; after 24 h it was slightly but significantly lower in the encephalectomized fetuses.

The thyroid gland of the decapitated fetuses showed the usual hypoplasia and lack of response to PTU.

These experiments suggest that the thyrostimulating activity of the fetal pituitary is to a large extent independent from the hypothalamus, and that the negative feedback mechanism of thyroid hormones is effected largely upon the pituitary gland itself. The slight difference in thyroid ^{131}I concentration, 24 h after ^{131}I injection was paralleled by a reduction in circulating thyroxine found in other experiments (Jost *et al.*, 1970). Some participation of the hypothalamus in the release of TSH is likely. This is in agreement with the presence of TRF in the hypothalamus of the rat fetus (Conklin *et al.*, 1973).

Mitskevich and Rumyantseva (1972) confirmed that encephalectomized rat and rabbit fetuses respond with a typical thyroid hypertrophy to antithyroid drugs; the response was weaker in encephalectomized guinea pig fetuses. In this species, the hypothalamus seems to exert a more important role in regulating the TSH secretion.

The nearly normal thyrostimulating activity of the pituitary of the rat fetus in the absence of the hypothalamus might result from the transfer of TRF from mother to fetus. Although D'Angelo *et al.* (1971), who injected very large doses of synthetic TRF into pregnant rats, observed some transfer from mother to fetus, it is not very likely that under more physiological conditions appreciable amounts of TRF pass from mother to fetus. A definitive proof still has to be given.

Secretion of growth hormone

The rather early appearance of growth hormone in the developing pig fetal pituitary was demonstrated as early as 1929 by Smith and Dortzbach. More recently, the presence of growth hormone in the pituitary gland and blood plasma of the human (Kaplan *et al.*, 1972) and sheep fetus (Bassett *et al.*, 1970) has been well documented. In the rat fetus growth hormone was detected in the pituitary (Contopoulos and Simpson, 1957) and in the plasma (Strosser and Mialhe, 1972; Rieutort, 1972) from day 19 onward; in the plasma it increases enormously during the next two days. Both in the sheep (Bassett *et al.*, 1970) and in the rat fetus (Rieutort, 1972) it has been demonstrated that the plasma growth hormone in the plasma originates from the fetal pituitary, since it is absent in hypophysectomized fetuses.

In human anencephalic fetuses, Grumbach and Kaplan (1973) observed a very low concentration of growth hormone (blood from the umbilical vein) in comparison to normal infants. This suggested a hypothalamic regulation of growth hormone secretion.

We made some preliminary observations on rat fetuses encephalectomized on day 19.5. On day 21.5, the mother animal was anesthetized with pentobarbital (4 mg/100 g) and laparotomized. Blood was obtained successively from several fetuses: each fetus was gently exteriorized from the uterine horn, so as to preserve the umbilical circulation through the placenta *in situ*. Blood was collected at the armpit. The fetus was then removed and dissected in order to verify that the hypothalamus had been removed and to secure the pituitary gland. Growth hormone determinations were made with a radio-immunological method described earlier (Rieutort, 1972), on plasma or on pituitary extracts.

The mean growth hormone content of the pituitary of 11 encephalectomized fetuses was only slightly lower than that of controls. In 13 encephalectomized fetuses, the plasma growth hormone was significantly reduced (Fig. 1). In 3 others, it was below the sensitivity of the method and as low as in decapitated fetuses; it could not be ascertained whether in these cases the pituitaries were in good condition.

In adult rats, pentobarbital anesthesia increases the plasma growth hormone (Takahashi *et al.*, 1971); if such an effect was responsible for the difference between encephalectomized and control fetuses, it would appear that the response is more pronounced in the presence than in the absence of the fetal hypothalamus. On the other hand, in adult rats various stresses decrease GH release (Schalch and Reichlin, 1968). One may wonder therefore whether encephalectomy acted as a stress and decreased in this way plasma GH. This has to be verified in sham operated fetuses; it seems not very likely as far as encephalectomy suppresses the depletion of ascorbic acid in the fetal adrenals in response to another stress, the injection of formaldehyde

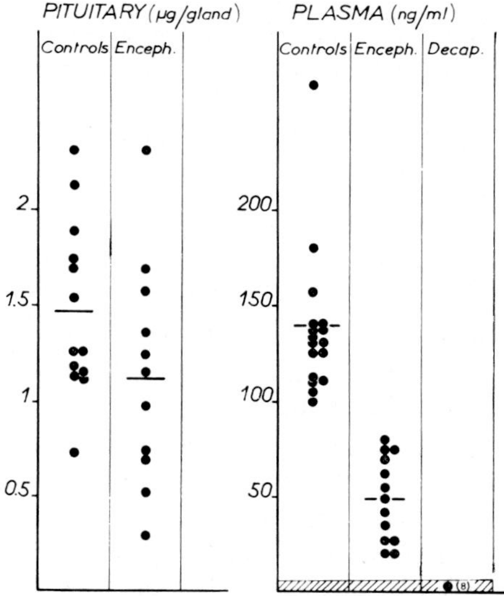

Fig. 1. Growth hormone contents in the pituitary of encephalectomized rat fetuses and littermate controls and in the plasma of the same fetuses and of decapitates. Individual values (except those of the decapitates) and means are shown. The limit of sensitivity of the method is indicated (hatched) for plasma values.

References p. 217–219

(Cohen *et al.*, 1968). These preliminary data suggest that in the rat fetus the hypothalamus plays a role in the release of growth hormone.

Pituitary corticostimulating activity

The role of the pituitary in controlling the size and function of the fetal adrenal cortex is well documented in the rat and in several other mammalian species (Jost, 1966a and b; Jost and Picon, 1970). In the absence of the pituitary gland, the growth of rat fetal adrenals is severely reduced and their effects on target organs are suppressed, for instance the induction of glycogen storage in the fetal liver. Interference with growth of the adrenals in the entire fetus can also result from a negative feedback mechanism, for example when the maternal plasma corticosteroids are elevated.

In the first series of experiments we compared the weight of the adrenals on day 21 in rat fetuses which were either decapitated or encephalectomized on day 19 and in littermate sham operated or intact controls. Growth of the adrenals of both the decapitated and encephalectomized fetuses was similarly stopped; however a purified preparation of beef CRF or rat hypothalamic extracts restored adrenal growth to normal in the encephalectomized but not in the decapitated fetuses; brain extracts had no effect (Jost, 1966b; Jost *et al.*, 1966). This gave evidence that the pituitary of the encephalectomized fetus lacked some hypothalamic influence. Adrenocortical atrophy after removal of the hypothalamus was confirmed in rats (Fujita *et al.*, 1970) and in rats, rabbits and guinea pigs (Mitskevich and Rumyantseva, 1972).

The absence of the fetal hypothalamus also prevents the adrenocortical response to stress on day 20. Pregnant rats were adrenalectomized; approximately 4 h later two fetuses were encephalectomized. One of the encephalectomized fetuses and one control were given subcutaneously 25 μl of a 2 % formaldehyde solution. Two hours after this stress factor a 21 % depletion of the ascorbic acid concentration had occurred in the adrenals of the entire fetuses, but not in those of the encephalectomized fetuses (Cohen *et al.*, 1968). On the other hand, Hiroshige and Sato (1971) observed an increase within 2 min of the CRF activity in the hypothalamus of fetuses recovered by cesarian section the day before birth and submitted to stress factors (ether + laparotomy).

In other series of experiments, still under progress, the hypothalamo–hypophyseal relations were studied according to the developmental stage. As was already mentioned, in fetuses encephalectomized on day 19 the adrenals were, on day 21, as reduced as in decapitated animals. On the contıary, in fetuses encephalectomized on day 17 and sacrificed on day 19, the adrenals were significantly heavier than those of the decapitated animals (Fig. 2a) (data from Cohen *et al.*, 1971). Some autonomous, *i.e.* not hypothalamic-dependent, corticostimulating activity of the pituitary seems to occur before day 19, but not after that stage. The difference between earlier and later stages might result either from the loss by the pituitary of the capacity for autonomous functioning, or from an increased negative feedback mechanism by maternal corticosteroids after day 19 or from an increased sensitivity of the pituitary to the feedback. In order to suppress the maternal corticosteroids, the mother was adrenal-

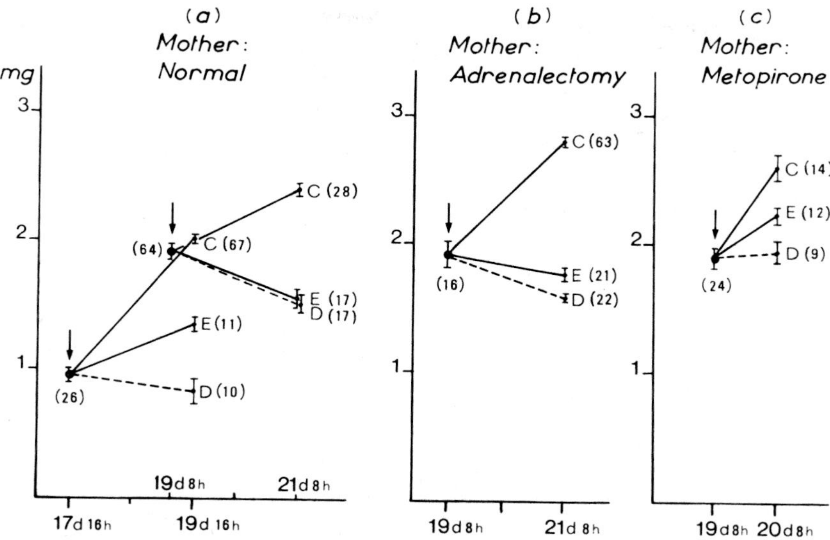

Fig. 2. Fetal adrenal weight under several experimental conditions. A straight line is drawn between the weight of the adrenals in controls on the day of surgery (arrows) and the final weight at the end of the experiment in controls (C), encephalectomized (E) and decapitated fetuses (D). The number of fetuses are given in brackets. Standard errors are indicated as vertical bars. Developmental stages expressed in days and hours from the time of fertilization, *i.e.* from 1 a.m. on the night of cohabitation (*cf.* Jost and Picon, 1970). (a) *in normal pregnant animals:* effects of encephalectomy and decapitation performed on either day 17 or day 19 (arrows) on the weight of the adrenals 2 days later (data from Cohen *et al.*, 1971); (b) a similar experiment done on *adrenalectomized females*, between days 19 and 21 (data from Dupouy and Jost, 1970); (c) a similar experiment done on a *metopirone treated female*, between days 19 and 20 (data from Dupouy, 1971).

ectomized on day 14, and the fetuses were either decapitated or encephalectomized on day 19 and sacrificed on day 21: the adrenals of the encephalectomized fetuses were now somewhat heavier than those of the decapitates (Fig. 2b). The difference was even more pronounced when surgery was performed on the fetus on day 18 (Fig. 3a); under these experimental conditions the autonomous pituitary corticostimulating activity was especially conspicuous between days 18 and 19. This is reflected in the marked difference in the glycogen contents of the liver in fetuses which were either decapitated or encephalectomized on day 18 (Fig. 3b). Recently we studied the corticosterone concentration in the adrenals of fetuses in mother animals adrenalectomized on day 14. In 9 fetuses encephalectomized on day 19, the corticosterone concentration on day 21 was intermediary (10.99 μg/g) between the 38 controls (14.90 μg/g) and the 13 decapitates (6.00 mg/g) (Dupouy *et al.*, 1973).

Besides the maternal corticosteroids there was still another possible source of corticosteroids to be considered: the adrenals of the littermate control fetuses. Kamoun (1970) showed that until day 16 of pregnancy, corticosterone is undetectable in the plasma of pregnant rats adrenalectomized the day before; on days 18 and 20, when the fetal adrenals have become active, significant amounts of corticosterone are

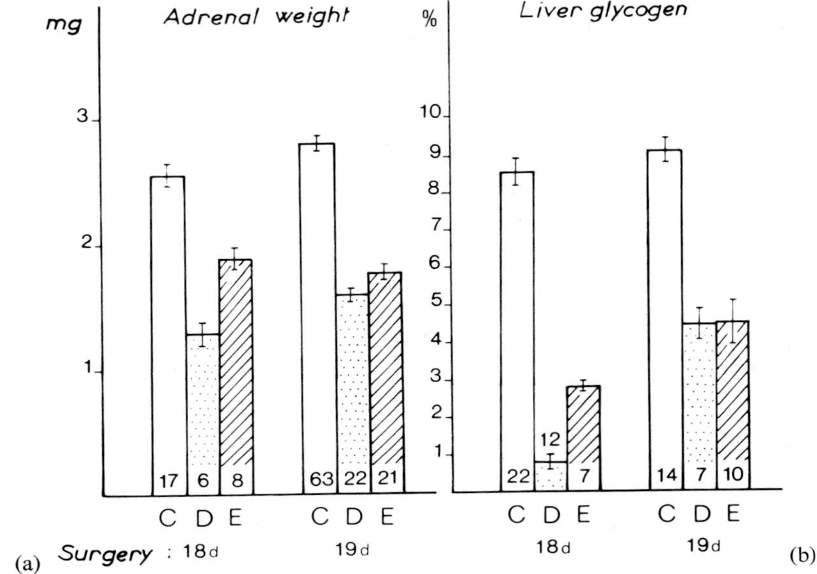

Fig. 3. Experiments carried out on pregnant rats, adrenalectomized on day 14. All fetuses were sacrificed on day 21. Effects of encephalectomy (E) and decapitation (D) performed either on day 18 or on day 19 on fetal adrenal weight, and on glycogen in the fetal liver (per cent of fresh tissue weight). C = controls. The number of fetuses and the standard errors are indicated (data from Dupouy and Jost, 1970).

found in the maternal plasma. We confirmed that fact in females adrenalectomized on day 14 and studied at different stages thereafter (Dupouy *et al.*, to be published).

In order to suppress the corticosteroids as much as possible in the whole system, pregnant rats were given metopirone (100 mg in two doses with glucose) and sacrificed 24 h later (the toxicity of the drug does not permit a more prolonged experiment); control animals received only glucose. A significant increase in adrenal weight was found in encephalectomized fetuses as compared with decapitates (Fig. 2c).

It thus would appear that the corticostimulating activity of the fetal pituitary of the rat to a very large extent depends upon the presence of the hypothalamus; but the fetal pituitary also has some autonomous non-hypothalamic-dependent corticostimulating activity which is easily stopped after day 19 by a negative feedback effect of corticosteroids. The reason why this negative feedback is more efficient on and after day 19 than before might depend either upon an increased sensitivity of the pituitary or on differences in the concentration of free corticosterone in the fetal plasma. This is currently being investigated.

CONCLUSIONS

The experiments summarized in the present paper indicate that in the rat fetus the TSH, GH and ACTH release by the pituitary depends, to a varying extent, upon the hypothalamus. Autonomous or non-hypothalamic-dependent hormone release can also be demonstrated to a varying extent. As a whole, the data obtained so far in the encephalectomized rat fetus duplicate the conditions observed in human anencephalic newborns (Jost and Geloso, 1967), insofar as the thyroid is well developed, the plasma growth hormone is reduced and the adrenal gland is strikingly underdeveloped.

SUMMARY

Only one of the several problems raised by the ontogenetic development of hypothalamo–hypophyseal relations is discussed in the present paper, namely the role of the hypothalamus in the production and release of pituitary hormones. Some techniques used in the field are surveyed, especially fetal encephalectomy, a procedure which permits removal of the hypothalamus with the brain and which allows the pituitary gland to remain *in situ*, and morphological or biochemical studies of the hypothalamus itself. Some experiments performed on rat fetuses are presented.

The thyrostimulating hypophyseal activity is to a very large extent independent of the hypothalamus and can be increased in encephalectomized fetuses submitted to propyl-thio-uracil.

The plasma concentration of growth hormone was found to be diminished in encephalectomized rat fetuses.

The corticostimulating activity is almost abolished in fetuses encephalectomized on day 19 and studied on day 21; it is only diminished in those fetuses which are encephalectomized on day 17 and studied on day 19. One reason — at least — of this difference is a negative feedback exerted directly on the pituitary by corticosteroids produced by the mother and by littermate fetuses.

The condition of the encephalectomized rat fetus resembles that prevailing in human anencephalics.

REFERENCES

Bassett, J. M., Thorburn, G. D. and Wallace, A. L. C. (1970) The plasma growth hormone concentration of the foetal lamb. *J. Endocrinol.*, **48**, 251–263.

Cohen, A., Pernot, J.-Cl. et Jost, A. (1968) Rôle de l'hypothalamus dans la réponse des surrénales du foetus de Rat à une agression par le formol. *C. R. Soc. Biol. (Paris)*, **162**, 2070–2073.

Cohen, A., Dupouy, J.-P. et Jost, A. (1971) Influence de l'hypothalamus sur l'activité corticostimulante de l'hypophyse foetale du Rat au cours de la gestation. *C. R. Acad. Sci. (Paris)*, **273**, 883–886.

Conklin, P. M., Schwindler, W. J. and Hull, S. F. (1973) Hypothalamic thyrotropin releasing factor activity and pituitary responsiveness during development in the rat. *Neuroendocrinology*, **11**, 197–211.

Contopoulos, N. and Simpson, M. E. (1957) Presence of trophic hormones in fetal rat pituitary. *Fed. Proc.*, **16**, 103.

Daikoku, S. (1966) A method of deencephalon in the fetal rat. *Okajimas Folia anat. jap.*, **42**, 39–49.

Daikoku, S., Kotsu, T. and Hashimoto, M. (1971) Electron microscopic observations on the development of the median eminence in perinatal rats. *Z. Anat. Entwickl.-Gesch.*, **134**, 311–327.

D'Angelo, S. A., Wall, N. R. and Bowers, C. Y. (1971) Maternal–fetal endocrine interrelations: demonstrations of TSH release from the fetal hypophysis in pregnant rats administered synthetic TRH. *Proc. Soc. exp. Biol. (N.Y.)*, **137**, 175–178.

DUPOUY, J. P. (1971) Réponse du complexe hypothalamo-hypophysaire du foetus de rat à un blocage de la biosynthèse des corticostéroïdes par la métopirone. Influence du cortisol. *C.R. Acad. Sci. (Paris)*, **273** 962–965.

DUPOUY, J. P., COFFIGNY, H. ET JOST, A. (1973) *Action de l'hypophyse et de l'hypothalamus sur la teneur en corticostérone et l'ultrastructure de la surrénale foetale du rat. 5e Colloque de Neuroendocrinologie, Marseille.*

DUPOUY, J. P., ET JOST, A. (1970) Activité corticotrope de l'hypophyse foetale du rat: influence de l'hypothalamus et des corticostéroïdes. *C.R. Soc. Biol. (Paris)*, **164**, 2422–2427.

ETKIN, W. (1963) Metamorphosis-activating system of the frog. *Science*, **139**, 810–814.

FINK, G. AND SMITH, G. C. (1971) Ultrastructural features of the developing hypothalamo–hypophyseal axis in the rat. *Z. Zellforsch.*, **119**, 208–226.

FUJITA, T., EGUCHY, Y., MORIKAWA, Y. AND HASHIMOTO, Y. (1970) Hypothalamic–hypophysial adrenal and thyroid systems: observations in fetal rats subjected to hypothalamic destruction, brain compression and hypervitaminosis A. *Anat. Rec.*, **166**, 659–671.

GELOSO, J. P. (1967) Fonctionnement de la thyroïde et corrélations thyréo-hypophysaires chez le foetus de rat, *Ann. Endocr. (Paris)*, **28**, Suppl. 1, 1–80.

GLYDON, R. ST. J. (1957) The development of the blood supply of the pituitary in the albino rat, with special reference to the portal vessels. *J. Anat. (Lond.)*, **91**, 237–244.

GRUMBACH, M. AND KAPLAN, S. L. (1973) Ontogenesis of growth hormone, insulin, prolactin and gonadotropin secretion in the human foetus. In *Foetal and Neonatal Physiology*, Proc. Sir Joseph Barcroft Centenary Symposium, K. S. COMLINE *et al.* (Eds.), Cambridge Univ. Press, Cambridge, pp. 462–487.

HALÁSZ, B., KOSARAS, B. AND LENGVÁRI, I. (1972) Ontogenesis of the neurovascular link between the hypothalamus and the anterior pituitary in the rat. In *Brain–Endocrine Interaction: Median Eminence: Structure and Function*, K. M. KNIGGE, D. E. SCOTT AND A. WEINDL (Eds.), *Int. Symp. Munich 1971*, Karger, Basel, pp. 27–34.

HIROSHIGE, T. AND SATO, T. (1971) Changes in hypothalamic content of corticotropin-releasing activity following stress during neonatal maturation in the rat. *Neuroendocrinology*, **7**, 257–270.

HYYPPÄ, M. (1969) A histochemical study of the primary catecholamines in the hypothalamic neurons of the rat in relation to the ontogenetic and sexual differentiation. *Z. Zellforsch.*, **98**, 550–560.

JOST, A. (1947) Expériences de décapitation de l'embryon de lapin. *C. R. Acad. Sci. (Paris)*, **225**, 322–324.

JOST, A. (1953) Problems of fetal endocrinology: the gonadal and hypophyseal hormones. *Rec. Progr. Horm. Res.*, **8**, 379–418.

JOST, A. (1957a) Action du propylthiouracile sur la thyroïde de foetus de rat intacts ou décapités. *C. R. Soc. Biol. (Paris)*, **151**, 1295–1298.

JOST, A. (1957b) Le problème des interrelations thyréo-hypophysaires chez le foetus et l'action du propylthiouracile sur la thyroïde foetale du rat. *Rev. suisse Zool.*, **64**, 821–832.

JOST, A. (1959) Action du propylthiouracile sur la thyroïde du foetus de lapin intact, décapité ou injecté de thyroxine. *C. R. Soc. Biol. (Paris)*, **153**, 1900–1902.

JOST, A. (1961) The role of fetal hormones in prenatal development. *The Harvey Lectures*, **55**, 201–226.

JOST, A. (1966a) Anterior pituitary function in foetal life. In *The Pituitary Gland, Vol. 2, Chap. 9*, G. W. HARRIS AND B. T. DONOVAN (Eds.), Butterworth, London, pp. 299–323.

JOST, A. (1966b) Problems on fetal endocrinology: the adrenal glands. *Rec. Progr. Horm. Res.*, **22**, 541–574.

JOST, A. (1970) Hormonal factors in the development of the male genital system. In *The Human Testis*, E. ROSEMBERG AND C. A. PAULSEN (Eds.), Plenum Press, New York, pp. 11–17.

JOST, A., DUPOUY, J. P. AND GELOSO-MEYER, A. (1970) Hypothalamo–hypophyseal relationships in the fetus. In *The Hypothalamus*, L. MARTINI, M. MOTTA AND F. FRASCHINI (Eds.), Academic Press, New York, pp. 605–615.

JOST, A., DUPOUY, J. P. ET MONCHAMP, A. (1966) Fonction corticotrope de l'hypophyse et hypothalamus chez le foetus de rat. *C. R. Acad. Sci. (Paris)*, **262**, 147–150.

JOST, A. ET GELOSO, A. (1967) Réponse de la thyroïde foetale du rat au propylthiouracile en l'absence d'hypothalamus. Remarques sur les glandes endocrines du foetus anencéphale humain. *C. R. Acad. Sci. (Paris)*, **265**, 625–627.

JOST, A. AND PICON, L. (1970) Hormonal control of fetal development and metabolism, *Advances in Metabolic Disorders, Vol. 4*, Academic Press, New York, pp. 123–184.

KAMOUN, A. (1970) Activité cortico-surrénale au cours de la gestation, de la lactation et du développement pré- et post-natal chez le rat. I. Concentration et cinétique de disparition de la corticostérone. *J. Physiol. (Paris)*, **62**, 5–32.

KAPLAN, S. L., GRUMBACH, M. M. AND SHEPARD, T. H. (1972) The ontogenesis of human fetal hormones. I. Growth hormone and insulin. *J. clin. Invest.*, **51**, 3080–3092.

LIGGINS, G. C. (1969) The foetal role in the initiation of parturition in the ewe. In *Foetal Autonomy*, G. E. W. WOLSTENHOLME AND M. O'CONNOR (Eds.), Churchill, London, pp. 218–244.

MESS, B. (1970) Intrahypothalamic localization and onset of production of Thyrotrophin Releasing Factor (TRF) in the albino rat. *Hormones*, **1**, 332–341.

MITSKEVICH, M. S. AND RUMYANTSEVA, O. N. (1972) A possible role of the hypothalamus in the control of adrenocortical and thyroid functions during fetal life. *Ontogenesis*, **3**, 376–384.

MITSKEVICH, M. S., RUMYANTSEVA, O. N., PROSHLYAKOVA, E. V. AND SERGEENKOVA, G. P. (1970) Encephalectomy in mammal embryos (rat, rabbit, guinea pig). *Ontogenesis*, **1**, 631–635.

NAKAI, T., SAKAMOTO, S. AND KIGAWA, T. (1972) Hypothalamic–pituitary testicular interrelations in fetal rats: effects of castration on ^{3}H-leucine uptake in pituitary basophilic cells. *Endocr. jap.*, **19**, 133–137.

RIEUTORT, M. (1972) Dosage radioimmunologique de l'hormone somatotrope de Rat à l'aide d'une nouvelle technique de séparation. *C. R. Acad. Sci. (Paris)*, **274**, 3589–3592.

SCHALCH, D. S. AND REICHLIN, S. (1968) Stress and growth hormone release. In *Growth Hormone*, A. PECILE AND E. E. MÜLLER (Eds.), Excerpta Medica, Amsterdam, pp. 211–225.

SMITH, P. E. AND DORTZBACH, C. (1929) The first appearance in the anterior pituitary of the developing pig foetus of detectable amounts of the hormone stimulating ovarian maturity and general body growth. *Anat. Rec.*, **43**, 277–297.

SMITH, G. C. AND SIMPSON, R. W. (1970) Monoamine fluorescence in the median eminence of fetal, neonatal and adult rats. *Z. Zellforsch.*, **104**, 541–556.

STROSSER, M. T. ET MIALHE, P. (1972) L'hormone de croissance plasmatique au cours du développement foetal et postnatal chez le rat. *Gen. comp. Endocr.*, **18**, 625.

TAKAHASHI, K., DAUGHADAY, W. H. AND KIPNIS, D. M. (1971) Regulation of immunoreactive growth hormone secretion in male rats. *Endocrinology*, **88**, 909–917.

DISCUSSION

DÖRNER: Is it justified to conclude, Dr. Jost, from your excellent presentation on the ontogenic development of the hypothalamus that there are various steps in the differentiation? The first being a differentiation of the peripheral gland, *i.e.* of the gonads or the thyroid, then a second differentiation step of the hypophysis, and as a third step, the differentiation of the hypothalamus.

JOST: There is no doubt that an early embryonic primordium can become, for example, a thyroid gland, which makes thyroxine, even in the absence of the pituitary. The pituitary is not necessary to cause the cellular differentiation of a peripheral gland such as the thyroid. But these glands never function as endocrine glands without the hypophysis, as far as they do not release enough hormones to affect their target organs. The problem whether the pituitary is an exception to this rule, and would secrete hormones even in the absence of the hypothalamus is not yet completely solved, because one has still to study this at very early stages. My surmise would be that appreciable amounts of some hormones, *e.g.* TSH, can be released from the pituitary even in the absence of the hypothalamus. But the hypothalamus soon controls the activity and the intensity of the release.

PILGRIM: In relation to this problem, I think that it is obvious from your experiments that the pituitary can function on a certain level without the hypothalamus, but probably only if the hypothalamus had been present already for some time.

JOST: One should study very early stages to know whether this possibility is true.

LEVINE: We do know that adrenocorticoids do exert a feedback action on the adult hypothalamus. Could you postulate whether the feedback site for adrenocorticoids will be the hypothalmus in the fetus as well?

JOST: Most of the feedback action of adrenocorticoids is probably on the fetal hypothalamus, and a small part of this action is on the pituitary, since in the absence of the hypothalamus a small feedback effect of corticosteroids directly on the pituitary was still demonstrable. For the regulation of the thyroid this is probably the reverse, here the feedback is mainly on the pituitary.

Environment Dependent Brain Organization and Neuroendocrine, Neurovegetative and Neuronal Behavioral Functions

G. DÖRNER

Institute of Experimental Endocrinology, Humboldt University, Berlin (G.F.R.)

INTRODUCTION

Fundamental processes of life, such as reproduction, metabolism and information processing, are regulated by means of the following feedback control systems.

(1) the hypothalamo–hypophyseal–gonadal system;

(2) other neuroendocrine and/or neurovegetative systems; and

(3) neuronal behavioral systems.

From extensive animal experiments and clinical studies (Dörner, 1972) one may regard the hypothalamo–hypophyseal–gonadal system as a model system and then the following ontogenetic organizational rules may be deduced (Fig. 1): (1) during a critical organization period of the brain an open-loop regulatory system is converted into a closed-loop feedback control system. The regulating variable and the regulated

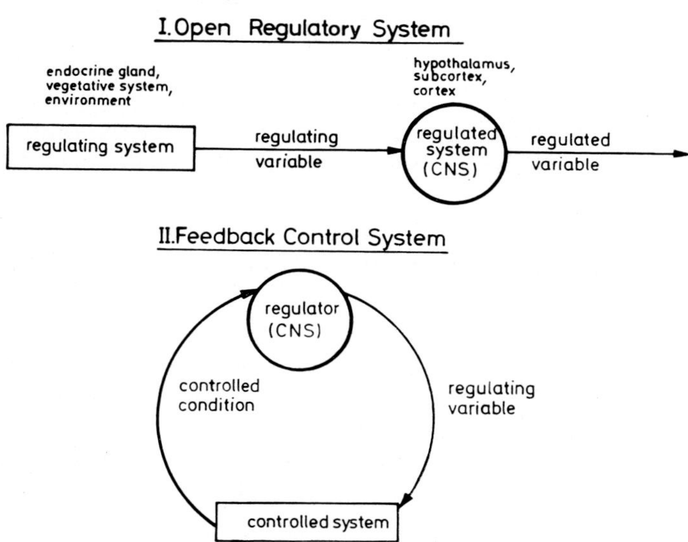

Fig. 1. Ontogenetic transformation of an open-loop regulatory system into a closed-loop feedback control system during organization of the central nervous system.

References p. 233–236

element of the open-loop system are then transformed into the controlled condition (homeostatic variable) and the central regulator, respectively, of the feedback control system (transformation rule); (2) the size of the primary regulating and secondary homeostatic variable determines, during a critical organization period of the CNS regulator, its quality and hence the functional and tolerance ranges of the whole feedback control system for life (determination rule).

This principle may be explained by enduring modifications of gene expressibility produced by the regulating (and/or homeostatic) variable during cytogenesis (and/or synaptogenesis) of specific brain centers. Thus, homeostatic variables which are to be established for the entire functional phase are preprogramming their own feedback control systems during a critical organization phase. In other words, what is to be regulated by the brain for life is already attempting to regulate itself during brain differentiation.

Several findings are presented which suggest that this elementary principle may be functioning, in fact, for the regulation of reproduction, metabolism and processing of information in the brain.

THE HYPOTHALAMO–HYPOPHYSEAL–GONADAL SYSTEM

The gonadotropic function of the pituitary gland was first demonstrated by Aschner (1912) who observed distinct gonadal atrophy in dogs following hypophysectomy. Subsequently, the gonadotropins were discovered by Aschheim and Zondek (1927) at our institute and by Smith and Engle (1927).

In 1932, Hohlweg and Junkmann envisaged the central nervous system as the regulator of the hypophyseal gonadotropic function, presenting for the first time a cybernetic functional unit of the brain–pituitary–gonadal axis. This concept was strengthened by the demonstration of a positive estrogen feedback (Hohlweg, 1934), when the corpus luteum formation was induced by a single injection of estrogen in intact immature female rats as opposed to hypophyseal stalk-sectioned or hypothalamus-lesioned animals (Westman and Jakobsohn 1938; Döcke and Dörner, 1965). Barraclough and Gorski (1961) distinguished a rostral sex center located in the preoptic suprachiasmatic region, that regulates the cyclic gonadotropin secretion in females, and a caudal center located in the ventromedial arcuate region, that is responsible for the tonic gonadotropin secretion in both sexes. All these regulatory mechanisms may be mediated by gonadotropin-releasing substances of the type postulated by Harris (1953), subsequently described by McCann and coworkers (1960) and finally isolated by Schally and coworkers (1971). In 1941, Brookhart and Dey demonstrated by means of intrahypothalamic lesions in guinea pigs a "mating center" responsible for sexual behavior patterns, which was not identical with the hypothalamic "sex center" postulated by Hohlweg and Junkmann (1932) that regulates gonadotropin secretion. Dörner *et al.* (1968) distinguished a "male mating center" located in the anterior hypothalamic area and a "female mating center" located in the ventromedial hypothalamic area. In rats of both sexes, predominantly male or female sexual behavior could be selectively stimulated or abolished either by intrahypothalamic sex hormone implants or by hypothalamic electrolytic lesions in these regions.

Today's knowledge regarding the hypothalamo–anterior pituitary–gonadal axis may be summarized as follows. In the so-called hypophyseotropic area of the basal central hypothalamus (Halász *et al.*, 1962) FSH- and LH-releasing substances are secreted under the influence of neurotransmitters (Kamberi, 1974). They are transported by the hypothalamo–hypophyseal portal vessels to the anterior pituitary, to stimulate the secretion of the appropriate gonadotropin. In females, an additional cyclic sex center is present in the rostral hypothalamus, responsible for the cyclic "over"-production of the LH-releasing substance. Periodically, an "over"-release of hypophyseal LH occurs, promoting the induction of ovulation.

The hypophyseal gonadotropic hormones regulate the generative gonadal functions as well as the secretion of sex hormones, which in turn regulate the physiological functions of the respective target organs. In addition, the sex hormones produce either a predominantly inhibitory (negative) or excitatory (positive) feedback effect on the hypothalamus and/or the hypophysis, depending on the combination of individual hormones (androgens, estrogens and gestagens) and also the concentration and duration of effect of these substances. Furthermore, the sex hormones sensitize the hypothalamic mating centers to sensory stimulations, which reach the diencephalon via pathways from the cerebral cortex.

Androgens play a dual role in this respect, since they are able to stimulate both the male and female centers, which are located in the anterior hypothalamus and the central hypothalamus, respectively. The sensitivity of the centers to androgens depends on the degree of differentiation. With a more highly differentiated male center the resulting sexual behavior will be male while female sexual behavior is seen if the corresponding center is more highly differentiated (Dörner, 1972).

SEX HORMONE DEPENDENT BRAIN ORGANIZATION AND REPRODUCTION

As early as 1936, Pfeiffer observed that in rats, independent of genetic sex, the lack of testes during a critical developmental phase resulted in the genesis of a cyclic hypophyseal gonadotropin release, whereas the presence of testes gave rise to a tonic (continuous) hypophyseal gonadotropin secretion. Thus, ovarian cycles could be induced in male rats castrated immediately after birth and implanted with ovaries when adult. In contrast, ovulations could be produced neither in female rats implanted with testes neonatally nor in male rats orchidectomized after the critical neonatal period and then implanted with ovaries.

Wilson and coworkers (1941) reported that female rats androgenized during the early neonatal period showed acyclic ovarian function without corpora lutea in adulthood. Barraclough and Gorski (1961) showed that the development of a cyclic or tonic hypophyseal gonadotropin secretion is determined by an androgen-dependent hypothalamic differentiation phase. A low androgen level during this critical phase

References p. 233–236

produced a cyclic hypophyseal gonadotropin release, whereas a high androgen level resulted in a tonic hypophyseal gonadotropin secretion. Takewaki (1962) confirmed the earlier findings of Wilson (1943) that a high estrogen level during the hypothalamic differentiation phase can also cause tonic gonadotropin secretion and acyclic function during postpuberal life.

With regard to sexual behavior, a remarkable observation was reported by Dantchakoff (1938) that was confirmed by Phoenix *et al.* (1959). Female guinea pigs, androgenized prenatally, exhibited strong male sexual behavior when the androgen treatment was repeated in adulthood. Based on these results, Phoenix and coworkers (1959) distinguished a prenatal differentiation period and a postpuberal activation period with respect to sexual behavior. During the first phase, differentiation of the hypothalamus is dependent on the androgen level, while in the second phase the hypothalamus is activated by either androgens or estrogens. Wilson *et al.* (1941) and Wilson (1943) observed reduced female sexual behavior in adult female rats which had been androgenized neonatally. Grady and Phoenix (1963) and Harris (1964) reported that male rats, orchidectomized shortly after birth, showed particularly strong female sexual behavior when treated with estrogen in adulthood. These observations, confirmed by other workers (Feder and Whalen, 1965; Neumann *et al.*, 1967, 1970), pointed to the possible significance of the androgen level during the critical hypothalamic differentiation phase for the development of the direction of sex drive.

The following findings were obtained in our laboratories concerning sex hormone dependent brain differentiation.

(1) Male rats castrated on the first day of life showed predominantly heterotypical, *i.e.* homosexual, behavior following androgen substitution in adulthood (Dörner, 1967, 1969, 1970, 1972). In other words, genetic males displaying a temporary androgen deficiency during the critical hypothalamic differentiation period, but normal or approximately normal androgen level in adulthood, were excited preferentially by partners of the same genetic sex.

(2) The neuroendocrine conditioned male homosexuality could be completely prevented by a single androgen injection during the perinatal hypothalamic differentiation period or suppressed by stereotaxic lesions in the hypothalamic ventromedial nuclei during adulthood (Dörner and Hinz, 1968; Dörner *et al.*, 1968).

(3) The higher the androgen level during the hypothalamic differentiation period, the stronger developed the male and the weaker the female sexual behavior during the postpuberal functional period, irrespective of the genetic sex. A complete inversion of sexual behavior was even observed in male and female rats following androgen deficiency in the males and androgen overdoses in the females during the hypothalamic differentiation period. Due to these findings, a neuroendocrine predisposition for primary hyposexuality, bisexuality and homosexuality may be based on different degrees of androgen deficiency in males and androgen (or even estrogen) overdoses in females (Dörner, 1969, 1970).

(4) The higher the androgen level during the critical hypothalamic differentiation period, the smaller the nuclear volumes of the nerve cells observed in the preoptic

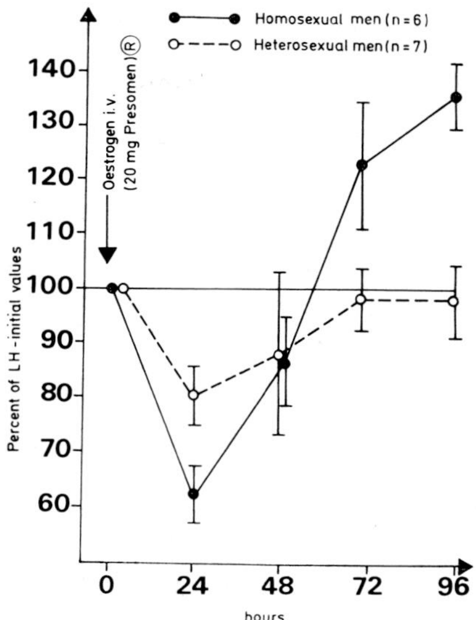

Fig. 2. Changes in LH values in serum of homosexual and heterosexual men after intravenous injection of estrogen (mean ± S.E.M.).

anterior hypothalamic area and in the ventromedial nucleus throughout life (Dörner and Staudt, 1969).

(5) In male rats castrated on the first day of life, a positive estrogen feedback-effect could be induced as well as in normal females, but not in males castrated later nor in neonatally androgenized (or estrogenized) females (Dörner and Döcke, 1964; Döcke and Dörner, 1966).

(6) A positive estrogen feedback-effect could also be induced in homosexual men in contrast to heterosexual men (Dörner *et al.*, 1972), as illustrated in Fig. 2. In the meantime, this finding was confirmed by comparing the LH-response to estrogen of 21 homosexual men with that of 20 heterosexual men (Dörner *et al.*, in preparation). The induction of a positive estrogen feedback-effect in women has been described by several authors during recent years (Van de Wiele *et al.*, 1970; Nillius and Wide, 1971; Tsai and Yen, 1971; Franchimont, 1972). Hence, our data suggest that homosexual men may possess, in fact, a predominantly female-differentiated brain.

(7) In female rats nonphysiologically high androgen and/or estrogen levels during the perinatal hypothalamic differentiation phase caused symptoms similar to the Stein–Leventhal syndrome with anovulatory sterility and polycystic ovaries, as already described by other authors (Barraclough and Gorski, 1961; Flerkó *et al.*, 1969; Lloyd and Weisz, 1969) and/or a neuroendocrine predisposition for female hypo-, bi- or homosexuality (Dörner and Fatschel, 1970; Dörner 1972; Dörner and Hinz, 1972).

(8) Very high androgen or estrogen doses administered during the critical hypothalamic differentiation period gave rise to hypogonadism in both sexes.

In summary, the following correlations were found between changes of the androgen

References p. 233–236

TABLE I

RELATIONSHIP BETWEEN CHANGES IN THE ANDROGEN AND/OR ESTROGEN LEVELS DURING THE HYPOTHALAMIC DIFFERENTIATION PHASE AND PERMANENT SEXUAL DISORDERS DURING THE HYPOTHALAMIC FUNCTIONAL PHASE

Hypothalamic differentiation phase	*Hypothalamic functional phase*
(1) Androgen deficiency in genetic males	Predisposition for male hypo-, bi- or homosexuality
(2) Moderate androgen or estrogen overdosage in genetic females	Stein–Leventhal-like syndrome; predisposition for female hypo-, bi- or homosexuality
(3) Very high androgen or estrogen levels in genetic males or females	Hypogonadotropic hypogonadism (idiopathic eunuchoidism; idiopathic hypothalamic insufficiency)

and/or estrogen levels during the hypothalamic differentiation phase and permanent sexual disorders during the postpuberal hypothalamic functional phase (Table I):

(1) In genetic males, an androgen deficiency during the hypothalamic organization phase results in a more or less female differentiation of the brain; *i.e.* a neuroendocrine predisposition for male hyposexuality, bisexuality or even homosexuality and the evocability of a positive estrogen feedback-effect.

(2) In genetic females, overdoses of androgen (or estrogen) during the hypothalamic organization phase lead to more or less male differentiation of the brain, *i.e.* a Stein–Leventhal-like syndrome and/or a neuroendocrine predisposition for female hypo-, bi- or homosexuality.

(3) Very high androgen and/or estrogen levels during hypothalamic differentiation give rise to hypogonadotropic hypogonadism in both sexes.

In view of these findings, important disturbances of sexual functions may be based on discrepancies between the genetic sex and the sex hormone level during the hypothalamic differentiation phase. A causal prophylaxis may become possible in the future by preventing such discrepancies during the time of sex-specific brain differentiation. Three preconditions towards this aim have already been achieved:

(1) Comparative studies of hypothalamic biomorphosis in 84 human fetuses and hundreds of rats have led to the conclusion that the critical hypothalamic differentiation period may be timed in the human between the 4th and 7th month of fetal life (Dörner and Staudt, 1972).

(2) A simple and reliable method for prenatal diagnosis of the genetic sex was developed using fluorescence microscopy of amniotic fluid cells (Dörner *et al.*, 1971b, 1973b).

(3) Significantly higher testosterone concentrations were found in amniotic fluids of male fetuses than in female fetuses (Dörner *et al.*, 1973c).

In view of these findings, preventive therapy for human subjects with disorders of sex hormone dependent brain organization may become possible in the future. Furthermore, the following ontogenetic organization rules (Dörner, 1973a) were deduced from animal experiments and clinical studies (Fig. 3):

(1) During a critical hypothalamic organization phase an open-loop system (*e.g.* placenta–fetal gonad–fetal hypothalamus) is converted into a feedback control system (hypothalamo–hypophyseal–gonadal system). The neuroendocrine contact between the hypothalamus and the anterior pituitary is established by the production of neurohormones and their transport via the developing hypophyseal portal vessels. The regulating variable and the regulated subsystem of the primary open-loop system (*e.g.* sex hormone and hypothalamus) are then transformed into the controlled condition (homeostatic variable) and the regulator, respectively, of the secondary feedback control system (transformation rule).

(2) The size of the primary regulating variable (*e.g.* sex hormone) during the critical organization phase determines the quality of the hypothalamic regulator during the functional phase (determination rule); *i.e.*, the rated (or reference) value of the regulator and subsequently the functional range of the whole neuroendocrine system is irreversibly established (*e.g.* normo- or hypogonadotropic; cyclic or acyclic; hetero-, hypo-, bi- or homosexual).

As a possible explanation of this irreversible determination at the molecular level, the following theory may be advanced. Three different periods of hormonal actions on the genetic material may be distinguished: (1) differentiation, (2) maturation, and (3) functional phase, although there exists some overlapping between these periods.

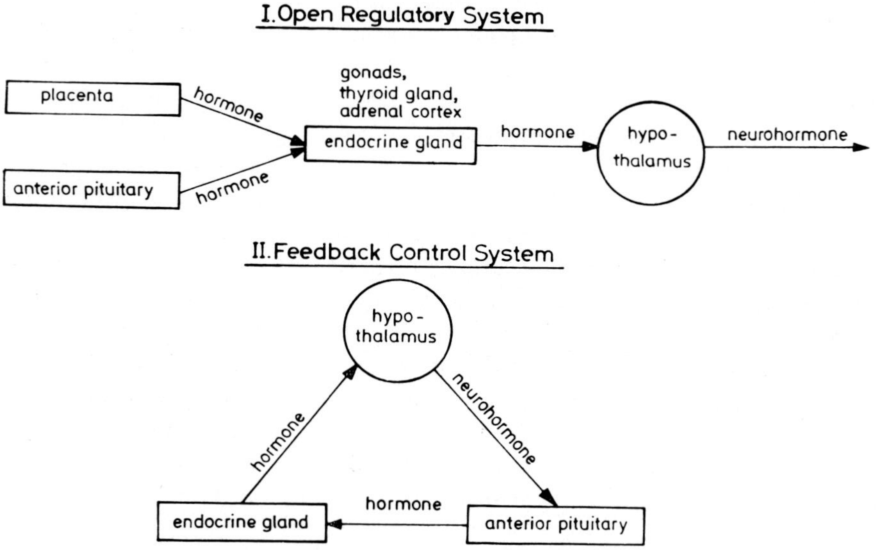

Fig. 3. Schematic representation of the ontogenetic organization of a neuroendocrine system: (a) hypothalamus–anterior pituitary–endocrine gland.

References p. 233–236

(1) During a critical organization phase, the specification, *i.e.* the capability of transcription (transcriptability) of specific genes in nerve cells is determined by the quantity of specific effectors (*e.g.* sex hormones).

(2) During the maturation phase, the transcriptability and/or translatability of these nerve cells may then be modulated (*i.e.*, changed quantitatively) in relation to these effectors, *e.g.*, by the development of specific receptors.

(3) During the functional phase, transcription and/or translation are then induced (or repressed) by these effectors.

HORMONE DEPENDENT BRAIN ORGANIZATION AND OTHER NEUROENDOCRINE FUNCTIONS

The ontogenetic organization rules deduced from differentiation processes in the hypothalamo–hypophyseal–gonadal system regulating the maintenance of species may also be valid for other neuroendocrine systems that are responsible for the maintenance of individual life. In this connection the following neuroendocrine systems may be mentioned:

(1) *Hypothalamo–hypophyseal–thyroid system with the control of the thyroid hormone level.* Thyroid hormone deficiency during the organization phases caused by lack of iodine gives rise to cretinism (Labhart, 1971). In animal experiments, irreversible structural changes of specific hypothalamic nuclei were observed following thyroxine deficiency during the differentiation phase (Szántó *et al.*, 1970). Finally, thyroid hormone overdoses during the hypothalamic organization phase result in secondary hypothyroidism for the entire life (Bakke *et al.*, 1972).

(2) *Hypothalamo–hypophyseal–adrenocortical system with the control of glucocorticoid level.* Glucocorticoid overdosage during the hypothalamic organization period leads to inhibition of development associated with hypothalamic disturbances (Sawano *et al.*, 1969). Furthermore, we have found that high doses of progestagens with glucocorticoid-like activities administered during the hypothalamic organization period give rise to permanent adrenal atrophy in rats (Dörner *et al.*, 1971a).

(3) *Neuroendocrine feedback system responsible for the carbohydrate and fat metabolism with the control of the insulin level.* Since disorders of the carbohydrate and fat metabolism and their complications belong to the most important diseases at all, knowledge about their etiopathogenesis is of particular relevance. A hypothalamic satiety center, located in the ventromedial nucleus, is provided with specific glucoreceptors. An elevation of glucose utilization results in an increased activity of this center, and hence of satiety. In contrast, diminution of glucose utilization leads to a decreased activity of this satiety center which, on its part, influences a feeding center located in the hypothalamic ventrolateral area; *i.e.*, a decreased activity of the satiety center results in an increased activity of the feeding center associated with hunger, whereas activation of the satiety center causes an inactivation of the feeding center (Anand *et al.*, 1962; Anand, 1971).

Besides the "satiety neurons", neurons are also located in the ventromedial hypothalamus, responsible for an increased production of growth hormone-releasing

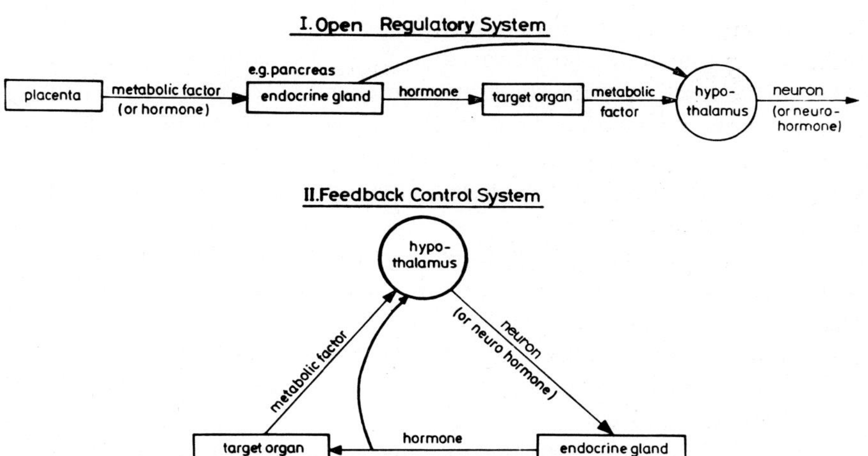

Fig. 4. Schematic representation of the ontogenetic organization of a neuroendocrine system: (b) hypothalamus–endocrine gland (*e.g.* pancreas)–target organ.

hormone (GH-RH) following a decrease of the glucose level in blood (Bernardis and Frohman, 1970). The GH-RH transported in the portal vessels of the hypophyseal stalk to the anterior pituitary stimulates the secretion of hypophyseal growth hormone (GH). GH, on its part, inhibits glucose utilization and enhances lipolysis. In addition, an activation of the ventromedial nucleus may lead to glycogenolysis and lipolysis through stimulation of the sympathetic nervous system associated with elevated epinephrine release (Shimazu *et al.*, 1966). A similar effect is exerted by glucagon produced in the pancreatic α-cells. A hypothalamic influence on the corticotropin-releasing factor, and hence on the release of ACTH which can raise the blood glucose by stimulating the secretion of cortisol, may also be considered. On the other hand, activation of the hypothalamic feeding center, located in the ventrolateral area by stimulating the vagal nerve, gives rise to an increased release of insulin from the pancreatic β-cells (Kaneto *et al.*, 1967). Insulin enhances glucose utilization leading to hypoglycemia and increased lipogenesis. The elevated glucose utilization exerts, in turn, a stimulating effect on the hypothalamic satiety center, while hypoglycemia causes a derepressing action on the GH-RH producing neurons. Thus, a self-regulating system responsible for carbohydrate and fat metabolism shows the properties of a closed-loop system.

This neuroendocrine feedback control system for carbohydrate and fat metabolism may develop again from an open-loop regulatory system (Fig. 4). In this case, the placental glucose transfer from the mother determines the quantity of insulin production of the fetal pancreas regulating glucose level and lipogenesis in the fetus (Steinke and Driscoll, 1965; Oakley *et al.*, 1972). Furthermore, insulin may represent a (direct or — by means of glucose utilization and/or glucose level — indirect) regulating variable for the fetal hypothalamus during its organization period.

The central question is as follows: is insulin as the regulating and then homeostatic variable capable of inducing irreversible changes during the hypothalamic organization

References p. 233–236

phase concerning the function and/or structure of the hypothalamic regulator responsible for carbohydrate and fat metabolism during the hypothalamic functional phase? In this context, the following experimental and clinical findings were obtained:

(1) In male rats, a temporary hyperinsulinism was produced during the hypothalamic differentiation phase by means of insulin administration followed by a decreased glucose tolerance, hypercholesterolemia and overweight in juvenile and/or adult life (Dörner and Götz, 1972).

(2) Moreover, these perinatally insulinized adult rats displayed a significant increase of the nuclear volumes of neurons in the hypothalamic ventromedial nucleus (Dörner and Staudt, unpublished data).

(3) In 3745 diabetics, relatives of the mother's side showed a clearcut higher occurrence of diabetes ($P < 0.001$) than those of the father's side (Dörner *et al.*, 1973a; Mohnike *et al.*, 1973).

(4) Human subjects born in periods with high food supply (*i.e.* carried by mothers with an increased tendency to hyperglycemia during pregnancy) had also a significantly increased frequency of diabetes in adulthood as compared to subjects of similar age but born in a post-war period with shortage of food supply (Dörner and Mohnike, 1973a).

(5) Children of mothers with glucosuria during pregnancy displayed a significantly higher occurrence of glucosuria ($P < 0.001$) than children of control mothers (Mohnike *et al.*, 1973).

(6) Children of mothers with glucosuria and decreased glucose tolerance during pregnancy showed a primary hyperinsulinism associated with diminished insulin response to glucose loading (Dörner *et al.*, 1973c). Similar responses were described in prediabetics (Cerasi and Luft, 1972). Hence, our findings suggest that fetal hyperinsulinism, accessible to a preventive therapy, may represent in fact a predisposing factor for the postnatal development of diabetes mellitus and obesity too.

Some additional experimental data in support of this theory are:

(1) the offspring of female rats treated with alloxan during pregnancy showed hyperglycemia and overweight in juvenile and adult life (Bartelheimer and Kloos, 1952);

(2) when alloxan diabetic rats became pregnant, many of their descendants showed spontaneous diabetes (Baranov, personal communication);

(3) Okamoto (1965) produced experimental diabetes in several successive parent generations of rats, rabbits and guinea pigs by use of alloxan and cortisol, respectively. In the F4–F5 generations of these animals, spontaneous diabetes occurred. This finding was confirmed by other authors (Görgen *et al.*, 1967; Kramer and Schulze, 1969) who observed a highly significant predominance with regard to the transmission of diabetes on the mother's side. In my opinion, genetic selections as well as perinatal enduring modifications for the development of diabetes mellitus may have been produced in these experiments.

As far as the pancreas was examined histologically, a marked hypertrophy of the Langerhans' islets was found in the newborn animals indicating in fact a hyperinsulinism existing during the critical hypothalamic differentiation period.

BRAIN ORGANIZATION AND NEUROVEGETATIVE OR NEURONAL BEHAVIORAL FUNCTIONS

It remains to be examined whether the ontogenetic organization rules mentioned above hold also true for neurovegetative and neuronal behavioral systems. As possible controlled conditions (homeostatic variables) of neurovegetative feedback control systems being able to act as determining regulating variables during critical brain differentiation periods may be mentioned: (1) food supply for the nutritional system, (2) O_2- and CO_2-tensions for the cardiovascular and respiratory systems, and (3) body temperature for the thermoregulatory feedback control system.

In an initial clinical study, it was found (Dörner, 1973b) that adult men born in periods with high food supply (1930–1939 and 1950–1953) showed, in fact, highly significantly increased body weights per body length as compared to men of similar age but born in war or post-war periods with shortage of food supply (1941–1946). The absolute difference of body weight between these two groups was 6 kg on the average.

Furthermore, we might confirm recent observations of Knittle (1972) made in rats that there exists a positive correlation between the quantity of perinatal food supply and the development of obesity in juvenile and/or adult life, which was explained, however, by the induction of fat cell hyperplasia during a critical phase of ontogeny. On the other hand, these findings suggest that the ontogenetic organization rules may also hold true, in fact, for neurovegetative systems, as illustrated in Fig. 5. But extensive experiments are yet to be done along this line in order to examine the possible validity of these rules for further neurovegetative systems regulated by subcortical brain regions, *e.g.* the thermoregulatory, cardiovascular and respiratory system. Thus, the idiopathic respiratory distress syndrome of newborns with prenatal anoxia

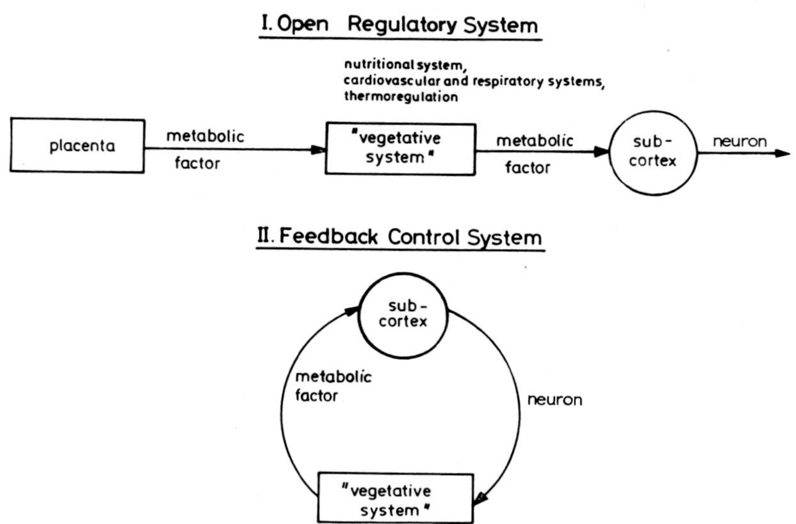

Fig. 5. Schematic representation of the possible ontogenetic organization of neurovegetative systems.

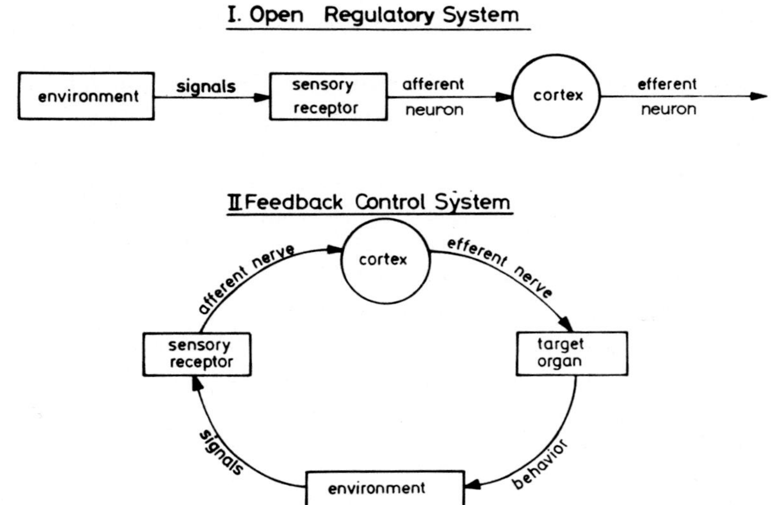

Fig. 6. Schematic representation of the ontogenetic organization of neuronal behavioral systems.

(Haworth and James, 1970) may be caused, for instance, by an increased P_{CO_2}, *i.e.* an abnormal concentration of the regulating and later on homeostatic variable CO_2 and pH, respectively, during the critical organization period of the subcortical respiratory center.

On the other hand, with regard to neuronal behavioral systems, regulated by the cerebral cortex the following theory is advanced. The cybernetic control of neuronal behavioral systems is — similar to that of neuroendocrine and neurovegetative systems — dependent on determining and modulating factors.

(1) Determinations may be caused again dependent of the genetic material by the internal and external environment during critical periods of brain differentiation, *i.e.* during cytogenesis (and/or synaptogenesis) of specific brain regions. In these critical periods, the transcriptability of neuronal DNA may be preprogrammed by environmental signals. Such permanent alterations may be caused during cell differentiations by the environment dependent production of specific regulatory proteins, as nuclear histones and/or acidic proteins, or even by changes in DNA sequences, *e.g.* translocation, paramutation or gene amplification (see Edelman and Gally, 1970). As shown in Fig. 6, an open-loop regulatory system may be converted again into a feedback control system regulating behavior. Recent publications of several authors speak in favor of the fact that environmental signals can act, indeed, as determining regulatory variable during critical organization periods for various neuronal behavioral systems (Henderson, 1967; Capretta, 1969; Galambos and Hillyard, 1970; Immelmann, 1970; Tembrock, 1971). Padilla described as early as 1935, that newly hatched chicks, if prevented from pecking for 2 weeks, can starve to death beside a pile of grain.

(2) Modulations of neuronal behavioral systems can then be caused by internal and/or external environmental signals during maturation and/or functional phases of the brain; *e.g.* (a) internal modulations through hormones and/or metabolic param-

eters, and (b) external motivations through habituation, conditioning, and conscious experiences.

CONCLUSIONS

Disorders of differentiation are either caused by primary genetic defects or by modifications acquired during critical organization periods of pre- and/or early post-natal life. All treatment of genetic diseases is purely symptomatic, and the search for early and lasting corrections of primary genetic defects has been unsuccessful so far. Genetic exchange, if possible as a therapeutic principle, will be restricted to a small number of selected patients (Fuhrmann, 1972). Thus, genetic counseling will remain more important than any genetic manipulation (engineering) or even interruptions of pregnancies in cases of prenatally diagnosed genetic defects.

In contrast, a causal preventive therapy seems to be possible in structural and/or functional disturbances established as modifications during critical ontogenetic organization periods. This principle has been used already most successful, as far as congenital malformation and/or mental defects are concerned caused by infections, radiation, immunological conflicts, oxygen deficiency or drugs during critical periods of pre- and/or perinatal life. On the other hand, in my opinion, it has not been realized sufficiently in order to prevent disorders of neuroendocrine, neurovegetative and neuronal behavioral systems. Thus, "structural teratology" must be completed by "functional teratology". In this connection, it should be emphasized that physiological effectors, *e.g.* hormones, metabolic products and presumably neurotransmitters too, can act as teratogens in nonphysiological concentrations and during critical organization periods, especially of the brain. Many of the so-called idiopathic, *i.e.* etiopathogenetically unknown functional disturbances, diseases and syndromes might be explained by this principle and even prevented in the future.

REFERENCES

Anand, B. K. (1971) Regulation of food and water intake. In *Proc. int. Union physiol. Sci., XXV Int. Cong., Vol. VIII, Munich*, pp. 11–12.

Anand, B. K., Chling, G. S. and Singh, B. (1962) Effect of glucose on the activity of hypothalamic "feeding centers". *Science*, **138**, 597–598.

Aschheim, S. und Zondek, B. (1927) Hypophysenvorderlappenhormon und Ovarialhormon im Harn von Schwangeren. *Klin. Wschr.*, **6**, 1322.

Aschner, B. (1912) Über die Beziehungen zwischen Hypophyse und Genitale. *Arch. Gynäk.*, **97**, 200–228.

Bakke, I. L., Lawrence, N. and Robinson, S. (1972) Late effects of thyroxine injected into the hypothalamus of the neonatal rat. *Neuroendocrinology*, **10**, 183–195.

Barraclough, C. A. and Gorski, R. A. (1961) Evidence that the hypothalamus is responsible for androgen-induced sterility in the female rat. *Endocrinology*, **68**, 68–79.

Bartelheimer, H. und Kloos, H. (1952) Die Auswirkung des experimentellen Diabetes auf Gravidität und Nachkommenschaft. *Z. ges. exp. Med.*, **119**, 246–265.

Bernardis, L. L. and Frohman, L. A. (1970) Effect of lesion size in the ventromedial hypothalamus on growth hormone and insulin levels in weanling rats. *Neuroendocrinology*, **6**, 319–328.

BROOKHART, J. M. AND DEY, F. L. (1941) Reduction of sexual behaviour in male guinea pigs by hypothalamic lesions. *Amer. J. Physiol.*, **133**, 551–554.

CAPRETTA, P. J. (1969) The establishment of food preferences in chicks *Gallus gallus*. *Anim. Behav.*, **17**, 229–231.

CERASI, E. AND LUFT, R. (1972) Clinical diabetes and theories of pathogenesis. In: *Handbook of Physiology, Vol. 1: Endocrine pancreas*, F. F. STEINER AND N. FREINKEL (Eds.), Williams and Wilkins, Baltimore, Md., pp. 627–640.

DANTCHAKOFF, V. (1938) Rôle des hormones dans la manifestation des instincts sexuels. *C. R. Acad. Sci. (Paris)*, **206**, 945–947.

DÖCKE, F. UND DÖRNER, G. (1966) Tierexperimentelle Untersuchungen zur Ovulationsauslösung mit Gonadotropinen und Östrogenen. 4. Mitt., Zur neurohormonalen Regulation der Ovulation. *Zbl. Gynäk.*, **88**, 273–282.

DÖRNER, G. (1967) Tierexperimentelle Untersuchungen zur Frage einer hormonalen Pathogenese der Homosexualität. *Acta biol. med. germ.*, **19**, 569–584.

DÖRNER, G. (1969) Zur Frage einer neuroendokrinen Pathogenese, Prophylaxe und Therapie angeborener Sexualdeviationen. *Dtsch. med. Wschr.*, **94**, 390–396.

DÖRNER, G. (1970) The influence of sex hormones during the hypothalamic differentiation and maturation phases on gonadal function and sexual behaviour during the hypothalamic functional phase. *Endokrinologie*, **56**, 280–291.

DÖRNER, G. (1972) *Sexualhormonabhängige Gehirndifferenzierung und Sexualität*, Gustav Fischer, Jena and Springer, Berlin.

DÖRNER, G. (1973a) Zur Bedeutung prä- oder perinataler Umweltsbedingungen für die postnatale Regelung neuroendokriner Systeme. *Endokrinologie*, **61**, 107–124.

DÖRNER, G. (1973b) Die mögliche Bedeutung der prä- und/oder perinatalen Ernährung für die Pathogenese der Obesitas. *Acta biol. med. germ.*, **30**, K 19–22.

DÖRNER, G. AND DÖCKE, F. (1964) Sex-specific reaction of the hypothalamo–hypophysial system of rats. *J. Endocr.*, **30**, 265–266.

DÖRNER, G., DÖCKE, F. UND HINZ, G. (1968) Entwicklung und Rückbildung neuroendokrin bedingter männlicher Homosexualität. *Acta biol. med. germ.*, **21**, 577–580.

DÖRNER, G. UND FATSCHEL, J. (1970) Wirkungen neonatal verabreichter Androgene und Antiandrogene auf Sexualverhalten und Fertilität von Rattenweibchen. *Endokrinologie*, **56**, 29–48.

DÖRNER, G. UND GÖTZ, F. (1972) Hyperglykämie und Übergewicht bei neonatal insulinisierten erwachsenen Rattenmännchen. *Acta biol. med. germ.*, **29**, 467–470.

DÖRNER, G. AND HINZ, G. (1968) Induction and prevention of male homosexuality by androgen. *J. Endocr.*, **40**, 387–388.

DÖRNER, G. UND HINZ, G. (1972) Neuroendokrin bedingte Prädisposition für weibliche Homosexualität bei erhaltener zyklischer Ovarialfunktion. *Endokrinologie*, **59**, 48–52.

DÖRNER, G. UND MOHNIKE, A. (1973a) Zur möglichen Bedeutung der prä- und/oder frühpostnatalen Ernährung für die Pathogenese des Diabetes mellitus. *Acta biol. med. germ.*, **31**, K7–K10.

DÖRNER, G., MOHNIKE, A., HONIGMANN, G., SINGER, P. UND PADELT, H. (1973b) Zur möglichen Bedeutung eines pränatalen Hyperinsulinismus für die postnatale Entwicklung eines Diabetes mellitus. *Endokrinologie*, **61**, 430–432.

DÖRNER, G., PHUONG, N. T. UND HINZ, G. (1971a) Einfluß von Chlormadinonazetat auf die Hypothalamusdifferenzierung männlicher Ratten. *Acta biol. med. germ.*, **26**, 105–114.

DÖRNER, G., ROHDE, W. UND BAUMGARTEN, G. (1971b) Bestimmung des genetischen Geschlechts des Foeten durch fluoreszenzmikroskopischen Nachweis des Y-Chromosoms in Fruchtwasserzellen. *Acta biol. med. germ.*, **26**, 1095–1098.

DÖRNER, G., ROHDE, W., BAUMGARTEN, G., HERTER, U., HALLE, H., GRUBER, G., RÖSSNER, P., BERGMANN, K. H., GÖTZ, F. UND ZILLMANN, R. (1973b) Zur pränatalen Geschlechtsbestimmung im Fruchtwasser und peripheren mütterlichen Blut durch fluoreszenzmikroskopischen Nachweis des Y-Chromosoms. *Zbl. Gynäk.*, **95**, 625–634.

DÖRNER, G., ROHDE, W. UND KRELL, L. (1972) Auslösung eines positiven Östrogenfeedback-Effekt bei homosexuellen Männern. *Endokrinologie*, **60**, 297–301.

DÖRNER, G., ROHDE, W., STAHL, F., KRELL, L. AND MASIUS, W. G. Neuroendocrine conditioned predisposition for homosexuality in men. *Arch. sex. Beh.*, in preparation.

DÖRNER, G., STAHL, F., ROHDE, W., HALLE, H., RÖSSNER, P., GRUBER, D. UND HERTER, U. (1973) Radioimmunologische Bestimmung des Testosterongehalts im Fruchtwasser männlicher und weiblicher Feten. *Endokrinologie*, **61**, 317–320.

DÖRNER, G. AND STAUDT, J. (1969) Structural changes in the hypothalamic ventromedial nucleus of the male rat, following neonatal castration and androgen treatment. *Neuroendocrinology*, **4**, 278–281.

DÖRNER, G. UND STAUDT, J. (1972) Vergleichende morphologische Untersuchungen der Hypothalamusdifferenzierung bei Ratte und Mensch. *Endokrinologie*, **59**, 152–155.

EDELMAN, G. M. AND GALLY, J. A. (1970) Arrangement and evolution of eukaryotic genes. In *The Neurosciences*, F. O. SCHMITT (Ed.), Rockefeller Univ. Press, New York, pp. 962–972.

FEDER, H. H. AND WHALEN, R. E. (1965) Feminine behavior in neonatally castrated and estrogen-treated male rats. *Science*, **147**, 306–307.

FLERKÓ, B., PETRUSZ, B. AND TIMA, L. (1969) On the mechanism of androgen sterilization. *Neuroendocrinology*, **4**, 164–169.

FRANCHIMONT, P. (1972) Human gonadotrophin secretion. *J. roy. Coll. Phy. Lond.*, **6**, 283–298.

FUHRMANN, W. (1972) Therapy of genetic diseases in man and the possible place of genetic engineering. In *Advances in the Biosciences, 8. Workshop on Mechanisms and Prospects of Genetic Exchange*, G. RASPÉ (Ed.), Pergamon, London, pp. 387–395.

GALAMBOS, R. AND HILLYARD, S. A. (1970) Determinants of neural and behavioral plasticity: an overview. In *The Neurosciences*, F. O. SCHMITT (Ed.), Rockefeller Univ. Press, New York, pp. 289–297.

GÖRGEN, H., KRAMER, M. AND SCHULZE, E. (1967) Production of spontaneous diabetes by the treatment of the parent generations with alloxan. *Naunyn-Schmiedeberg's Arch. exp. Path. Pharmak.*, **259**, 173–174.

GRADY, K. L. AND PHOENIX, C. H. (1963) Hormonal determinants of mating behavior by adult male rats castrated neonatally. *Amer. Zool.*, **3**, 482–487.

HALÁSZ, B., PUPP, L. AND UHLARIK, S. (1962) Hypophysiotrophic area in the hypothalamus. *J. Endocr.*, **25**, 147–154.

HARRIS, G. W. (1953) Die Physiologie des Hypothalamus und der Hypophyse in Beziehung zur Gynäkologie. *Geburtsh. u. Frauenheilk.*, **13**, 70–71.

HARRIS, G. W. (1964) Sex hormones, brain development and brain function. *Endocrinology*, **75**, 627–648.

HAWORTH, S. G. AND JAMES, L. S. (1970) The respiratory distress syndrome. In *Scientific Foundation of Obstetrics and Gynaecology*, E. E. PHILLIP, J. BARNES AND M. NEWTON (Eds.), Heinemann, London, pp. 411–421.

HENDERSON, N. D. (1967) Prior treatment effects on open field behaviour of mice — a genetic analysis, *Anim. Behav.*, **15**, 364–376.

HOHLWEG, W. (1934) Veränderungen des Hypophysenvorderlappens und des Ovariums nach Behandlung mit großen Dosen von Follikelhormon. *Klin. Wschr.*, **13**, 92–95.

HOHLWEG, W. UND JUNKMANN, K. (1932) Die hormonal-nervöse Regulierung der Funktion des Hypophysenvorderlappens. *Klin. Wschr.*, **11**, 321–323.

IMMELMANN, K. (1970) Lernen durch Prägung. *Naturw. u. Med.*, **7**, 15–29.

KAMBERI, I. (1974) Role of brain monoamines in regulation of the secretion of hypophysiotropic hormones of the hypothalamus. In *Endocrinology of Sex*, G. DÖRNER (Ed.), Barth, Leipzig, in press.

KANETO, A., KOSAKA, K. AND NAKOA, K. (1967) The effect of stimulation of the vagus nerve on insulin secretion. *Endocrinology*, **80**, 530–536.

KNITTLE, J. L. (1972) Maternal diet as a factor in adipose tissue cellularity and metabolism in the young rat. *J. Nutr.*, **102**, 427–434.

KRAMER, M. AND SCHULZE, E. (1969) Erzeugung eines Spontan-Diabetes bei Ratten durch Behandlung der Elterngeneration mit Alloxan. In *Advances in the Biosciences, 1. Schering Symposium on Endocrinology*, G. RASPÉ (Ed.), Pergamon, London, pp. 222–230.

LABHART, A. (1971) *Klinik der inneren Sekretion*, Springer, Berlin, pp. 215–224.

LLOYD, C. W. AND WEISZ, J. (1969) Interrelationships of steroid and neural mechanisms. In *Proc. 2nd Internat. Congr. Hormonal Steroids*, L. MARTINI, F. FRASCHINI AND M. MOTTA (Eds.), Excerpta Medica, Amsterdam, pp. 917–924.

MCCANN, S. M., TALEISNIK, M. S. AND FRIEDMAN, H. M. (1960) LH-releasing activity in hypothalamic extracts. *Proc. Soc. exp. Biol. (N.Y.)*, **104**, 432–434.

MOHNIKE, A., STEINDEL, E. UND DÖRNER, G. (1973) Graviditätsglukosurie, ein intrauteriner Diabetesrisikofaktor? *Dtsch. Gesundh.-Wes.*, **28**, 279–283.

NEUMANN, F., VON BERSWORDT-WALLRABE, R., ELGER, W., STEINBECK, H., HAHN, J. D. AND KRAMER, M. (1970) Aspect of androgen-dependent events as studied by antiandrogens. *Rec. Progr. Hormone Res.*, **26**, 337–410.

NEUMANN, F., ELGER, W. UND VON BERSWORDT-WALLRABE, R. (1967) Intersexualität männlicher Feten und Hemmung androgenabhängiger Funktionen bei erwachsenen Tieren durch Testosteronblocker. *Dtsch. med. Wschr.*, **92**, 360–366.

NILLIUS, S. J. AND WIDE, L. (1971) Induction of a midcycle-like peak of luteinizing hormone in young women by exogenous oestradiol-17β. *J. Obstet. Gynaec. Brit. Cwlth.*, **78**, 822–827.

OAKLEY, N. W., BEARD, K. W. AND TURNER, R. C. (1972) Effect of substained maternal hyperglycaemia on the fetus in normal and diabetic pregnancies. *Brit. med. J.*, **1**, 466–469.

OKAMOTO, K. (1965) Induction of diabetic disposition and of spontaneous diabetes in the descendants of diabetic animals. *Proc. 2nd Int. Congr. Endocrinology, London, 1964*, Excerpta Medica, Amsterdam, p. 1018.

PADILLA, S. G. (1935) Further studies on the delayed pecking of chicks. *J. comp. Psychol.*, **20**, 413–443.

PFEIFFER, C. A. (1936) Sexual differences of the hypophyses and their determination by the gonads. *Amer. J. Anat.*, **58**, 195–225.

PHOENIX, C. H., GOY, R. W., GERALL, A. A. AND YOUNG, W. C. (1959) Organizing action of prenatally administered testosterone propionate on the tissues mediating mating behavior in the female guinea pig. *Endocrinology*, **65**, 369–382.

SAWANO, S., ARIMURA, A., SCHALLY, A. V., REDDY, T. W. AND SCHAPIRO, S. (1969) Neonatal corticoid administration: effect upon adult pituitary growth hormone releasing hormone activity. *Acta endocr. (Kbh.)*, **61**, 57–67.

SCHALLY, A. V., KASTIN, A. J. AND ARIMURA, A. (1971) Hypothalamic follicle stimulating hormone (FSH) and luteinizing hormone (LH)-regulating hormone: structure, physiology and clinical studies. *Fertil. and Steril.*, **22**, 703–721.

SHIMAZU, T., FUKUDA, A. AND BAN, T. (1966) Reciprocal influence of the ventromedial and lateral nuclei on blood glucose level and liver glycogen. *Nature (Lond.)*, **210**, 1178–1179.

SMITH, P. E. AND ENGLE, E. T. (1927) Experimental evidence regarding the role of the anterior pituitary in the development and regulation of the genital system. *Amer. J. Anat.*, **40**, 159–217.

STEINKE, J. AND DRISCOLL, G. (1965) The extractable insulin content of pancreas from fetuses and infants of diabetic and control mothers. *Diabetes*, **14**, 573–578.

SZANTÓ, L., MESS, B. AND REVICZKY, A. (1970) Interrelations of thyroid and diencephalic activity. *Acta med. Acad. Sci. hung.*, **27**, 235–242.

TAKEWAKI, K. (1962) Some aspects of hormonal mechanism involved in persistent estrus in the rat. *Experientia (Basel)*, **18**, 1–6.

TEMBROCK, G. (1971) *Grundlagen der Tierpsychologie*, Pergamon, Oxford, und Vieweg, Braunschweig.

TSAI, C. C. AND YEN, S. S. C. (1971) Acute effects of intravenous infusion of 17β-estradiol on gonadotrophin release in pre- and post-menopausal women. *J. clin. Endocr.*, **32**, 766–771.

VAN DE WIELE, K. L., BOGUMIL, F., DYRENFURTH, F., FERIN, M., JEWELEWICZ, R., WARREN, M., RIZKALIAH, T. AND MIKHAIL, G. (1970) Mechanisms regulating the menstrual cycle in women. *Recent Progr. Hormone Res.*, **26**, 63–95.

WESTMAN, A. UND JAKOBSOHN, D. (1938) Über die luteinisierende Wirkung des Follikelhormons. *Acta obstet. gynec. scand.*, **18**, 115–123.

WILSON, J. G. (1943) Reproductive capacity in adult female rats treated prepuberally with estrogenic hormone. *Anat. Rec.*, **86**, 341–359.

WILSON, J. G., HAMILTON, J. B. AND YOUNG, W. C. (1941) Influence of age and presence of the ovaries on reproductive function in rats injected with androgens. *Endocrinology*, **29**, 784–787.

DISCUSSION

VAN DER SCHOOT: In view of the far reaching consequences of the principles that you introduced in your presentation I should like to ask you a question. The picture that you presented about the development of homosexuality fits indeed very well for rats. One wonders, however, how far this general picture of sexual differentiation on the behavior and ovulation is applicable to other mammals, and in particular to human. The congenital adrenogenital syndrome in human *e.g.* is compatible with normal ovulation and pregnancy, and in monkeys it is possible to obtain a positive feedback in both males and females by means of estrogen injections. Homosexuality in men seems moreover to have a very complex genesis. It cannot only be a matter of hormones, also *e.g.* social circumferences will play a role.

DÖRNER: I would agree indeed that we have to differentiate between various forms of homosexuality. Most sexuologists agree, however, that there is a type of genuine or inborn homosexuality. In this type social circumferences cannot play an essential role by definition. Only in these cases, who never displayed an excitability by the other sex, we have shown this positive estrogen feedback effect. In heterosexual or bisexual men we could not find it.

DREWETT: The critical thing for a positive feedback in monkeys was the fact that they had to be castrated. This could mean that if you would compare the testosterone levels in homosexual and heterosexual groups, that there might be a difference.

DÖRNER: We have measured the testosterone levels in homosexual and heterosexual men by radioimmunoassay and there was no significant difference. Nevertheless, we have found a significant difference in the evokability of the positive estrogen feedback effect. In transsexual men there may be some reduction in testosterone levels, but in the homosexual men these levels were within the normal range in blood and urine.

LEVINE: I have some general comments to make. I agree with Dr. Dörner on the probability that testosterone has an effect on the sexual differentiation during ontogenesis. There is however, much basic disagreement about the concept of homosexuality. The concept of homosexuality in animals has never been established in a sense that they function in a sexual way. There is one critical question about human homosexuality in this concept: what does the human homosexual use to achieve his sexual satisfaction? He uses his penis, testosterone, and ejaculates. So he is using all the male mechanisms. His only difference is that he likes boys in stead of girls.

DÖRNER: The male animals I spoke about are preferentially, *i.e.* significantly, more excited by a partner of the same sex than by one of the other sex, that is the same as in human homosexuality. But when the animal is mounted it shows lordosis; thus female sexual behavior. The male animals I spoke about were neonatally castrated and became implanted testes or obtained testosterone again in the adult. It is not so surprising that these animals with a female differentiated hypothalamus show some female behavior, and some lordosis, but it is very surprising that they show predominantly female-like behavior and excitability, even after androgen treatment. Since androgens are not only activating the male mating center, but also the female mating center in adulthood, it is therefore important to test both, the male and the female behavior after the same manipulation, in order to see what the prefential behavior and excitability is.

LEVINE: Androgens given neonatally can have an effect on ovulation without affecting the sexual behavior.

DÖRNER: We have found that the differentiation of the mating centers and the differentiation of the gonadotropic hormone regulating centers overlap. But the differentiation is not identical. If you give androgen at the beginning of the hypothalamic differentiation period the gonadotropic hormone regulating centers are strongly influenced, but if you give androgens at the end of the hypothalamic differentiation period, *e.g.* in rats on day 10 of postnatal life, there is no more marked influence on the gonadotropic hormone regulating centers, but you do get an influence on mating centers. Then, it is possible to obtain dissociations between gonadotropic hormone secretion on the one side, and sexual behavior on the other side. We have also seen from a morphological point of view that the differentiation of the preoptic area with the cyclic center *e.g.* is completed at about 6 days after birth. There are no more proliferating cells in this region after this period. But in the ventromedial nucleus region, that is an area regulating female sexual behavior, are still proliferating cells. The differentiation of this region is only complete at day 16 of postnatal life. Therefore, if we give androgens or estrogens on day 10 of life, we can see approximately normal estrous cycles associated with clearcut changes of sexual behavior in adulthood.

The Role of the Hypothalamus in Puberty

B. T. DONOVAN

Department of Physiology, Institute of Psychiatry, London (Great Britain)

INTRODUCTION

Few would question the high importance of the hypothalamus in the control of sexual development, although progress in understanding the fundamental mechanism has been slow. The evidence for this postulate has been surveyed recently (Donovan and van der Werff ten Bosch, 1965; Critchlow and Bar-Sela, 1967) and is based upon observations of the following kind: The gonadotropin necessary for development of the gonads is secreted by the pars distalis, in that hypophysectomy arrests sexual development, and gonadotropin administration to juveniles causes sexual precocity. The secretion of the necessary follicle-stimulating hormone (FSH) and luteinizing hormone (LH) is promoted by releasing factors from the hypothalamus and in their absence is minimal, so that the operation of a drive from the hypothalamus is indicated. Accordingly, damage to the hypophyseal portal vessels carrying the releasing factors to the pars distalis depresses the output of gonadotropin and results in gonadal atrophy. Further, damage to the hypothalamus, which does not involve the portal vessels, may cause sexual precocity, as may damage to other parts of the brain. Involvement of gonadal hormones in the mechanisms controlling puberty has been suggested by the finding that gonadectomy in infancy increases the output of gonadotropin, that the systemic administration of estrogen to immature rats can precipitate gonadotropin release and advance puberty, and that the application of estrogen to the hypothalamo–hypophyseal region acts likewise. It is also almost axiomatic that the gonads are relatively quiescent during infancy and that a hypothalamic drive toward gonadotropin release, as seen in the adult, is not in evidence before puberty. Recent advances in knowledge of the changing secretion of gonadotropin secretion during development prompt a re-examination of some of the generalizations that have just been stated and an appraisal of such concepts provides the substance of the following pages. Discussion will be confined to the situation in the female, while exhaustive coverage of the topic is not practicable on this occasion.

THE SECRETION OF GONADOTROPIN

Work with radioimmunoassay techniques has recently shown that substantial amounts of gonadotropin are in circulation very early in life. The concentration of FSH and

References p. 249–252

LH in the serum of human fetuses has been followed by Grumbach and Kaplan (1973), who found maximal concentrations of serum FSH between 100 and 140 days, with an abrupt fall to basal level by 160 days. A similar peak in the serum concentration of LH in the fetus was recorded and it is noteworthy that the highest figures were obtained at a time when maternal HCG secretion was low. Grumbach and Kaplan (1973) interpreted their findings as indicative of an autonomous secretion of FSH and LH by the pituitary of the fetus between 100 and 140 days, or of the relatively unrestrained secretion at this time of the hypothalamic gonadotropin releasing factor. Later in fetal development the inhibitory feedback mechanism matures so that a hypothalamic restraint upon gonadotropin discharge becomes operative. The hypothalamo–hypophyseal system could well be functional during the period of high gonadotropin output, for monoamine fluorescence appears in the hypothalamus during the tenth week of fetal life and in the median eminence during the thirteenth week (Hyyppä, 1972) although the level of activity of the system remains to be determined. McArthur (1972) and Blizzard *et al.* (1972) have reviewed the subsequent course of gonadotropin secretion during childhood and it is clear that LH secretion may continue for about 6 months after delivery. Both FSH and LH are always detectable in the serum before a slow rise in concentration begins about 4 years before puberty. Luteinizing hormone is present in the blood of both male and female lamb fetuses from day 55 to day 120 of gestation, but from approximately day 120 to term the concentration is very low (Foster *et al.*, 1972c). The administration of hypothalamic gonadotropin releasing factor to sheep fetuses during the last quarter of gestation elicited the secretion of LH, and the magnitude of the response of 3- or 11-day-old lambs was less than that of 126- to 138-day-old fetuses (Foster *et al.*, 1972b). Neonatal gonadectomy did not enhance the increase of serum LH recorded in female lambs from about 10 days of age, whereas a marked increase was observed in male lambs from about 8 days of age (Foster *et al.*, 1972a). However, when lambs spayed at 5 days of age were followed for a longer time, a steep rise in serum LH became apparent from about 10 days of age which was suppressed by the implantation of estradiol at operation (Liefer *et al.*, 1972). These findings were taken to indicate that the testes exert a negative feedback action on gonadotropin secretion during the early neonatal period, whereas the establishment of a negative feedback system in the female occurs much later in postnatal life. High concentrations of luteinizing hormone have been detected in the blood of the fetal guinea-pig, for Donovan *et al.* (1973) have found that the immunoassayable LH concentration in the plasma of fetuses rose to a peak at about 65 days, fell sharply at birth, and then rose to a second peak about 5 days later. From this time LH declined to reach a basal level at about 20 days of postnatal age, when the concentration was approximately equal to that of adults. There were also indications of a sex difference, with a tendency for there to be more LH in the plasma of males than in females up to about 20 days of postnatal life. The guinea-pig differs from the sheep in that removal of the gonads on days 0, 5, 10, 15, 25 and 35 of postnatal life, followed by blood collection at autopsy 10 days later, caused a significant rise in the plasma LH content at all ages studied.

The factors underlying the substantial degree of gonadotropin secretion evident at some time during fetal life in the human, sheep and guinea-pig are not understood. If, as might be expected, the hypothalamus is responsible for the high level of gonadotropin secretion, then what is the cause of the inhibition or depression of hormone output after this phase? Alternatively, if gonadotropin can be secreted autonomously for a while by the pituitary gland of the fetus, how does this ability come to be lost? It is important to bear in mind that definitive proof that the immunoassayable hormone is, or can be, biologically active is desirable. Diebel *et al.* (1973) have drawn attention to major discrepancies between the FSH activity measured in rat sera and pituitary extracts by bioassay and that determined in radioimmunoassays, as well as to small, but consistent, discrepancies in the results from the 3 radioimmunoassay systems tested. They suggested that gonadal steroids might influence not only the amount, but the kind(s) of tropic hormone(s) secreted by the hypophysis. In the female rat, the concentration of immunoassayable FSH rises from around day 5 to a peak at about 15 days before declining to the concentrations found in the adult; luteinizing hormone is released in a similar manner during infancy (Weisz and Ferin, 1970; Kragt and Dahlgren, 1972; Ojeda and Ramirez, 1972; Meijs-Roelofs *et al.* 1973b), although Swerdloff *et al.* (1972) reported a near constant level. In view of the relatively high concentration of gonadotropic hormone in circulation early in infancy it is perhaps surprising that ovarian tissue from adults becomes non-functional when grafted into immature females, as was shown many years ago by Foà (1900). When the ovaries of 3-week-old rats were transplanted to a subcutaneous location in newborn animals the follicular apparatus degenerated and at the end of a week was similar to that of the host. But when gonadotropin was given the follicles of the host responded while those of the grafts remained unaltered (Picon, 1956). The interpretation of such work can be complicated (Ben-Or, 1970) but such findings have been taken to indicate that gonadotropin secretion is minimal in infancy; this conclusion is no longer satisfactory. It is still possible that the pituitary gland secretes a hormone that antagonizes the action of gonadotropin upon the ovaries in infancy (Woods and Simpson, 1961). Alternatively, is the immunoassayable hormone detectable soon after birth only part of the complete molecule, with the full synthetic capacity of the hypophysis being gained somewhat later in infancy, perhaps under the influence of gonadal hormone? On the other hand, treatment of infant mice from birth with an antiserum prepared in rabbits against rat gonadotropin inhibited ovarian growth and had disturbed granulosa cell proliferation in animals killed at 14 days of age. These changes were reversed when human FSH or human menopausal gonadotropin was given (Eshkol *et al.*, 1970; Eshkol and Lunenfeld, 1972).

GONADAL HORMONES AND GONADOTROPIN SECRETION

Gonadotropin secretion in infancy is affected by gonadal hormones (Donovan and van der Werff ten Bosch, 1965; Critchlow and Bar-Sela, 1967). Treatment of infant rats with small amounts of estrogen can advance puberty in that ovulation and estrous cycles set in earlier than otherwise anticipated (Ramirez, 1964; Ramirez and Sawyer,

References p. 249–252

1965). This reaction is apparent from about 25 days of age and can be elicited by single injections of hormone (Ying and Greep, 1971). It shows, like the Hohlweg effect (Hohlweg, 1934), that the mechanism controlling the adult type or pattern of gonadotropin secretion is fully differentiated well before it is normally put into operation. When two injections of estradiol benzoate, spaced 2 days apart, were given to immature rats, the second caused the release of LH as early as 22 days (Caligaris *et al.*, 1972). This response is an example of the positive feedback of gonadal steroids upon gonadotropin secretion; recent demonstrations of negative feedback can also be cited. Both estradiol benzoate and testosterone propionate depressed the rise in FSH and LH concentration in the serum of 6- to 9-day-old male rats that normally followed castration, and FSH secretion in spayed 5- to 10-day-old females was also inhibited (Goldman and Gorski, 1971). Caligaris *et al.* (1972) recorded an increase in the serum concentration of LH in female rats 10 days after removal of the ovaries on the day of birth or 10 days later. Taking their study of the positive feedback action of estrogen into account, these workers suggested that the failure of ovarian steroids to activate the release of LH before 20–22 days of age might indicate that some neural structure involved in the events leading to the phasic surge of LH was still immature, while the fact that serum LH had increased after ovariectomy by 10 days of age showed that the mechanism for the constant or tonic release was already functional. As the release of LH by progesterone in estrogen-primed rats depended upon the integrity of the rostral afferents to the hypothalamus (Taleisnik *et al.*, 1970), it seemed likely that the extrahypothalamic structures involved in the phasic response might not be fully developed. Thus Caligaris *et al.* (1972) distinguished between the operation of tonic and phasic mechanisms in the control of gonadotropin release during infancy and pointed to the phasic process as being the key system in the timing of puberty.

Direct assays of the amount of estrogen in circulation during infancy are of more recent date, although Kelch *et al.* (1972) have established a close correlation between the concentration of plasma estradiol and bone age, chronological age and the stage of pubertal development in children. Meijs-Roelofs *et al.* (1973a) found a remarkable parallelism in rats over days 5–21 between the curves of changing blood content of estradiol and FSH, which rose to a peak concentration around 15 days and then fell steeply to basal levels around day 25. The simultaneous presence of high concentrations of estradiol and FSH would seem to indicate that there was no estrogenic inhibition of gonadotropin release, yet ovariectomy on day 13 increased the blood content of FSH as observed 2 days later, and ovariectomy on day 15 prevented the fall seen over the next 10 days in intact females. Further, treatment of spayed immature rats with near-physiological amounts of estrogen, as judged by the maintenance of uterine weight, prevented the postgonadectomy rise in serum FSH. As these workers note, such remarkable changes might be accounted for on the basis of a change in the amount of estrogen-binding protein present in the blood of the infant animals (Raynaud *et al.*, 1971). The protein was present in the plasma of 20-day fetal rats and gradually disappeared over the first 4 postnatal weeks. It was present in the blood of both sexes and readily bound estradiol and estrone, but not testosterone or androstenedione. Comparison of the effect upon uterine weight of estradiol and a synthetic estrogen (R 2858), which was not bound by the protein, clearly showed that estradiol action was impeded by the protein, for the potency of a threshold amount (0.05 μg) of estradiol increased with age from 5 to 28 days, while R 2858 was highly active throughout this period (Raynaud, 1973). Thus, by protein binding, at least part of the estradiol produced by the ovaries soon after birth may be prevented from acting upon the accessory sex organs, or the hypothalamo–hypophyseal system. However, treatment of postnatal female rats with an anti-estradiol serum for 4-day periods at 2, 6, 10, 14 or 18 days of age depressed the gain in ovarian and uterine weight and decreased follicular diameter and the number of granulosa cells (Reiter *et al.*, 1972), so that some active estrogen is in circulation. Ramaley (1971) has followed the binding of estradiol, progesterone, testosterone and androsterone to serum proteins in growing rats and observed a fall in estrogen binding and an increase in testosterone and progesterone binding at puberty. No changes occurred with androsterone and similar patterns were found after gonadectomy, so that the binding of gonadal steroids did not seem to be an important factor in the control of sexual maturation. Quite different work involving a comparison of the effects of ovarian tissue or estradiol upon ovarian growth after implantation into the spleen or kidney in infantile rats (Donovan and O'Keeffe, 1966; Donovan *et al.*, 1967) indicates that the hormones produced by the ovaries should not be equated with estradiol. Further, since the protein that binds estrogen does not bind testosterone, the high level of gonadotropin secretion in the male cannot be accounted for on the simple basis of inactivation of androgen.

Attempts to correlate the changes in the biological indices of gonadotropin action with hormone assay findings have not proved rewarding. Removal of one testis from newborn rats caused a rapid rise in plasma FSH, and a delayed increase in LH content while hemiovariectomy did not change FSH secretion in females for at least 10 days, when compensatory hypertrophy of the remaining ovary became apparent (Ojeda and Ramirez, 1972). Other workers disagree over the occurrence of compensatory hypertrophy after removal of one ovary in the immature rat (Baker and Kragt, 1969; Dunlap *et al.*, 1972; Dunlap and Gerall, 1973) or mouse (Peters and Braathen, 1973). Hemiovariectomy of 15-, 25- or 35-day-old guinea pigs did not elicit the compensatory hypertrophy evident in adults, and the implantation of estrogen into hemispayed immature females did not inhibit ovarian growth, although luteinized follicles were produced (Donovan and Lockhart, 1972). As noted earlier, complete removal of the ovaries from infant guinea-pigs caused a rise in LH secretion (Donovan *et al.*, 1973).

THE BRAIN AND GONADOTROPIN SECRETION

The observations just described are not readily reconciled with the view that the hypothalamus is inhibited by the small amounts of gonadal hormone produced during infancy, and that puberty sets in as the neural mechanism matures and more gonadal hormone is required to sustain the inhibition; particularly when the occurrence of "sub-threshold" fluctuations in gonadotropin secretion (Donovan and van der Werff

References p. 249–252

ten Bosch, 1965; McArthur, 1972) and of a pulsatile pattern of release of gonadotropic hormone before puberty in children (Boyar *et al.*, 1972) are borne in mind.

There is fair agreement that damage to the hypothalamus may advance the onset of puberty, although there is less concordance over the mode of action or precise location of the effective region (Donovan and van der Werff ten Bosch, 1965; Critchlow and Bar-Sela, 1967). The long-term effects upon ovarian function in the rat are variable, in that normal estrous cycles may ensue or an arrest of ovulation with persistent vaginal estrus set in. The development of persistent vaginal estrus has been associated with large lesions in the anterior hypothalamus, whereas lesions in the middle hypothalamus did not lead to a suspension of ovulation (Meijs-Roelofs and Moll, 1972). Stimulation of the anterior hypothalamus or arcuate nuclear region has enhanced gonadotropin secretion and accelerated sexual maturation in the rat in that after the first ovulation regular estrous cycles continued (Meijs-Roelofs, 1972; Meijs-Roelofs and Uilenbroek, 1973), but the expectation that hypothalamic lesions also cause gonadotropin secretion has been called into question by Ramirez (1973) who found no increase in the blood content of FSH and LH in 21-day-old rats, after placement of lesions in the hypothalamus at 16 days, although precocious puberty was induced.

A fully satisfactory explanation of the mode of action of hypothalamic lesions in hastening sexual development has yet to be advanced. The view of Hohlweg and Junkmann (1932) that a sex center in the brain was exquisitely sensitive to gonadal hormones in infancy and so inhibited gonadotropin secretion has proved most attractive. With increasing age and the progressive maturation of the brain the center would become progressively less depressed by estrogen, for example, so that gradually more gonadotropin would be released and puberty set in. However, examples of the positive feedback of estrogen in inducing gonadotropin secretion have already been cited, while Smith and Davidson (1968) have shown that the application of estrogen to the hypothalamus of rats for 48 h can advance puberty. They postulated the existence of two sensitive areas in the hypothalamus: a "negative feedback controller" in the median eminence, with a threshold to inhibition by estrogen that increases during development and so produces a gradual rise in the circulating level of ovarian hormone, and a "positive feedback controller" in the anterior hypothalamus that responds to the raised blood concentration of estrogen by promoting the secretion of gonadotropin. This concept may be likened to the tonic and phasic mechanism mentioned earlier.

Just as the hypothalamic control of the secretion of anterior pituitary hormones is modulated by the limbic system in adults, so experimental interference with the amygdala and hippocampus can alter the timing of sexual maturation (Donovan, 1971). Unhappily, diverse findings concerning the effect of lesions in these regions have been obtained. Thus, Elwers and Critchlow (1960) found that lesions in the medial part of the amygdaloid nuclei placed at 18–20 days of age accelerated puberty whereas Relkin (1971) found that similar lesions made at 4 days delayed this event, a difference which perhaps suggests that the age at which the lesions were made influenced the response. No change in the day of vaginal opening was observed with amygdaloid lesions by Bloch and Ganong (1971), who prepared their animals at 21–23 days, although all these investigators found that lesions in the antero-basal hypothalamus accelerated vaginal canalization. In part, these discordant results may be attributed to the use on some occasions of stainless steel electrodes and of platinum on others. The interpretation of the effects of brain lesions can be hazardous (Donovan, 1966), and Velasco (1972) has recently shown that bilateral lesions in the medial amygdaloid nuclei made with steel electrodes caused early vaginal opening, whereas those made with platinum electrodes did not. It seemed that the advanced ovulation resulted from an irritative side effect of the iron deposited from the steel electrode, rather than from a specific loss of tissue.

Hippocampal damage in the infant rat has not altered sexual development in a consistent manner, for vaginal opening has remained unaffected (Kling, 1964), been delayed (Zarrow *et al.*, 1969) or been advanced (Brown-Grant and Raisman, 1972).

Whatever the precise role of the limbic system may be, it is evidently capable of modulating the secretion of gonadotropin necessary for sexual development. Critchlow and Bar-Sela (1965) pointed out that exposure to estrogen favored the promotion of gonadotropin secretion by the amygdaloid nuclei, while the feedback action of estrogen upon gonadotropin release may be exerted through the limbic system (Lawton and Sawyer, 1970; Stumpf, 1971). In quite recent work, Kawakami and Terasawa (1972) found that electrical stimulation of the arcuate nuclei of the hypothalamus in 27-day-old rats caused a rise in serum LH, while stimulation of the hippocampus increased the serum FSH concentration. However, daily stimulation of these structures over days 27, 28 and 29 did not produce a rise in gonadotropin secretion in animals killed immediately after the third stimulation, although excitation of the medial preoptic area and of the amygdala raised the serum content of FSH. Pretreatment with estrogen facilitated the release of FSH following stimulation of the medial preoptic area.

THE METABOLIC CONTROL OF PUBERTY

That the control of puberty should be considered in the widest possible context is indicated by the "critical weight" hypothesis of Frisch and Revelle (1971). Many studies of the relationship between the rate of growth, size and body build at the age of menarche lead to the conclusion that a high growth rate and an early menarche are linked (Tanner, 1962; Crisp, 1970). The better nourished children of the professional or managerial classes have, in the past, experienced menarche at an earlier age than those of unskilled parents, while the obese girl tends to reach menarche earlier than her lighter counterpart. Frisch and Revelle (1971) used longitudinal growth data and determined the height and weight of children at several well-defined stages of adolescent development. They found (see Frisch, 1973) that the mean weight of girls at the time of initiation of the adolescent growth spurt (30 kg), at the time of maximum rate of weight gain (39 kg), and at menarche (48 kg) did not differ for early or late maturers although the mean height at each of these events increased significantly with age. When this information was set alongside other observations indicating that puberty is

References p. 249–252

more closely related to weight than to age in rats, mice, pigs and cattle, it was reasonable to propose that a direct relationship between a critical body weight, representing a critical metabolic rate, and menarche exists. It was assumed that the attainment of the critical weight causes a change in metabolic rate per unit mass (or surface area) which in turn affects the hypothalamo–pituitary–ovarian feedback mechanisms by decreasing its sensitivity to estrogen, with the feedback being reset at a level high enough to induce sexual maturation. The evidence for the association is convincing, such as a comparison of the weight at menarche of 30 undernourished girls (43.5 ± 0.92 (SEM) kg at 14.4 ± 0.17 years) with that for 30 well-nourished girls (44.6 ± 1.2 kg at 12.5 ± 0.17 years) (Frisch, 1972), but which is cause and which is effect is less clear. Could the process governing sexual development collaterally influence body weight and metabolism? Frisch *et al.* (1973) have related the changes in body weight to changes in body composition. When total water, lean body weight and fat were calculated for each of 169 girls, the total water (26.2 ± 0.18 l) did not differ at menarche in early and late maturing children, although early maturing girls showed a greater increase in fat per kilogram than later maturers. When the ratio of total water to body weight was calculated, the coefficient of variability of the ratio at menarche was 55% that of the weight at menarche. Such observations show that a consistent metabolic pattern is established at puberty, and direct attention toward various body compartments.

A relationship between rate of growth and age at puberty has been established in rats and guinea-pigs (Widdowson and McCance, 1960; Mills and Reed, 1971) and has been explored in some detail by Kennedy and Mitra (1963). They used the simple expedient of taking pairs of litters of 9–10 pups born on the same day and redistributing them so that 3 sucklings were given to one mother, and the rest to the other. When weaned at 21 days the average weight of the optimally fed young was over 40 g, whereas those reared in large litters weighed between 15 and 25 g. From weaning all animals were given an unlimited supply of food. Vaginal opening in the optimally fed rats occurred significantly earlier (38.1 ± 5 days) than in the retarded females (43.4 ± 4 days). Vaginal opening was more closely related to weight than to age, and mating occurred at virtually identical weights in both groups (102.4 ± 15 g; 99.1 ± 21 g). Kennedy and Mitra (1963) and Kennedy (1966) have pointed out that the weanling rat, like many young animals, eats far more than it needs to build new tissue. This is because of the high cost of thermoregulation in the infant, partly as a result of the fact that the surface area through which heat is lost is large compared with the weight of tissue producing heat, and partly because infants resist falls in body temperature by producing more heat rather than restricting heat loss. During the short period when body size is changing very rapidly from about 50 to 150 g, food intake varies in proportion to surface area and is virtually independent of age, indicating that thermostasis is the dominant factor controlling food intake at this time of life and that food intake or its correlate metabolic rate may act as the normal signal to initiate puberty. Kennedy (1966) goes on to remark that some developmental milestones, such as skeletal epiphyseal changes, and the onset of estrous cycles, are influenced more by body size than by age. "It is perhaps more than coincidence that thermostasis, thyroid regulation, the onset of puberty and the control of ovulation are all disturbed by lesions in virtually the same part of the anterior hypothalamus".

That the anterior hypothalamus is generally considered to be concerned with the control of heat loss lends plausibility to a thermogenic view, for lesions in this region disrupt heat loss mechanisms (Myers, 1969). If hyperthermia was produced in an infant by such a lesion, the resultant acceleration in metabolic turnover and disposal of hormones might, almost inadvertently, lead to precocious puberty. However, this mechanism is not likely since mice reared at a temperature of −3 °C, in which metabolic turnover would be expected to be increased, showed vaginal opening at 33 days, while this event occurred in animals reared at 21 °C at 22 days. Despite this marked difference in timing, sexual maturation occurred at the same body weight under both conditions (Barnett and Coleman, 1959). Additionally, the metabolic clearance rate of estrone and estradiol in 27- to 29-day immature rats was the same as that of 2.5- to 3-month-old adults when related to body size (De Hertog *et al.*, 1970). Further, dysthermia is seldom reported in cases of precocious puberty (Bauer, 1954).

The critical weight hypothesis is less satisfactory when experimental sexual precocity is taken into account, for accelerated growth to a "pubertal" weight would be expected in treated infant animals before sexual maturation occurred. This is not commonly observed. For example, lesions in the anterior hypothalamus which advanced puberty in infant rats did not cause a marked gain in weight and the animals were smaller at puberty than controls (Donovan and van der Werff ten Bosch, 1959). Exposure of the anterior hypothalamus or preoptic area to estrogen for two days can advance puberty without markedly affecting body weight (Smith and Davidson, 1968).

The "critical weight" concept has been developed from studies of the human situation and should be applied to other species with caution, particularly in view of the long delay between brain maturation and the onset of sexual function in the primate. It is also a peculiarity of the primate that treatment of the prepuberal female with estrogen fails to elicit a surge of gonadotropin release (Knobil *et al.*, 1972).

Nevertheless, the suggestion that under normal circumstances changes in water and fat metabolism may be important, has its attractions, particularly as these functions come under hypothalamic control. At the very least, it becomes difficult to ignore the fact, too often neglected, that many physiological processes, besides that of sexual development, are controlled by the hypothalamus and that it is unrealistic to attempt to understand one to the exclusion of others. The gonadal hormones, for example, exert a feedback action upon the brain that influences more than sexual behavior and gonadotropin secretion. Estrogens promote motor activity, depress food intake and cause a loss in body weight. Different regions of the hypothalamus are involved in these reactions, for implants of estradiol benzoate into the preoptic area increased locomotor activity, but did not depress the food intake of spayed female rats, whereas similar implants into the ventromedial hypothalamus depressed food intake without changing locomotor activity (Wade and Zucker, 1970). Rothchild (1967) has attempted to account for the inhibition of ovulation by progesterone on the comparable basis of a generalized response on the part of the hypothalamus to the hormone. In alluding to the prepubertal rat he comments: "I am aware of no information on body temperature

References p. 249–252

changes or on prolactin secretion before and during puberty, but the changes in the modalities we have discussed certainly agree with a process in which activity in the ventromedial nucleus increases and activity in the far-lateral area decreases in association with the transition to sexual maturity, *i.e.* with a decrease in inhibition over folliculotropin secretion."

The monoamines present in the hypothalamus are known to be concerned in some way with the control of gonadotropin secretion (Fuxe and Hökfelt, 1970; Glowinski, 1970; Anton-Tay and Wurtman, 1971; Coppola, 1971; McCann *et al.*, 1972) and the likely involvement of adrenaline, nor-adrenaline and 5-hydroxytryptamine in the control of body temperature (Cremer and Bligh, 1969; Feldberg, 1970) might be more than coincidental. When injected into the cerebral ventricle or anterior hypothalamus, catecholamines caused a fall in body temperature in the cat, dog and monkey, a rise in the rabbit and sheep, a dose-dependent rise or fall in rats and mice and have no effect in oxen or goats, whereas 5-hydroxytryptamine generally caused a fall in body temperature in all species. Body temperature has been regarded as being set by the relative activities of the catecholamines and 5-hydroxytryptamine within the hypothalamus, so that the gradual fall in body temperature during infancy and adolescence in children (Tanner, 1962) could reflect a steadily increasing monoaminergic tonus within the hypothalamus. Such a monoaminergic tonus might provide the control setting around which a variety of hypothalamic functions become balanced, and through which some degree of integration is achieved. Depression of hypothalamic nor-adrenaline content by treatment with reserpine has delayed both rate of body growth and vaginal opening in rats, but the hypothalamic content of nor-adrenaline in pair-fed control females that reached puberty at the same body weight and age was normal. Hypothalamic nor-adrenaline could also be reduced in amount by treatment with α-methyl-*p*-tyrosine without affecting the onset of puberty (Weiner and Ganong, 1971). However, nor-adrenaline might not be the important monoamine, while the hypothalamic content may not truly reflect functional activity.

The limbic system can also be incorporated in a "metabolic" explanation of the control of sexual maturation. The amygdala and hippocampus influence eating and drinking besides modulating endocrine function, and lesions in these structures can elicit a hyperphagia as marked and persistent as occurs after damage to the hypothalamus (Grossman, 1967).

CONCLUDING REMARKS

Despite the current and welcome interest in the control of puberty it is clear that our understanding of the mechanism concerned has not deepened in the last decade: indeed the newer information has cast doubt upon the validity of long-established ideas. Attention has been focussed on but a few of the problems awaiting resolution; others could not be covered on this occasion. However, the delineation of a problem is always the first step in an analysis, while the answers themselves may emerge quite soon.

REFERENCES

ANTON-TAY, F. AND WURTMAN, R. J. (1971) Brain monoamines and endocrine function. In *Frontiers in Neuroendocrinology*, L. MARTINI AND W. F. GANONG (Eds.), Oxford Univ. Press, New York, pp. 45–66.

BAKER, F. D. AND KRAGT, C. L. (1969) Maturation of the hypothalamic–pituitary–gonadal negative feedback system. *Endocrinology*, **85**, 522–527.

BARNETT, S. A. AND COLEMAN, E. M. (1959) The effect of low environmental temperature on the reproductive cycle of female mice. *J. Endocr.*, **19**, 232–240.

BAUER, H. G. (1954) Endocrine and other clinical manifestations of hypothalamic disease. *J. clin. Endocr.*, **14**, 13–31.

BEN-OR, S. (1970) Development of the ovary under different experimental conditions. In *Gonadotrophins and Ovarian Development*, W. R. BUTT, A. C. CROOKE AND M. RYLE (Eds.), Livingstone, Edinburgh, pp. 266–271.

BLIZZARD, R. M., PENNY, R., FOLEY, T. P., BAGHDASSARIAN, A., JOHANSON, A. AND YEN, S. S. C. (1972) Pituitary–gonadal interrelationships in relation to infancy. In *Gonadotropins*, B. B. SAXENA, C. G. BELING AND H. M. GANDY (Eds.), Wiley-Interscience, New York, pp. 502–523.

BLOCH, G. J. AND GANONG, W. F. (1971) Lesions of the brain and the onset of puberty in the female rat. *Endocrinology*, **89**, 898–901.

BOYAR, R., FINKELSTEIN, J., ROFFWARG, H., KAPEN, S., WEITZMAN, E. AND HELLMAN, L. (1972) Synchronization of augmented luteinizing hormone secretion with sleep during puberty. *New Engl. J. Med.*, **287**, 582–586.

BROWN-GRANT, K. AND RAISMAN, G. (1972) Reproductive function in the rat following selective destruction of afferent fibres to the hypothalamus from the limbic system. *Brain Res.*, **46**, 23–42.

CALIGARIS, L., ASTRADA, J. J. AND TALEISNIK, S. (1972) Influence of age on the release of luteinizing hormone induced by oestrogen and progesterone in immature rats. *J. Endocr.*, **55**, 97–103.

COPPOLA, J. A. (1971) Brain catecholamines and gonadotropin secretion. In *Frontiers in Neuroendocrinology*, L. MARTINI AND W. F. GANONG (Eds.), Oxford Univ. Press, New York, pp. 129–143.

CREMER, J. E. AND BLIGH, J. (1969) Body-temperature and responses to drugs. *Brit. med. Bull.*, **25**, 299–306.

CRISP, A. H. (1970) Premorbid factors in adult disorders of weight, with particular reference to primary anorexia nervosa (weight phobia). A literature review. *J. psychosom. Res.*, **14**, 1–22.

CRITCHLOW, V. AND BAR-SELA, M. E. (1967) Control of the onset of puberty. In *Neuroendocrinology, Vol. 2*, L. MARTINI AND W. F. GANONG (Eds.), Academic Press, New York, pp. 101–162.

DIEBEL, N. D., YAMAMOTO, M. AND BOGDANOVE, E. M. (1973) Discrepancies between radioimmunoassays and bioassay for rat FSH: evidence that androgen treatment and withdrawal can alter bioassay-immunoassay ratios. *Endocrinology*, **92**, 1065–1078.

DONOVAN, B. T. (1966) Experimental lesions of the hypothalamus. *Brit. med. Bull.*, **22**, 249–253.

DONOVAN, B. T. (1971) The extra-hypothalamic control of gonadotrophin secretion. In *Control of Gonadal Steroid Secretion*, D. T. BAIRD AND J. A. STRONG (Eds.), Univ. Press, Edinburgh, pp. 1–12.

DONOVAN, B. T. AND LOCKHART, A. N. (1972) Ovarian–pituitary interaction in the immature female guinea-pig. *J. Endocr.*, **57**, lvi.

DONOVAN, B. T., LOCKHART, A. N., MACKINNON, P. C. B., MATTOCK, J. M. AND PEDDIE, M. J. (1973) Changes in the plasma concentration of luteinizing hormone during development in the guinea-pig. *J. Endocr.*, **59**, xxxviii.

DONOVAN, B. T. AND O'KEEFFE, M. C. (1966) The liver and the feedback action of ovarian hormones in the immature rat. *J. Endocr.*, **34**, 469–478.

DONOVAN, B. T., O'KEEFFE, M. C. AND O'KEEFFE, H. T. (1967) Intrasplenic implants of oestradiol and uterine growth in infantile and pubertal rats. *J. Endocr.*, **37**, 93–98.

DONOVAN, B. T. AND VAN DER WERFF TEN BOSCH, J. J. (1959) The hypothalamus and sexual maturation in the rat. *J. Physiol. (Lond.)*, **147**, 78–92.

DONOVAN, B. T. AND VAN DER WERFF TEN BOSCH, J. J. (1965) *Physiology of Puberty*, *Monog. Physiol. Soc. No. 15*, Edward Arnold, London.

DUNLAP, J. L. AND GERALL, A. A. (1973) Compensatory ovarian hypertrophy can be obtained in neonatal rats. *J. Reprod. Fertil.*, **32**, 517–519.

DUNLAP, J. L., PREIS, L. K. AND GERALL, A. A. (1972) Compensatory ovarian hypertrophy as a function of age and neonatal androgenization. *Endocrinology*, **90**, 1309–1314.

ELWERS, M. AND CRITCHLOW, V. (1960) Precocious ovarian stimulation following hypothalamic and amygdaloid lesions in rats. *Amer. J. Physiol.*, **198**, 381–385.

ESHKOL, A. AND LUNENFELD, B. (1972) Gonadotropic regulation of ovarian development in mice during infancy. In *Gonadotropins*, B. B. SAXENA, C. G. BELING AND H. M. GANDY (Eds.), Wiley-Interscience, New York, pp. 335–346.

ESHKOL, A., LUNENFELD, B. AND PETERS, H. (1970) Ovarian development in infant mice. Dependence on gonadotropic hormones. In *Gonadotrophins and Ovarian Development*, W. R. BUTT, A. C. CROOKE AND M. RYLE (Eds.), Livingstone, Edinburgh, pp. 249–258.

FELDBERG, W. (1970) Monoamines of the hypothalamus as mediators of temperature response. In *The Hypothalamus*, L. MARTINI, M. MOTTA AND F. FRASCHINI (Eds.), Academic Press, London, pp. 213–232.

FOA, C. (1900) La greffe des ovaires, en relation avec quelques questions de biologie générale. *Arch. ital. Biol.*, **34**, 43–73.

FOSTER, D. L., COOK, B. AND NALBANDOV, A. V. (1972a) Regulation of luteinizing hormone (LH) in the fetal and neonatal lamb: effect of castration during the early postnatal period on levels of LH in sera and pituitaries of neonatal lambs. *Biol. Reprod.*, **6**, 253–257.

FOSTER, D. L., CRUZ, T. A. C., JACKSON, G. L., COOK, B. AND NALBANDOV, A. V. (1972b) Regulation of luteinizing hormone in the fetal and neonatal lamb. III. Release of LH by the pituitary *in vivo* in response to crude ovine hypothalamic extract or purified porcine gonadotrophin releasing factor. *Endocrinology*, **90**, 673–683.

FOSTER, D. L., KARSCH, F. J. AND NALBANDOV, A. V. (1972c) Regulation of luteinizing hormone (LH) in the fetal and neonatal lamb. II. Study of placental transfer of LH in sheep. *Endocrinology*, **90**, 589–592.

FRISCH, R. E. (1972) Weight at menarche: similarity for well-nourished and undernourished girls at differing ages, and evidence for historical constancy. *Pediatrics*, **50**, 445–450.

FRISCH, R. E. (1973) The critical weight at menarche and the initiation of the adolescent growth spurt, and the control of puberty. In *The Control of the Onset of Puberty*, M. GRUMBACH, G. GRAVE AND F. MAYER (Eds.), Wiley-Interscience, New York, in press.

FRISCH, R. E. AND REVELLE, R. (1971) Height and weight at menarche and a hypothesis of menarche. *Arch. Dis. Childh.*, **46**, 695–701.

FRISCH, R. E., REVELLE, R. AND COOK, S. (1973) Components of weight at menarche and the initiation of the adolescent growth spurt in girls: estimated total water, lean body weight and fat. *Human Biol.*, **45**, 469–483.

FUXE, K. AND HÖKFELT, T. (1970) Central monoaminergic systems and hypothalamic function. In *The Hypothalamus*, L. MARTINI, M. MOTTA AND F. FRASCHINI (Eds.), Academic Press, London, pp. 123–138.

GLOWINSKI, J. (1970) Metabolism of catecholamines in the central nervous system and correlation with hypothalamic functions. In *The Hypothalamus*, L. MARTINI, M. MOTTA AND F. FRASCHINI (Eds.), Academic Press, London, pp. 139–152.

GOLDMAN, B. D. AND GORSKI, R. A. (1971) Effects of gonadal steroids on the secretion of LH and FSH in neonatal rats. *Endocrinology*, **89**, 112–115.

GROSSMAN, S. P. (1967) *A Textbook of Physiological Psychology*, John Wiley, New York.

GRUMBACH, M. M. AND KAPLAN, S. L. (1973) Ontogenesis of growth hormone, insulin, prolactin and gonadotropin secretion in the human foetus, *Foetal and Neonatal Physiology. Proceedings of the Sir Joseph Barcroft Centenary Symposium*, University Press, Cambridge, pp. 462–487.

DE HERTOG, R., EKKA, E., VAN DER HEYDEN, I. AND HOET, J. J. (1970) Metabolic clearance rates and the interconversion factors of estrone and estradiol-17β in the immature and adult female rat. *Endocrinology*, **87**, 874–880.

HOHLWEG, W. (1934) Veränderungen des Hypophysenvorderlappens und des Ovariums nach Behandlung mit grossen Dosen von Follikelhormon. *Klin. Wschr.*, **13**, 92–95.

HOHLWEG, W. AND JUNKMANN, K. (1932) Die Hormonal-Nervöse Regulierung der Funktion des Hypophysenvorderlappens. *Klin. Wschr.*, **11**, 321–323.

HYYPPÄ, M. (1972) Hypothalamic monoamines in human fetuses. *Neuroendocrinology*, **9**, 257–266.

KAWAKAMI, M. AND TERASAWA, E. (1972) Electrical stimulation of the brain on gonadotropin secretion in the female prepuberal rat. *Endocr. jap.*, **19**, 335–347.

KELCH, R. P., GRUMBACH, M. M. AND KAPLAN, S. L. (1972) Studies on the mechanism of puberty in man. In *Gonadotropins*, B. B. SAXENA, C. G. BELING AND H. M. GANDY (Eds.), Wiley-Interscience, New York, pp. 524–534.

KENNEDY, G. C. (1966) Food intake, energy balance and growth. *Brit. med. Bull.*, **22**, 216–220.

KENNEDY, G. C. AND MITRA, J. (1963) Body weight and food intake as initiating factors for puberty in the rat. *J. Physiol. (Lond.)*, **166**, 408–418.

KLING, A. (1964) Effects of rhinencephalic lesions on endocrine and somatic development in the rat. *Amer. J. Physiol.*, **206**, 1395–1400.

KNOBIL, E., DIERSCHKE, D. J., YAMAJI, T., KARSCH, F. J., HOTCHKISS, J. AND WEICK, R. F. (1972) Role of estrogen in the positive and negative feedback control of LH secretion during the menstrual cycle of the rhesus monkey. In *Gonadotropins*, B. B. SAXENA, C. G. BELING AND H. M. GANDY (Eds.), Wiley-Interscience, New York, pp. 72–86.

KRAGT, C. L. AND DAHLGREN, J. (1972) Development of neural regulation of follicle-stimulating hormone (FSH) secretion. *Neuroendocrinology*, **9**, 30–40.

LAWTON, I. E. AND SAWYER, C. H. (1970) Role of amygdala in regulating LH secretion in the adult female rat. *Amer. J. Physiol.*, **218**, 622–626.

LIEFER, R. W., FOSTER, D. L. AND DZIUK, P. J. (1972) Levels of LH in the sera and pituitaries of female lambs following ovariectomy and administration of estrogen. *Endocrinology*, **90**, 981–985.

MCARTHUR, J. W. (1972) Gonadotropins in relation to sexual maturity. In *Gonadotropins*, B. B. SAXENA, C. G. BELING AND H. M. GANDY (Eds.), Wiley-Interscience, New York, pp. 487–501.

MCCANN, S. M., KALRA, P. S., KALRA, S. P., DONOSO, A. O., BISHOP, W., SCHNEIDER, H. P. G., FAWCETT, C. P. AND KRULICH, L. (1972) The role of monoamines in the control of gonadotropin and prolactin secretion. In *Gonadotropins*, B. B. SAXENA, C. G., BELING AND H. M. GANDY (Eds.), Wiley-Interscience, New York, pp. 49–60.

MEIJS-ROELOFS, H. M. A. (1972) Effect of electrical stimulation of the hypothalamus on gonadotrophin release and the onset of puberty. *J. Endocr.*, **54**, 277–284.

MEIJS-ROELOFS, H. M. A. AND MOLL, J. (1972) Differential effects of anterior and middle hypothalamic lesions on vaginal opening and cyclicity. *Neuroendocrinology*, **9**, 297–303.

MEIJS-ROELOFS, H. M. A. AND UILENBROEK, J. TH. J. (1973) Gonadotropin release and follicular development after electrical stimulation of the hypothalamus in the immature female rat. In *The Development and Maturation of the Ovary and its Functions, Int. Congr. Series No. 267*, H. PETERS (Ed.), Excerpta Medica, Amsterdam, pp. 117–123.

MEIJS-ROELOFS, H. M. A., UILENBROEK, J. TH. J., DE JONG, F. H. AND WELSCHEN, R. (1973a) Plasma oestradiol-17β and its relation to serum follicle-stimulating hormone in immature female rats. *J. Endocr.*, **59**, 295–312.

MEIJS-ROELOFS, H. M. A., UILENBROEK, J. TH. J., OSMAN, P. AND WELSCHEN, R. (1973b) Serum levels of gonadotropins and follicular growth in prepuberal rats. In *The Development and Maturation of the Ovary and its Functions, Int. Congr. Series No. 267*, H. PETERS (Ed.), Excerpta Medica, Amsterdam, pp. 3–11.

MILLS, P. G. AND REED, M. (1971) The onset of first oestrus in the guinea-pig and the effects of gonadotrophins and oestradiol in the immature animal. *J. Endocr.*, **50**, 329–337.

MYERS, R. D. (1969) Temperature regulation: neurochemical systems in the hypothalamus. In *The Hypothalamus*, W. HAYMAKER, E. ANDERSON AND W. J. H. NAUTA (Eds.), Thomas, Springfield, Ill., pp. 506–523.

OJEDA, S. R. AND RAMIREZ, V. D. (1972) Plasma level of LH and FSH in maturing rats: response to hemigonadectomy. *Endocrinology*, **90**, 466–472.

PETERS, H. AND BRAATHEN, B. (1973) The effect of unilateral ovariectomy in the neonatal mouse on follicular development. *J. Endocr.*, **56**, 85–89.

PICON, L. (1956) Sur le rôle de l'age dans la sensibilité de l'ovaire à l'hormone gonadotrope chez le rat. *Arch. Anat. micr. Morph. exp.*, **45**, 311–341.

RAMALEY, J. A. (1971) Steroid binding to serum proteins in maturing male and female rats. *Endocrinology*, **89**, 545–552.

RAMIREZ, V. D. (1964) Advancement of puberty in the rat by estrogen. *Anat. Rec.*, **148**, 325.

RAMIREZ, V. D. (1973) In discussion. In *The Control of the Onset of Puberty*, M. GRUMBACH, G. GRAVE AND F. MAYER (Eds.), Wiley-Interscience, New York, in press.

RAMIREZ, V. D. AND SAWYER, C. H. (1965) Advancement of puberty in the female rat by estrogen. *Endocrinology*, **76**, 1158–1168.

RAYNAUD, J.-P. (1973) Influence of rat estradiol binding plasma protein (EBP) on uterotrophic activity. *Steroids*, **21**, 249–258.

RAYNAUD, J.-P., MERCIER-BODARD, C. AND BAULIEU, E. E. (1971) Rat estradiol binding plasma protein (EBP). *Steroids*, **18**, 767–788.

REITER, E. O., GOLDENBERG, R. L., VAITUKAITIS, J. L. AND ROSS, G. T. (1972) A role for endogenous estrogen in normal ovarian development in the neonatal rat. *Endocrinology*, **91**, 1537–1539.

RELKIN, R. (1971) Relative efficacy of pinealectomy, hypothalamic and amygdaloid lesions in advancing puberty. *Endocrinology*, **88**, 415–418.

ROTHCHILD, I. (1967) The neurologic basis for the anovulation of the luteal phase, lactation and pregnancy. In *Reproduction in the Female Mammal*, G. E. LAMMING AND E. C. AMOROSO (Eds.), Butterworths, London, pp. 30–54.

SMITH, E. R. AND DAVIDSON, J. M. (1968) Role of estrogen in the cerebral control of puberty in female rats. *Endocrinology*, **82**, 100–108.

STUMPF, W. E. (1971) Autoradiographic techniques and the localization of estrogen, androgen, and glucocorticoid in the pituitary and brain. *Amer. Zool.*, **11**, 725–739.

SWERDLOFF, R. S., JACOBS, H. S. AND ODELL, W. D. (1972) Hypothalamic–pituitary–gonadal interrelationships in the rat during sexual maturation. In *Gonadotropins*, B. B. SAXENA, C. G. BELING AND H. M. GANDY (Eds.), Wiley-Interscience, New York, pp. 546–561.

TALEISNIK, S., VELASCO, M. E. AND ASTRADA, J. J. (1970) Effect of hypothalamic deafferentation on the control of luteinizing hormone secretion. *J. Endocr.*, **46**, 1–7.

TANNER, J. M. (1962) *Growth and Adolescence, 2nd edition*, Blackwell, Oxford.

VELASCO, M. E. (1972) Opposite effects of platinum and stainless-steel lesions of the amygdala on gonadotropin secretion. *Neuroendocrinology*, **10**, 301–308.

WADE, G. N. AND ZUCKER, I. (1970) Modulation of food intake and locomotor activity in female rats by diencephalic hormone implants. *J. comp. physiol. Psychol.*, **72**, 328–336.

WEINER, R. I. AND GANONG, W. F. (1971) Effect of the depletion of brain catecholamines on puberty and the estrous cycle in the rat. *Neuroendocrinology*, **8**, 125–135.

WEISZ, J. AND FERIN, M. (1970) Pituitary gonadotrophins and circulating LH in immature rats — a comparison between normal females and males and females treated with testosterone in neonatal life. In *Gonadotrophins and Ovarian Development*, W. R. BUTT, A. C. CROOKE AND M. RYLE (Eds.), Livingstone, Edinburgh, pp. 339–350.

WIDDOWSON, E. M. AND MCCANCE, R. A. (1960) Some effects of accelerating growth. I. General somatic development. *Proc. roy. Soc. B*, **152**, 188–206.

WOODS, M. C. AND SIMPSON, M. E. (1961) Characterization of the anterior pituitary factor which antagonizes gonadotrophins. *Endocrinology*, **68**, 647–661.

YING, S.-Y. AND GREEP, R. O. (1971) Effect of age of rat and dose of a single injection of estradiol benzoate (EB) on ovulation and the facilitation of ovulation by progesterone (P). *Endocrinology*, **89**, 785–790.

ZARROW, M. X., NAQVI, R. H. AND DENENBERG, V. H. (1969) Androgen-induced precocious puberty in the female rat and its inhibition by hippocampal lesions. *Endocrinology*, **84**, 14–19.

DISCUSSION

MEIJS-ROELOFS: I should like to ask you a question concerning the experiments in which hypothalamic lesions were placed in mammals inducing precocious puberty without inducing an increase in the gonadotropic hormones. The lesions were placed at day 16, but the gonadotropic hormone levels were only measured at day 21. What are the consequences of waiting 5 days for the results?

DONOVAN: I would agree that if lesions have been placed in the hypothalamus, serial determinations of the gonadotropic hormone levels are desirable. This is, however, very difficult in baby rats. Another point that needs to be made, is that we need to take the existence of surges of gonadotropic hormone secretion into account. It is known that children get nocturnal spurts of gonadotropic hormones as puberty approaches so that it may be unrealistic to take just one sample at one time of the day, and draw conclusions from that single determination of gonadotropin. Serial studies over the course of the day are needed.

MEIJS-ROELOFS: Another question that I would like to ask is whether you favor the hypothesis of Smith and Davidson* that puberty occurs by an estrogen stimulation of the preoptic area. In our estrogen determinations we could not find any increase during the last period before puberty.

* SMITH, E. R. AND DAVIDSON, J. M. (1968) Role of estrogen in the cerebral control of puberty in female rats. *Endocrinology*, **82**, 100–108.

DONOVAN: We need more measurements of estrogen blood levels to judge this hypothesis. A point that also needs to be made in this connection is that we have not so far mentioned a change in the sensitivity of the pituitary gland to releasing factors. There is evidence that the sensitivity of the sheep pituitary gland to releasing factor changes during development. And surprisingly enough, the pituitary gland of the fetus is more sensitive to releasing factors than the pituitary gland of the early postnatal animal. What underlies this difference, I don't know. Dr. Guillemin was mentioning similar information for the human, so perhaps he may be able to enlighten us.

GUILLEMIN: This refers to measurements made in collaboration with Grumbach and Kaplan during the last few months, in which a constant dose of 150 μg of pure synthetic LRF was given to a very large number of normal children from infancy to early and later adulthood. The effects on the plasma levels of LH and FSH are rather striking. There is, as Dr. Donovan just mentioned for sheep, a striking increase in the pituitary responsiveness to LRF in terms of LH secretion, with very little responsiveness in the infant, and increasing responsiveness to the prepuberal period, reaching a plateau between puberty and early adulthood. On the other hand, the responsiveness to that same dose of LRF in terms of FSH secretion is absolutely identical in a 6 month infant as it is in a 19-year-old teenager. So there is a dissociation in the responsiveness of the pituitary to the LRF in terms of the two gonadotropic hormones.

If I may, I will add one more comment in relation to an earlier statement made by Dr. Donovan. I do agree with him that it is really necessary in experiments or clinical observations to measure any of these pituitary hormone levels at a single time during the day. There is also no doubt that the secretion of LH in man and in the monkey is essentially modulated by a series of extremely short spurts of release, and also modulated by circadian rhythms. So one has to be very careful with the interpretation of a single measurement at a given time.

JOST: Would it be possible to measure the binding sites for LRF in the pituitary, in the course of development?

GUILLEMIN: To my knowledge it has not been done. It will not be simple, but the technology is available. We have already reported on the process of purification of a LRF receptor in normal pituitaries from adult rats. We do find at least two populations of LRF receptors, a classical observation, one with high affinity and low capacity, and the opposite for the other. So I do agree that this is a very interesting possibility now.

SALAMAN: I wonder if I could ask Dr. Donovan to comment on the experiments by MacKinnon and ter Haar in Oxford** regarding the changes in protein synthesis of the brain around puberty. They found marked rises in overall protein synthesis in areas such as the amygdala and the preoptic area, just before and over puberty. Even more amazing is the diurnal cycle that they describe of protein synthesis, particularly in the preoptic area. In the male and in the female at the time of puberty there was a differentiation into two diametrically opposed cycles. I wonder if these changes are primary or secondary to the process of puberty, or do you think they could in any way be causally related to the onset?

DONOVAN: The observations of MacKinnon and ter Haar are certainly important. But what we don't know is whether these are cause or effect. It is a great problem to distinguish between the mechanisms which are driving gonadotropic hormone secretion and mechanisms which are, as it were, responding to changes in the secretion of hormones.

** TER HAAR, M. B. AND MACKINNON, P. C. B. (1972) An investigation of cerebral protein synthesis in various states of neuroendocrine activity. In *Topics in Neuroendocrinology (Progr. in Brain Research, Vol. 38)*, J. ARIËNS KAPPERS AND J. P. SCHADÉ (Eds.), Elsevier, Amsterdam, pp. 211–223.

The Role of the Fetal Hypothalamus in Development of the Feto-Placental Unit and in Parturition

D. F. SWAAB AND W. J. HONNEBIER

Netherlands Central Institute for Brain Research, and Wilhelmina Gasthuis, Department of Obstetrics and Gynaecology, University of Amsterdam, Amsterdam (The Netherlands)

INTRODUCTION

Fetal growth is known to be influenced by many maternal and environmental factors (*e.g.* Knobil and Caton, 1953; Gruenwald, 1966; Kloosterman, 1966; Thomson *et al.*, 1968; Hoet, 1969). Moreover, maternal neuroendocrine factors are supposed to be involved in the initiation of parturition (Lincoln, 1971, 1974; Swaab, 1972).

Recent studies indicate that the fetal brain also plays an active role in fetal body and placental growth (Honnebier and Swaab, 1973a; Swaab and Honnebier, 1973). In addition, the fetus is considered to be involved in the timing of the onset of delivery (Liggins *et al.*, 1967; Liggins, 1968, 1969; Chard *et al.*, 1970).

The present paper will review the experimental and clinical basis for the assumption of an active role of the fetal hypothalamus in intrauterine growth and parturition.

INFLUENCE OF FETAL BRAIN LESIONS UPON DEVELOPMENT OF THE FETO-PLACENTAL UNIT

In order to study the influence of the fetal brain on intrauterine growth of the fetus and the placenta, various experimental techniques (see below) have been applied to eliminate the fetal brain or the fetal hypophysis. Observations on human anencephalics were also considered for this purpose since the hypothalamus is either missing or grossly disturbed in this condition (Tuchmann-Duplessis and Gabe, 1960; Eguchi, 1969; Grumbach and Kaplan, 1973).

Influence upon fetal growth

In mice, fetal body growth was inhibited after X-ray destruction of the fetal hypophyseal area (Raynaud, 1950).

In rat, reduction in the average body weight after fetal decapitation was reported to be 5–15% (Jost, 1966a) or 20% (Heggestad and Wells, 1965). A reduction in body weight of 12% (Fujita *et al.*, 1970) or 22% (Swaab and Honnebier, 1973) was observed after electrocoagulation of the fetal hypothalamus or brain respectively, while

References p. 275–280

7% reduction was observed after compression of the fetal rat brain (Fujita *et al.*, 1970). We used a simple technique in removing fetal rat brain by aspiration. This procedure appeared to have various advantages over fetal decapitation. By means of this procedure a reduction of 25–33% of the fetal body weight was observed (Swaab and Honnebier, 1973). Since reduction of fetal body weight has been reported after destroying the hypothalamus, but not after destroying other brain areas (Fujita *et al.*, 1970), the primary cause of this growth reduction is likely to be situated in or near the hypothalamus.

Decapitation of fetal mice (Eguchi, 1961) and rabbits (Jost, 1966a) did not cause any growth inhibition. Important for the interpretation of the difference between these results and our data except for possible species differences, however, is the fact that with our procedure all the fetuses in a litter are brainless. This excludes the possibility

TABLE I

EARLY GROWTH INHIBITION IN PREGNANCIES WITH ANENCEPHALY

Moment at which the uterine size was considered to be too small for gestation length in records of pregnancies with anencephaly, described by Honnebier and Swaab (1973a) and in one patient (*) sent to us by Dr. W. H. M. van der Velden (see Honnebier *et al.*, 1974).

Gestation length (in weeks) that uterus was found to be too small for date for the first time	*Estimated uterine size (in weeks) at that moment*	*Gestation length at which parturition took place (weeks)*
13	12	40
13	12	38
15	12	44
15	10–12	33
16	14	37
17	15	35
18	16	31
18	14	39
19	18	42
20*	18*	35*
20	18	41
20	18	36
22	20	34
23	22	37
24	23	43
24	22	46
25	22	33
26	22	51
26	25	34
26	24	47
26	24	52
28	24	43
29	26	36
29	27	43
34	32	45

of the transfer of substances from the fetal brain or hypophysis of unoperated littermates to the operated fetuses via the maternal circulation.

In sheep, coagulation of the fetal hypophyseal area inhibited fetal body growth and bone-age development (Liggins and Kennedy, 1968). A decreased birth weight was also observed after fetal decapitation in sheep (Lanman and Schaffer, 1968) and in Guernsey calf fetuses suffering from an inherited adenohypophyseal aplasia (Kennedy *et al.*, 1957). Even after prolonged gestation this last malformed fetus is reported to remain small and to have a premature appearance (Kennedy *et al.*, 1957).

Hutchinson *et al.* (1962) do not give sufficiently exact data to judge the existence of an effect of fetal hypophysectomy upon intrauterine growth in monkey.

A reduced growth exists in human anencephalics (Hoet, 1969; Milic and Adamsons, 1969; Kučera and Doležalová, 1972; Honnebier and Swaab, 1973a). The average ponderal growth of anencephalics was estimated to be 97 g/week against 143 g/week in controls. The anencephalics reached therefore only far beyond 40 weeks the birth weight of the control group at term (Honnebier and Swaab, 1973a).

Our data on anencephalic birth weights (Honnebier and Swaab, 1973a) start at 28 weeks of pregnancy. Clinical observations on these patients indicate, however, that growth retardation starts already much earlier (Table I).

However, other serious congenital malformations are also reported to have lower birth weights (Battaglia, 1970; Kučera and Doležalová, 1972; Schutt, 1965). Therefore the data of our anencephalic group were compared with those of children having other kinds of serious congenital malformations (Table II). Birth weight of anencephalic children appeared to be definitely less than that of the group with other congenital anomalies (Fig. 1).

TABLE II

CONGENITAL ANOMALIES DESCRIBED IN FIGS. 1, 2, 3

Congenital anomalies of the children presented in Figs. 1, 2 and 3. Classification according to: International statistical classification of diseases, injuries and causes of death 1965, detailed list (three-digit categories, see C. B. S., 1972).

Classification	*Number*
745; Cong. anom. of ear, face and neck	2
746; Cong. anom. of heart	26
748; Cong. anom. of respiratory system	1
750; Other cong. anom. of upper alimentary tract	17
751; Other cong. anom. of digestive system	11
753; Cong. anom. of urinary system	29
755; Other cong. anom. of limbs	1
756; Other cong. anom. of musculoskeletal system	9
757; Cong. anom. of skin, hair and nails	2
758; Other and unspecified cong. anom.	2
759; Cong. syndromes affecting multiple systems	11
Total	111

References p. 275–280

Twin pregnancies, discordant for anencephaly, show that the decreased birth weight in anencephaly is not caused by an alteration in maternal environment (Honnebier and Swaab, 1973a).

Children with a congenital aplasia of the hypophysis are reported to have normal

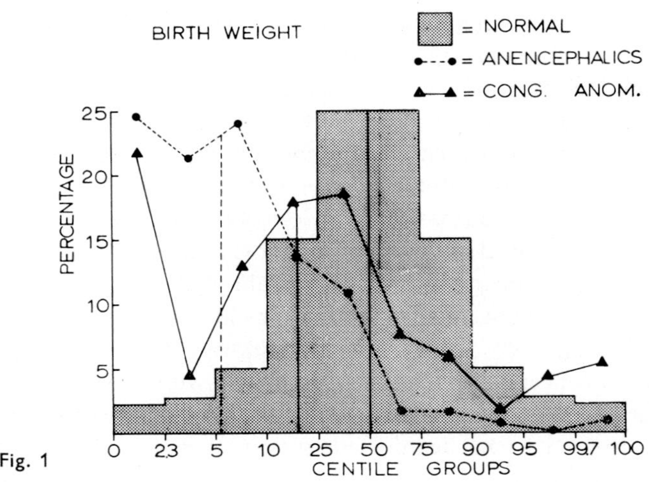

Fig. 1

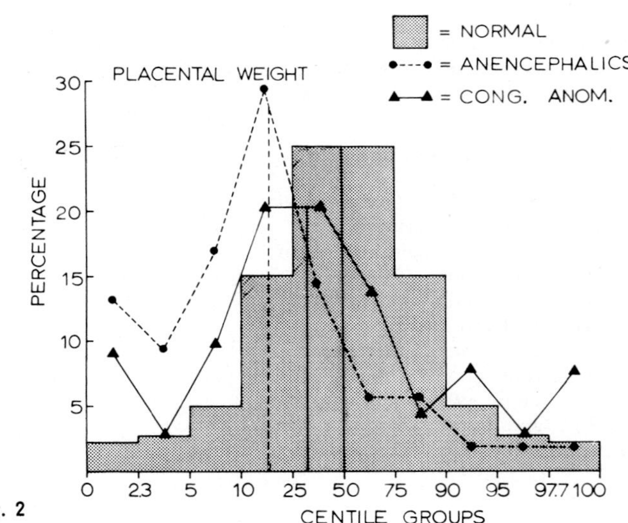

Fig. 2

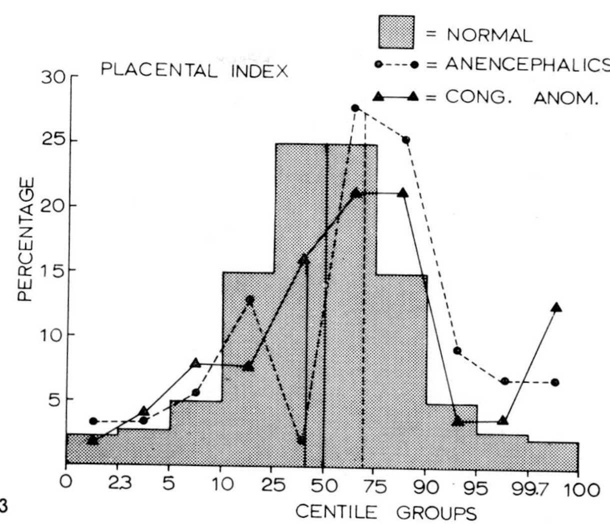

Fig. 3

Figs. 1–3. The distribution of birth weight, placental weight and placental index over the centile groups of the controls is shown in Figs. 1, 2 and 3 respectively. The data of the "normal" group obtained by Kloosterman (1970) are compared with those of 122 anencephalics (Honnebier and Swaab, 1973a) and 111 children with serious congenital anomalies (except for anencephaly) (for details see Honnebier and Swaab, 1973a). The anencephalic birth weights have been corrected approximately for their absence of brain weight. Appropriate brain weights according to gestation length were obtained by interpolation using the data of Gruenwald (1963) and Blinkov and Glezer (1968). These brain weights were added to the birth weights of the anencephalic children. The group of children having other serious congenital anomalies was collected out of the 147 children dying perinatally between 1948 and 1957 in the Training School for Midwives in Amsterdam, and between 1958 and 1971 in the Department of Obstetrics and Gynaecology, University of Amsterdam. From this group 36 children were excluded which had congenital anomalies of the central nervous system (22 cases of congenital hydrocephalus (category according to C.B.S., 1972: 742), 1 case of spina bifida (category: 741) and 13 other congenital anomalies of the nervous system (category: 743)). The distribution of the remaining 111 children over the various categories is given in Table II. From the distribution of the data over the centile groups and the place of the median (vertical lines, calculated according to De Jonge, 1963, chapt. 2.2.2.), it appears that the correlated anencephalics show a lower birth weight and placental weight and a higher placental index (as compared to the normal group as well as to the group having congenital anomalies). The group with congenital malformations shows birth and placental weights lying in between the normal group and the anencephalics, while its placental index was less than that of the control group.

birth weights (Table III). The primary cause of the reduced growth in anencephalics is therefore likely to be situated at the level of the hypothalamus. This relative unimportance of the presence of the hypophysis for the stimulation of intrauterine growth is supported by the data on anencephalics (see above) and by the work of Dupouy and Jost (1970) who observed an even greater decrease in body weight in encephalectomized rats that have a hypophysis than in decapitated animals.

References p. 275–280

TABLE III

CHILDREN WITH PITUITARY APLASIA

Literature reference	*Gestation length (weeks)*	*Sex*	*Birth weight (g)*	*Pituitary*	*Age at death*
Blizzard and Alberts (1956)	40	♂	4400 97.7–100 centile*	No adenohypophysis No pituitary stalk No neurohypophysis	18 h
Brewer (1957)	34.5	♀	1970 10–25 centile	No adenohypophysis Normal neurohypophysis	4.75 h
Reid (1960)	40	♂	4050 90–95 centile	No adenohypophysis No neurohypophysis No pituitary stalk	19 h
Dunn (1966)	44	♀	5464 97.7–100 centile	Only neurohypophyseal tissue	18 h
Johnson *et al.* (1973)	41.5	♀	3400 25–50 centile	No adenohypophysis Small neurohypophysis	21 days

* According to Kloosterman (1970).

Influence upon placental growth

A strongly positive relation seems to exist between placental and fetal weight (Kloosterman and Huidekoper, 1954; Dawes, 1968; Solomon and Friesen, 1968). Removal or destruction of a part of the placenta reduces fetal growth (Dawes, 1968; Gammal, 1973), while removal of the fetus inhibits placental growth (Van Wagenen and Catchpole, 1965).

In rat, placental weight decreased after aspiration of the fetal brain, while the placental index (*i.e.* placental weight divided by fetal weight) appeared to be increased (Swaab and Honnebier, 1973).

In sheep, microscopical placental changes were reported after fetal hypophysectomy (Liggins and Kennedy, 1968); placental weights were unfortunately not mentioned.

In human anencephalics a decreased placental weight and increased placental index were also observed (Milic and Adamsons, 1969; Honnebier and Swaab, 1973a). Reduction in placental weight of anencephalics appeared to be much more pronounced than that of other congenital anomalies (Fig. 2). The placental index of the last group is even less than in the controls while in the anencephalics it is increased (Fig. 3).

INFLUENCE OF HYPOPHYSEAL HORMONES AND OTHER COMPOUNDS UPON DEVELOPMENT OF THE FETO-PLACENTAL UNIT

As has been mentioned before, destruction of the fetal brain or hypophyseal area induces growth retardation of the fetus and of the placenta in various species. In order to determine factors responsible for these inhibitions, the influence of some hypophyseal compounds, brain extracts and insulin on intrauterine development were determined, and are discussed in the following paragraphs.

Growth hormone

Growth hormone has been demonstrated in the fetal pituitary of various species (Bassett *et al.*, 1970; Franchimont, 1971; Grumbach and Kaplan, 1973) as well as in the fetal circulation (Bassett *et al.*, 1970). Although rat plasma growth hormone decreases after encephalectomy and disappears after decapitation (Jost *et al.*, 1974), and is lower in anencephalics than in normal newborns (Grumbach and Kaplan, 1973) the role of this hormone in fetal growth seems, however, not to be important.

Placental transfer of growth hormone is supposed to be unimportant (Jost and Picon, 1970; Bassett *et al.*, 1970), yet effects were observed in the fetus after administration of the hormone to the mother. Administration of crude growth hormone preparations (Phyol, Antuitrin-G) to the mother rat induced in some groups a slight increase in full term body weight (Hultquist and Engfeldt, 1949), while pure growth hormone induced in some groups a slight increase—but in other groups even a decrease—in body weight (Engfeldt and Hultquist, 1953). Literature data about these experiments are difficult to interpret because of the possible effects of impurities in

References p. 275–280

the crude growth hormone preparations that were used. This problem is illustrated by the fact that Zamenhof *et al.* (1966) could not find an increased fetal body weight after either s.c. or i.v. administration of NIH-growth hormone (3 mg/day) to the mother, although a pronounced increase in fetal brain weight was observed.

Administration of growth hormone directly to the fetus failed to give clear-cut results. That administration of bovine growth hormone to decapitated rats should increase fetal weight cannot be maintained, since only 5 experimental fetuses were studied (Heggestad and Wells, 1965). In addition, we were unable to confirm this growth promoting effect (Tables VI, VII). Also our preliminary studies in brain-aspirated fetuses, injecting 10 μg NIH-GH-S 9 (ovine) in 50 μl 0.9% NaCl and 10 μg or 50 μg of this hormone in 50 μl oleum rapae per fetus, were negative in this respect. In decapitated rabbit fetuses, only a trend towards increase in size was observed after growth hormone administration (Jost, 1966a). Injection of growth hormone into the intact rat fetus gave either slightly decreased birth weights (Hwang and Wells, 1959; Heggestad and Wells, 1965) or no effect at all (Table VII).

There is also no clinical evidence at present for a role of growth hormone in human fetal growth. Congenital aplasia of the fetal pituitary goes together with normal birth weight (Table II). In addition, ateleotic dwarfs (McKusick and Rimoin, 1967; Rimoin *et al.*, 1966a, b) and acromegalic women usually have children with normal birth weights (Solomon and Friesen, 1968; Hoet, 1969). In addition, no relation was found between maternal growth hormone levels and birth weight (Parekh *et al.*, 1973), while even a negative relation exists between fetal growth hormone levels and birth weight after 20–21 weeks of pregnancy (Cramer *et al.*, 1971; Kaplan *et al.*, 1972).

A remarkable increase in placental weight was found after growth hormone injection into brain-aspirated fetuses (Table VIII) as well as into intact ones (Table IX). This effect was also obtained by administration of cyclic-AMP (Table VIII) to brain-aspirated fetuses. The latter observation and the finding that this effect could not be obtained in intact fetuses cannot be interpreted in the present stage of the study. The lower placental weights found in brain-aspirated fetuses (Swaab and Honnebier, 1973) and in anencephalic infants (Honnebier and Swaab, 1973a), together with the heavier placentas after growth hormone treatment, suggest that growth hormone may play a role in the development of the placenta. Also interesting in this respect is the trend towards heavier placentas found after growth hormone administration to pregnant rats (Croskery *et al.*, 1973).

Adrenocorticotropic hormone (ACTH)

The fetal adrenal has been shown to be stimulated by the fetal hypothalamo–hypophyseal system in man, as well as in various animals (Jost, 1966b; Jost *et al.*, 1970; Honnebier *et al.*, 1974). ACTH that has been identified in sheep fetal pituitaries and blood (Alexander *et al.*, 1973) does not pass the placenta (Alexander *et al.*, 1971; Honnebier *et al.*, 1974). Transplacental passage of adrenal steroids does take place, however (Angervall and Martinsson, 1969; Jost and Picon, 1970).

Corticotropin administration to pregnant rats was found to decrease fetal body weight (Hultquist and Engfeldt, 1949; Noumura, 1959). Although cortisone injection in the pregnant mother rat has been reported to increase fetal growth (Angervall and Martinsson, 1969), others found a slight decrease in fetal growth (Noumura,

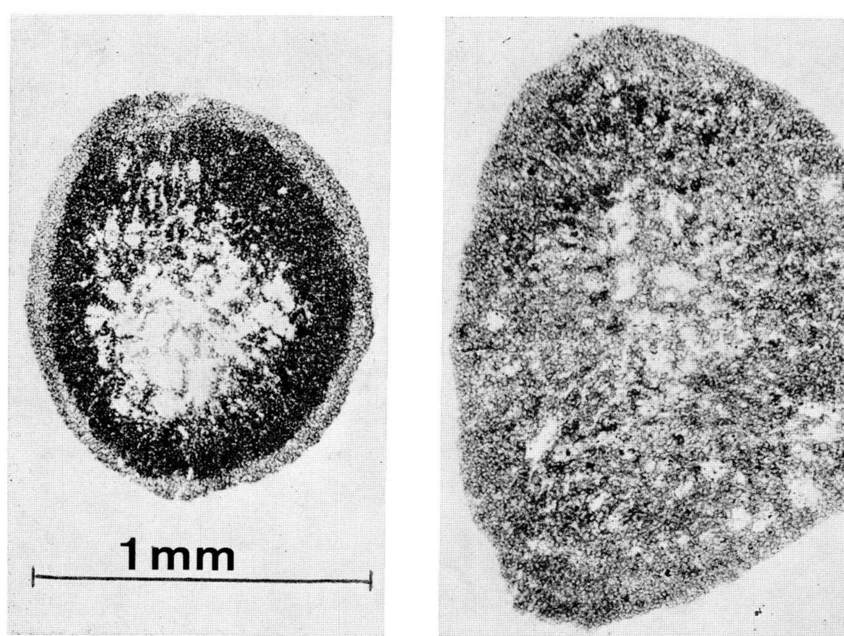

Fig. 4. Stimulation of fetal adrenal size by ACTH. For description of the experimental procedure and the Sudan-black-B staining of the sections, see legend of Table IV. Left: control adrenal (brain-aspirated fetus receiving 0.9% NaCl); right: ACTH-stimulated adrenal (brain-aspirated fetus receiving ACTH).

1959). Moreover, injections of corticotropins as well as of adrenal steroids into intact and decapitated fetuses were reported to inhibit fetal growth (Noumura, 1959; Kitchell and Wells, 1952). In brain-aspirated fetuses a wide lipid-free layer was found in the outer zone of the adrenals (Fig. 4), that has also been described after fetal decapitation (Noumura, 1959). After ACTH treatment (Cortrosyn-depot®) this outer zone of the adrenals also contained lipids (Fig. 4). Adrenal length and width were 32% and 42% enlarged respectively (Table IV). Neither body weight nor placental weight could however be restored, either by short- or by long-acting ACTH (Tables VI, VIII).

Since also in humans, birth weight of children with congenital adrenal hyperplasia (Price *et al.*, 1971) and hypoplasia (O'Donohoe and Holland, 1968) is normal, the fetal adrenal cannot be important for fetal growth.

References p. 275–280

TABLE IV

EFFECT OF ACTH ADMINISTRATION TO BRAINLESS FETUSES ON FETAL ADRENAL SIZE

In the present experiment, the 5 control and 5 Cortrosyn-depot treated animals of Table VI were studied. The adrenals of the two fetuses situated at the ovarian ends of the left uterine horns were frozen in liquid-nitrogen-cooled Freon-12. Serial sections (16 μm) were made on a cryostat. The sections were post-fixed in an ethanol–formalin–acetic acid mixture and alternately stained by means of thionin (Windle *et al.*, 1943) or Sudan-black-B (Noumura, 1959). Of each fetal adrenal the largest diameter ("length") and the diameter perpendicular to this ("width") were measured under a microscope. The increase in length as well as in width after ACTH administration is highly significant (Student t-test: $P < 0.001$; De Jonge, 1963, chapt. 13).

	Control fetuses	*ACTH treated fetuses*
Treatment of the fetus	Brain aspiration day 19 and administration of 50 μl 0.9% NaCl s.c. to the fetus.	Brain aspiration day 19 and administration of 50 μl Cortrosyn-depot® s.c. to the fetus.
Adrenal length (in mm)	1.34 (S.E.M. 0.030)	1.77 (S.E.M. 0.024)
Adrenal width (in mm)	0.95 (S.E.M. 0.042)	1.35 (S.E.M. 0.060)

Thyreotropic hormone (TSH)

The rat fetal thyroid is regulated near term by the fetal pituitary (Jost, 1966a; Jost *et al.*, 1970). TSH has been identified in the fetal pituitary (Conklin *et al.*, 1973) and serum (Erenberg and Fisher, 1973). TSH is not supposed to pass the placenta (Kajihara *et al.*, 1972; Furth and Pagliara, 1971; Knobil and Josimovich, 1959; Jost *et al.*, 1974). In contrast, TSH-RF (Kajihara *et al.*, 1972; D'Angelo and Wall, 1972), iodine and some organic iodine-containing compounds cross the placenta readily (Furth and Pagliara, 1971). Administration of thyreotropic hormone to the pregnant mother rat (Hultquist and Engfeldt, 1949) was found to reduce fetal body weight slightly.

Hopkins and Thorburn (1972) observed a decreased body weight and bone-age in athyroid lambs. In addition, the same phenomenon was seen in fetal lambs from ewes given methylthiouracil (Lascelles and Setchell, 1959). These experiments indicate a role of TSH in fetal growth in sheep. Thyroidectomized rabbit fetuses, however, did not show any significant reduction in fetal weight (Jost, 1971). Moreover, no increase in fetal body weight was observed after administration of TSH or thyroid hormones to the rat fetus (Hwang and Wells, 1959; the present study: Table VI). No positive relation exists between thyroid weight and body weight in rat fetuses (Angervall and Lundin, 1963). Also the placental weight was not found to change after TSH administration (Table VIII).

As congenital hypothyroidism is reported to be accompanied by normal birth weight

(Page, 1969; Paine, 1971), also in the human the thyroid does not seem to be important for intrauterine growth.

Gonadotropic hormones

Gonadotropic hormones are present in the fetal pituitary (Jost, 1966a; Ingleton *et al.*, 1971; Grumbach and Kaplan, 1973) and in the fetal circulation (Levina, 1972; Grumbach and Kaplan, 1973). Prolactin (Delost, 1971), LH (Foster *et al.*, 1972) and HCG (Honnebier *et al.*, 1974) were found not to cross the placenta.

Prolactin (Delost, 1971), chorionic gonadotropins and PMS gonadotropins (Hultquist and Engfeldt, 1949), injected into pregnant animals, did not influence fetal body weight. Prolactin administration to intact mouse fetuses (Jean and Delost, 1969) and prolactin, LH or FSH to brain-aspirated rat fetuses (Table VI) did not change either fetal body weight or placental weight (Table VIII). Furthermore, after chorionic gonadotropins or PMS given to decapitated or intact rat fetuses (Noumura, 1959) or HCG administration to brain-aspirated fetuses (Table VI) no increase in body weight was observed. Administration of chorionic gonadotropins to the fetuses did not change placental weight (Table VIII). Gonadotropic hormones thus do not have any stimulating effect on intrauterine growth.

Sex hormones, *i.e.* estrogens (Delost, 1971; Kitchell and Wells, 1952; Noumura, 1959), testosterone (Noumura, 1959), progesterone (Kitchell and Wells, 1952), administered to the rat fetus, or testosterone-propionate (Slob, 1972), androstenediol-dipropionate, methyl-androstenediol-dipropionate (Jost, 1954), administered to the mother failed to increase fetal body weight. Moreover, injection of a progesterone–estrogen combination into intact pregnant rabbit fetuses does not influence placental weight (Butterstein and Leathem, 1970). And so neither gonadotropic hormones nor sex hormones stimulate intrauterine growth.

Posterior and intermediate lobe hormones

Vasopressin, oxytocin and neurophysins are present in the fetal hypothalamo–neurohypophyseal system (HNS) (Dicker and Tyler, 1953; Roffi, 1958; Alexander *et al.*, 1973). In addition, these hormones have been found in the fetal blood (Alexander *et al.*, 1973; Chard, 1972; Edwards, 1971; Robinson *et al.*, 1971).

No difference in pup weight was observed in Brattleboro rats being homozygous or heterozygous for hypothalamic diabetes insipidus (Boer *et al.*, 1974). Therefore, a lack of active maternal and fetal vasopressin does not influence fetal growth. This agrees with the results of two sets of preliminary experiments (Swaab and Veerkamp, unpublished data) in which vasopressin was administered to pregnant rats (Table V). Parturition occurred in all these groups within the same period. As is shown in Table V, vasopressin treatment of the mother did not increase pup weight.

During the administration of oxytocin to the litter, the uterus became clearly pale and contracted. However, the litters were not born before the day of weighing (*i.e.* day 21). Oxytocin (Piton-S) slightly decreased the weight of the brain-aspirated fetuses

References p. 275–280

TABLE V

INFLUENCE ON PUP WEIGHT OF VASOPRESSIN ADMINISTRATION TO THE MOTHER

Pregnant rats obtained the following treatments from day 7 of pregnancy (the day that spermatozoa were found in the vaginal smear being taken as day 0). Ia: 9 controls were injected with arachnoid oil in the same volume as the experimental rats. Ib: 9 rats obtained vasopressin (Pitressin tannate in oil, Parke-Davis, 5 pressor units/ml) in a dose of 1 I.U. daily from day 7 of pregnancy and of 2 I.U. daily from day 14 of pregnancy. Ic: 7 rats were given the same treatment as in b, but the drinking water contained 1% NaCl. IIa: 7 controls obtained a treatment as in Ia. IIb: 4 animals were given vasopressin in a dose of 5 I.U. daily from day 7.

Group	*Number of mothers*	*Mean number of pups per litter*	*Weight of the whole litter (g)*	*Mean body weight per pup (g)*
Ia	9	11.3	59.5	5.3
Ib	9	11.4	57.7	5.0
Ic	7	11.7	60.7	5.2
IIa	7	11.1	59.4	5.3
IIb	4	6.4	31.9	5.3

(Table VI), but had no influence on placental weight (Table VIII). Since anti-oxytocin administration to pregnant rats did not change pup weight (Kumaresan *et al.*, 1971), maternal oxytocin, which is able to cross the placenta (Chard, 1972), cannot be important for intrauterine growth.

An increased fetal body weight (Table VI) as well as placental weight (Table VIII) was found after administration of α-MSH to brain-aspirated fetuses. Much more work needs to be done on the intermediate lobe hormone after this surprising finding, especially since MSH has been determined at an early stage in the pituitary glands of human fetuses (Kastin *et al.*, 1968).

Hypothalamic extracts

Four Sephadex G-25 fractions (A, B, C and E) of the entire base of the brain of 3–4-month-old calves were kindly supplied by Dr. A. Witter (Utrecht). These fractions have been prepared according to De Wied *et al.* (1969), except for the HCl extraction, which was performed ultrasonically at 0 °C for 5 min, and the separation on Sephadex which was performed at 4 °C. Solutions of the 4 fractions in physiological saline were neutralized to pH 7 by means of 0.5 *N* NaOH.

In preliminary experiments, 10 or 100 μg of fraction C in physiological saline and 100 or 500 μg of fraction C in oleum rapae did not change either body or placental weight in brain-aspirated fetuses. In intact fetuses, 5 mg of fractions B, C and E per fetus did not increase body or placental weight. Also 5 mg of fraction C per fetus at pH 5, 6 or 8 did not induce body weight or placental weight changes (Tables VI–IX).

Since 5 mg of fraction A per fetus caused death of the pregnant rats, this extract has been administered in a lower dose (50 μg). This dose did not cause a significant increase in fetal body weight (Table VII). It did, however, increase placental weight in the experiments reported in Table IX. However, in later experiments, this effect of fraction A did not appear to be reproducible.

Insulin

Although no specific hypothalamic factor seems to control the endocrine pancreas, insulin secretion is influenced by various neuroendocrine and neural mechanisms (Ensink and Williams, 1972; Malaisse, 1972). The excessive fetal growth during gestational diabetes mellitus is supposed to be causally related to a fetal hyperinsulinism (Hoet, 1969; Jost and Picon, 1970). To obtain this increased fetal growth, a normal hypothalamo–hypophyseal system should be necessary (De Gasparo and Hoet, 1971; Van Assche *et al.*, 1970), although an observation of Frandsen and Stakemann (1964) is conflicting in this respect. Since the endocrine pancreas in anencephalics is found to be less developed (Van Assche, 1968), the hypothalamo–hypophyseal system may thus be necessary for complete morphological and functional maturation of the pancreatic endocrine tissue (Van Assche *et al.*, 1970). Moreover, rat fetal plasma insulin increases 5-fold (Freie and Bouman, 1973) in the same period as we observed an increased fetal growth rate in intact fetuses (Swaab and Honnebier, 1973). These data, pointing to a functional connection between hypothalamus and pancreas made it of interest in the framework of the present study to investigate the influence of insulin upon fetal and placental development.

Insulin has been found in fetal pancreas (Jack and Milner, 1972; Fujimoto and Williams, 1972; Freie and Bouman, 1973) and in blood (Freie and Bouman, 1973; Grumbach and Kaplan, 1973). The placental block to insulin transfer (Jost and Picon, 1970) between the mother and the fetus is probably only partial (Solomon and Friesen, 1968; Knobil and Josimovich, 1959; Spellacy, 1971; Gitlin *et al.*, 1965).

Insulin tardum (Organon), injected in a dose of 2 I.U. into 3 litters of brain-aspirated rat fetuses on day 19, caused maternal death in 2 cases. From the one living animal the 5 live fetuses (of the total litter of 11) did not show increased growth, and nor did the placenta. Lower doses were not able to increase either fetal weight (Tables VI, VII) or placental weight (Table VIII). Picon (1967) obtained an increased fetal weight by means of insulin administration to intact fetuses, but he treated only half the litter. The question arises therefore whether this scheme of insulin, administered to the whole litter, would also cause death of the mother, and could thus be considered to be an unphysiological dose. Since decapitation of rabbit fetuses increases fetal plasma insulin level, and glucose induces increased insulin release from the pancreas of these animals (Jack and Milner, 1973), while the body weight of decapitated rabbit fetuses does not change (Jost, 1966a), the role of insulin in intrauterine growth of rabbit does not seem to be important.

In the adult situation a complex relationship exists between insulin and growth hormone. In adult rats (Salter and Best, 1953) and human ateleotic dwarfs (Editorial, 1953) insulin shows a growth promoting effect and a nitrogen retention effect respectively. In addition, growth hormone was reported to stimulate insulin secretion (Martin

TABLE VI

INFLUENCE ON FETAL BODY WEIGHT

Tables VI–IX. Influence of various substances on rat fetal body weight and placental weight. The general animal treatment, operation procedure and brain-aspiration were performed according to Swaab and Honnebier (1973). All animals were operated ("brain-aspiration") or sham operated ("intact") on the morning of day 19 of pregnancy. The animals received their hormones in a volume of 50 μl (*cf.* Heggestad and Wells, 1965; Noumura, 1959) of sterile physiological saline (0.9% NaCl) subcutaneously into the back by means of a 0.50 mm $\times$ 16 mm needle, unless mentioned otherwise in the tables. Doses were given per fetus, and the whole litter was treated in the same way. The controls received 50 μl of 0.9% NaCl only, using the same procedure. In order to correct the fetal weights for differences in cranial filling, all fetal weights reported in these tables were measured after emptying the cranial cavity by suction (Swaab and Honnebier, 1973). The differences between the groups were tested by Student *t*-test (De Jonge, 1963, chapt. 13) and $P < 0.05$ was considered to be statistically significant.

Treatment of brain-aspirated fetuses	*n mothers*	*n fetuses*	*Experimental group body weight (and S.E.M.)*	*n mothers*	*n fetuses*	*Control group body weight (and S.E.M.)*	*Student t-test*	*Result*
Rat growth hormone; NIAMD-RAT GH-B-1; 150 μg	6	45	2.65 (0.063)	5	51	2.87 (0.053)	$P<0.02$	↓
ACTH: Cortrosyn (Organon); 150 μg	5	37	3.11 (0.095)	5	51	3.09 (0.070)	$0.80<P<0.90$	—
ACTH: Cortrosyn-Depot (Organon); 50 μl depot containing 1 mg/ml	5	47	3.09 (0.058)	5	56	3.44 (0.042)	$P<0.001$	↓
TSH; NIH-TSH-S6 (ovine); 150 μg	5	39	2.95 (0.067)	5	43	2.97 (0.090)	$0.60<P<0.70$	—
Prolactin; NIH-P-S9 (ovine); 150 μg	5	52	3.07 (0.056)	5	41	3.20 (0.097)	$0.20<P<0.30$	—
LH; NIH-LH-S18 (ovine); 150 μg	4	39	3.23 (0.072)	5	26	3.23 (0.100)	$P>0.9995$	—
FSH; NIH-FSH-S9 (ovine); 150 μg	5	42	3.13 (0.069)	5	29	3.18 (0.098)	$0.80>P>0.70$	—
HCG: Pregnyl (Organon); 10 I.U.	4	37	3.23 (0.044)	5	51	3.17 (0.048)	$0.30<P<0.40$	—
Piton-S; (Organon, 10 U/ml) 0.5 U	5	40	3.22 (0.050)	5	48	3.41 (0.042)	$P<0.01$	↓
Hypothalamic extract Fraction-C, 5 mg	5	38	2.69 (0.083)	5	48	3.41 (0.042)	$P<0.001$	↓
Insulinum Tardum; 0.4 I.U. (Organon) 40 I.U./ml diluted 1/5 with phys. sal.	5	32	3.18 (0.067)	5	39	3.28 (0.053)	$0.20<P<0.30$	—
Insulinum Tardum; 0.2 I.U. (Organon) 40 I.U./ml diluted 1/10 with phys. sal.	5	44	2.90 (0.069)	5	39	3.28 (0.053)	$P<0.001$	↓
Placenta Suspensata-depot (Schwarzhaupt); 50 μl	5	45	3.54 (0.055)	5	48	3.65 (0.038)	$0.10<P<0.20$	—
Cyclic AMP: N^6-2′ O-Dibutyryl-adenosine-3′: 5′-monophosphate, cyclic, monosodium salt (Boehringer), 150 μg	5	40	3.57 (0.062)	5	41	3.58 (0.060)	$0.90<P<0.95$	—
α-MSH; 150 μg	9	84	3.38 (0.043)	8	75	2.92 (0.039)	$p<0.001$	↑

TABLE VII

INFLUENCE ON FETAL BODY WEIGHT

Treatment of intact fetuses	*n mothers*	*n fetuses*	*Experimental group body weight (and S.E.M.)*	*n mothers*	*n fetuses*	*Control group body weight (and S.E.M.)*	*Student t-test*	*Result*
Rat growth hormone (see Table VI); 150 μg	5	41	4.02 (0.082)	5	50	4.13 (0.055)	$0.10 < P < 0.20$	—
Hypothalamic extract Fraction A; 50 μg	5	45	3.98 (0.065)	5	56	3.84 (0.059)	$0.20 > P > 0.10$	—
Hypothalamic extract Fraction B; 5 mg	5	35	3.67 (0.072)	5	44	3.69 (0.059)	$0.90 > P > 0.80$	—
Hypothalamic extract Fraction C; 500 μg	5	50	3.66 (0.066)	5	51	3.93 (0.037)	$P < 0.001$	↓
Hypothalamic extract Fraction C; 5 mg	8	67	3.71 (0.048)	8	81	3.87 (0.041)	$0.01 < P < 0.02$	↓
Hypothalamic extract Fraction E; 5 mg	5	41	3.71 (0.106)	5	50	3.84 (0.069)	$0.40 > P > 0.30$	—
Insulinum Tardum (see Table VI); 0.2 I.U./fetus	6	57	3.65 (0.056)	5	51	3.81 (0.054)	$0.05 < P < 0.10$	—
Cyclic AMP (see Table VI); 150 μg	8	76	3.69 (0.053)	8	81	3.81 (0.034)	$0.05 < P < 0.10$	—

References p. 275–280

TABLE VIII

INFLUENCE ON PLACENTAL WEIGHT

Treatment of brain-aspirated fetuses	*n mothers*	*n fetuses*	*Experimental group placental weight (and S.E.M.)*	*n mothers*	*n fetuses*	*Control group placental weight (and S.E.M.)*	*Student t-test*	*Result*
ACTH: Cortrosyn (Organon); 150 μg	5	37	0.51 (0.021)	5	51	0.48 (0.008)	$0.10 < P < 0.20$	—
ACTH: Cortrosyn-Depot (Organon); 50 μl depot containing 1 mg/ml	5	47	0.50 (0.013)	5	56	0.50 (0.013)	$0.95 < P < 0.975$	—
TSH; NIH-TSH-S6; 150 μg	5	39	0.48 (0.012)	5	43	0.49 (0.012)	$0.60 < P < 0.70$	—
Prolactin NIH-P-S9; 150 μg	5	52	0.50 (0.009)	5	41	0.49 (0.011)	$0.60 < P < 0.70$	—
LH; NIH-LH-S18 (ovine); 150 μg	4	39	0.54 (0.013)	5	26	0.53 (0.022)	$0.70 > P > 0.60$	—
FSH; NIH-FSH-S9 (ovine); 150 μg	5	42	0.53 (0.011)	5	29	0.52 (0.021)	$0.70 > P > 0.60$	—
HCG: Pregnyl (Organon); 10 I.U.	4	37	0.50 (0.010)	5	51	0.49 (0.010)	$0.95 < P < 0.975$	—
Piton-S (Organon 10 U/ml); 0.5 U	5	40	0.50 (0.010)	5	48	0.51 (0.010)	$0.60 < P < 0.70$	—
Hypothalamic extract Fraction C; 5 mg	5	38	0.48 (0.012)	5	48	0.51 (0.010)	$0.10 < P < 0.20$	—
Insulinum Tardum (see before); 0.4 I.U.	5	32	0.48 (0.010)	5	39	0.49 (0.009)	$0.50 < P < 0.60$	—
Insulinum Tardum (see before); 0.2 I.U.	5	44	0.50 (0.010)	5	39	0.49 (0.009)	$0.30 < P < 0.40$	—
Placenta Suspensata-depot (Schwarzhaupt); 50 μl	5	45	0.49 (0.011)	5	48	0.54 (0.012)	$P < 0.01$	↓
Rat growth hormone (see Table VI); 150 μg	6	45	0.54 (0.014)	5	51	0.49 (0.011)	$P < 0.02$	↑
α-MSH; 150 μg	9	84	0.51 (0.007)	8	75	0.45 (0.006)	$P < 0.001$	↑
Cyclic AMP (see Table VI); 150 μg	5	40	0.57 (0.013)	5	41	0.51 (0.009)	$P < 0.001$	↑

TABLE IX

INFLUENCE ON PLACENTAL WEIGHT

Treatment of intact fetuses	n mothers	n fetuses	Experimental group placental weight (and S.E.M.)	n mothers	n fetuses	Control group placental weight (and S.E.M.)	Student t-test	Result
Hypothalamic extract Fraction B; 5 mg	5	35	0.53 (0.012)	5	44	0.55 (0.014)	$0.30 > P > 0.20$	—
Hypothalamic extract Fraction C; 500 μg	5	50	0.52 (0.012)	5	51	0.51 (0.007)	$0.40 < P < 0.50$	—
Hypothalamic extract Fraction C; 5 mg	8	67	0.54 (0.012)	8	81	0.53 (0.007)	$0.60 < P < 0.70$	—
Hypothalamic extract Fraction E; 5 mg	5	50	0.54 (0.012)	5	41	0.52 (0.008)	$0.50 > P > 0.40$	—
Insulinum Tardum (see Table VI); 0.2 I.U.	5	57	0.50 (0.010)	5	51	0.51 (0.009)	$0.30 < P < 0.40$	—
Cyclic AMP (see Table VI); 150 μg	8	76	0.51 (0.011)	8	81	0.52 (0.006)	$0.50 < P < 0.60$	—
Rat growth hormone (see Table VI); 150 μg	5	41	0.55 (0.012)	5	50	0.51 (0.009)	$P < 0.02$	↑
Hypothalamic extract Fraction A; 50 μg	5	45	0.57 (0.010)	5	56	0.52 (0.009)	$P < 0.005$	↑

References p. 275–280

and Gagliardino, 1967), and was found to be unable to cause anabolism of protein in the absence of insulin (Editorial, 1953). Hypophysectomized rats had a lower insulin content and production of the pancreas, while the normal reaction to glucose administration was not found. This reaction was restored by growth hormone (Hoet, 1969). Whether or not a functional relationship between insulin and growth hormone also exists in the fetus has still to be studied.

At present, therefore, there are not sufficient data to justify calling insulin "the fetal growth hormone", as has been done in the literature (Hoet, 1969).

THE ROLE OF THE FETAL PITUITARY IN PARTURITION

The idea that the fetal pituitary could play a role in the initiation of parturition has arisen from observations of congenital malformations. Predisposition for prolonged pregnancy was reported in anencephaly without hydramnios (Malpas, 1933). In addition, a prolonged gestation occurred in Guernsey cattle having an inherited adenohypophyseal aplasia (Kennedy *et al.*, 1957) and in cyclopian sheep fetuses. The latter anomaly, caused by the ingestion of *Veratrum californicum*, is accompanied by absence or displacement of the fetal pituitary (Binns *et al.*, 1964). From the already classical experiments performed mainly in sheep and involving adrenalectomy, fetal hypophysectomy and administration of ACTH or adrenal glucocorticoids, the role of the fetal hypothalamo–hypophyseal–adrenal system in the initiation of labor has obtained experimental support (Drost and Holm, 1968; Liggins *et al.*, 1967; Liggins and Kennedy, 1968; Liggins, 1969). Additional evidence is that about the time of delivery, fetal plasma cortisol (Bassett and Thorburn, 1969; Nathanielsz *et al.*, 1972) and plasma ACTH (Alexander *et al.*, 1971, 1973) increase. In Ango goats habitual abortion was found to be accompanied by adrenal hyperplasia (Osinga, 1969). In addition, the finding of remarkably high posterior lobe hormone and neurophysin levels in fetal blood, apparently produced by the fetus (Chard *et al.*, 1970, 1971; Chard, 1971; Hoppenstein *et al.*, 1968; Robinson *et al.*, 1971), pointed to a possible involvement of the fetal hypothalamo–neurohypophyseal system in parturition.

However, observations in other animals and in man make the role of the fetal hypophysis questionable, at least as a general mechanism in the animal kingdom. Fetal adrenalectomy in rats gave varying results being partly against the "Liggins" hypothesis mentioned above. Josimovich (1969, and personal communication) found that after adrenalectomy on day 19¼ and 19¾ of pregnancy, an increasing proportion of pregnancies lasted a good day beyond term (nothing is known however, about the dispersion in gestation length of the fetuses born earlier!), while the same operation carried out between day 18½ and 19⅛ did not produce a significant amount of increased gestation lengths. Kirsch (1938) could not demonstrate that the rat fetuses would be essential for the determination of gestation length. In rats bearing brainless fetuses, gestation length did not appear to be prolonged (Swaab and Honnebier, 1973). Moreover, in contradistinction to the sheep (see above), fetal adrenal activity seems to decrease before birth in rat (Jost, 1966b, 1973; Cohen, 1963). In addition, after injection (on day 19) of ACTH (Cortrosyn®, Cortrosyn-depot®) in brain-aspirated fetuses (see Table VI) parturition never occurred within the next 48 h, although the fetal adrenals were clearly activated (Table V).

In a preliminary experiment on rabbits we observed that after fetal brain aspiration on day 25 (day of copulation is day 1 of pregnancy) the experimental animal as well as the sham operated control delivered on the expected time (day 32), while in the experimental animal two well-formed and well-aspirated fetuses were recovered and one macerated fetus. The mother probably ate the other 10 fetuses (Nathanielsz and Swaab, unpublished observation, Cambridge 1973). This is in agreement with the normal gestation length of decapitated rabbit fetuses in reduced litters (Jost, 1973). These observations certainly need confirmation since they are in disagreement with the "Liggins" hypothesis that has obtained additional experimental support in this species by the finding that cortisone can initiate parturition (Kendall and Liggins, 1972).

Donovan and Peddie (1973) showed that damage of the hypothalamus of the guinea-pig even causes abortion instead of the expected prolonged gestation. Increased fetal adrenal activity was not the sole course of parturition in this species (Donovan and Peddie, 1974).

Fetal hypophysectomy in the monkey, performed by placement of radioactive sources in the sella turcica, was associated with prolongation of gestation (Chez *et al.*, 1970). However, Mueller-Heubach *et al.* (1972) observed that after fetal adrenalectomy, labor was initiated at or near term in the majority of instances. Another argument against the "Liggins" hypothesis in this species is the finding of Van Wagenen and Catchpole (1965) that after removal of the fetus the placenta was born at normal term. This phenomenon has also been reported for rat (Kirsch, 1938) and mouse (Newton, 1935).

The few observations of Hutchinson *et al.* (1962) in pregnant monkeys suggested that delivery at normal term was possible after fetal hypophysectomy.

Human anencephalics do not have a normal functional hypothalamus (see above). In spite of the general opinion in the literature (*e.g.* Malpas, 1933; Jost, 1966a; Liggins and Kennedy, 1968; Milic and Adamsons, 1969) the supposed prolonged gestation length in anencephaly without hydramnios was not confirmed by our observations. The mean gestation length (36.6 weeks) in this anomaly was even much shorter than that of the control group (39.6 weeks). Spontaneously born anencephalics from pregnancies where there was a normal amount of amniotic fluid had a similar mean gestation length (39.7 weeks) as had the control group (39.6 weeks) although a high percentage of both pre- and post-maturely born anencephalics was observed (Honnebier and Swaab, 1973a). In spite of the abrupt rise in fetal cortisol observed in spontaneously born children, again suggesting a role of the adrenal in the initiation of parturition (Murphy, 1973), congenital adrenal hyperplasia is reported not to influence gestation length (Price *et al.*, 1971). These arguments show that the relationship between adrenal weight and pregnancy length, as proposed by Anderson *et al.* (1971), does not exist in human. Four children reported in the literature have a congenital aplasia of the hypophysis, and do not show a prolonged gestation length

References p. 275–280

(Table III). In addition, initiation of parturition could not be obtained by transabdominal injections of corticotropins, HCG or posterior lobe hormones (Honnebier and Swaab, 1973b; Honnebier *et al.*, 1974).

Therefore, strong evidence for a role of the fetal pituitary in the initiation of parturition only exists in sheep. However, since fetal hypophysectomy in this species has necessarily to be executed rather a long time (6 days) before term (Liggins, 1969), the influence of growth inhibition (*cf.* Liggins and Kennedy, 1968) certainly needs to be investigated before drawing a firm conclusion even here.

CONCLUSIONS

Elimination of the fetal brain or fetal hypophyseal area causes a reduction of fetal ponderal growth in various animal species. The same phenomenon is found in the human anencephalus. In addition, absence of fetal brain in rat and human is accompanied by a decreased placental growth. The primary cause of the ponderal growth inhibition will probably be situated in the hypothalamo–hypophyseal system.

The influence of hypophyseal hormones, hypothalamic extracts and related compounds on the development of the feto-placental unit were therefore tested by injection of these compounds directly into intact or brain-aspirated fetuses. Results of a pilot study are reported in the present study. Several compounds of hypophyseal origin were found to increase placental weight (*i.e.*, growth hormone and α-MSH), while α-MSH also increases fetal body weight. More work is needed to see how these actions are interrelated and what their physiological function might be.

Although observations on congenital anomalies and experiments on sheep suggested an essential role for the fetal hypothalamo–hypophyseal system in the initiation of labor, several observations plead against such a role in other species at present. Particularly in rat and human, the moment of initiation of parturition was found to be independent of the fetal hypothalamus. Since the delivery phase of brain-aspirated fetuses was considerably protracted (Swaab *et al.*, 1973), the fetal brain may rather play a role in the *course* of labor than in its *initiation*.

ACKNOWLEDGEMENTS

The authors are greatly indebted to Miss J. W. L. Nolten and Mr. B. Fisser for their valuable technical assistance, to Miss B. L. Huidekoper for her help with the collection of clinical data, to Dr. M. A. Corner for the correction of the manuscript, and to Miss J. van der Velden for her secretarial work.

We are grateful to the National Institute of Arthritis and Metabolic Diseases for NIH-GH, TSH, Prolactin, FSH and LH, to CIBA-GEIGY for α-MSH and to Dr. A. Witter for the hypothalamic fractions A, B, C and E.

REFERENCES

ALEXANDER, D. P., BRITTON, H. G., FORSLING, M. L., NIXON, D. A. AND RATCLIFFE, J. G. (1971) The concentration of adrenocorticotrophin, vasopressin and oxytocin in the foetal and maternal plasma of the sheep in the latter half of gestation. *J. Endocr.*, **49**, 179–180.

ALEXANDER, D. P., BRITTON, H. G., FORSLING, M. L., NIXON, D. A. AND RATCLIFFE, J. G. (1973) Adrenocorticotrophin and vasopressin in foetal sheep and the response to stress. In *The Endocrinology of Pregnancy and Parturition. Experimental Studies in the Sheep, Proc. Int. Symp.*, C. G. PIERREPONT (Ed.), Alpha Omega Alpha Publishing, Cardiff, pp. 112–125.

ANDERSON, A. B. M., LAURENCE, K. M., DAVIES, K., CAMPBELL, H. AND TURNBULL, A. C. (1971) Fetal adrenal weight and the cause of premature delivery in human pregnancy. *J. Obstet. Gynaec. Brit. Cwlth*, **78**, 481–488.

ANGERVALL, L. AND LUNDIN, P. M. (1963) Hypophysectomy in pregnant rats with special reference to the endocrine organs of the offspring. *Acta endocr. (Kbh.)*, **42**, 591–600.

ANGERVALL, L. AND MARTINSSON, A. (1969) Overweight in offspring of cortisone-treated pregnant rats. *Acta endocr. (Kbh.)*, **60**, 36–46.

BASSETT, J. M. AND THORBURN, G. D. (1969) Foetal plasma corticosteroids and the initiation of parturition in sheep. *J. Endocr.*, **44**, 285–286.

BASSETT, J. M., THORBURN, G. D. AND WALLACE, A. L. C. (1970) The plasma growth hormone concentration of the foetal lamb. *J. Endocr.*, **48**, 251–263.

BATTAGLIA, F. C. (1970) Intrauterine growth retardation. *Amer. J. Obstet. Gynec.*, **106**, 1103–1114.

BINNS, W., JAMES, L. F. AND SHUPE, J. L. (1964) Toxicosis of *Veratrum californicum* in ewes and its relationship to a congenital deformity in lambs. *Ann. N.Y. Acad. Sci.*, **111**, 571–576.

BLINKOV, S. M. AND GLEZER, I. I. (1968) *The Human Brain in Figures and Tables. A Quantitative Handbook*, Basic Books, Plenum Press, New York, p. 336, Table 115.

BLIZZARD, R. M. AND ALBERTS, M. (1956) Hypopituitarism, hypoadrenalism, and hypogonadism in the newborn infant. *J. Pediat.*, **48**, 782–792.

BOER, K., BOER, G. J. AND SWAAB, D. F. (1974) Does the hypothalamo–neurohypophysial system play a role in gestation length or parturition? In *Integrative Hypothalamic Activity, Progress in Brain Research, Vol. 41*, D. F. SWAAB AND J. P. SCHADÉ (Eds.), Elsevier, Amsterdam, pp. 307–320.

BREWER, D. B. (1957) Congenital absence of the pituitary gland and its consequences. *J. Path. Bact.*, **73**, 59–67.

BUTTERSTEIN, G. M. AND LEATHEM, J. H. (1970) Ribosomes in normal and giant placentae. *J. Endocr.*, **48**, 473–474.

C.B.S. (Centraal Bureau voor de Statistiek) (1972) Maandstatistiek van bevolking en volksgezondheid, Jaarg. 20, Suppl. In *Jaaroverzicht Bevolking en Volksgezondheid 1971*, Staatsuitgeverij, 's-Gravenhage, p. 62.

CHARD, T. (1971) Recent trends in the physiology of the posterior pituitary. In *Current Topics in Experimental Endocrinology, Vol. 1*, L. MARTINI AND V. H. T. JAMES (Eds.), Academic Press, New York, pp. 81–120.

CHARD, T. (1972) The posterior pituitary in human and animal parturition. *J. Reprod. Fertil.*, Suppl. **16**, 121–138.

CHARD, T., BOYD, N. R. H., FORSLING, M. L., MCNEILLY, A. S. AND LANDON, J. (1970) The development of a radioimmunoassay for oxytocin: the extraction of oxytocin from plasma, and its measurement during parturition in human and goat blood. *J. Endocr.*, **48**, 223–234.

CHARD, T., HUDSON, C. N., EDWARDS, C. R. W. AND BOYD, N. R. H. (1971) Release of oxytocin and vasopressin by the human foetus during labour. *Nature (Lond.)*, **234**, 352–354.

CHEZ, R. A., HUTCHINSON, D. L., SALAZAR, H. AND MINTZ, D. H. (1970) Some effects of fetal and maternal hypophysectomy in pregnancy. *Amer. J. Obstet. Gynec.*, **108**, 643–650.

COHEN, A. (1963) Corrélation entre l'hypophyse et le cortex surrénal chez le foetus de rat. Le cortex surrénal du nouveau-né. *Arch. Anat. micr. Morph. exp.*, **52**, 279–289.

CONKLIN, P. M., SCHINDLER, W. J. AND HULL, S. F. (1973) Hypothalamic thyrotropin releasing factor. *Neuroendocrinology*, **11**, 197–211.

CRAMER, D. W., BECK, P. AND MAKOWSKI, E. L. (1971) Correlation of gestational age with maternal human chorionic somatomammotropin and maternal and fetal growth hormone plasma concentrations during labor. *Amer. J. Obstet. Gynec.*, **109**, 649–655.

CROSKERRY, P. G., SMITH, G. K., SHEPARD, B. J. AND FREEMAN, K. B. (1973) Perinatal brain DNA in the normal and growth hormone-treated rat. *Brain Res.*, **52**, 413–418.

D'ANGELO, S. A. AND WALL, N. R. (1972) Maternal–fetal endocrine interrelations: effects of synthetic thyrotrophin releasing hormone (TRH) on the fetal pituitary–thyroid system of the rat, *Neuro-endocrinology*, **9**, 197–206.

DAWES, G. S. (1968) The placenta and foetal growth. In *Foetal and Neonatal Physiology*, G. S. DAWES (Ed.), Yearbook Medical Publishers, Chicago, Ill., pp. 42–59.

DE GASPARO, M. AND HOET, J. J. (1971) Normal and abnormal foetal weight gain. In *Excerpta Medica International Congress Series No. 231, Proc. VII Congr. Intern. Diabetes Federation, Buenos Aires, 23–28 August, 1970*, K. R. RODRIGUEZ AND J. VALLANCE-OWEN (Eds.), pp. 667–677.

DE JONGE, H. (1963) *Inleiding tot de Medische Statistiek, Vol. I and II*, Nederlands Instituut voor Preventieve Geneeskunde, Leiden.

DELOST, P. (1971) Fetal endocrinology and the effect of hormones on development. In *The Bio-psychology of Development*, E. TOBACH, L. R. ARONSON AND E. SHAW (Eds.), Academic Press, New York, pp. 195–229.

DE WIED, D., WITTER, A., VERSTEEG, D. H. G. AND MULDER, A. H. (1969) Release of ACTH by substances of central nervous system origin. *Endocrinology*, **85**, 561–569.

DICKER, S. E. AND TYLER, C. (1953) Vasopressor and oxytocic activities of the pituitary glands of rats, guinea pigs and cats and of human foetuses. *J. Physiol. (Lond.)*, **121**, 206–214.

DONOVAN, B. T. AND PEDDIE, M. J. (1973) Foetal hypothalamic and pituitary lesions, the adrenal glands and abortion in the guinea pig. In *Foetal and Neonatal Physiology*, K. S. COMLINE *et al.* (Eds.), Cambridge Univ. Press, Mass., pp. 603–605.

DONOVAN, B. T. AND PEDDIE, M. J. (1974) The adrenal glands, oestrogen, and the control of parturition in the guinea pig. In *Integrative Hypothalamic Activity, Progress in Brain Research, Vol. 41*, D. F. SWAAB AND J. P. SCHADÉ (Eds.), Elsevier, Amsterdam, pp. 281–288.

DROST, M. AND HOLM, L. W. (1968) Prolonged gestation in ewes after foetal adrenalectomy. *J. Endocr.*, **40**, 293–296.

DUNN, J. M. (1966) Anterior pituitary and adrenal absence in a live-born normocephalic infant, *Amer. J. Obstet. Gynec.*, **96**, 893–894.

DUPOUY, J.-P. ET JOST, A. (1970) Activité corticotrope de l'hypophyse foetale du rat: influence de l'hypothalamus et des corticostéroides. *C. R. Soc. Biol. (Paris)*, **164**, 2422–2427.

EDITORIAL (1953) Insulin and growth. *Brit. med. J.*, **II**, 383–384.

EDWARDS, C. R. W. (1971) Measurement of plasma and urinary vasopressin by immunoassay. *Proc. roy. Soc. Med.*, **64**, 842–844.

EGUCHI, Y. (1961) Atrophy of the fetal mouse adrenal following decapitation in utero, *Endocrinology*, **68**, 716–719.

EGUCHI, Y. (1969) Interrelationships between the fetal and maternal hypophyseal–adrenal axes in rats and mice. In *Physiology and Pathology of Adaptation Mechanisms: Neural–Neuroendocrine–Humoral*, E. BAJUSZ (Ed.), Pergamon Press, Oxford, pp. 3–27.

ENGFELDT, B. AND HULTQUIST, G. T. (1953) Administration of crystalline growth hormone to pregnant rats. *Acta endocr. (Kbh.)*, **14**, 181–188.

ENSINCK, J. W. AND WILLIAMS, R. H. (1972) Hormonal and nonhormonal factors modifying man's response to insulin. In *Handbook of Physiology, Vol. I, Endocrinology*, R. O. GREEP, E. B. ASTWOOD, D. F. STEINER, N. FREINKEL AND S. R. GEIGER (Eds.), American Physiological Society, Williams and Wilkins, Baltimore, Md., pp. 665–684.

ERENBERG, A. AND FISHER, D. A. (1973) Thyroid hormone metabolism in the foetus. In *Foetal and Neonatal Physiology, Proc. Sir Joseph Barcroft Centenary Symposium*, K. S. COMLINE *et al.* (Eds.), Cambridge Univ. Press, Mass., pp. 508–526.

FOSTER, D. L., KARSCH, F. J. AND NALBANDOV, A. V. (1972) Regulation of luteinizing hormone (LH) in the fetal and neonatal lamb. II. Study of placental transfer of LH in the sheep. *Endocrinology*, **90**, 589–592.

FRANCHIMONT, P. (1971) *Sécrétion Normale et Pathologique de la Somatotrophine et des Gonado-trophines Humaines*, Masson, Paris, p. 306.

FRANDSEN, V. A. AND STAKEMANN, G. (1964) The site of production of oestrogenic hormones in human pregnancy. III. Further observations on the hormone excretion in pregnancy with an-encephalic foetus. *Acta endocr. (Kbh.)*, **47**, 265–276.

FREIE, H. M. P. EN BOUMAN, P. R. (1973) Perinatale spiegels van insuline en glucose bij de rat. *14e Federatieve Vergadering van medisch–biologische Verenigingen, Groningen, 26–28 april, 1973*, Abstract No. **125**.

FUJIMOTO, W. Y. AND WILLIAMS, R. H. (1972) Insulin release from cultured fetal human pancreas. *Endocrinology*, **91**, 1133–1136.

FUJITA, T., EGUCHI, Y., MORIKAWA, Y. AND HASHIMOTO, Y. (1970) Hypothalamic–hypophysial adrenal and thyroid systems: observations in fetal rats subjected to hypothalamic destruction, brain compression and hypervitaminosis A. *Anat. Rec.*, **166**, 659–672.

FURTH, E. D. AND PAGLIARA, A. S. (1971) Thyroid hormones. In *Endocrinology of Pregnancy*, F. FUCHS AND A. KLOPPER (Eds.), Harper and Row, London, pp. 216–234.

GAMMAL, E. B. (1973) Effects of placental lesions on foetal growth in rats. *Experientia (Basel)*, **29**, 201–203.

GITLIN, D., KUMATE, J. AND MORALES, C. (1965) On the transport of insulin across the human placenta. *Pediatrics*, **35**, 65–69.

GRUENWALD, P. (1963) Chronic fetal distress and placental insufficiency. *Biol. Neonat. (Basel)*, **5**, 215–265.

GRUENWALD, P. (1966) Growth of the human fetus. II. Abnormal growth in twins and infants of mothers with diabetes, hypertension, or isoimmunization. *Amer. J. Obstet. Gynec.*, **94**, 1120–1132.

GRUMBACH, M. M. AND KAPLAN, S. L. (1973) Ontogenesis of growth hormone, insulin, prolactin and gonadotropin secretion in the human foetus. In *Foetal and Neonatal Physiology, Proc. Sir Joseph Barcroft Centenary Symposium*, K. S. COMLINE *et al.* (Eds.), Cambridge Univ. Press, Mass., pp. 462–487.

HEGGESTAD, C. B. AND WELLS, L. J. (1965) Experiments on the contribution of somatotrophin to prenatal growth in the rat. *Acta anat. (Basel)*, **60**, 348–361.

HOET, J. J. (1969) Normal and abnormal foetal weight gain. In *Ciba Foundation Symposium on Foetal Autonomy*, G. E. W. WOLSTENHOLME AND M. O'CONNOR (Eds.), Churchill, London, pp. 186–217.

HONNEBIER, W. J. AND SWAAB, D. F. (1973a) The influence of anencephaly upon intrauterine growth of fetus and placenta and upon gestation length. *J. Obstet. Gynaec. Brit. Cwlth*, **80**, 577–588.

HONNEBIER, W. J. AND SWAAB, D. F. (1973b) The role of the human foetal brain in the onset of labour. *J. Endocrinol.*, **57**, xxx–xxxi.

HONNEBIER, W. J., JÖBSIS, A. C. AND SWAAB, D. F. (1974) The effect of hypophysial compounds and Human Chorionic Gonadotrophin (HCG) upon the anencephalic fetal adrenal and parturition in the human. *J. Obstet. Gynaec. Brit. Cwlth*, in press.

HOPKINS, P. S. AND THORBURN, G. D. (1972) The effects of foetal thyroidectomy on the development of the ovine foetus. *J. Endocr.*, **54**, 55–66.

HOPPENSTEIN, J. M., MILTENBERGER, F. W. AND MORAN, W. H. (1968) The increase in blood levels of vasopressin in infants during birth and surgical procedures. *Surg. Gynec. Obstet.*, **127**, 966–974.

HULTQUIST, G. T. AND ENGFELDT, B. (1949) Giant growth of rat fetuses produced experimentally by means of administration of hormones to the mother during pregnancy. *Acta endocr. (Kbh.)*, **3**, 365–376.

HUTCHINSON, D. L., WESTOVER, J. L. AND WILL, D. W. (1962) The destruction of the maternal and fetal pituitary glands in subhuman primates. *Amer. J. Obstet. Gynec.*, **83**, 857–865.

HWANG, U. K. AND WELLS, L. J. (1959) Hypophysis–thyroid system in the fetal rat: thyroid after hypophyseopriva, thyroxin, triiodothyronine, thyrotrophin and growth hormone. *Anat. Rec.*, **134**, 125–141.

INGLETON, P. M., ENSOR, M. AND HANCOCK, M. P. (1971) Identification of prolactin in foetal rat pituitary. *J. Endocr.*, **51**, 799–800.

JACK, P. M. B. AND MILNER, R. D. G. (1972) Effect of foetal decapitation on the development of insulin secretion in the rabbit. *J. Endocr.*, **55**, xxvi.

JACK, P. M. B. AND MILNER, R. D. G. (1973) Effect of decapitation on the development of insulin secretion in the foetal rabbit. *J. Endocr.*, **57**, 23–31.

JEAN, CH. ET DELOST, P. (1969) Prolactine et embryogenèse mammaire chez la souris. *J. Physiol. (Paris)*, **61**, 323–324.

JOHNSON, J. D., HANSEN, R. C., ALBRITTON, W. L., WERTHEMANN, U. AND CHRISTIANSEN, R. O. (1973) Hypoplasia of the anterior pituitary and neonatal hypoglycemia. *J. Pediat.*, **82**, 634–641.

JOSIMOVICH, J. B. (1969) The foetal role in the initiation of parturition in the ewe. In *Foetal Autonomy*, G. E. W. WOLSTENHOLME AND M. O'CONNOR (Eds.), Churchill, London, pp. 231–232 (discussion).

JOST, A. (1954) Hormonal factors in the development of the fetus. *Cold Spr. Harb. Symp. quant. Biol.*, **19**, 167–181.

JOST, A. (1966a) Anterior pituitary function in foetal life. In *The Pituitary Gland*, G. W. HARRIS AND B. T. DONOVAN (Eds.), Butterworths, London, pp. 299–323.

JOST, A. (1966b) Problems of fetal endocrinology: the adrenal glands. In *Recent Progress in Hormone Research, Vol. 22*, G. PINCUS (Ed.), Academic Press, New York, pp. 541–574.

JOST, A. (1971) Hormones in development: past and prospects. In *Hormones in Development*, M. HAMBURG AND E. J. W. BARRINGTON (Eds.), Appleton-Century-Crofts, New York, pp. 1–18.

JOST, A. (1973) Does the foetal hypophyseal–adrenal system participate in delivery in rats and in rabbits? In *Foetal and Neonatal Physiology, Proc. Sir Joseph Barcroft Centenary Symposium*, K. S. COMLINE *et al.* (Eds.), Cambridge Univ. Press, Mass., pp. 589–593.

JOST, A., DUPOUY, J. P. AND GELOSO-MEYER, A. (1970) Hypothalamo–hypophyseal relationships in the fetus. In *The Hypothalamus*, L. MARTINI, M. MOTTA AND FRASCHINI (Eds.), Academic Press, New York, pp. 605–615.

JOST, A., DUPOUY, J. P. AND RIEUTORT, M. (1974) The ontogenetic development of hypothalamo–hypophyseal relations. In *Integrative Hypothalamic Activity, Progress in Brain Research, Vol. 41*, D. F. SWAAB AND J. P. SCHADÉ (Eds.), Elsevier, Amsterdam, pp. 209–219.

JOST, A. AND PICON, L. (1970) Hormonal control of fetal development and metabolism. In *Advances in Metabolic Disorders, Vol. 4*, R. LEVINE AND R. LUFT (Eds.), Academic Press, New York, pp. 123–184.

KAJIHARA, A., KOJIMA, A., ONAYA, T., TAKEMURA, Y. AND YAMADA, T. (1972) Placental transport of thyrotropin releasing factor in the rat. *Endocrinology*, **90**, 592–594.

KAPLAN, S. L., GRUMBACH, M. M. AND SHEPARD, T. H. (1972) The ontogenesis of human fetal hormones I. Growth hormone and insulin. *J. clin. Invest.*, **51**, 3080–3093.

KASTIN, A. J., GENNSER, G., ARIMURA, A., CLINTON MILLER, III, M. AND SCHALLY, A. V. (1968) Melanocyte-stimulating and corticotrophic activities in human foetal pituitary glands. *Acta endocr. (Kbh.)*, **58**, 6–10.

KENDALL, J. Z. AND LIGGINS, G. C. (1972) The effect of dexamethasone on pregnancy in the rabbit. *J. Reprod. Fertil.*, **29**, 409–413.

KENNEDY, P., KENDRICK, J. W. AND STORMONT, C. (1957) Adenohypophyseal aplasia, an inherited defect associated with abnormal gestation in Guernsey cattle. *Cornell Vet.*, **47**, 160–178.

KIRSCH, R. E. (1938) A study on the control of length of gestation in the rat with notes on maintenance and termination of gestation. *Amer. J. Physiol.*, **122**, 86–93.

KITCHELL, R. L. AND WELLS. L. J. (1952) Functioning of the hypophysis and adrenals in fetal rats: effects of hypophysectomy, adrenalectomy, castration, injected ACTH and implanted sex hormones. *Anat. Rec.*, **112**, 561–585.

KLOOSTERMAN, G. J. (1966) Prevention of prematurity, *Ned. T. Verlosk.*, **66**, 361–379.

KLOOSTERMAN, G. J. (1970) On intrauterine growth. *Int. J. Gynaec. Obstet.*, **8**, 895–912.

KLOOSTERMAN, G. J. AND HUIDEKOPER, B. L. (1954) The significance of the placenta in "Obstetrical mortality". *Gynaecologia (Basel)*, **138**, 529–550.

KNOBIL, E. AND CATON, W. L. (1953) The effect of hypophysectomy on fetal and placental growth in the rat. *Endocrinology*, **53**, 198–201.

KNOBIL, E. AND JOSIMOVICH, J. B. (1959) Placental transfer of thyrotropic hormone, thyroxine, triiodothyronine, and insulin in the rat. *Ann. N.Y. Acad. Sci.*, **75**, 895–904.

KUČERA, J. AND DOLEŽALOVÁ, V. (1972) Prenatal development of malformed fetuses at 28–42 weeks of gestational age. *Biol. Neonat. (Basel)*, **20**, 253–261.

KUMARESAN, P., KAGAN, A. AND GLICK, S. M. (1971) Oxytocin antibody and lactation and parturition in rats. *Nature (Lond.)*, **230**, 468–469.

LANMAN, J. T. AND SCHAFFER, A. (1968) Gestational effects of fetal decapitation in sheep. *Fertil. and Steril.*, **19**, 598–605.

LASCELLES, A. K. AND SETCHELL, B. P. (1959) Hypothyroidism in sheep. *Aust. J. biol. Sci.*, **12**, 455–465.

LEVINA, S. E. (1972) Times of appearance of LH and FSH activities in human fetal circulation. *Gen. comp. Endocr.*, **19**, 242–246.

LIGGINS, G. C. (1968) Premature parturition after infusion of corticotrophin or cortisol into foetal lambs. *J. Endocr.*, **42**, 323–329.

LIGGINS, G. C. (1969) Premature delivery of foetal lambs infused with glucocorticoids. *J. Endocr.*, **45**, 515–523.

LIGGINS, G. C. AND KENNEDY, P. C. (1968) Effects of electrocoagulation of the foetal lamb hypophysis on growth and development. *J. Endocr.*, **40**, 371–381.

LIGGINS, G. C., KENNEDY, P. C. AND HOLM, L. W. (1967) Failure of initiation of parturition after electrocoagulation of the pituitary of the fetal lamb. *Amer. J. Obstet. Gynec.*, **98**, 1080–1086.

LINCOLN, D. W. (1971) Labour in the rabbit: Effect of electrical stimulation applied to the infundibulum and median eminence. *J. Endocr.*, **50**, 607–618.

LINCOLN, D. W. (1974) Maternal hypothalamic control of labor. In *Integrative Hypothalamic Activity, Progress in Brain Research, Vol. 41*, D. F. SWAAB AND J. P. SCHADÉ (Eds.), Elsevier, Amsterdam, 1974, pp. 289–306.

MALAISSE, W. J. (1972) Hormonal and environmental modification of islet activity. In *Handbook of Physiology, Sect. 7, Vol. 1, Endocrinology*, R. O. GREEP, E. B. ASTWOOD, D. F. STEINER, N. FREINKEL AND S. R. GEIGER (Eds.), Amer. Physiol. Soc., Williams and Wilkins, Baltimore, Md., pp. 237–260.

MALPAS, P. (1933) Postmaturity and malformations of the foetus. *J. Obstet. Gynaec. Brit. Emp.*, **40**, 1046–1053.

MARTIN, J. M. AND GAGLIARDINO, J. J. (1967) Effect of growth hormone on the isolated pancreatic islets of rat in vitro. *Nature (Lond.)*, **213**, 630–631.

MCKUSICK, V. A. AND RIMOIN, D. L. (1967) General Tom thumb and other midgets, *Sci. Amer.*, **217**, 102–110.

MILIC, A. B. AND ADAMSONS, K. (1969) The relationship between anencephaly and prolonged pregnancy. *J. Obstet. Gynaec. Brit. Cwlth*, **76**, 102–111.

MUELLER-HEUBACH, E., MYERS, R. E. AND ADAMSONS, K. (1972) Effects of adrenalectomy on pregnancy length in the rhesus monkey. *Amer. J. Obstet. Gynec.*, **112**, 221–226.

MURPHY, B. E. P. (1973) Does the human fetal adrenal play a role in parturition? *Amer. J. Obstet. Gynec.*, **115**, 521–525.

NATHANIELSZ, P. W., COMLINE, R. S., SILVER, M. AND PAISEY, R. B. (1972) Cortisol metabolism in the fetal and neonatal sheep. *J. Reprod. Fertil.*, Suppl. **16**, 39–59.

NEWTON, W. H. (1935) "Pseudo-parturition" in the mouse, and the relation of the placenta to post-partum oestrus. *J. Physiol. (Lond.)*, **84**, 196–207.

NOUMURA, T. (1959) Development of the hypophyseal–adrenocorticol system in the rat embryo in relation to the maternal system. *Jap. J. Zool.*, **10**, 279–300.

O'DONOHOE, N. V. AND HOLLAND, P. D. J. (1968) Familial congenital adrenal hypoplasia. *Arch. Dis. Childh.*, **43**, 717–723.

OSINGA, A. (1969) De "foeto-placental unit" en het geboorteproces. *T. Diergeneesk.*, **94**, 768–776.

PAGE, E. W. (1969) Human fetal nutrition and growth. *Amer. J. Obstet. Gynec.*, **104**, 378–387.

PAINE, B. G. (1971) Growth and development. In *Endemic Cretinism, Proc. Symp. 1971, Inst. of Human Biology (Monograph Ser. No. 2)*, B. S. HETZEL AND P. O. PHAROAH (Eds.), Surrey Beatty, Chipping Norton, N.S.W., pp. 94–100.

PAREKH, M. C., BENJAMIN, F. AND GASTILLO, N. (1973) The influence of maternal human growth hormone secretion on the weight of the newborn infant. *Amer. J. Obstet. Gynec.*, **115**, 197–201.

PICON, L. (1967) Effect of insulin on growth and biochemical composition of the rat fetus. *Endocrinology*, **81**, 1419–1521.

PRICE, H. V., CONE, B. A. AND KEOGH, M. (1971) Length of gestation in congenital adrenal hyperplasia. *J. Obstet. Gynaec. Brit. Cwlth*, **78**, 430–434.

RAYNAUD, A. (1950) Recherches expérimentales sur le développement de l'appareil génital et le fonctionnement des glandes endocrines des foetus de souris et de mulot, *Arch. Anat. micr. Morph. exp.*, **39**, 518–576.

REID, J. D. (1960) Congenital absence of the pituitary gland. *J. Pediat.*, **56**, 658–664.

RIMOIN, D. L., MERIMEE, T. J. AND MCKUSICK, V. A. (1966a) Growth hormone deficiency in man: an isolated recessively inherited defect. *Science*, **152**, 1635–1637.

RIMOIN, D. L., MERIMEE, T. J. AND MCKUSICK, V. A. (1966b) Sexual ateliotic dwarfism: A recessively inherited isolated deficiency of growth hormone. *Trans. Ass. Amer. Phycns*, **79**, 297–311.

ROBINSON, A. G., ZIMMERMAN, E. A. AND FRANTZ, A. G. (1971) Physiologic investigation of posterior pituitary binding proteins neurophysin I and neurophysin II. *Metabolism*, **20**, 1148–1155.

ROFFI, J. (1958) Dosage de la vasopressine dans l'hypophyse du foetus de rat en fin de gestation. *C. R. Soc. Biol. (Paris)*, **152**, 741–743.

SALTER, J. AND BEST, C. H. (1953) Insulin as a growth hormone. *Brit. med. J.*, **II**, 353–356.

SCHUTT, W. (1965) Foetal factors in intrauterine growth retardation. In *Clinics in Developmental Medicine*, M. DAWKINS AND B. MACGREGOR (Eds.), The Spastics Society Medical Education and Information Unit in ass. with Heinemann, London, No. **19**, pp. 1–9.

SLOB, A. K. (1972) *Perinatal Endocrine and Nutritional Factors Controlling Physical and Behavioural Development in the Rat*, Med. Thesis, Offset Printing Bureau V.C.V.O., Vlaardingen.

SOLOMON, S. AND FRIESEN, H. G. (1968) Endocrine relations between mother and fetus. *Ann. Rev. Med.*, **19**, 399–430.

SPELLACY, W. N. (1971) Insulin. In *Endocrinology of Pregnancy*, F. FUCHS AND A. KLOPPER (Eds.), Harper and Row, New York, pp. 197–215.

SWAAB, D. F. (1972) The hypothalamo–neurohypophysial system and reproduction. In *Topics in Neuroendocrinology, Progress in Brain Research, Vol. 38*, J. ARIËNS KAPPERS AND J. P. SCHADÉ (Eds.), Elsevier, Amsterdam, pp. 225–244.

SWAAB, D. F. AND HONNEBIER, W. J. (1973) The influence of removal of the fetal rat brain upon intrauterine growth of the fetus and the placenta and on gestation length. *J. Obstet. Gynaec. Brit. Cwlth*, **80**, 589–597.

SWAAB, D. F., HONNEBIER, W. J. AND BOER, K. (1973) The role of the foetal rat brain in labour. *J. Endocr.*, **57**, xxxi–xxxii.

THOMSON, A. M., BILLEWICZ, W. Z. AND HYTTEN, F. E. (1968) The assessment of fetal growth. *J. Obstet. Gynaec. Brit. Cwlth*, **75**, 903–916.

TUCHMANN-DUPLESSIS, H. ET GABE, M. (1960) Absence de produit de neurosécrétion dans la post-hypophyse des anencephales. *Bull. Acad. Méd. (Paris)*, **144**, 102–104.

VAN ASSCHE, F. A. (1968) A morphological study of the Langerhans' islets of the fetal pancreas in late pregnancy. *Biol. Neonat. (Basel)*, **12**, 331–342.

VAN ASSCHE, F. A., GEPTS, W. AND DE GASPARO, M. (1970) The endocrine pancreas in anencephalics. A histological, histochemical and biological study. *Biol. Neonat. (Basel)*, **41**, 374–388.

VAN WAGENEN, G. AND CATCHPOLE, H. R. (1965) Growth of the fetus and placenta of the monkey. *Amer. J. Phys. Anthrop.*, **23**, 23–34.

WINDLE, W. F., RHINES, R. AND RANKIN, J. (1943) A Nissl method using buffered solutions of thionin, *Stain Technol.*, **18**, 77–86.

ZAMENHOF, S., MOSLEY, J. AND SCHULLER, E. (1966) Stimulation of the proliferation of cortical neurons by prenatal treatment with growth hormone. *Science*, **152**, 1396–1397.

DISCUSSION

JOST: I was very interested in your paper of course, especially in what you said about anencephalic infants having a normal mean gestation length. My limited observations on rabbits would, in agreement with your preliminary experiment, suggest that the fetal pituitary has no important effect upon the length of pregnancy or the initiation of delivery. The point you discussed about the role of the fetal hypothalamus in body growth seems also of importance. It is our experience that the encephalectomized young are consistently smaller than decapitated fetuses. My explanation was that the surgery we have to do for encephalectomizing the young is much more stressful to the uterus than for decapitation, which is a very simple operation. Were you suggesting that fetal growth hormone could control the size of the placenta, and thus the size of the fetus?

SWAAB: I did not mean to say that growth inhibition after brain aspiration is caused by the placenta that lacks growth hormone. Growth of the fetus and placenta are to a certain degree two separate entities. We can restore the placental weight after brain aspiration again by injecting growth hormone into the fetus, without increasing the fetal body weight.

LEQUIN: In the human it has been shown that the placental weight is correlated with HCS-levels. I just wonder why the placental lactogen was not mentioned here, since HCS-levels are very high.

SWAAB: We are of course mostly interested in hypophyseal and hypothalamic factors involving fetal growth. We tested, however, one preparation, placenta suspensata, that should contain HCS. It did not increase fetal body weight or placental weight. Perhaps we have to try a better HCS preparation.

The Adrenal Glands, Estrogen, and the Control of Parturition in the Guinea Pig

B. T. DONOVAN AND M. J. PEDDIE

Department of Physiology, Institute of Psychiatry, London (Great Britain)

INTRODUCTION

Marked species individuality is apparent in the hormonal changes at parturition and reflected by maternal plasma steroid hormone concentrations. In women the content of estrogen and progesterone rises steadily from the third month of gestation and remains high till term; in the cow the plasma content of progesterone falls near term whereas that of estrogen rises to a peak just before delivery; while in the guinea-pig maternal plasma progesterone and estrogen both reach high levels during mid-gestation and tend to decline slightly a few days before parturition (Bedford *et al.*, 1972). The changes in the secretion of steroids by the fetus have been less extensively investigated, although the output of adrenal hormones has been followed. In the sheep, a dramatic increase in fetal plasma cortisol content during the last few days of gestation occurs before any change in maternal hormone levels (Bassett and Thorburn, 1969; Comline *et al.*, 1973); such changes are much less marked in the calf or foal (Comline *et al.*, 1973). Fetal plasma cortisol levels rise abruptly during the last week of gestation in the guinea-pig, but in this case only after a rise in maternal plasma cortisol (Jones, 1973, personal communication).

Alterations in adrenal activity in the fetus merit examination in view of the possible influence of the fetus in triggering the onset of parturition. Much evidence for this activity has come from work in the sheep, where fetal adrenalectomy, or destruction of the hypophysis, delays or inhibits parturition (Liggins *et al.*, 1967; Drost and Holm, 1968). The infusion of ACTH or cortisol into the ovine fetus during the last third of pregnancy causes premature delivery, while the same amount of hormone given to the mother is ineffective (Liggins, 1968). However, large doses of the fluorinated steroid with a highly potent corticoid-like action, dexamethasone, have precipitated parturition in the ewe (Fylling, 1971; Bosc, 1972a).

Interaction between the fetal and maternal endocrine systems also takes place, with the fetus probably acting as instigator. Liggins (1968) reported that it was necessary to hypophysectomize both partners in twin pregnancies in order to delay delivery. Destruction of the pituitary gland of the fetal lamb also prevented the premature parturition otherwise caused by maternal treatment with dexamethasone (Bosc, 1972b). It might be supposed that the rise in the secretion of cortisol by the fetal

References p. 287–288

adrenals follows an increased release of ACTH from the hypophysis of the fetus (Nathanielsz *et al.*, 1972), although the adrenal glands could also show an enhanced response to the tropic hormone (Anderson *et al.*, 1972). But the prime cause of the increased secretion of ACTH remains in question. Metabolic stress does not seem to be the responsible factor since plasma Po_2, Pco_2 and pH remain unaltered during the last 10 days of gestation and during delivery (Comline and Silver, 1970, 1972).

In some way the brain of the guinea-pig fetus can modify the length of intra-uterine life. Lesions placed in the central hypothalamus of a single fetus midway through gestation were frequently associated with abortion of the entire litter around day 50 of pregnancy, whereas sham operations, or damage of the pituitary or stalk–median eminence region, did not have this effect (Donovan and Peddie, 1973). The basis of the reaction to hypothalamic lesions in the fetus is not understood although adrenal function could be altered. Destruction of the fetal hypophysis on day 40 of gestation retarded adrenal growth as determined on days 50 or 60, but as stalk–median eminence lesions did not have the same effect it appears that the secretion of ACTH is autonomous and does not require promotion by the hypothalamus. Nevertheless, since it is possible that delivery is triggered at term in the guinea-pig by an increased output of ACTH from the fetal hypophysis, the effect upon gestation of ACTH administration to the fetus has been examined. The response to dexamethasone treatment has also been followed, and attempts made to delay parturition by maternal administration of estrogen and progesterone. Since the work is still in hand the following account is in the nature of a progress report.

METHODS

ACTH administration

Guinea-pig fetuses were injected on day 50 of gestation with 0.1 mg synthetic ACTH (0.1 ml Synacthen®, Ciba, long acting zinc suspension, 100 I.U./ml). To do this the pregnant female was anesthetized with Nembutal® (Abbott Laboratories, 40 mg/kg, i.p.), a midline ventral laparotomy incision made and some or all of the fetuses in the litter injected subcutaneously through the uterine wall with the hormone. As the needle was withdrawn the uterine wall at this point was gently clasped with forceps and a ligature tied around the puncture to avoid loss of amniotic fluid. Fifteen pregnancies have been studied to date. In 4 all the fetuses were injected with ACTH and in another 4 some of the fetuses were left undisturbed. These 8 sows were autopsied at 55 days of gestation. One pregnant guinea-pig in which part of the litter was injected with ACTH was killed at 65 days and 3 similarly treated females were allowed to go to term, as were 3 further females in which all the fetuses had been injected. For control purposes the fetuses of 8 pregnant females were injected with 0.1 ml sterile water at 50 days; 4 were autopsied at 55 days and 4 allowed to continue to term.

Estrogen treatment

A pellet of approximately 15 mg estradiol was implanted subcutaneously on day 50 into each of 19 pregnant females. Four were killed on day 60, 4 on day 70, 2 on day 72 and one each on days 75 and 80. The estradiol pellet was removed from 4 females on day 60 and the 3 remaining sows were allowed to go to term. A further 8 pregnant animals were given a pellet of estradiol on day 40; 4 were killed on day 50 and 4 on day 60. Additionally, a pellet of estradiol was implanted into each of 20 guinea-pigs on the day of birth, with 10 animals being killed 10 days later and 10 after 20 days. Equal numbers of males and females were used.

Progesterone and dexamethasone treatment

Twenty milligrams progesterone was implanted subcutaneously into each of 4 females on day 50 of pregnancy and they were kept until delivery occurred. Eight females were injected with 2.5 mg dexamethasone (Organon) twice daily from day 50 of pregnancy. Four were killed on day 60 and treatment of the remaining 4 was continued until parturition.

All pregnant females were kept under close observation during the experimental period and examined daily for signs of impending vaginal opening and pelvic relaxation. At autopsy the mother was anesthetized with Nembutal, the uterus exposed and the fetuses removed in turn and weighed. The adrenals, gonads, and uterus or seminal vesicles were recovered, weighed and fixed for histological examination.

Differences were tested, using Student's *t*-test. A level of $P < 0.05$ was considered to be significant.

RESULTS

Parturition in our colony normally occurs between days 67 and 69 of gestation in animals carrying 3–6 young, or on days 70 or 71 in females carrying only one or two fetuses.

Effect of ACTH

None of the animals injected with ACTH on day 50 showed signs of abortion by 55 days, although the adrenal glands of all the fetuses autopsied at this time were enlarged whether or not all the fetuses in a litter were injected (23.62 $\pm$ 0.092 mg (S.E.M.); 19 animals). The weights of the adrenals of the blank-injected fetuses (10.65 $\pm$ 1.00 mg; 5 animals) did not differ from those of the untreated littermate controls (10.52 $\pm$ 0.37 mg; 12 animals). The adrenal glands of the fetuses given ACTH at 50 days were at 55 days approximately equal in size to those of controls on day 68 of gestation (22.93 $\pm$ 1.96 mg; $n = 27$) or day 0 postnatally (26.15 $\pm$ 2.75 mg; $n = 12$).

When litters injected with ACTH were followed until delivery took place, parturition

References p. 287–288

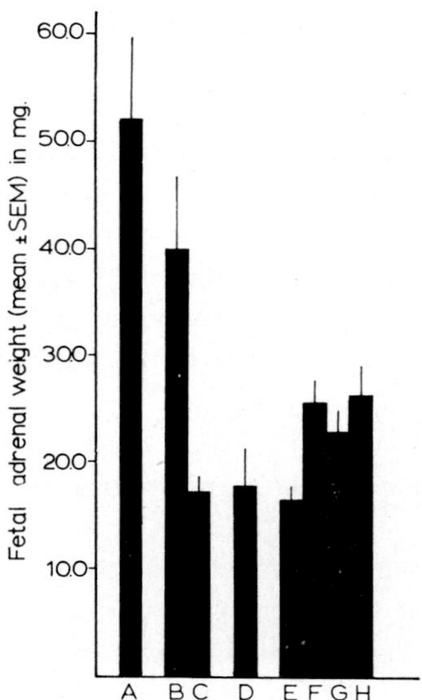

Fig. 1. The adrenal weights (mean ± S.E.M.) of fetuses from litters in which all individuals were injected with ACTH at 50 days with subsequent premature delivery (A), or from litters in which only some individuals were treated with ACTH at 50 days (B). In the latter case the adrenal weights of the non-injected fetuses are included separately (C). For comparison, the adrenal weights of blank-injected (D) or untreated fetuses at various times during gestation (E: 60 days; F: 65 days; G: 68 days) or on the day of birth (H) are included. The number of animals in each group is: A, 9; B, 9; C, 5; D, 13; E, 22; F, 21; G, 27; H, 12.

was found to be advanced by 3–5 days in those females in which all the fetuses had been injected. Thus, delivery occurred at 62, 63 and 64 days in 3 animals, when the adrenals of the injected fetuses were substantially larger (51.96 ± 7.66 mg, n = 9) than those of control fetuses of similar age or older (60 days: 16.49 ± 1.37 mg, n = 22; 65 days: 25.84 ± 1.50 mg, n = 21), or of those of fetuses similarly injected with ACTH and killed at 55 days. In these mothers vaginal changes and pelvic relaxation occurred between 12 and 24 h before delivery. Vaginal changes and pelvic relaxation were observed around 63 days in the females in which only some fetuses had been injected with ACTH at 50 days, but parturition did not occur until day 68 (2 litters), or day 70 (1 litter). The adrenal weights of these ACTH-treated fetuses were 38.98 ± 6.41 mg (n = 8), compared with 19.94 ± 1.80 mg (n = 5) for their littermate controls. In the animals carrying blank-injected fetuses pelvic relaxation and vaginal changes again occurred about 24 h before delivery, which took place at the normal time (68 days, 1 litter; 69 days, 2 litters; 70 days, 1 litter), and with no evident increase in adrenal weight (17.84 ± 3.29 mg; n = 13) (see also Fig. 1).

Effect of estradiol

The high dosage of estradiol employed caused substantial uterine hypertrophy and vaginal opening in the female fetuses. Thus the 8 uteri of the fetuses from pregnant

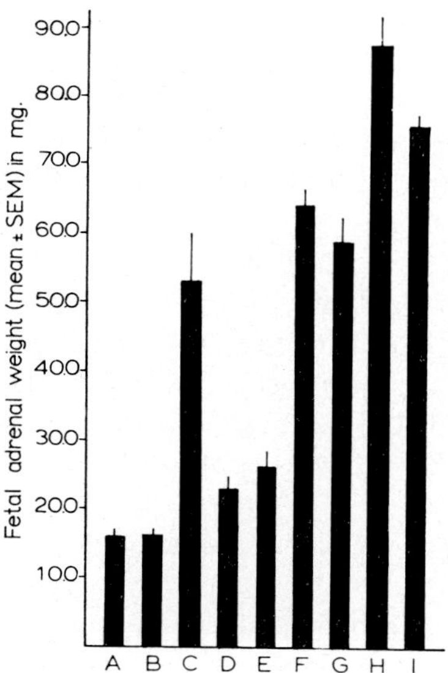

Fig. 2. Adrenal weights (mean ± S.E.M.) of fetuses from mothers implanted with 15 mg estradiol at 50 days of gestation and autopsied at 60 (A) or 70 days (C), compared with control litters at 60 days (B), 68 days (D), and on the day of birth (E). The adrenal weights of animals given 15 mg estradiol implants on day 0, for either 10 (F) or 20 days (H), and untreated immature guinea-pigs of 10 (G) or 20 days (I) are included. The number of animals in each group is: A, 16; B, 22; C, 11; D, 27; E, 12; F, 10; G, 20; H, 10; I, 10.

females treated with estradiol from days 50 to 60 weighed 209.1 ± 14.9 mg, compared with 104.2 ± 11.4 mg in 6 controls of the same age.

Estrogen treatment delayed or inhibited delivery in that it occurred at 70 days in two cases and at 72 days in another. In the absence of parturition 4 animals were autopsied at 70 days, 2 animals at 72 days, 1 at 75 days and 1 at 80 days. Vaginal opening or pelvic relaxation was not apparent at autopsy. When 4 pregnant guinea-pigs were treated with estradiol over days 50–60, when the pellet was removed, vaginal softening, pelvic relaxation and delivery occurred at the normal time in 3 of them. In the fourth animal, which carried only one fetus, delivery occurred 48 h after removing the estradiol implant.

Adrenal hypertrophy was striking in those fetuses exposed to estrogen near term (Fig. 2). The mean adrenal weight of 11 fetuses autopsied at 70 days was 52.40 ± 5.78 mg, compared with 22.99 ± 1.15 mg (n = 27) and 26.93 ± 1.96 mg (n = 12) for fetuses on day 68 of gestation and on the day of birth respectively. The adrenal glands were also enlarged in 11 fetuses from estrogen-treated females which delivered at 70 or 72 days (53.53 ± 4.79 mg). When pregnant females were autopsied at 50 or 60 days after exposure to estrogen for 10 or 20 days from day 40 the fetal adrenal weights (6.81 ± 0.30 mg, n = 20; 11.21 ± 0.36 mg, n = 15, respectively) were low compared with untreated controls of the same age (50 days: 11.96 ± 0.37 mg, n = 35; 60 days: 16.49 ± 1.37 mg, n = 22). However, the body-weights were also reduced (19.11 ± 0.53 g, n = 20; 31.72 ± 1.61 g, n = 15) when compared with the untreated controls

References p. 287–288

(50 days: 30.24 ± 1.47 g, n = 35; 60 days: 45.94 ± 7.91 g, n = 22). When estrogen treatment was begun at 50 days body and adrenal growth were not affected at 60 days, the weights being 45.15 ± 3.38 g, n = 15 and 16.05 ± 1.16 mg, n = 15 respectively. No change in adrenal weight was evident in the fetuses delivered at the usual time by mothers provided with estradiol implants over days 50–60, although uterine hypertrophy was manifest in the fetuses.

The implantation of a 15 mg pellet of estradiol into neonatal guinea-pigs did not alter adrenal size over a 10- or 20-day period. At 10 days of age the adrenal weight of treated animals was 61.06 ± 2.12 mg (n = 10) compared with 57.02 ± 2.53 mg (n = 20) for untreated young of the same age, and at 20 days the adrenal weights were 87.08 ± 4.50 mg (n = 10) and 75.40 ± 2.33 mg (n = 10) respectively. These differences are not significant, while body growth was not affected.

Effect of progesterone

The 20 mg pellet of progesterone implanted subcutaneously on day 50 of gestation did not affect the onset or completion of delivery, which occurred at 67 days (2 animals) and 68 days (2 animals). The uterine weights of the fetuses were not altered (119.6 ± 14.3 mg; n = 5), nor were the adrenal weights (20.27 ± 1.65 mg; n = 8).

Effect of dexamethasone

The daily injection of 5 mg dexamethasone into the pregnant female did not advance delivery in the 8 animals so treated; parturition occurred normally at 68 or 69 days in the 4 animals left until term. The adrenal glands of the 13 newborn animals were of normal size (30.38 ± 2.05 mg, compared with 26.93 ± 1.93 mg in 12 untreated newborn controls). Similarly, in the 4 animals autopsied at 60 days of gestation the weight of the fetal adrenal glands remained unchanged (15.66 ± 1.24 mg for 19 fetuses from treated dams *versus* 16.49 ± 1.37 mg for 22 controls).

DISCUSSION AND CONCLUSIONS

Unlike the sheep, in the guinea-pig parturition was not induced within a few days by the administration of ACTH to the fetus or by treatment of the pregnant female with large amounts of dexamethasone from day 50, although abortion could readily be induced at this age by other means involving manipulation of the fetus (see Donovan and Peddie, 1973). There were indications that a single injection of ACTH given at 50 days exerted a long lasting action and that parturition could be induced a few days earlier than normal. For this to occur, it seems necessary to treat all the fetuses in a litter; the presence of but one uninjected fetus *in utero* prevented the effect, even though the pelvic signs of hormonal change within the mother developed. In view of the response to ACTH, it is curious that dexamethasone did not affect the length of gestation. Ash *et al.* (1973) found that dexamethasone, given as a single or repeated dose from day 57 of pregnancy, failed to induce parturition when injected into the dam (up to 8 mg/day for at least 7 days) or fetus (up to 100 μg/day for at least 7 days). It is noteworthy that 10 mg/day dexamethasone caused premature parturition in the sheep (Bosc, 1972a), while 20 mg dexamethasone daily sufficed to cause delivery in the cow (Adams and Wagner, 1970). It is possible that dexamethasone failed to cross the placenta in adequate amount to affect the fetus, for maternally administered dexamethasone did not depress fetal adrenal weight, possibly because the concentration of corticosteroid-binding globulin in the pregnant guinea-pig is very high (Rosenthal *et al.*, 1969). However, direct treatment of the fetus with dexamethasone was not more effective than maternal administration in inducing premature delivery (Ash *et al.*, 1973) and dexamethasone failed to suppress ACTH secretion in the neonatal guinea-pig (D'Angelo, 1966).

Estradiol delayed or prevented parturition and caused adrenal enlargement in the members of the litter. The latter reaction was evidently not a direct response to the presence of estradiol, since it could not be elicited earlier in pregnancy or after delivery, but was associated with the delay in the onset of labor. It is possible that the stressful nature of this situation brought about an increased secretion of ACTH in the fetuses. There was, for example, little or no amniotic fluid around the fetuses autopsied at 70 days or later, whereas it was still present in the amniotic sacs of fetuses within 24 h of parturition at 68 days of gestation.

In accord with the findings of Porter (1970), who gave daily injections of methyl progesterone or introduced progesterone beneath the myometrium, progesterone did not influence the course of parturition in the guinea-pig. By contrast, treatment of the pregnant rabbit with progesterone inhibits both normal delivery, and that otherwise hastened by concurrent administration of dexamethasone or cortisol (Heckel and Allen, 1937; Nathanielsz *et al.*, 1973).

The present observations give rise to more questions than answers. Although the evidence for the participation of the adrenal glands in the control of parturition is much less convincing in the guinea-pig than in the sheep, the hastening of delivery by ACTH in animals near term may be of significance. The need for the treatment of all fetuses points to the operation of at least one additional factor, the nature of which remains to be explained.

ACKNOWLEDGEMENTS

We wish to thank the Medical Research Council for supporting this work, and Maureen Harrison for her excellent technical assistance.

REFERENCES

ADAMS, W. M. AND WAGNER, W. C. (1970) The role of corticoids in parturition. *Biol. Reprod.*, **3**, 223–228.

ANDERSON, A. B. M., PIERREPOINT, C. G., TURNBULL, A. C. AND GRIFFITHS, K. (1972) Steroid investigation in the developing sheep foetus. In *The Endocrinology of Pregnancy and Parturition*, C. G. PIERREPOINT (Ed.), Alpha-Omega-Alpha Publishing, Cardiff, pp. 23–39.

ASH, R. W., CHALLIS, J. R. G., HARRISON, F. A., HEAP, R. B., ILLINGWORTH, D. V., PERRY, J. S. AND POYSER, N. L. (1973) Hormonal control of pregnancy and parturition; a comparative analysis. In *Foetal and Neonatal Physiology*, R. S. COMLINE, K. W. CROSS, G. S. DAWES AND P. W. NATHANIELSZ (Eds.), Cambridge Univ. Press, Cambridge, pp. 551–561.

BASSETT, J. M. AND THORBURN, G. D. (1969) Foetal plasma corticosteroids and the initiation of parturition in sheep. *J. Endocr.*, **44**, 285–286.

BEDFORD, C. A., CHALLIS, J. R. G., HARRISON, F. A. AND HEAP, R. B. (1972) The role of oestrogens and progesterone in the onset of parturition in various species. *J. Reprod. Fertil.*, **Suppl. 16**, 1–24.

BOSC, M. J. (1972a) The induction and synchronization of lambing with the aid of dexamethasone. *J. Reprod. Fertil.*, **28**, 347–358.

BOSC, M. J. (1972b) Effects of maternal or foetal hypophysectomy on the parturition of dexamethasone treated sheep. *C. R. Acad. Sci. (Paris)*, **274**, 93–96.

COMLINE, R. S. AND SILVER, M. (1970) Daily changes in foetal and maternal blood of conscious pregnant ewes, with catheters in umbilical and uterine vessels. *J. Physiol. (Lond.)*, **209**, 567–586.

COMLINE, R. S. AND SILVER, M. (1972) The composition of foetal and maternal blood during parturition in the ewe. *J. Physiol. (Lond.)*, **222**, 233–256.

COMLINE, R. S., SILVER, M., NATHANIELSZ, P. W. AND HALL, L. W. (1973) Parturition in the larger herbivores. In *Foetal and Neonatal Physiology*, R. S. COMLINE, K. W. CROSS, G. S. DAWES AND P. W. NATHANIELSZ (Eds.), Cambridge Univ. Press, Cambridge, pp. 606–612.

D'ANGELO, S. A. (1966) *Functional Maturation of the Pituitary–Thyroid and Adrenal Systems in the Guinea Pig, Proc. 2nd Int. Congr. Hormonal Steroids, Milan*, Excerpta Medica ICS No. 132, Amsterdam, 901.

DONOVAN, B. T. AND PEDDIE, M. J. (1973) Foetal hypothalamic and pituitary lesions, the adrenal glands and abortion in the guinea pig. In *Foetal and Neonatal Physiology*, R. S. COMLINE, K. W. CROSS, G. S. DAWES AND P. W. NATHANIELSZ (Eds.), Cambridge Univ. Press, Cambridge, pp. 603–605.

DROST, M. AND HOLM, L. W. (1968) Prolonged gestation in ewes after foetal adrenalectomy. *J. Endocr.*, **40**, 293–296.

FYLLING, P. (1971) Premature parturition following dexamethasone administration to pregnant ewes. *Acta Endocr. (Kbh.)*, **66**, 289–295.

HECKEL, G. P. AND ALLEN, W. M. (1937) Prolongation of pregnancy in the rabbit by injection of progesterone. *Amer. J. Physiol.*, **119**, 330–331.

LIGGINS, G. C. (1968) Premature parturition after infusion of corticotrophin or cortisol into foetal lambs. *J. Endocr.*, **42**, 323–329.

LIGGINS, G. C., KENNEDY, P. C. AND HOLM, L. W. (1967) Failure of initiation of parturition after electrocoagulation of the pituitary of the foetal lamb. *Amer. J. Obstet. Gynec.*, **98**, 1080–1086.

NATHANIELSZ, P. W., COMLINE, R. S., SILVER, M. AND PAISEY, R. B. (1972) Cortisol metabolism in the foetal and neonatal sheep. *J. Reprod. Fertil.*, **Suppl. 16**, 39–59.

NATHANIELSZ, P. W., ABEL, M. AND SMITH, G. W. (1973) Hormonal factors in parturition in the rabbit. In *Foetal and Neonatal Physiology*, R. S. COMLINE, K. W. CROSS, G. S. DAWES AND P. W. NATHANIELSZ (Eds.), Cambridge Univ. Press, Cambridge, pp. 594–602.

PORTER, D. G. (1970) The failure of progesterone to affect myometrial activity in the guinea pig. *J. Endocr.*, **46**, 425–434.

ROSENTHAL, M. E., SLAUNWHITE, W. R. AND SANDBERG, A. A. (1969) Transcortin: a corticosteroid-binding protein of plasma. XI. Effects of estrogens on pregnancy in guinea pigs. *Endocrinology*, **85**, 825–830.

DISCUSSION

VAN DELFT: You have obtained marked adrenal changes after estradiol injections. Estradiol has been shown to affect the hypothalamo–hypophyseal–adrenal axis at several levels. It has of course an effect on the adrenal, but also important effects on both the median eminence and the pituitary. One useful way of looking whether an effect is at the pituitary or adrenal level would be to assay blood-ACTH levels. Have you looked at this ?

PEDDIE: We should like to assay fetal plasma ACTH levels after estradiol treatment, but we first have to obtain the suitable antibodies.

Maternal Hypothalamic Control of Labor

DENNIS W. LINCOLN

Department of Anatomy, The Medical School, Bristol (Great Britain)

INTRODUCTION

Hippocrates expressed remarkable foresight when he postulated a role for the fetus in the organization of birth. It has taken some 2000 years, however, to provide the current but still superficial evidence to show that the fetus or feto-placental unit determines the day of birth in sheep. The physical task of delivering the young is still the prerogative of the mother. Her role is far from "passive" and control is exercised to determine the exact hour of birth on that final day of gestation. Such dual control embodies a most sound rationale. The fetus is the best judge of its own maturity (or viability); the mother has the faculties to assess the environmental factors which will govern survival at the time of birth. I shall consider the maternal control, as expressed through the activities of the hypothalamus and other neural structures. The role of the fetus will be discussed in a subsequent paper (Swaab and Honnebier, 1974).

Much of the evidence for the maternal control of labor is circumstantial. Every experimental investigation is confounded by the dominating role of the fetus and the fact that maternal control is not restricted to one readily defined neural or neuroendocrine system. There are, however, many elementary questions which suggest a high degree of maternal involvement. What mechanism, for example, determines the "explosive" pattern of delivery displayed by several species, such that when conditions are favorable the mother can give birth and escape before being taken by predators? The rabbit will give birth to 10 or more pups in as many minutes (Cross, 1958a; Fuchs, 1964; Porter and Schofield, 1966). When circumstances are less than adequate the birth of the young may be delayed for several hours. What determines the circadian periodicity in the time of birth? Even women, who are perhaps less tied to environmental cues than most species, tend to give birth at night (Conroy and Mills, 1970; Alison, 1972). Experimentally, our evidence is scant. Lesions of the hypothalamus of cat, guinea-pig, rat and rabbit disturb the normal processes of labor, creating dystocia (Fisher *et al.*, 1938; Gale and McCann, 1961; Nibbelink, 1961). But, what structures have been destroyed? Ablation of the paraventricular nuclei does not necessarily abolish the synthesis and release of oxytocin. The supraoptic nuclei contain as much oxytocin, or more (Lederis, 1961; Dyer *et al.*, 1973), and 50% of the neurons are involved in oxytocin release as judged from studies of milk ejection (Wakerley and Lincoln, 1973a). Deafferentation of the paraventricular nuclei is also without effect in

References p. 301–305

terms of the milk-ejection reflex (Voloschin and Tramezzani, 1973). Furthermore, it is impossible in the destruction of the paraventricular nuclei to avoid damage of anterior hypothalamic tissue, and the consequences of such accidental destruction must not be disregarded. Lesions of the anterior hypothalamus disturb sleep and circadian rhythms, and lead to changes in the function of the autonomic nervous system (Nauta, 1946; Hernándéz-Péon and Chávez-Ibarra, 1963; Lincoln *et al.*, 1972). Electrical stimulation is more satisfactory, at least it attempts to elicit an electrophysiological response from the nervous tissue. Stimulation of the median eminence and infundibulum of the rabbit at 30 or 31 days of gestation promotes within minutes the "normal" delivery of the young (Cross, 1958a; Lincoln, 1971). It is far more difficult to use electrical stimulation to explore the opposing functions, *i.e.* mechanisms inhibitory to the processes of labor. We are left, therefore, with a mass of indirect and circumstantial evidence on which to formulate our discussions.

Three links between the maternal hypothalamus and the uterine contractions of labor will be established. Whilst they are discussed separately, their interaction must not be overlooked.

(1) anterior pituitary, ovary, estrogen and progesterone,
(2) posterior pituitary, oxytocin and the Ferguson reflex,
(3) autonomic nervous system, catecholamines and stress.

UTERINE ACTIVITY: IN SUMMARY

The contractions of the uterus, as expressed in labor, are subject to a variety of physical, hormonal and neural constraints. These will be discussed only in so far as they set the scene for the operation of maternal influences.

Birth is brought about by the co-ordinated contractions of the myometrium; the smooth muscle of the uterine wall. Such muscle displays "intrinsic contractility", and so effective are these rhythmical contractions that the uterus can expel its contents when removed from the body (Kurdinowski, 1904). The myometrium functions as an electrical syncytium (Tomita, 1966; Barr *et al.*, 1968) where cells are joined by intracellular coupling — the nexus or nexal junction (Dewey and Barr, 1964) and operated electrophysiologically in multicellular units (Abe and Tomita, 1968; Anderson, 1969). The resting membrane potential of the myometrial cell is about -40 to -70 mV, depending on the species and the levels of estrogen and progesterone (Carsten, 1968), and electrical excitation is sodium and calcium dependent (Kao *et al.*, 1961; Anderson, 1969; Kleinhaus and Kao, 1969; Anderson *et al.*, 1971). Depolarization of this resting potential leads in many instances to spike (action potential) activity and the propagation of a wave of contractions across the uterine wall. A directional property is conferred on these rhythmical contractions by the presence of pacemaker regions (Fuchs, 1971). These are not anatomically distinct but, by virtue of a lower threshold for excitation, they generate a propagated contraction which in labor runs from the oviductal junction towards the cervix.

ANTERIOR PITUITARY, OVARY, ESTROGEN AND PROGESTERONE

A number of humoral factors (estrogens, progestogens and prostaglandins) of both placental and ovarian origin profoundly influence uterine contractility, and for many years attempts have been made to formulate a universal theory of parturition based on their actions. The prostaglandins are perhaps our most potent stimulators of uterine activity and they could function as the final denominator by which other control mechanisms are expressed. At this time, however, we have no evidence that prostaglandins of uterine origin influence the hypothalamus, or *vice-versa*.

The estrogen-dominated uterus displays a high level of activity. Removal of the estrogen or the addition of progesterone (by physiological routes) abolishes the co-ordinated contractions of the uterus, leaving weak contractions devoid of spike activity (Carsten, 1968). This "progesterone-block" of the myometrial contraction has been the subject of much criticism, but has never been satisfactorily disproved. Too much weight has been placed on variations in the plasma levels of estrogen and progesterone; not enough attention has been given to the kinetics of hormone action within the myometrium. (For recent reviews: Bedford *et al.*, 1972; Thomas, 1973.) The placenta is the major source of progesterone and estrogen during pregnancy, particularly in the larger species, and the availability of these hormones to the adjacent myometrium is more likely to relate to local factors than to levels in peripheral plasma. Furthermore, the myometrium appears to contain a binding-protein which has the ability during gestation to hold progesterone at a level in excess of the plasma concentration

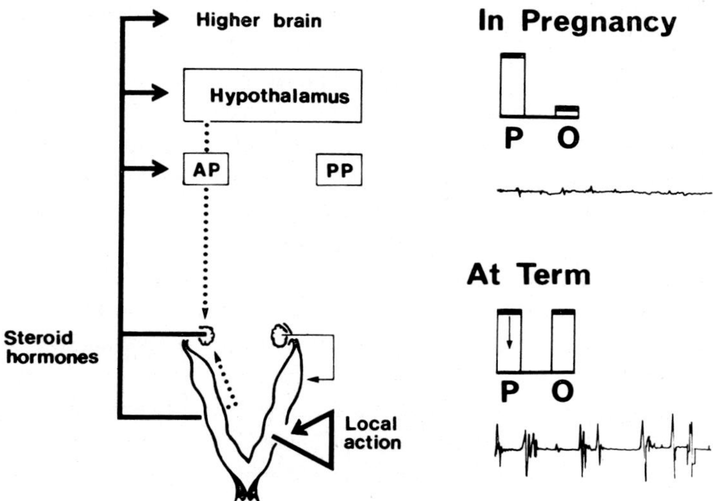

Fig. 1. Pathways of steroid action and control in labor. Two modes of action are expressed: (i) local action within the confines of the uterus, and (ii) action on the brain and pituitary. Pathways by which these steroids may be controlled, via pituitary or uterine trophic and lytic factors, are shown by the discontinuous lines. AP, anterior pituitary; PP, posterior pituitary. The right-hand part of the figure illustrates the activities of progesterone (P) and estrogen (O) during pregnancy and at term, (i) in the form of a histogram depicting the plasma levels of hormone and (ii) in terms of uterine contractions (time being expressed on the horizontal axis).

References p. 301–305

(Davies and Ryan, 1973). For such reasons, therefore, the circulating levels of progesterone and estrogen may be of less consequence to the uterus than they are to their more distant target organs: the hypothalamus and mammary gland (Fig. 1).

There are 3 aspects of the progesterone–estrogen story which are of particular relevance to this discussion.

(1) Are the circulating levels of estrogen and progesterone, and their changes at the time of labor, under hypothalamic control?

(2) Does the feedback of these hormones to the brain influence pituitary function and/or behavior?

(3) To what extent is the myometrial response to oxytocin governed by estrogen and progesterone?

Hypothalamic control

To many species the corpora lutea contribute a major and indispensable source of progesterone (and estrogen) throughout gestation. In other species, ovariectomy whilst not leading to abortion may severely disrupt the normal process of labor. Ovariectomy of the rat for example at day 15 of gestation or later prolongs pregnancy and leads to a protracted delivery of the young (Csapo and Wiest, 1969). Estrogen has been implicated in these observations for there is a sharp rise in the estrogen production of the ovary on days 20–21 (Yoshinaga *et al.*, 1969; Shaika, 1971) and estrogen administration to the gonadectomized animal causes the delivery of the young at the normal time (Csapo, 1969). What, we must ask, determines this change in steroid biosynthesis? Is it caused by the withdrawal of a placental luteotropin or the production of a lytic factor in the prostaglandin group, or does it relate to changes in pituitary function? We have no answer. The rabbit is another of our ovarian dependent species, and is probably the finest example of the "progesterone-block" hypothesis (Csapo, 1961; 1969). The withdrawal of progesterone by ovariectomy promotes abortion. Conversely, gestation is prolonged by exogenous progesterone or by the maintenance of luteal activity. However, it is difficult to reconcile the slow withdrawal of progesterone from day 28 of gestation with the explosive delivery of the young on day 31 or 32 (Cross, 1958a; Fuchs, 1964; Csapo and Takeda, 1965). Once again, we do not know how much hypothalamic control is involved in the maintenance of luteal activity in the rabbit. Hypophysectomy during pregnancy has yielded conflicting results, but it is a gross manipulation to apply under the best of circumstances. In women, successful pregnancy and parturition has been observed after hypophysectomy (Jorgensen *et al.*, 1973). This does not mean that the pituitary–ovarian axis is without importance, especially in those species in which pregnancy is ovarian dependent. Pencharz and Long (1931) and Selye *et al.* (1933) were amongst the first to show that hypophysectomy of the rat changed the normal course of parturition. But the question remains, were their observations due to changes in ovarian function?

Feedback

A rise in circulating estrogen and/or a fall in progesterone could account for many of the behavioral manifestations observed at about the time of parturition. The post-partum estrus and ovulation is a clear case where — if we are to use the basic determinants of the estrous cycle — the changes in steroid activity within the hypothalamus and pituitary have precipitated a behavioral estrus and the ovulatory release of LH. The development of maternal behavior and the onset of lactation may be further examples of action of hormones on the brain. The changes in steroid production in the hours surrounding labor are probably greater than those of the estrous cycle, but in comparison they are grossly understudied.

The feedback of steroids is also important in the operation of the posterior pituitary. The oxytocin content of the posterior lobe fluctuates considerably during the estrous cycle and levels are highest during proestrus and estrus (Heller, 1959; König and Böttcher, 1966). A parallel situation is observed in the enzymic activity of the neurosecretory nuclei of the hypothalamus (Swaab and Jongkind, 1970). The biological significance of these observations is obscure for the posterior pituitary carries a great reserve of hormone relative to the likely demands of labor or lactation. Jones and Pickering (1972) have calculated a basal turnover of oxytocin in the male rat in water balance (a non-stimulated situation) at 18 mU/day. Under the stimulated conditions of lactation in the rat, we have calculated a physiological demand for about 100 mU/day (Lincoln *et al.*, 1973). Even this is a small figure when compared with a pituitary content of oxytocin which may exceed 500 mU. It has been suggested that the changes in oxytocin content are estrogen dependent (König and Ehlers, 1968), but activity also increases after gonadectomy (Zambrano and De Robertis, 1968). The level of pituitary oxytocin has been measured during pregnancy (Heller, 1959) and in the rat at least it has been shown to fall sharply over the course of parturition (Fuchs and Saito, 1971). No study has determined whether the changes in steroid production in the hours prior to parturition have any relevance to the synthesis, storage or release of the posterior pituitary peptides.

The "Ferguson reflex" (Ferguson, 1941), to which further reference will be made in the following section, is steroid dependent. Distension of the reproductive tract of the conscious ewe or goat promotes as a reflex function the release of oxytocin (Roberts and Share, 1968, 1970), and the response is enhanced by estrogen administration and eliminated by progesterone (Roberts, 1971). Where, we should like to know, are these steroids acting? Progesterone may modify the reflex at a hypothalamic level. The infusion of progesterone into the lateral ventricle is more effective than a systemic infusion in terms of the oxytocin release following vaginal distension (Roberts, 1971). For estrogen, we have no information.

Uterine activity

Estrogen and progesterone influence the ionic permeability of the myometrial cell in such a way that contractile activity is enhanced by estrogen and reduced by pro-

References p. 301–305

gesterone (Kao, 1961, 1967; Kao and Nishiyama, 1964). In addition, estrogen increases and progesterone abolishes the uterine response to oxytocin. It is difficult, however, to account fully for such basic observations. Hyperpolarization of the myometrial cell was one of the earliest suggestions to account for the inactivity of the progesterone-dominated uterus (Daniel and Singh, 1958), but this view has been questioned in subsequent studies and has failed to gain much support (Kao, 1967; Saldivar and Melton, 1966). A somewhat more tenable explanation is that progesterone reduces the rate at which the myometrial cell undergoes depolarization, possibly due to the firm binding of calcium ions within the myometrial cell.

The control of the uterine response to oxytocin, by progesterone, is of considerable significance to the maintenance of pregnancy and the control of labor. If such a mechanism were not available, abortion might well result from the release of oxytocin during nursing or coitus. The development of oxytocin sensitivity at term does not occur abruptly (the guinea-pig may be an exception, Porter, 1970). With the rabbit, oxytocin has no effect on uterine activity until day 29 of gestation, from then onwards the sensitivity increases to peak at the time of parturition on day 31–32. This parallels the withdrawal of ovarian progesterone and luteal regression (Csapo and Takeda, 1965). The timing of this change is fascinating in another respect. The rabbit mated on the post-partum estrus and allowed to become pregnant will lactate until day 26–27 of gestation, and not a day longer. If nursing were to continue (and any approach from the pups is violently resisted) the suckling-induced release of oxytocin would precipitate labor. With women the change in oxytocin sensitivity is, in some ways, exponential. A gradual increase in uterine sensitivity is seen as term approaches with a massive increase occurring during the hours of labor (Smyth, 1958; Csapo and Sauvage, 1968; Theobald *et al.*, 1969). With the rabbit, progesterone and its withdrawal provided a tenable explanation for the change in uterine sensitivity. With women the situation is far from clear since progesterone levels in the plasma do not fall until the terminal stages of delivery (Short, 1961; Yannone *et al.*, 1969; Yoshimi *et al.*, 1969).

POSTERIOR PITUITARY, OXYTOCIN AND THE FERGUSON REFLEX

Oxytocin is indispensible for milk ejection (Cross and Harris, 1952). The same cannot be said of labor, though oxytocin is amongst our most potent stimulators of myometrial activity and is invariably found in the blood at some stage during the delivery of the young. Its role in labor is probably to augment the intrinsic contractile activity of the uterus. By facilitating the rapid expulsion of the young in multitocus species (rabbit) or strengthening the uterine contractions during the terminal stages of delivery in larger species (man), oxytocin could contribute significantly to the birth of live young.

Oxytocin and vasopressin (and probably other posterior pituitary peptides) are synthesized in *both* the paraventricular and supraoptic nuclei of the hypothalamus. These octapeptide hormones are there attached to a neurophysin carrier protein and packaged into neurosecretory granules for transport to and storage within the neurosecretory terminals of the neural lobe (Morris and Cannata, 1973; Cannata and Morris, 1973). As early as 1938, Haterius and Ferguson demonstrated that electrical stimulation of the neural stalk evoked the release of posterior pituitary factors. This has since been confirmed, both *in vivo* (Cross and Harris, 1952; Harris *et al.*, 1969) and *in vitro* (Ishida, 1970; Dreifuss *et al.*, 1971; Nordmann and Dreifuss, 1972), and consistently it has been observed that stimulation of the neurosecretory fibers or the posterior pituitary has to exceed a critical frequency of about 30 pulses/sec for oxytocin release to occur. Stimulation, however, provides only indirect evidence *viz.* the relationship between action potentials and hormone release and tells us nothing about the patterns of electrical activity which actually occur under physiological conditions. Paraventricular and supraoptic neurons, reliably identified by antidromic activation, display a background of spike activity which occasionally attains rates of 8–10/sec (Dyball and Koizumi, 1969; Sundsten *et al.*, 1970; Negoro and Holland, 1972; Wakerley and Lincoln, 1973b). We are still uncertain whether this background activity, which on average runs at about 2–4 spikes/sec, causes any release of oxytocin; it is far below the 30/sec threshold observed with electrical stimulation. Our doubts are further increased by our recent discovery that the firing rate of the neurosecretory cell reaches 50–100 spikes/sec during the release of oxytocin at milk ejection (Lincoln and Wakerley, 1972; Wakerley and Lincoln, 1973a, b). An example from one of our recordings of a supraoptic neuron during reflex milk ejection in the rat is shown in an earlier paper in this volume (Dyer, 1974). For each release of oxytocin, about 50% of the neurosecretory cells accelerate in a stereotyped manner to 50–100 spikes/sec for 2–4 sec. This uniform activation of the neurosecretory system precipitates the release of a standard quantum of about 1 mU oxytocin, and the phenomenon repeats itself every 5–15 min despite the continuous sucking of the young. Whilst these results relate to milk ejection, they illustrate a number of principles concerning the control and operation of the neurosecretory system which could apply to the release of oxytocin during labor.

(1) The afferent input (the suckling stimulus) is *continuous*; the output of the neurosecretory system is *pulsatile*.

(2) The response of the neurosecretory cell at oxytocin release is a *stereotyped, all-or-none*, period of activation.

(3) *Uniform* pulses of oxytocin are released in response to each period of activation, each release lasting 2–4 sec. The response recurs at *regular* intervals of 5–15 min.

(4) The overall process is *frequency modulated*. Thus, the efficiency of milk ejection depends not on the size of one oxytocin release but on the number of pulses discharged/unit time.

For further details see Lincoln *et al.* (1973).

Oxytocin release in labor

Ablation of the neurosecretory system, at least in some of the laboratory animals, creates a prolonged and protracted labor (Fisher *et al.*, 1938; Nibbelink, 1961). This, coupled with the fact that pituitary stores of oxytocin fall considerably during labor,

References p. 301–305

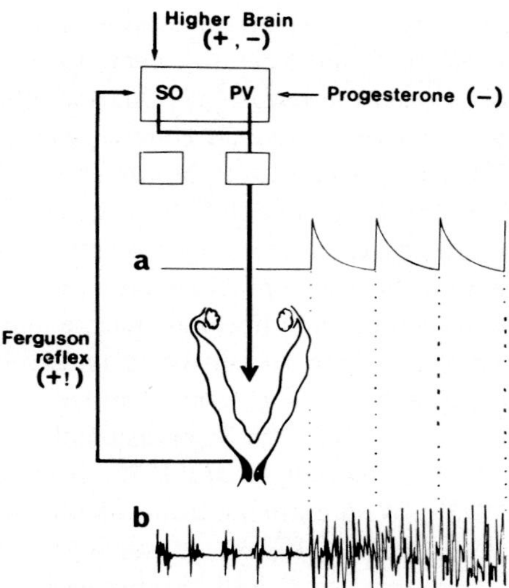

Fig. 2. Oxytocin release and its control during labor. The main components of this diagram follow the design of Fig. 1. SO, supraoptic nuclei; PV, paraventricular nuclei. The release of oxytocin, under the positive and negative influences of higher neural centers, the Ferguson reflex and the circulating steroids (see Fig. 1), is expressed as an intermittent function (a) involving the regular release of a uniform pulse of hormone. The response of the estrogenized-uterus (b) is continuous activation, with contractions continuing from one oxytocin pulse to the next.

suggests some neurosecretory involvement with the normal process of delivery (Acher *et al.*, 1956; Fuchs and Saito, 1971). There is at least one genetic anomaly in which we can observe the effects of profound changes in posterior pituitary function. The "Brattleboro strain" of Long–Evans rat fails to produce vasopressin and, probably as a result of extreme stimulation of the neurosecretory system in an attempt to restore water balance, the oxytocin levels of the pituitary are depleted (Valtin *et al.*, 1965; Burford *et al.*, 1971). Parturition of the homozygous animals is prolonged (Boer *et al.*, 1974). Diabetes insipidus in women, however, is consistent with normal labor (Hendricks, 1954; Cobo *et al.*, 1972).

The measurement of oxytocin in plasma is a more definitive index of oxytocin release, but the results have been notoriously inconsistent. It would be a fair comment to say that most studies have detected oxytocin at some stage of labor and levels have usually been highest during the terminal stages of delivery (Chard, 1972). There are two notable features about the oxytocin levels which have been published. The range of the oxytocin titer has always been large, and the values have tended to decline over the years. The variability of the oxytocin titer is indicative of a "surge-like" release of hormone (Fox and Knaggs, 1969; McNeilly, 1972), though measurements have never been made at sufficiently short intervals to fully establish the pattern of release. With a half-life of 2–4 min (Fabian *et al.*, 1969; Chard *et al.*, 1970) and a pulse which may be released in a few seconds, measurements of plasma oxytocin have to be very frequent — every 10–20 sec.

A single pulse of oxytocin promotes an increase in uterine contractility for a period which considerably exceeds the circulatory half-life of the hormone. In the rabbit at 31 days gestation a single intravenous injection of 50–200 mU oxytocin will maintain uterine activity for 30 min and promote the delivery of the full litter (Cross, 1958a; Fuchs, 1964). However, it is possible that oxytocin persists in the blood for longer than the original calculations of the half-life suggested, for Gibbens *et al.* (1972) have found evidence for a double exponential in the rate of oxytocin decay such that lower concentrations of hormone are removed at a reduced rate. Thus, an occasional release of oxytocin, which might be detected in only one sample in 10, may be of major significance to the normal course of parturition. A pulsatile release of hormone has other attractive features. For example, if the pulses are correctly timed, one can maintain and predict the response of the target tissue.

A number of attempts have been made to capitalize on the milk ejection response as an on-line measure of oxytocin release in labor. Milk ejection has been observed in women though it is not a frequent event (Gunther, 1948; Cobo, 1968). Observations on other species, where we have more evidence for an oxytocin involvement, have been hampered by the fact that anesthesia — a necessary prerequisite for the measurement of intramammary pressure — blocks the reflex release of this hormone.

Induction of labor

Posterior pituitary extracts have been used from the early part of this century for the induction of birth (Blair-Bell, 1909; Hofbauer, 1911) and, until the advent of prostaglandin therapy, synthetic oxytocin has played a major part in obstetrical work. Experimentally, electrical stimulation can be used to release posterior pituitary hormones and in the rabbit this is a most effective way of inducing labor (Cross, 1958a; Lincoln, 1971). When stimulation is applied to the pituitary stalk, 50 pulses/sec for 20 sec, at 31–32 days of gestation labor commences within minutes and continues for up to half-an-hour (Fig. 3). Spontaneous delivery follows a very similar pattern. Uterine contractions remain weak until almost the time of delivery. There is then a sudden increase in contractile activity and the young are rapidly expelled (Fuchs, 1964). To simulate these events one has to administer a large pulse of oxytocin (50–200 mU), slow infusions of oxytocin do not precipitate the rapid delivery of the young. Electrical stimulation for 20 sec, as applied above, would release a similar pulse of hormone and this could account for the remarkable parallel between the electrically induced labor and the spontaneous event.

There are some minor but most interesting differences between spontaneous labor in the rabbit and that induced by exogenous oxytocin and electrical stimulation. In the first place, labor always ceased about 25–30 min after electrical stimulation irrespective of how many young had been delivered (Fig. 3). One would have expected, had there been an effective Ferguson reflex operated by the distension of the lower parts of the reproductive tract during the expulsion of the young, that labor would always have continued to completion. Both Fuchs (1964) in the measurement of uterine activity and Haldar (1970) in the measurement of plasma levels of oxytocin failed to observe any effective release of oxytocin in association with the delivery of individual

References p. 301–305

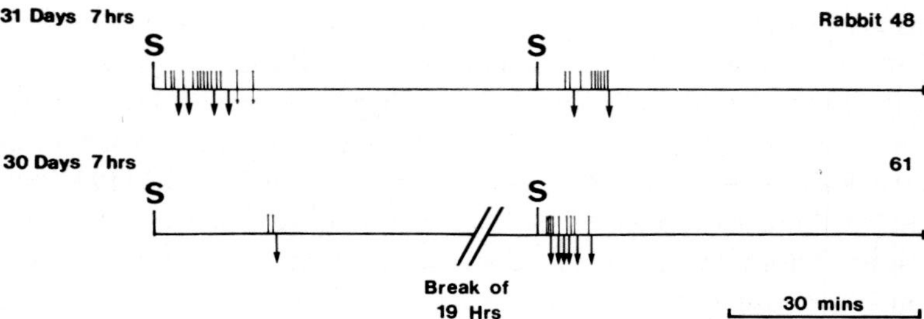

Fig. 3. Induction of labor in two conscious rabbits by electrical stimulation of the pituitary stalk. Stimulation (S) was applied through chronically implanted platinum–rhodium electrodes; stimulation parameters: 20 sec duration, 50 pulses/sec, 1 msec biphasic square waves, 1 mA peak to peak current. The rabbits were of the 'Californian strain' and weighed approximately 3.5 kg. Abdominal straining actions are shown by the marks placed vertically above each trace. The delivery of a pup is shown by each large arrow; the expulsion of a separated placenta is shown by the smaller arrow. The stage of gestation at which stimulation was applied is recorded in days and hours from the time of mating. Spontaneous delivery would have occurred between days 32 and 33. Note: (i) effective labor ceased within 30 min of stimulation, (ii) events were restarted by further stimulation after 1 and/or 19 h, and (iii) stimulation was more effective in terms of the number of pups delivered at 31 days 7 h than 24 h earlier.

young. It would appear, therefore, that the rabbit releases one large pulse of oxytocin and this is adequate to complete delivery under normal circumstances. If labor has not been completed within the allotted 30 min, uterine activity ceases until further hormone release occurs some hours later. There was no apparent neurosecretory deficiency following the electrical induction of labor for further stimulation 1, 2 or even 24 h later induced effective contractions of the uterus and the birth of further live pups (Fig. 3). The effectiveness of electrical stimulation increased towards term, *i.e.* stimulation on day 31 led to the birth of few young, stimulation on day 32 usually induced the delivery of the full litter — particularly in multiparous does. What stimulus triggers the apparent neurosecretory activation prior to the onset of spontaneous delivery remains something of a mystery.

Electrical stimulation was invariably more effective in the induction of labor than treatment with exogenous oxytocin (i.v.), for the administration of exogenous hormone was either ineffective or events moved to the other extreme and delivery was explosive and most traumatic. There are several explanations that one could apply. Stimulation may have released oxytocin in amounts or in a form which was more physiological to the animal than the synthetic oxytocin preparation "Syntocinon". Alternatively, other hormonal factors may have been involved. Vasopressin was undoubtedly released by electrical stimulation (Harris, 1947); this could have accounted for the difference observed in the degree of hemorrhage in the mother associated with delivery. The levels of "oxytocic-activity" in plasma have shown a tendency to decline in recent years with improvements in the specificity of the assay systems, and it is possible that the current radio-immunoassays are not measuring what is biologically important to the uterus. The possibility of there being unknown posterior pituitary factors is supported by the recent discovery of "coherin", a peptide which controls the motility of the gastrointestinal tract (Goodman and Hiatt, 1972).

AUTONOMIC NERVOUS SYSTEM, CATECHOLAMINES AND STRESS

The uterus, in common with most of the abdominal viscera, receives an autonomic innervation, the sympathetic function of which is supported by circulating catecholamines of adrenal origin. The entire system is under hypothalamic control, but it is difficult to isolate the medullary and spinal pathways along which these autonomic functions are propagated or to uphold the classical separation into sympathetic and parasympathetic components. Considerably more is known about the peripheral distribution of the nerves and a number of studies have considered the uterine innervation in detail (Krantz, 1959; Owman and Sjoberg, 1966; Nakanishi and Wood, 1971) (Fig. 4).

The autonomic innervation of the uterus is not essential, for denervation of the uterus does not lead to muscle degeneration and contractile activity continues. Likewise denervation of the uterus and adrenalectomy (to remove the major source of circulating catecholamines) does not abolish labor (Carlson and De Foe, 1965; Fuchs, 1971). The noradrenaline content of the uterus, and this is the only catecholamine synthesized in the uterus, is low but cannot be accounted for on the basis of vascular innervation alone (Wurtman *et al.*, 1964). The adrenaline and noradrenaline content of the uterus increase during pregnancy (Spratto and Miller, 1968). Whilst this

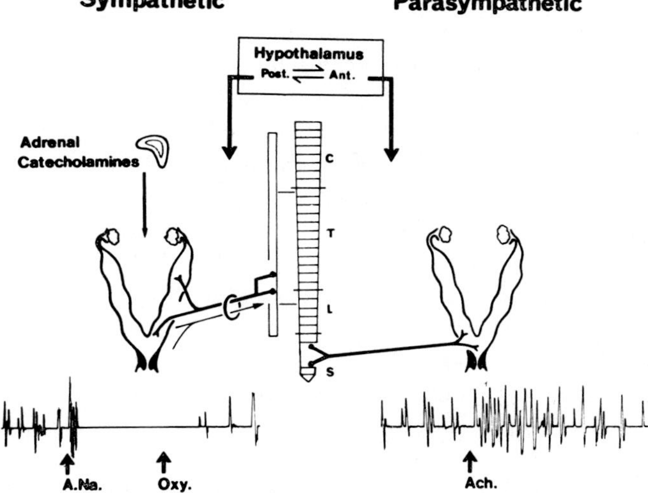

Fig. 4. Autonomic nervous control of uterine contractility. The basic components of this diagram follow the design of Fig. 1, with a spinal cord added to illustrate the segmental origin of the autonomic innervation. The sympathetic (adrenergic) innervation, controlled by the more caudal parts of the hypothalamus and augmented in its action by catecholamines of adrenal origin, is shown to the left. The somewhat more restricted parasympathetic innervation is shown on the other side. The action of catecholamines (A, adrenaline; Na, noradrenaline) on the uterus is shown as mild activation with prolonged inhibition and blockade of the oxytocin (Oxy) response. For further information on these effects and the excitatory response to acetylcholine (Ach) see the text. Visceral afferent fibers from the uterus are shown running in association with the sympathetic outflow from the lower thoracic and upper lumbar segments. Possible functions: relay of uterine pain, autonomic integration, afferents to neuroendocrine (oxytocin release) and spinal reflexes.

increase in catecholamine content is gradual and parallels the increase in the size of the uterus, the fall at parturition is dramatic — the rat uterus loses 80% of its adrenaline in the hours surrounding parturition (Wurtman *et al.*, 1964).

Electrical stimulation of the hypogastric and splanchnic nerves of estrogen-treated post-parturient rabbit profoundly effects uterine contractility. There is an initial increase in activity followed by quiescence and relaxation (Schofield, 1952; Setekleiv, 1964). A similar response is observed after electrical stimulation of the sympathetic centers in the dorsomedial hypothalamus (Cross, 1958b). These effects are readily simulated by the injection of adrenaline and noradrenaline and the use of these catecholamines and their antagonists has been the subject of considerable study. The response of the uterus is species and hormone dependent, and the direction of the response can be explained in a "receptor-theory". The α-receptors of the myometrium respond to noradrenaline (and to adrenaline to a lesser degree) and are excitatory to the contractile activity. The β-receptors responding primarily to adrenaline are inhibitory. Both receptors have been identified within the myometrium, but their "availability" is steroid dependent. Estrogen favors α-receptor activity and progesterone β-activity (Miller, 1967; Diamond and Brody, 1966). For these reasons, therefore, diametrically opposed responses will be encountered according to whether the animal is pregnant or non-pregnant. For example, adrenaline infusion reduces or blocks the spontaneous or oxytocin-induced contractions of the pregnant human uterus (Pose *et al.*, 1962). The opposite effect, an increase in activity, is observed following adrenaline administration during the first half of the menstrual cycle (Cieciorowska and Telko, 1961). It remains to be resolved whether this balance in adrenergic receptor activity has any functional significance to the normal process of labor. Propranolol, a β-blocking agent, has been used to precipitate labor; presumably by eliminating β-receptors the bias in receptor activity swings in favor of the excitatory α-units (Shabanah *et al.*, 1968).

The cholinergic innervation of the uterus is less extensive than the adrenergic system, and is confined to the cervical region (Coupland, 1969; Nakanishi and Wood, 1971). The myometrium contracts in response to acetylcholine, but the response is considerably attenuated during pregnancy (Sandberg *et al.*, 1958; Sala and Fish, 1965; Nakanishi and Wood, 1971).

As a general rule, it may be said that the myometrium is protected from autonomic influences during pregnancy (Nakanishi *et al.*, 1969). However, the autonomic nervous system must not be regarded as a system which is only activated on demand, it displays continuous activity (autonomic tone), the level of which varies hour by hour and from one individual to another. It could in effect operate "set-point" governing the contractile activity of the uterus at given levels of estrogen, progesterone and oxytocin.

The sympathetic nerves are associated with visceral afferents carrying sensory information from the uterus. The central projection of this information is unknown, but it undoubtedly reaches both thalamic (pain perception) and hypothalamic levels (Abrahams *et al.*, 1964; Lincoln, 1969). The hypothalamic projection could represent the afferent component of the Ferguson reflex; alternatively, it could function in the integration of autonomic activity.

CONCLUSIONS AND SUMMARY

The maternal control of labor may not do more than modulate by a few hours the inherent course of events, but it must be given careful consideration. The evidence presented in the foregoing discussion suggests, most strongly, that maternal mechanisms contribute significantly to the promotion of the live birth and/or the delivery of the young at a time which favors survival.

The 3 maternal interactions with uterine activity, established on the basis of rather fragmentary evidence, provide all the control that would be necessary to explain the labor of any species, but it would be foolish to construct from these mechanisms a general scheme for the control of labor. Species diversity is so great. There is, however, one general principle that we may apply. Throughout the major part of gestation, everything is organized to safeguard pregnancy, sometimes to the point that efforts are duplicated. Take the release of oxytocin, for example. During pregnancy the Ferguson reflex is blocked, but should oxytocin be released by any other stimulus (*e.g.* coitus or nursing) the uterus would still fail to respond. Stress inhibits the release of oxytocin at a neural level (Aulsebrook and Holland, 1969a, b) and, in addition, abolishes uterine activity by adrenergic inhibition. With the onset of labor, "everything changes", and once more there is "a time to be born".

REFERENCES

ABE, Y. AND TOMITA, T. (1968) Cable properties of smooth muscle. *J. Physiol. (Lond.)*, **196**, 87–100.

ABRAHAMS, V. C., LANGWORTH, E. P. AND THEOBALD, G. W. (1964) Potential evoked in the hypothalamus and cerebral cortex by electrical stimulation of the uterus. *Nature(Lond.)*, **203**, 654–656.

ACHER, R., CHAUVET, J. ET OLIVRY, G. (1956) Sur l'existence éventuelle d'une hormone unique neurohypophysaire. *Biochim. biophys. Acta (Amst.)*, **22**, 428–433.

ALISON, J. (1972) Hour of birth in primates and man. *Folia primatol.*, **18**, 108–121.

ANDERSON, N. C. (1969) Voltage-clamp studies on uterine smooth muscle. *J. gen. Physiol.*, **54**, 145–165.

ANDERSON, N. C., RAMON, F. AND SNYDER, A. (1971) Studies on calcium and sodium in uterine smooth muscle excitation under current-clamp and voltage-clamp condition. *J. gen. Physiol.*, **58**, 322–339.

AULSEBROOK, L. H. AND HOLLAND, R. C. (1969a) Central regulation of oxytocin release with and without vasopressin release. *Amer. J. Physiol.*, **216**, 818–829.

AULSEBROOK, L. H. AND HOLLAND, R. C. (1969b) Central inhibition of oxytocin release. *Amer. J. Physiol.*, **216**, 830–842.

BARR, L., BERGER, W. AND DEWEY, M. M. (1968) Electrical transmission at the nexus between smooth muscle cells. *J. gen. Physiol.*, **51**, 347–369.

BEDFORD, C. A., CHALLIS, J. R. G., HARRISON, F. A. AND HEAP, R. B. (1972) The role of oestrogens and progesterone in the onset of parturition in various species. *J. Reprod. Fertil.*, **Suppl. 16**, 1–23.

BLAIR-BELL, W. (1909) The pituitary body and the therapeutic value of the infundibular extract in shock, uterine atony, and intestinal paresis. *Brit. med. J.*, **ii**, 1609–1613.

BOER, K., BOER, G. J. AND SWAAB, D. F. (1974) Does the hypothalamo–neurohypophyseal system play a role in gestation length or parturition? In *Integrative Hypothalamic Activity, Progr. Brain Res., Vol. 41*, D. F. SWAAB AND J. P. SCHADÉ (Eds.), Elsevier, Amsterdam, pp. 307–320.

BURFORD, G. D., JONES, C. W. AND PICKERING, B. T. (1971) Tentative identification of a vasopressin-neurophysin and an oxytocin-neurophysin in the rat. *Biochem. J.*, **124**, 809–813.

CANNATA, M. A. AND MORRIS, J. F. (1973) Changes in the appearance of hypothalamo–neurohypophysial neurosecretory granules associated with their maturation. *J. Endocr.*, **57**, 531–538.

CARLSON, R. R. AND DE FOE, V. J. (1965) Role of the pelvic nerve *versus* the abdominal sympathetics in the reproductive functions of the female rat. *Endocrinology*, **77**, 1014–1022.

CARSTEN, M. E. (1968) Regulation of myometrial composition, growth and activity. In *Biology of Gestation*, N. S. ASSALI (Ed.), Academic Press, London, pp. 355–425.

CHARD, T. (1972) The role of the posterior pituitary in human and animal parturition. *J. Reprod. Fertil.*, **Suppl. 16**, 121–138.

CHARD, T., BOYD, N. R. H., FORSLING, M. L., MCNEILLY, A. S. AND LANDON, J. (1970) The development of a radio-immunoassay for oxytocin. The extraction of oxytocin from plasma, and its measurement during parturition in human and goat blood. *J. Endocr.*, **48**, 223–234.

CIECIOROWSKA, A. AND TELKO, M. (1961) The effects of adrenaline and noradrenaline on the contractions of the human uterus. *Gynaecologia (Basel)*, **152**, 39–49.

COBO, E. (1968) Uterine and milk-ejection activities during human labour. *J. appl. Physiol.*, **24**, 317–323.

COBO, E., DE BERNAL, M. AND GAITAN, E. (1972) Low oxytocin secretion in diabetes insipidus associated with normal labour. *Amer. J. Obstet. Gynec.*, **114**, 861–866.

CONROY, R. T. W. L. AND MILLS, J. N. (1970) Circadian rhythm in birth. The development, synchronization and maintenance of circadian rhythms. In *Human Circadian Rhythms*, Churchill, London, pp. 112–114.

COUPLAND, R. E. (1969) The distribution of cholinergic and other nerve fibres in the human uterus. *Postgrad. med. J.*, **45**, 78–79.

CROSS, B. A. (1958a) On the mechanism of labour in the rabbit. *J. Endocr.*, **16**, 261–276.

CROSS, B. A. (1958b) The motility and reactivity of the oestrogenized rabbit uterus *in vivo:* with comparative observations on milk ejection. *J. Endocr.*, **16**, 237–260.

CROSS, B. A. AND HARRIS, G. W. (1952) The role of the neurohypophysis in the milk-ejection reflex. *J. Endocr.*, **8**, 148–161.

CSAPO, A. (1961) Progesterone and the defence mechanism of pregnancy. In *Ciba Foundation Study Group No. 9*, G. E. W. WOLSTENHOLME AND M. P. CAMERON (Eds.), Churchill, London, pp. 3–12.

CSAPO, A. (1969) The four direct regulatory factors of myometrial function. In *Ciba Foundation Study Group No. 34*, G. E. W. WOLSTENHOLME AND J. KNIGHT (Eds.), Churchill, London, pp. 13–42.

CSAPO, A. AND SAUVAGE, J. (1968) The evolution of uterine activity during human pregnancy. *Acta obstet. gynec. scand.*, **47**, 181–212.

CSAPO, A. AND TAKEDA, H. (1965) Effect of progesterone on the electrical activity and intrauterine pressure of pregnant and parturient rabbits. *Amer. J. Obstet. Gynec.*, **91**, 221–231.

CSAPO, A. AND WIEST, W. G. (1969) An examination of the quantitative relationship between progesterone and the maintenance of pregnancy. *Endocrinology*, **85**, 735–746.

DANIEL, E. AND SINGH, H. (1958) The electrical properties of smooth muscle cell membrane. *Canad. J. Biochem.*, **36**, 959–975.

DAVIES, I. J. AND RYAN, K. J. (1973) The modulation of progesterone concentration in the myometrium of the pregnant rat by changes in cytoplasmic 'receptor' protein activity. *Endocrinology*, **92**, 394–401.

DEWEY, M. M. AND BARR, L. (1964) A study of the structure and distribution of the nexus. *J. Cell Biol.*, **23**, 553–585.

DIAMOND, J. AND BRODY, T. M. (1966) Hormonal alteration of the response of the rat uterus to catecholamines. *Life Sci.*, **5**, 2187–2193.

DREIFUSS, J. J., KALNINS, I., KELLY, I. S. AND RUF, K. B. (1971) Action potentials and release of neurohypophyseal hormones *in vitro*. *J. Physiol. (Lond.)*, **215**, 805–817.

DYBALL, R. E. J. AND KOIZUMI, K. (1969) Electrical activity in the supraoptic and paraventricular nuclei associated with neurohypophysial hormone release. *J. Physiol. (Lond.)*, **201**, 711–722.

DYER, R. G. (1974) The electrophysiology of the hypothalamus and its endocrine implications. In *Integrative Hypothalamic Activity*, *Progr. Brain Res.*, *Vol. 41*, D. F. SWAAB AND J. P. SCHADÉ (Eds.), Elsevier, Amsterdam, pp. 133–147.

DYER, R. G., DYBALL, R. E. J. AND MORRIS, J. F. (1973) The effect of hypothalamic deafferentation upon the ultrastructure and hormone content of the paraventricular nucleus. *J. Endocr.*, **57**, 509–516.

FABIAN, M., FORSLING, M. L., JONES, J. J. AND PRYOR, J. S. (1969) The clearance and antidiuretic potency of neurohypophyseal hormones in man and their plasma binding and stability. *J. Physiol. (Lond.)*, **204**, 653–658.

FERGUSON, J. K. W. (1941) Study of motility of intact uterus at term. *Surg. Gynec. Obstet.*, **73**, 359–366.

FISHER, C., MAGOUN, W. H. AND RANSON, S. W. (1938) Dystocia in diabetes insipidus. *Amer. J. Obstet. Gynec.*, **36**, 1–9.

FOX, C. A. AND KNAGGS, G. S. (1969) Milk-ejection activity (oxytocin) in peripheral venous blood of man during lactation and in association with coitus. *J. Endocr.*, **45**, 145–146.

FUCHS, A. R. (1964) Oxytocin and the onset of labour in rabbits. *J. Endocr.*, **30**, 217–224.

FUCHS, A. R. (1971) Uterine activating hormones. In *Endocrinology of Pregnancy*, F. FUCHS AND A. KLOPPER (Eds.), Harper and Row, New York, pp. 286–305.

FUCHS, A. R. AND SAITO, S. (1971) Pituitary oxytocin and vasopressin content of pregnant and parturient rats before, during, and after parturition. *Endocrinology*, **88**, 574–578.

GALE, C. C. AND MCCANN, S. M. (1961) Hypothalamic control of pituitary gonadotrophins. *J. Endocr.*, **22**, 107–117.

GIBBENS, D., BOYD, N. R., CROCKER, S., BAUMBER, S. AND CHARD, T. (1972) Plasma oxytocin levels following administration of Syntometrine in the third stage of labour. *J. Obstet. Gynaec. Brit. Cwlth*, **79**, 644–646.

GOODMAN, I. AND HIATT, R. B. (1972) Coherin: a new peptide of the bovine neurohypophysis with activity on gastrointestinal motility. *Science*, **178**, 419–421.

GÜNTHER, M. (1948) Posterior pituitary and labour. *Brit. Med. J.*, **i**, 567.

HALDAR, J. (1970) Independent release of oxytocin and vasopressin during parturition in the rabbit. *J. Physiol. (Lond.)*, **206**, 723–730.

HARRIS, G. W. (1947) The innervation and actions of the neurohypophysis; an investigation using the method of remote-control stimulation. *Phil. Trans. B*, **232**, 385–441.

HARRIS, G. W., MANABE, Y. AND RUF, K. B. (1969) A study of the parameters of electrical stimulation of unmyelinated fibres in the pituitary stalk. *J. Physiol. (Lond.)*, **203**, 67–81.

HATERIUS, H. O. AND FERGUSON, J. K. W. (1938) Evidence for the hormonal nature of the oxytocic principle of the hypophysis. *Amer. J. Physiol.*, **124**, 314–321.

HELLER, H. (1959) The neurohypophysis during the estrous cycle, pregnancy and lactation. In *Recent Progress in the Endocrinology of Reproduction*, C. W. LLOYD (Ed.), Academic Press, London, pp. 365–385.

HENDRICKS, C. H. (1954) The neurohypophysis in pregnancy. *Surg. Gynec. Obstet.*, **9**, 323.

HERNÁNDEZ-PÉON, R. AND CHÁVEZ-IBARRA, G. (1963) Sleep induced by electrical or chemical stimulation of the forebrain. *Electroenceph. clin. Neurophysiol.*, **Suppl. 24**, 188–198.

HOFBAUER, J. (1911) Hypophysenextrakt als Wehenmittel. *Zbl. Gynäk*, **35**, 137.

ISHIDA, A. (1970) The oxytocin release and the compound action potential evoked by electrical stimulation of the isolated neurohypophysis of the rat. *Jap. J. Physiol.*, **20**, 84–96.

JONES, C. W. AND PICKERING, B. T. (1972) Intra-axonal transport and turnover of neurohypophysial hormones in the rat. *J. Physiol. (Lond.)*, **227**, 553–564.

JORGENSEN, P. I., SELE, V., BUUS, O. AND DAMKJAER, M. (1973) Detailed hormonal studies during and after pregnancy in a previously hypophysectomized patient. *Acta endocrinol. (Kbh.)*, **73**, 117–132.

KAO, C. Y. (1961) Contents and distributions of potassium, sodium and chloride in uterine smooth muscle. *Amer. J. Physiol.*, **201**, 717–722.

KAO, C. Y. (1967) Ionic basis of electrical activity in uterine smooth muscle. In *Cellular Biology of the Uterus*, R. M. WYNN (Ed.), North Holland, Amsterdam, pp. 386–448.

KAO, C. Y. AND NISHIYAMA, A. (1964) Ovarian hormones and resting potential of rabbit uterine smooth muscle. *Amer. J. Physiol.*, **207**, 793–799.

KAO, C. Y., ZAKIM, D. AND BRONNER, F. (1961) Sodium influx and excitation in uterine smooth muscle. *Nature (Lond.)*, **192**, 1189–1190.

KLEINHAUS, A. L. AND KAO, C. Y. (1969) Electrophysiological actions of oxytocin on the rabbit myometrium. *J. gen. Physiol.*, **53**, 758–780.

KÖNIG, A. U. UND BÖTTCHER, K. (1966) Der Hormongehalt des Hypophysenhinterlappens während des Sexualcyclus von Wistar-Ratten. *Arch. Gynäk.*, **203**, 485–490.

KÖNIG, A. AND EHLERS, B. (1968) Effect of stilboestrol on the hormonal activity of the posterior pituitary lobe of mature female rats. *Rass. Neurol. veg.*, **22**, 193–199.

KRANTZ, K. E. (1959) Uterine contractions; Innervation of the human uterus. *Ann. N.Y. Acad. Sci.*, **75**, 770–784.

KURDINOWSKI, E. M. (1904) Physiologische und pharmakologische Versuche an der isolierten Gebärmutter. *Arch. Anat. Physiol., Lpz.*, **28**, 323.

LEDERIS, C. (1961) Vasopressin and oxytocin in the mammalian hypothalamus. *Gen. comp. Endocr.*, **1**, 80–89.

LINCOLN, D. W. (1969) Response of hypothalamic units to stimulation of the vaginal cervix: specific *versus* non-specific effects. *J. Endocr.*, **43**, 683–684.
LINCOLN, D. W. (1971) Labour in the rabbit: effect of electrical stimulation applied to the infundibulum and median eminence. *J. Endocr.*, **50**, 607–618.
LINCOLN, D. W., HILL, A. AND WAKERLEY, J. B. (1973) The milk ejection reflex of the rat: an intermittent function not abolished by surgical levels of anaesthesia. *J. Endocr.*, **57**, 459–476.
LINCOLN, D. W., SCHOOT, P., VAN DER, AND ZEILMAKER, G. H., (1972) EEG arousal, sympathetic activation and the induction of ovulation by electrochemical stimulation of the rat hypothalamus. *J. Endocr.*, **57**, xix–xx.
LINCOLN, D. W. AND WAKERLEY, J. B. (1972) Accelerated discharge of paraventricular neurosecretory cells correlated with reflex release of oxytocin. *J. Physiol. (Lond.)*, **222**, 23–24.
MCNEILLY, A. S. (1972) The blood levels of oxytocin during suckling and hand milking in the goat with some observations on the pattern of hormone release. *J. Endocr.*, **52**, 177–188.
MILLER, J. W. (1967) Adrenergic receptors in the myometrium. *Ann. N.Y. Acad Sci.*, **139**, 788–798.
MORRIS, J. F. AND CANNATA, M. A. (1973) Ultrastructural preservation of the dense core of posterior pituitary neurosecretory granules and its implications for hormone release. *J. Endocr.*, **57**, 517–529.
NAKANISHI, H., MACLEAN, J., WOOD, C. AND BURNSTOCK, G. (1969) The role of sympathetic nerves in control of the nonpregnant and pregnant uterus. *J. Reprod. Med.*, **2**, 20–23.
NAKANISHI, H. AND WOOD, C. (1971) Cholinergic mechanisms in the human uterus. *J. Obstet. Gynaec. Brit. Cwlth.*, **78**, 716–723.
NAUTA, W. J. H. (1946) Hypothalamic regulation of sleep. An experimental study. *J. Neurophysiol.*, **9**, 285–316.
NEGORO, H. AND HOLLAND, R. C. (1972) Inhibition of unit activity in the hypothalamic paraventricular nucleus following antidromic activation. *Brain Res.*, **42**, 385–402.
NIBBLELINK, D. W. (1961) Paraventricular nuclei, neurohypophysis and parturition. *Amer. J. Physiol.*, **200**, 1229–1232.
NORDMANN, J. J. AND DREIFUSS, J. J. (1972) Hormone release evoked by electrical stimulation in the absence of action potentials. *Brain Res.*, **45**, 604–607.
OWMAN, C. AND SJÖBERG, N. O. (1966) Adrenergic nerves in the female genital tract of the rabbit with remarks on cholinesterase containing structures. *Z. Zellforsch.*, **74**, 182.
PENCHARZ, R. I. AND LONG, J. A. (1931) The effects of hypophysectomy on gestation in the rat. *Science*, **74**, 206.
PORTER, D. G. (1970) The failure of progesterone to affect myometrial activity in the guinea-pig. *J. Endocr.*, **46**, 425–434.
PORTER, D. G. AND SCHOFIELD, B. M. (1966) Intra-uterine pressure changes during pregnancy and parturition in rabbits. *J. Endocr.*, **36**, 291–299.
POSE, S. V., GIBILS, L. A. AND ZUSPAN, F. P. (1962) Effect of L-epinephrine infusion on uterine contractility and cardiovascular system. *Amer. J. Obstet. Gynec.*, **84**, 297–306.
ROBERTS, J. S. (1971) Progesterone-inhibition of oxytocin release during vaginal distension evidence for a central site of action. *Endocrinology*, **89**, 1137–1141.
ROBERTS, J. S. AND SHARE, L. (1968) Oxytocin in plasma of pregnant lactating and cycling ewes during vaginal stimulation. *Endocrinology*, **83**, 272–278.
ROBERTS, J. S. AND SHARE, L. (1970) Inhibition by progesterone of oxytocin secretion during vaginal stimulation. *Endocrinology*, **87**, 812–815.
SALA, N. L. AND FISH, L. (1965) Effect of acetylcholine and atropine upon uterine contractility in pregnant women. *Amer. J. Obstet. Gynec.*, **91**, 1069–1075.
SALDIVAR, J. T. JR. AND MELTON, C. E. JR. (1966) Effects *in vitro* and *in vivo* of sex steroids on rat myometrium. *Amer. J. Physiol.*, **211**, 835–843.
SCHOFIELD, B. M. (1952) The innervation of cervix and cornu in the rabbit. *J. Physiol. (Lond.)*, **117**, 317–328.
SELYE, H., COLLIP, J. B. AND THOMSON, D. L. (1933) Effect of hypophysectomy upon pregnancy and lactation. *Proc. Soc. exp. Biol. (N.Y.)*, **30**, 589–590.
SETEKLEIV, J. (1964) Uterine motility of the estrogenized rabbit. III. Response to hypogastric and splanchnic nerve stimulation. *Acta physiol. scand.*, **62**, 137–149.
SHABANAH, E. H., TOTH, A., CARASSAVAS, M. D. AND MAUGHAN, G. B. (1968) The role of the autonomic nervous system in uterine contractility and blood flow. *Amer. J. Obstet. Gynec.*, **100**, 974–980.

SHAIKA, A. A. (1971) Estrone and estradiol levels in the ovarian venous blood from rats during the estrous cycle and pregnancy. *Biol. Reprod.*, **5**, 297–307.
SHORT, R. V. (1961) Progesterone. In *Hormones in the Blood*, C. H. GRAY AND A. L. BACHARACH (Eds.), Academic Press, London, pp. 379–437.
SMYTH, C. N. (1958) Uterine irritability. *Lancet*, **i**, 237.
SANDBERG, F., INGLEMAN-SANDBERG, A., LINDGREN, L. AND RYDEN, G. (1958) The effect of adrenaline, noradrenaline, and acetylcholine on the spontaneous motility in different parts of the pregnant and non-pregnant uterus. *J. Obstet. Gynaec. Brit. Emp.*, **65**, 965–972.
SPRATTO, G. AND MILLER, J. W. (1968) An investigation of the mechanism by which estradiol - 17β elevates the epinephrine content of the rat uterus. *J. Pharmacol. exp. Ther.*, **161**, 7–13.
SUNDSTEN, J. W., NOVIN, D. AND CROSS, B. A. (1970) Identification and distribution of paraventricular units excited by stimulation of the neural lobe of the hypophysis. *Exp. Neurol.*, **26**, 95–102.
SWAAB, D. F. AND JONGKIND, D. F. (1970) The hypothalamic neurosecretory activity during the oestrous cycle, pregnancy, parturition, lactation and persistent oestrus, and after gonadectomy in the rat. *Neuroendocrinology*, **6**, 133–145.
SWAAB, D. F. AND HONNEBIER, W. (1974) The role of the fetal hypothalamus in the development of the feto-placental unit and in parturition. In *Integrative Hypothalamic Activity*, *Progr. Brain Res.*, *Vol. 41*, D. F. SWAAB AND J. P. SCHADÉ (Eds.), Elsevier, Amsterdam, pp. 255–280.
THEOBALD, G. W., ROBERDS, M. F. AND SUTER, R. E. N. (1969) Changes in myometrial sensitivity to oxytocin in man during the last six weeks of pregnancy. *J. Obstet. Gynaec. Brit. Cwth.*, **76**, 385–393.
THOMAS, P. J. (1973) Steroid hormones and their receptors. *J. Endocr.*, **57**, 333–359.
TOMITA, T. (1966) Electrical responses of smooth muscle to external stimulation in hypertonic solution. *J. Physiol. (Lond.)*, **183**, 450–468.
VALTIN, H., SAWYER, W. H. AND SOKOL, H. W. (1965) Neurohypophyseal principles in rats homozygous and heterozygous for hypothalamic diabetes insipidus. *Endocrinology*, **77**, 701–706.
VOLOSCHIN, L. M. AND TRAMEZZANI, J. H. (1973) The neural input of the milk-ejection reflex on the hypothalamus. *Endocrinology*, **92**, 973–983.
WAKERLEY, J. B. AND LINCOLN, D. W. (1973a) Unit activity in the supraoptic nucleus during reflex milk ejection. *J. Endocr.*, **59**, xvi–xvii.
WAKERLEY, J. B. AND LINCOLN, D. W. (1973b) The milk-ejection reflex of the rat: a 20- to 40-fold acceleration in the firing of paraventricular neurones during the release of oxytocin. *J. Endocr.*, **57**, 477–493.
WURTMAN, R. J., AXELROD, J. AND POTTER, L. T. (1964) The disposition of catecholamines in the rat uterus and the effect of drugs and hormones. *J. Pharmacol. exp. Ther.*, **144**, 150–155.
YANNONE, M. E., MUELLER, J. AND OSBORNE, R. H. (1969) Protein binding of progesterone in the peripheral plasma during pregnancy and labour. *Steroids*, **13**, 773–781.
YOSHIMI, T., STROTT, C. A., MARSHALL, J. R. AND LIPSETT, M., (1969) Corpus luteum function in early pregnancy. *J. clin. Endocr.*, **29**, 225–230.
YOSHINAGA, K., HAWKINS, R. A. AND STOCKER, J. F. (1969) Estrogen secretion by the rat ovary *in vivo* during the estrous cycle and pregnancy. *Endocrinology*, **85**, 103–111.
ZAMBRANO, D. AND DE ROBERTIS, E. (1968) The ultrastructural changes in the neurohypophysis after destruction of the paraventricular nuclei in normal and castrated rats. *Z. Zellforsch.*, **88**, 496–510.

DISCUSSION

JOST: In connection with the possible role of the maternal hypothalamus in the initiation of parturition, I should like to ask you a few questions. What is the normal length of pregnancy in the rabbits, and how early can you produce delivery by stimulating the hypothalamus ?

LINCOLN: The normal gestation length of the rabbits, being animals from a local supply of a Californian strain, is slightly over 32 days. The earliest day on which parturition could be induced was 27.5 days, but that was an exception. Twenty-nine days was much more common, and it coincides with the time at which oxytocin itself would have an effect, and with the decline in the progesterone production. These facts fit in very well, but throughout these studies stimulation was always more effective than oxytocin itself — a finding that I cannot explain. Perhaps we are also causing a release of other hormones, perhaps it is only because we are releasing oxytocin in a form or in an amount that is physiological for that animal. Another difference between stimulation and oxytocin is that stimulation induces a very normal pattern of delivery, whereas when you give oxytocin it causes a very traumatic, explosive pattern of delivery with a great deal of hemorrhage.

FORBES: I would like to ask if it is possible to arrange to have a lactating rabbit in late pregnancy so that you could measure intramammary pressure as a direct sort of auto-assay during parturition. This would be possible in the cow, but I don't know if it is possible in the rabbit.

LINCOLN: This has been performed in women. The results, as far as oxytocin release during labor is concerned are, however, rather negative. When you start to apply this procedure to laboratory species, you generally have to apply anesthesia in order to record the intramammary pressure. The anesthesia itself tends to disrupt, however, the normal pattern of parturition, so it will be difficult to obtain results in rabbits using this procedure.

Does the Hypothalamo–Neurohypophyseal System play a role in Gestation Length or the Course of Parturition?

K. BOER, G. J. BOER AND D. F. SWAAB

The Netherlands Central Institute for Brain Research, Amsterdam (The Netherlands)

INTRODUCTION

During parturition the rat hypothalamo–neurohypophyseal system (HNS) shows signs of activation at the level of the supraoptic (SON) and paraventricular nuclei (PVN) (Swaab and Jongkind, 1970) and at the level of the neurohypophysis (Stutinsky, 1957, for reference see Sloper, 1966; Boer *et al.*, 1973). Moreover, parturition is accompanied by a release of oxytocin and vasopressin (for reviews see Heller and Ginsburg, 1966, and Swaab, 1972). Whether this activation of the HNS causes the initiation of parturition, or is itself caused by the process of parturition is still a matter of dispute.

Parturition can be induced by oxytocin infusion (*e.g.* Fuchs and Problete, 1970; Fuchs and Saito, 1971) in rat, or by electrical stimulation of the infundibulum and median eminence (Cross, 1958, Lincoln, 1971) in the rabbit. Such observations, however, do not give much information about the normal activities of the HNS in parturition. In addition, neither mechanical (Gale and McCann, 1961; *c.f.* Fitzpatrick, 1966) nor immunological elimination of HNS function (Kumaresan *et al.*, 1971) gave conclusive evidence that the HNS initiates parturition. Assays suggest that the endogenous blood levels of oxytocin are very low in human during spontaneous labor (Chard *et al.*, 1970; Boyd and Chard, 1973).

In order to test the hypothesis that the HNS is involved in the initiation or further course of parturition, delivery and its distribution during 24-h periods has been observed in rats after various treatments. The HNS was activated by water deprivation, while its inhibition was induced by alcohol administration (see below). Parturition was also studied in the Brattleboro strain of Long–Evans rats, since these animals display a hereditary hypothalamic diabetes insipidus involving an absence as well as an activation of certain HNS functions. The influence of two anesthetics on parturition was also examined. In the first place a review will be presented of the current state of knowledge concerning each of the conditions used in the present investigation.

HNS and alcohol

Alcohol induces a diuresis (Van Dyke and Ames, 1951; Fig. 2) which is supposed to be caused by inhibition of the HNS (Millet *et al.*, 1968; Cobo and Quintero, 1969; Fuchs and Wagner, 1963a, b). On the other hand, a low blood level of alcohol seems to have a stimulatory effect on oxytocin release (Cobo and Quintero, 1969). In support of the action of alcohol on the HNS, changes were found in nuclear size of the neurons of the SON and PVN and in the neurohypophyseal content of neurosecretory material (NSM) (Hirvonen *et al.*, 1966). The parameters used in that study, however, do not allow the conclusion that alcohol inhibits hormone synthesis in the SON or PVN (*e.g.* Swaab, 1970). Preliminary data obtained in our department (Van Leeuwen, unpublished) failed to show any inhibition of neurosecretory activity, as measured by the TPP-ase distribution (Swaab and Jongkind, 1970), following various schedules of alcohol administration.

Effects of ethanol on HNS activity were generally found at blood concentrations which approach those found in alcoholic coma (Cobo and Quintero, 1969). Ethanol might therefore not cause a *specific* inhibition of the HNS (Cobo and Quintero, 1969), as has often been supposed (Millet *et al.*, 1968), but will probably affect other central and peripheral neural systems as well (Eidelberg and Wooley, 1970).

Effects of ethanol upon parturition have been reported for rabbit and human. In rabbit it was possible to postpone delivery for about 30 h, while uterine sensitivity for oxytocin was unimpaired (Fuchs, 1966a). In human ethanol was effective in treatment of premature parturition, possibly owing to its inhibitory effect on uterine activity during labor (Zlatnik and Fuchs, 1972; Bieniarz *et al.*, 1971; Luukkainen *et al.*, 1967). No information is available for the rat in this respect. The dose administered in the present study by means of a stomach tube is reported to cause complete inhibition of milk ejection in rats when given intraperitoneally (Fuchs, 1969), although this is in dispute again (Lincoln, 1973).

HNS and water deprivation

Water deprivation is a strong stimulus for the activation of the SON, PVN and the neurohypophysis, as has been shown by morphological and enzymatic studies (*e.g.* Ortmann, 1951; Zambrano and De Robertis, 1966; Jongkind, 1969; Boudier *et al.*, 1970; Swaab, 1970; Boer *et al.*, 1973).

In the rat, water deprivation induces an increased protein synthesizing capacity in the perikarya of the SON (Norström, 1971) and also an increased transport of NSM along the stalk to the infundibular process (Norström and Sjöstrand, 1972). Neurohypophyseal content of vasopressin and oxytocin decreases during water deprivation (Jones and Pickering, 1969) while the plasma level of vasopressin (Little and Radford, 1964), as well as urinary excretion of vasopressin (Noble and Taylor, 1953) increase. On the basis of these results it would appear that the serum oxytocin level will also be elevated during water deprivation, but so far no direct data are available concerning blood and urine levels of oxytocin after such a treatment.

Jones and Pickering (1969) reported a 30% lower level of both nonapeptides in the neural lobe within 24 h after the start of water deprivation, while after 5 days the content was almost depleted. Moreover, Miller and Moses (1971) observed that, on the very first day of water deprivation, urinary vasopressin excretion in the rat reaches already its highest level, and remains so throughout at least a 4-day period. Therefore water deprivation periods of 1, 2 and 4 days were used in the present study.

HNS and Brattleboro rats

Diabetes insipidus (DI) in the Brattleboro rat is not due to absence of the HNS. Both the magnocellular nuclei and the neurohypophysis are morphologically intact (Sokol and Valtin, 1965; Scott, 1968; Kalimo and Rinne, 1972). Morphological and enzymatic studies showed in fact a highly activated HNS, just like that in normal rats deprived of water (Sokol and Valtin, 1965, 1967; Scott, 1968; Kalimo and Rinne, 1972; Swaab *et al.*, 1973). Together with a lower neurohypophyseal oxytocin content (Valtin *et al.*, 1965) and a higher neurophysin serum level (Cheng *et al.*, 1972) this suggests an increased synthesis and release of HNS products. Although this has not been measured directly, the serum level of oxytocin in the Brattleboro strain has been suggested to be elevated as compared to that in normal Long–Evans rats (Valtin *et al.*, 1965; Cheng *et al.*, 1972). With respect to the characteristics mentioned above, Brattleboro rats heterozygous for DI are intermediate between normal Long–Evans or Wistar rats and homozygous DI animals (Sokol and Valtin, 1965; Cheng *et al.*, 1972; Swaab *et al.*, 1973). The electrical activity of the SON, however, appears to be the same for both heterozygous and homozygous Brattleboro rats (Dyball, 1973).

In one of the early papers on the Brattleboro rats by Valtin and Schroeder (1964), reproductive abnormalities were mentioned for these animals. Although a lessened reproductive capability is well-documented for 5 generations of Brattleboro's by Saul *et al.* (1968), no explanation for this is known at present.

MATERIALS AND METHODS

For the present study virgin female rats weighing approximately 200 g were used: Wistar rats and Brattleboro rats, homo- and heterozygous for diabetes insipidus (DI), all obtained from T.N.O. (Zeist). The Brattleboro's were all of the same age (*ca.* 3 months). The animals were kept in individual cages at 25 °C and exposed to 12 h light daily (from 7 a.m. to 7 p.m.). They received tap water and standard chow (Hope Farms) *ad libitum*. Pregnant animals were obtained by mating females overnight: Wistar, homo- and heterozygous Brattleboro females with corresponding males. The day on which spermatozoa were observed in the morning vaginal smears was called day zero of pregnancy. After pregnancy had been confirmed by palpation, the Wistar rats were divided at random into the experimental groups.

Twenty-four hour observation of parturition was started at the latest by 9 a.m. on day 21. Red light was used during the dark period (*c.f.* König and Martin, 1968). For each rat the exact time of each delivery of a pup was noted, the pups were removed and weighed. Observations ended with the last delivery.

The first series of experiments included 4 groups of 8 rats each: one control group

References p. 317–319

(C1), two water-deprived groups (W4D and W2D, that did not receive water respectively from midday of day 17 or day 19 until the last pup was born), and one group that received ethanol (10 ml 10% v/v in tap water) every 12 h by a stomach tube as long as no pup was delivered, starting at 9 a.m. on day 21 (ALC 12/9). Beside these Wistar groups, the first series included the two Brattleboro groups: 6 rats homozygous for DI (HOM DI) and 7 rats heterozygous for DI (HET DI). Whether they were homo- or heterozygous for DI was determined prior to pregnancy by their daily water intake (Swaab *et al.*, 1973).

The second series of experiments, performed approximately one month later, included 4 groups of Wistar rats: a control group of 8 rats (C2), a water-deprived group of 7 rats (from midday of day 20 until the last pup was born; W1D), and 2 groups of 7 rats that received ethanol (10 ml 10% v/v in tap water) every 6 (ALC 6/3) or 12 h (ALC 12/3) for as long as no pup was born. Ethanol was administered by means of a stomach tube, beginning at 3 a.m. of day 21.

In addition to these experiments, two groups (each consisting of 5 female Wistar rats) received, at 9 a.m. on day 21 of pregnancy, a single s.c. injection of either urethane (UR) (1.2 g/kg) or sodium pentobarbital (PENT) (0.12 g/kg). Another group of 6 Wistar rats served as control (C3). Here the observations were made at 2-h intervals. Urine output per hour was measured, using metabolism cages (Jongkind, 1964), in 5 non-pregnant female Wistar rats which received ethanol (10 ml 10% v/v in tap water) twice, by means of a stomach tube, at an interval of 6 h. Whenever the amount of urine was adequate, the osmolality of the urinary sample was determined by means of freeze-point determination (Knauer Type M).

Statistical differences between the various observations were tested by the Student-*t*-test (De Jonge, 1963). A level of $P < 0.05$ was considered to be statistically significant.

RESULTS

Time distribution (Fig. 1)

The distribution of the onset of parturition in time is given for the two control groups (C1) and (C2). Thirteen of the 16 animals started delivery between 0.50 and 6.30 p.m. on day 21 (Fig. 1).

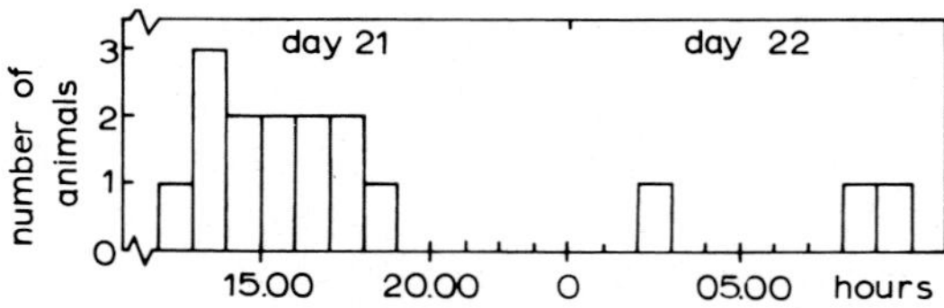

Fig. 1. Distribution of the onset of parturition in the Wistar controls (C1 and C2) over the 24 h of a day.

Alcohol (Figs. 2, 3 and Table I)

After administration of 10 ml 10% ethanol by stomach tube, an increased diuresis starts during the second hour and declines within the following 2 h. The urine osmolality gave a reciprocal picture (Fig. 2). Repeating the ingestion of alcohol 6 h after the first one, the rats reached a level of anesthesia in which they did not react to painful stimuli such as foot-pinching.

Only in the alcohol group ALC 12/3 a significant increase in gestation length was observed. This change was also significant as compared to animals from the ALC 6/3

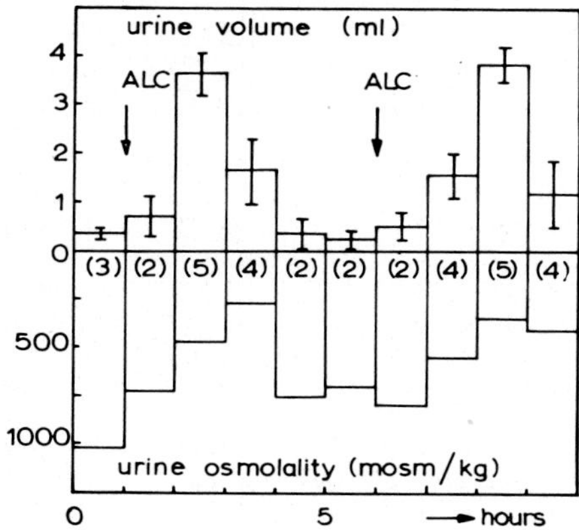

Fig. 2. Urine production and osmolality of Wistar females after ethanol ingestions (ALC: 10 ml 10% in tap water). The first volume was given at 09.00 a.m. Vertical lines indicate ± S.E.M. (5 animals). The number of the animals that produced enough urine per hour for osmolality measurement is given between brackets.

TABLE I

BIRTH WEIGHT OF PUPS FROM WISTAR RATS ON DIFFERENT SCHEDULES OF ALCOHOL TREATMENT AND DIFFERENT PERIODS OF WATER DEPRIVATION

	*Code**	*Pup weight (g)***	*Change****
Alcohol treatment	C1	5.12 ± 0.05 (89)	
	ALC 12/9	5.17 ± 0.06 (74)	n.s.
	C2	5.15 ± 0.04 (84)	
	ALC 12/3	5.16 ± 0.05 (57)	n.s.
	ALC 6/3	4.80 ± 0.07 (33)	− 7%
Water deprivation	C2	5.15 ± 0.04 (84)	
	W1D	5.02 ± 0.05 (62)	n.s.
	C1	5.12 ± 0.05 (89)	
	W2D	4.73 ± 0.05 (74)	− 8%
	W4D	3.16 ± 0.05 (81)	−38%

* For code see legend Fig. 3.
** Expressed in grams ± S.E.M. (number of pups).
*** Only mentioned if statistically significantly different at the $P < 0.05$ level as compared to their appropriate control (C). n.s. = not significantly different.

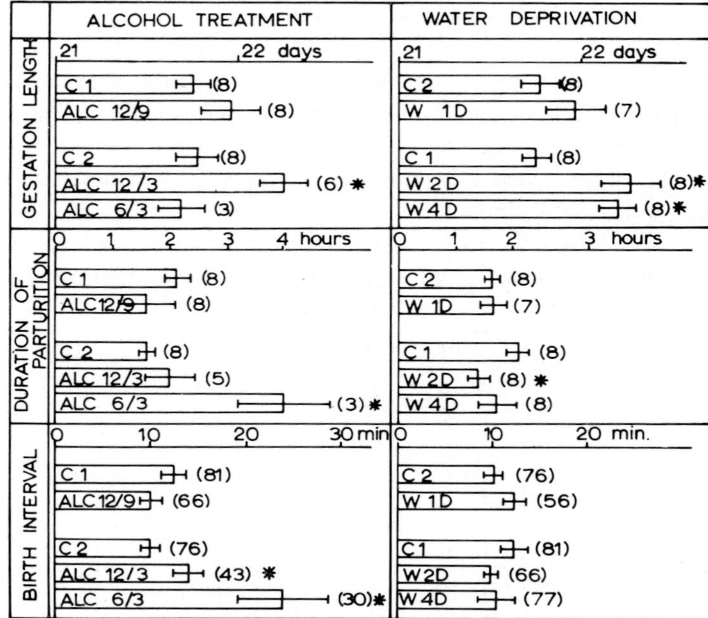

Fig. 3. Gestation length, duration of parturition and interval between births for Wistar rats on different schedules of *alcohol treatment* and different periods of *water deprivation*. Alcohol was given (10 ml 10% v/v by stomach tube) every 12 h starting at day 21, 3 a.m. (ALC 12/3) and 9 a.m. (ALC 12/9) *or* every 6 h starting at day 21, 3 a.m. (ALC 6/3). Water deprivation periods respectively 1 (W1D), 2 (W2D) and 4 days (W4D) before midday of day 21 of pregnancy. Controls in the two series of experiments are C1 and C2. Horizontal lines indicate ± S.E.M. Numbers between brackets are the the numbers of females and number of intervals respectively.
* Difference statistically significant at the $P < 0.05$ level as compared to the controls of the same series.

group ($0.01 < P < 0.02$). Three animals of the ALC 6/3 group delivered before the third alcohol ingestion. Of the other 4 animals in this group which were unconscious, 2 died after 4, and the others after 9 alcohol administrations. Immediate post-mortem examination of the animals showed that all the pups were still alive. In order to calculate gestation length in these cases the time at which the mother died was regarded as the end of gestation. Combining the values for gestation length of all 7 females of the ALC 6/3 group, a mean gestation length of 530.1 h (SEM 5.2) could be calculated, which is not significantly different ($0.10 < P < 0.20$) from the control value of 522.0 h (SEM 2.1).

By administration of alcohol every 6 h (ALC 6/3) the duration of parturition and the birth interval was increased. In the group ALC 12/3, on the other hand, only the birth interval was increased.

Water deprivation (Fig. 3 and Table I)

One day of water deprivation (W1D) prior to the mean delivery time of the control animals had no effect upon any of the parameters tested. Two and 4 days of water deprivation (W2D and W4D) postponed delivery by 12.5 and 11 h respectively. No differences in duration of parturition and birth intervals were seen between the water-deprived and the control groups, except for a shorter duration of parturition in the W2D group. A gradual decrease in pup weight was observed between 1 and 4 days of water deprivation, up to 38 % in the W4D group.

Brattleboro rats (Fig. 4 and Table II)

No difference in gestation length could be observed between heterozygous and homozygous Brattleboro rats. The same holds true for duration of parturition, possibly caused by the great variance in the HOM DI group. The mean birth interval for HOM DI was increased by about 70 % as compared to HET DI.

One homozygous DI female which started parturition failed to deliver a second pup within the following 24 h. Cesarian section at that time revealed only one more still living pup (weight 7.38 g). This animal was excluded from the calculations for mean

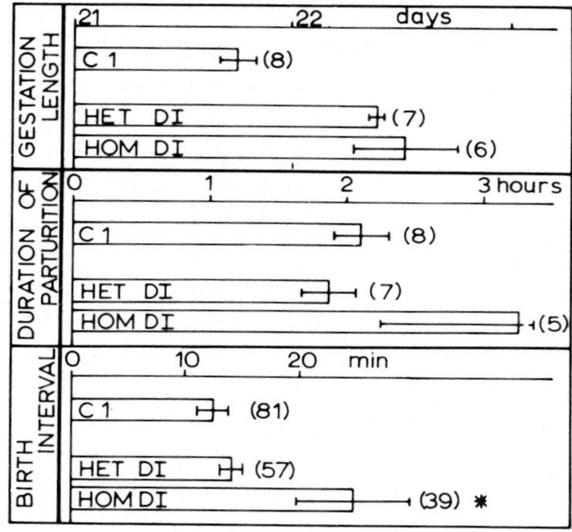

Fig. 4. Gestation length, duration of parturition and interval between births for Wistar (C1) and heterozygous (HET DI) and homozygous (HOM DI) Brattleboro rats. Horizontal lines indicate ± S.E.M. Numbers between brackets are respectively the numbers of females and the intervals between births. * Difference statistically significant at the $P < 0.05$ level as compared to HET DI.

TABLE II

MEAN PUP BIRTH WEIGHT AND LITTER SIZE FOR WISTAR (C1) AND FOR HETEROZYGOUS (HET DI) AND HOMOZYGOUS (HOM DI) BRATTLEBORO RATS

	Pup weight *grams ± S.E.M. (number of pups)*	*Litter size* *± S.E.M. (number of females)*
Wistar (C1)	5.12 ± 0.05 (89)	11.1 ± 0.5 (8)
HET DI	5.75 ± 0.04 (64)	9.1 ± 1.1 (7)
HOM DI	5.89 ± 0.07 (45)	7.7 ± 1.7 (6)

References p. 317–319

duration of parturition and mean birth interval as given in Fig. 3 and Table II. Including these 24 h in the values for HOM DI, the mean duration of parturition was calculated at 6.53 h (SEM 3.59), and the birth interval at 296 min (SEM 29.4). Using these values too, the duration of parturition did not differ significantly from the HET DI level ($0.2 < P < 0.3$, instead of $0.1 < P < 0.2$), while the significance of the birth interval increased ($P < 0.001$, instead of $0.01 < P < 0.02$).

No difference in pup weight was found between the two groups. Litter size was significantly lower for homozygous as compared to heterozygous Brattleboro rats.

Anesthesia (Table III)

No influence upon gestation length was caused by pentobarbital; urethane in contrast postponed delivery for about 18 h.

TABLE III

INFLUENCE OF SODIUM-PENTOBARBITAL (PENT) AND URETHANE (UR) UPON GESTATION LENGTH AS COMPARED TO CONTROLS (C3)

	Gestation length (expressed in 2 h intervals)
C3	261.2 ± 1.5 (6)*
PENT	261.0 ± 1.3 (5)
UR	270.2 ± 0.9 (5)**, §

* Mean ± S.E.M. (number of animals).
** Including two animals which had not delivered on day 22 at 6 p.m.
§ $P < 0.001$.

DISCUSSION

The concept that the HNS, and in particular oxytocin, is involved in the onset or course of parturition is based upon indirect evidence (Fuchs, 1966a; Fitzpatrick, 1969; Fuchs and Poblete, 1970; Burton and Forsling, 1971; Chard *et al.*, 1971; Fuchs and Saito, 1971; Lincoln, 1971). Until now, even the measurement of oxytocin levels in the blood during parturition (McNeilly *et al.*, 1972; Burton *et al.*, 1972) failed to give a decisive answer in this respect, due to the low levels (Chard *et al.*, 1971) and the intermittent spurt-like release (Gibbens *et al.*, 1972) of oxytocin. The determination of the start of the contraction phase by visual observation, as described by Rosenblatt and Lehrman (1963), appeared to be too inaccurate to rely upon (Fuchs, 1966b). In the present study the onset of the first delivery was therefore chosen for the onset of parturition.

The distribution per 24 h of the parturition in the two combined control groups

(C1 and C2) is closely related to the daily fluctuation of the neurohypophyseal oxytocin content, found in the *male* Wistar rat by König and Martin (1968). Most parturitions occurred at the time of day when these authors had measured a low oxytocin content. Although the relationship is very indirect, this agrees with the possibility of a supposed triggering role of oxytocin with respect to parturition. In contradiction to the expectation, 2 and 4 days of water deprivation did not shorten gestation length, but caused a postponement of the first delivery. Since, however, fetal weight influences the length of gestation (Csapo, 1969), one might be inclined to ascribe this increase in gestation length directly to the decreased fetal weights. This cannot be the only explanation, however, since the mean birth weight of the pups in the 4 day water-deprived group was 33 % lower than that in the 2 day water-deprived group, whereas their mean gestation lengths were not different at all.

Alcohol too caused increased gestation length in some of the schedules used. In spite of a different method used for administration of ethanol and the other species in this study, the results obtained in the present study confirm those of Fuchs (1966a) in the rabbit. Under alcoholic conditions too, however, lower fetal weight might play some role, since the birth weights of pups in the group receiving alcohol every 6 h (ALC 6/3) were lower than normal. This effect of alcohol agrees with the growth-inhibiting action of alcohol in the post-natal period (Ratcliffe, 1972). The group receiving alcohol every 12 h from 3 a.m. on day 21 on (ALC 12/3) had pups of normal weight, but the gestation length was prolonged. This also points to a relatively lower intra-uterine growth rate during the alcohol period as compared to normal.

Although ethanol is claimed to specifically inhibit the HNS (Millet *et al.*, 1968), it definitely has a severe anesthetic effect. Each successive alcohol ingestion further increased the anesthetic level, which resulted in complete unresponsiveness to painful stimuli such as foot-pinching, and finally in the death of 4 out of 7 rats from the group that received alcohol every 6 h. Therefore it was investigated whether anesthesia by pentobarbital (Nembutal®) or urethane was effective in postponing delivery. It appeared that only urethane gave a similar postponement as occurred in the ALC 6/3 group. The ineffectiveness of pentobarbital might be ascribed to the shorter anesthesia as compared to urethane (10 and 27 h respectively if one considers as the end of anesthesia the time that the animals were able to move after handling). Preliminary results obtained from 4 rats revealed that injection of the same amount of pentobarbital, now given at midday of day 21, gave prolonged gestation for about 21 h.

These data suggest that the anesthetic effect of alcohol is an important factor in the prolongation of gestation in this experiment. This is also supported by the fact that the diuresis after ingestion of ethanol (10 ml 10 % in tap water) is only increased for 2 h and the urine is hypotonic only for 1 h. This means that if release of hormones from the neural lobe is necessary for the onset of parturition, its possible inhibition by alcohol is too short to cause a postponement. In the Brattleboro rats no difference was found in gestation length between the rats homo- or heterozygous for DI. There was, however, a striking difference between them and normal Wistars ($P < 0.01$). In order to know whether this difference has any significance, this study needs to be repeated using Brattleboro rats, showing no hereditary DI.

References p. 317–319

An acceleration of the course of parturition was caused only by water deprivation lasting 2 days, 1 and 4 days water deprivation failing to produce any effect. This difference does not seem to be important, however, since it is ruled out when the litter size is included by calculating the mean interval between births. No influence of water deprivation upon the interval between births was found in any of the groups.

In some of the groups ethanol prolonged the duration of parturition, which was even more obvious when the interval between births was calculated. This is in agreement with the idea that oxytocin is needed for the normal course of delivery (Kumaresan *et al.*, 1971). On the other hand, the mothers were rather dull during parturition, which means that some influence of anesthesia cannot be ruled out.

The Brattleboro groups, whether homo- or heterozygous for DI, had the same duration of parturition. The mean interval between births, however, is longer in the homozygous than in the heterozygous group. This difference is accompanied by a decreased litter size in the homozygous group, but no relationship seems to exist between litter size and duration of parturition in this group of animals (unpublished data). The smaller litter size for the homozygous rats confirms the findings of Saul *et al.* (1968). The longer interval between births might be explained by a hypertonia of the uterus, caused by a high level of oxytocin (Poseiro and Noriega Guerra, 1961), which would be expected in the Brattleboro rat (Valtin *et al.*, 1965).

Recapitulating, we must conclude that the results of our study are more in conflict with, than in support of the hypothesis that posterior lobe hormones are important for the onset or course of parturition in rats.

SUMMARY

In the present study the hypothesis was tested that the rat hypothalamo–neurohypophyseal system (HNS) is involved in the initiation or further course of parturition. For this purpose the influence upon length of gestation and duration of parturition following water deprivation, alcohol administration, Nembutal and urethane anesthesia was investigated. All these treatments are presumed to influence HNS activity. For the same reason, gestation length and duration of parturition were studied in rats of the "Brattleboro-strain", both homo- and heterozygous for hypothalamic familiar diabetes insipidus.

None of these treatments resulted in a shortening of gestation length. Alcohol administration in one of the schedules used as well as both anesthetics caused a postponement of delivery. Gestation length in the Brattleboro strain was about one day longer than in the Wistar rats used in the rest of the study. No difference was observed, however, between Brattleboro's which were homo- or heterozygous for diabetes insipidus.

Duration of parturition, when corrected for litter size, was not shortened by any of the experimental situations. The duration was longer in 2 out of the 3 alcohol groups, however. The homozygous Brattleboro's were found to have a longer duration of parturition than the heterozygous ones.

These results are more in conflict with than in favor of the concept that the posterior lobe of the pituitary is important for the onset of parturition in rats.

REFERENCES

BIENIARZ, J., BURD, L., MOTEW, M. AND SCOMMEGNA, A. (1971) Inhibition of uterine contractility in labor. *Amer. J. Obstet. Gynec.*, **111** (7), 874–885.

BOER, G. J., SWAAB, D. F. AND JONGKIND, J. F. (1973) Changes in enzyme activities of the neurohypophysis after influencing the hypothalamo–neurohypophysial system of the rat. *J. Endocr.*, **57**, XVIII.

BOUDIER, J. L., BOUDIER, J. A. ET PICARD, D. (1970) Ultrastructure du lobe postérieur de l'hypophyse du rat et ses modifications au cours de l'excrétion de vasopressine. *Z. Zellforsch.*, **108**, 357–379.

BOYD, N. R. H. AND CHARD, T. (1973) Human urine oxytocin levels during pregnancy and labor. *Amer. J. Obstet. Gynec.*, **115** (6), 827–829.

BURTON, A. M. AND FORSLING, M. L. (1971) Hormone content of the neurohypophysis in foetal, new-born and adult guinea-pigs. *J. Physiol. (Lond.)*, **221**, 6–7P.

BURTON, A. M., CHALLIS, J. R. G. AND ILLINGWORTH, D. V. (1972) Oxytocin in the plasma of conscious guinea-pigs during parturition. *J. Physiol. (Lond.)*, **226**, 94–95P.

CHARD, T., BOYD, N. R. H., FORSLING, M. L., MCNEILLY, A. S. AND LANDON, J. (1970) The development of a radioimmuno-assay for oxytocin: the extraction of oxytocin from plasma, and its measurement during parturition in human and goat blood. *J. Endocr.*, **48**, 223–234.

CHARD, T., HUDSON, C. N., EDWARDS, C. R. W. AND BOYD, N. R. H. (1971) Release of oxytocin and vasopressin by the human foetus during labour. *Nature (Lond.)*, **234** (5328), 352–354.

CHENG, K. W., FRIESEN, H. G. AND MARTIN, J. B. (1972) Neurophysin in rats with hereditary hypothalamic diabetes insipidus (Brattleboro strain). *Endocrinology*, **90** (4), 1055–1063.

COBO, E. AND QUINTERO, C. A. (1969) Milk-ejection and antidiuretic activities under neurohypophyseal inhibition with alcohol and water overload. *Amer. J. Obstet. Gynec.*, **105**, 877–887.

CROSS, B. A. (1958) On the mechanism of labour in the rabbit. *J. Endocr.*, **16**, 261–276.

CSAPO, A. (1969) The four direct regulatory factors of myometrial function. In *Progesterone: Its Regulatory Effect on the Myometrium*, G. E. W. WOLSTENHOLME AND J. KNIGHT (Eds.), Churchill, London, pp. 13–42.

DE JONGE, H. (1963) *Inleiding tot de Medische Statistiek, Vol. II*, Ned. Inst. voor Preventieve Geneeskunde, Leiden, chapter 13.

DYBALL, R. E. J. (1973) Single unit activity in the supraoptic nucleus of Brattleboro rats. *J. Physiol. (Lond.)*, **231**, 39–40P.

EIDELBERG, E. AND WOOLEY, D. F. (1970) Effects of ethyl alcohol upon spinal cord neurones. *Arch. int. Pharmacodyn.*, **185** (2), 388–396.

FITZPATRICK, R. J. (1966) The posterior pituitary gland and the female reproductive tract. In *The Pituitary Gland, Vol. 3*, G. W. HARRIS AND B. T. DONOVAN (Eds.), Butterworths, London, pp. 453–504.

FITZPATRICK, R. J. (1969) *In Progesterone: Its Regulatory Effect on the Myometrium*, G. E. W. WOLSTENHOLME AND J. KNIGHT (Eds.), Churchill, London, p. 172 (discussion).

FUCHS, A. R. (1966a) The inhibitory effect of ethanol on the release of oxytocin during parturition in the rabbit. *J. Endocr.*, **35**, 125–134.

FUCHS, A. R. (1966b) Studies on the control of oxytocin release at parturition in rabbits and rats. *J. Reprod. Fertil.*, **12**, 418.

FUCHS, A. R. (1969) Ethanol and the inhibition of oxytocin release in lactating rats. *Acta Endocr. (Kbh.)*, **62**, 546–554.

FUCHS, A. R. AND PROBLETE, V. F. (1970) Oxytocin and uterine function in pregnant and parturient rats. *Biol. Reprod.*, **2** (2), 387–400.

FUCHS, A. R. AND SAITO, S. (1971) Pituitary oxytocin and vasopressin content of pregnant rats before, during, and after parturition. *Endocrinology*, **88** (3), 574–578.

FUCHS, A. R. AND WAGNER, G. (1963a) Effect of alcohol on release of oxytocin. *Nature (Lond.)*, **196**, 92–94.

FUCHS, A. R. AND WAGNER, G. (1963b) The effect of ethyl alcohol on the release of oxytocin in rabbits. *Acta Endocr. (Kbh.)*, **44**, 593–605.

GALE, C. C. AND MCCANN, S. M. (1961) Hypothalamic control of pituitary gonadotrophins. Impairment in gestation, parturition and milk ejection following hypothalamic lesions. *J. Endocr.*, **22**, 107–117.

GIBBENS, D., BOYD, N. R. H. AND CHARD, T. (1972) Spurt-like release of oxytocin during human labour. *J. Endocr.*, **53**, LIV–LV.

HELLER, H. AND GINSBURG, M. (1966) Secretion, metabolism and fate of the posterior pituitary hormones. In *The Pituitary Gland, Vol. 3*, G. W. HARRIS AND B. T. DONOVAN (Eds.), Butterworths, London, pp. 330–373.

HIRVONEN, J. I., KARLSSON, L. K. J. AND VIRTANEN, K. S. I. (1966) Inhibition of the secretory function of the hypothalamo–hypophyseal system by ethanol in the rat. *Ann. Med. exp. Fenn.*, **44**, 52–57.

JONES, C. W. AND PICKERING, B. T. (1969) Comparison of the effects of water deprivation and sodium chloride inhibition on the hormone content of the neurohypophysis of the rat. *J. Physiol. (Lond.)*, **203**, 449–458.

JONGKIND, J. F. (1964) *De Papilloze Rattenier. (Een Onderzoek naar Bouw en Restfunctie)*. Thesis, University of Amsterdam, Rototype, Amsterdam.

JONGKIND, J. F. (1969) Quantitative histochemistry of hypothalamus, II. Thiamine pyrophosphatase, nucleoside diphosphatase and acid phosphatase in the activated supraoptic nucleus of the rat. *J. Histochem. Cytochem.*, **17** (1), 23–29.

KALIMO, H. AND RINNE, U. (1972) Ultrastructural studies on the hypothalamic neurosecretory neurons of the rat. II. The hypothalamo–neurohypophyseal system in rats with hereditary hypothalamic diabetes insipidus. *Z. Zellforsch.*, **134**, 205–225.

KÖNIG, A. UND MARTIN, B. (1968) Tagesperiodische Schwankungen des neurohypophysären Oxytocingehaltes bei männlichen Wistarratten. *Acta Endocr. (Kbh.)*, **58**, 98–100.

KUMARESAN, P., KAGAN, A. AND CLICK, S. M. (1971) Oxytocin antibody and lactation and parturition in rats. *Nature (Lond.)*, **230** (5294), 468–469.

LINCOLN, D. W. (1971) Labour in the rabbit: effect of electrical stimulation applied to the infundibulum and median eminence. *J. Endocr.*, **50**, 607–618.

LINCOLN, D. W. (1973) Milk ejection during alcohol anaesthesia in the rat. *Nature (Lond.)*, **243**, 227–229.

LITTLE, J. B. AND RADFORD, E. P. JR., (1964) Bio-assay for antidiuretic activity in blood of undisturbed rats. *J. appl. Physiol.*, **19**, 179–186.

LUUKKAINEN, T., VÄISTÖ, L. AND JÄRVINEN, P. A. (1967) The effect of oral intake of ethyl alcohol on the activity of the pregnant human uterus. *Acta obstet. gynec. scand.*, **46**, 486.

MCNEILLY, A. S., MARTIN, M. J., CHARD, T. AND HART, I. C. (1972) Simultaneous release of oxytocin and neurophysin during parturition in the goat. *J. Endocr.*, **52**, 213–214.

MILLER, M. AND MOSES, A. M. (1971) Radioimmunoassay of urinary antidiuretic hormone with application to study of the Brattleboro rat. *Endocrinology*, **88**, 1389–1398.

MILLET, Y. A., TIEFFENBACH, L. ET LEUNG, J. (1968) Influence de l'alcool sur la réponse antidiurétique du noyau supra-optique à la stimulation électrique chez le Cobaye. *C.R. Soc. Biol. (Paris)*, **162** (7), 1437–1442.

NOBLE, R. L. AND TAYLOR, N. B. G. (1953) Antidiuretic substances in human urine after haemorrhage, fainting, dehydration and acceleration. *J. Physiol. (Lond.)*, **122**, 220–237.

NORSTRÖM, A. (1971) A functional study of the hypothalamo–neurohypophyseal system of the rat, with the use of a newly developed method for localized administration of labelled precursor. *Brain Res.*, **28**, 131–142.

NORSTRÖM, A. AND SJÖSTRAND, J. (1972) Effect of salt-loading, thirst and water-loading on transport and turnover of neurohypophyseal proteins of the rat. *J. Endocr.*, **52**, 87–105.

ORTMANN, R. (1951) Ueber experimentelle Veränderungen der Morphologie des Hypophysen-Zwischenhirn-Systems und die Beziehung der sog. 'Gomorisubstanz' zum Adiuretin. *Z. Zellforsch.*, **36**, 92–140.

POSEIRO, J. J. AND NORIEGA GUERRA, L. (1961) Dose-response relationships in uterine effects of oxytocin infusions. In *Oxytocin*, R. CALDEYRO-BARCIA AND H. HELLER (Eds.), Pergamon, Oxford, pp. 158–176.

RATCLIFFE, F. (1972) The effect of chronic ethanol administration on the growth of rats. *Arch. int. Pharmacodyn.*, **197**, 19–30.

ROSENBLATT, J. S. AND LEHRMAN, D. S. (1963) Maternal behavior of the laboratory rat. In *Maternal Behavior in Mammals*, H. L. RHEINGOLD (Ed.), John Wiley, New York, pp. 8–21.

SAUL, G. B., GARRITY, E. B. AND VALTIN, H. (1968) Inherited hypothalamic diabetes insipidus in the Brattleboro strain of rats. *J. Heredity*, **59**, 113–117.

SCOTT, D. E. (1968) Fine structural features of the neural lobe of the hypophysis of the rat with homozygous diabetes insipidus (Brattleboro strain). *Neuroendocrinology*, **3**, 156–176.

SLOPER, J. C. (1966) The experimental and cytopathological investigation of neurosecretion in the hypothalamus and pituitary. In *The Pituitary Gland, Vol. 3*, G. W. HARRIS AND B. T. DONOVAN (Eds.), Butterworths, London, pp. 131–239.

SOKOL, H. W. AND VALTIN, H. (1965) Morphology of the neurosecretory system in rats homozygous and heterozygous for hypothalamic diabetes insipidus (Brattleboro strain). *Endocrinology*, **77**, 692–700.

SOKOL, H. W. AND VALTIN, H. (1967) Evidence for the synthesis of oxytocin and vasopressin in separate neurons. *Nature (Lond.)*, **24**, 314–316.

SWAAB, D. F. (1970) *Factors Influencing Neurosecretory Activity of the Supraoptic and Paraventricular Nuclei in Rat, a Histochemical and Cytochemical Study*. Thesis, University of Amsterdam, Nooy, Purmerend.

SWAAB, D. F. AND JONGKIND, J. F. (1970) The hypothalamic neurosecretory activity during the oestrous cycle, pregnancy, parturition, lactation, and persistent oestrus, and after gonadectomy, in the rat. *Neuroendocrinology*, **6**, 133–145.

SWAAB, D. F. (1972) The hypothalamo-neurohypophysial system and reproduction. In *Topics in Neuroendocrinology, Progr. in Brain Res., Vol. 38*, J. ARIËNS KAPPERS AND J. P. SCHADÉ (Eds.), Elsevier, Amsterdam, pp. 133–145.

SWAAB, D. F., BOER, G. J. AND NOLTEN, J. W. L. (1973) The hypothalamo–neurohypophyseal system (H.N.S.) of the Brattleboro rat. *Acta Endocr. (Kbh.)*, **Suppl. 117**, 80.

VALTIN, H. AND SCHROEDER, H. A. (1964) Familial hypothalamic diabetes insipidus in rats. *Amer. J. Physiol.*, **206** (2), 425–430.

VALTIN, H., SAWYER, W. H. AND SOKOL, H. W. (1965) Neurohypophyseal principles in rats homozygous and heterozygous for hypothalamic diabetes insipidus (Brattleboro strain). *Endocrinology*, **77**, 701–706.

VAN DYKE, H. B. AND AMES, R. G. (1951) Alcohol diuresis. *Acta Endocr. (Kbh.)*, **7**, 110–121.

ZAMBRANO, D. AND DE ROBERTIS, E. (1966) The secretory cycle of supra-optic neurons in the rat. A structural–functional correlation. *Z. Zellforsch.*, **73**, 414–431.

ZLATNIK, F. J. AND FUCHS, F. (1972) A controlled study of ethanol in threatened premature labor. *Amer. J. Obstet. Gynec.*, **112** (5), 610–612.

DISCUSSION

FORBES: The observations on parturition during anesthesia have confused me. You have not mentioned the possibility of an influence of the fetuses being under anesthesia either by one of the anesthetics or by alcohol. Under anesthesia the fetuses are possibly not able to respond in a normal way, that will presumably initiate parturition in the first place.

G. BOER: As Dr. Swaab has demonstrated *(this volume)*, we do not think any more that the rat fetuses have a role in the initiation of parturition. Moreover, in the animals which died after several alcohol injections, immediate autopsy revealed that the pups were alive and in good condition.

LINCOLN: In our colony of Brattleboro rats in Bristol we have also noticed a lot of difficulties in the course of parturition of the homozygous Brattleboro rats. Have you already tried whether vasopressin treatment will restore the normal pattern of parturition ?

G. BOER: No, that is what we intend to do, but it is a lot of work because the 24 h of observation are a powerful stress for the observer.

K. BOER: We want to wait for the ‘normal’ Brattleboro, that means, Long–Evans rats having no diabetes insipidus, since these are in our opinion the only good control animals.

Sexual Differentiation of Behavior in Rat

HUIB VAN DIS AND NANNE E. VAN DE POLL

Netherlands Central Institute for Brain Research, Amsterdam (The Netherlands)

INTRODUCTION

The concept of psychosexual differentiation essentially refers to two intertwined research approaches to sex-related behavior, one originating from studies demonstrating sex differences in various behavior patterns between male and female animals, the other founded in the study of ontogenetic processes in which morphological differentiation has an impact upon the adult male–female differences in behavior.

In most species sexual dimorphism between adult males and females exists not only in physical appearance and physiological mechanisms but also in various aspects of their behavioral repertoire. Some of these male–female differences, for example in aggressive behavior, nest-building behavior or vocal repertoire, have been extensively studied experimentally or in field work.

The process of sexual differentiation has *e.g.* genetic, morphological, neuroendocrine and behavioral aspects. The normal ontogenetic development of an individual into masculine or feminine direction is basically dependent upon the individual's genotype. In a genetical male the primitive gonads will develop in the direction of testes; under the influence of products from the primitive testes the development of the Mullerian duct system is inhibited, and the Wolffian duct system is stimulated to develop into masculine internal genitalia. Fetal testicular androgen production further leads to the development of masculine external genitalia from the genital tubercle, to a tonic hypothalamic–hypophyseal–gonadal axis and differentiates the brain such that it is able to execute typical male behavioral activities in adulthood.

In the genetical female, because of absence of products from the developing testes, normal ontogenetic development will lead to feminine internal and external genitalia, a cyclic functioning hypothalamic–hypophyseal–gonadal axis, and a brain more ready to perform typical feminine behavior patterns in adulthood (Whalen, 1968; Jost, 1970; Dörner, 1974).

The effects of gonadal hormones on physiological processes and behavioral activities can be divided into "activational effects" and "organizational effects" (Young, 1965). The activational effects of gonadal hormones are reflected in the physiological and behavioral changes that take place when adult animals are castrated, ovariectomized or treated with male or female gonadal steroids. Sexual activity in the male rat, for example, diminishes following castration and is restored again following androgen

References p. 328–329

replacement therapy. Ovariectomy of the female rat results in an immediate loss of the cyclic feminine mating behavior, whereas subsequent treatment with estrogen and progesterone induces normal behavioral estrous. The organizational effects of gonadal hormones are reflected in morphological and behavioral changes due to the presence or absence of gonadal hormones during circumscribed "critical periods" in ontogenetic development. These changes are assumed to be irreversible. Some of them, however, are only exhibited in adulthood, when the gonadal hormones exhibit their activational effects upon the involved neural and peripheral tissues.

DIFFERENTIATION OF SEXUAL BEHAVIOR

Male and female mating patterns in the rat

When a receptive female is presented to a sexually active male rat, he first will follow her, sniff her genitals, and then mount her; some of the mountings will be accompanied by vaginal intromissions. After several mounts and intromissions ejaculation occurs and is followed by a post-ejaculatory "refractory" period lasting approximately 5 min, after which the whole sequence is repeated. Frequencies and latencies of mounts, intromissions and ejaculations are the measures generally used to quantify this behavior.

A receptive female, when presented to a sexually active male, will do her best to seduce him. She runs around, makes hopping movements and wiggles her ears. When she is mounted by the male she shows the lordosis posture. If she is insufficiently receptive, on the other hand, the female rat shows the typical "refusal" behavior. The feminine mating pattern is usually quantified by the frequencies of lordosis and refusals in response to a mounting male, and is commonly expressed as lordosis-quotient.

Heterotypical mating patterns

Male and female rats predominantly execute the specific elements of their sex-typical behavior. Under certain circumstances, however, they can exhibit mating patterns typical for the opposite sex.

Intact male rats only rarely show fragments of the female mating pattern. Stone (1924) and Beach (1938, 1945) reported some cases of animals showing this kind of behavior spontaneously. Aren-Engelbrektsson *et al.* (1970) showed that daily administration of estrogen produced low lordosis-quotients in a high percentage of male animals. Davidson and Bloch (1969) found that only large doses of estrogen were able to achieve high percentages of lordosis responses, and only after a latency of several days of treatment. In both studies, progesterone failed to facilitate the occurrence of the female mating pattern.

Heterotypical mating behavior is more commonly shown by female rats than by males (Beach and Rasquin, 1942). Treatment with either estrogen or testosterone moreover stimulates the execution of both mounts and intromission patterns in ovariectomized rats (Södersten, 1972).

Although some authors use the term "homosexual" behavior patterns for the execution of heterotypical elements of sexual behavior (*e.g.* Dörner, 1974), it seems to us better to preserve this word to the sexual preference patterns.

The effect of perinatal hormones upon the organization of adult sexual behavior

It thus appears that both male and female rats possess the potential to show heterotypical as well as homotypical mating patterns. There are, however, clear sex differences in the probability of occurrence and the displayed intensities of the various behavioral elements. In rats the critical developmental period for the emergence of this male–female difference appears to extend from some days before until some days after birth. Manipulation of the androgen levels during this period, either by neonatal castration or by perinatal administration of antiandrogens in the case of male animals or by perinatal androgen-treatment of female animals, has a drastic effect upon the mating behavior patterns displayed in adulthood.

Contemporary research can be briefly summarized as follows: genetically male rats, castrated neonatally and treated with testosterone in adulthood, will exhibit mounting behavior. However, the mating pattern is both qualitatively and quantitatively different from that in male rats castrated in adulthood and subsequently given testosterone-replacement therapy. Thus, few or no intromissions or ejaculations are displayed by neonatally castrated animals despite high mounting rates. When these animals are treated with estrogen, either alone or followed by progesterone, high levels of feminine mating behavior can be observed. Male rats castrated as adults, on the other hand, show only minimal levels of feminine mating behavior following such treatment (Grady *et al.*, 1965; Whalen and Edwards, 1967). Similar changes are seen when antiandrogens are administered perinatally (Neumann *et al.*, 1970).

Neonatal administration of androgens to genetically female rats results in a masculinization of the mating patterns induced by hormone administration in adulthood. Thus, in comparison with control females, neonatally androgenized female rats exhibit less hormone-induced feminine sexual behavior. When treated in adulthood with testosterone, furthermore, neonatally androgenized females show the intromission pattern more frequently than do control females. Neonatal estrogen administration to genetically female rats also has masculinizing effects upon adult behavior (Whalen and Edwards, 1967; Beach, 1968; Södersten, 1973).

Morphological and physiological mechanisms underlying sexual differentiation of behavior

The male–female differences at the behavioral level have their counterparts in qualitative and quantitative morphological and physiological sex differences, each of which could possibly contribute to certain aspects of the observed differences in behavior. The explanation of the differences only in terms of central nervous system mechanisms seems to be a too narrow approach, at the present stage of knowledge, because some

References p. 328–329

peripheral changes could also contribute to these sex differences. In the case of sexual behavior, for instance, the organizational effect of gonadal hormones upon mating patterns could conceivably be due on the one hand to effects upon the development of the penis, pheromonal factors or perception, and on the other to effects upon the central nervous system.

The penis. Following an undisturbed ontogenic development a male rat will have a phallus of normal size and diameter with functional receptors on the glans penis. These penile receptors will atrophy when male rats are castrated in adulthood (Beach and Levinson, 1950). The importance of the stimulation of these receptors during sexual activity for the total display of male sexual behavior has been shown by experiments in which the glans penis was desensitized by local application of anesthetics (Carlsson and Larsson, 1964; Adler and Bermant, 1966). Such anesthesia reduces the occurrence of intromissions and ejaculations, although these animals mount vigorously. The behavioral deficiency resulting from this treatment thus appears to be very much like that in neonatally castrated male rats. It has been shown recently that the penis only develops properly when androgens are present during the critical perinatal period (Beach *et al.*, 1969). The deficient male sexual behavior of neonatally castrated male rats might therefore be supposed to be due simply to these qualitative changes in the development of the penis. The increased levels of feminine sexual behavior exhibited by these neonatally castrated males, however, cannot of course be explained by these changes in peripheral receptors.

The spinal cord. Erection and ejaculation are responses based upon spinal reflexes. Male rats with complete mid-thoracic spinal transection continue to show penile reflexes in response to genital stimulation of the glans (Hart, 1968a). The occurrence of these reflexes is dependent upon the presence of androgen, as shown by castration, by systemic administration of testosterone, and by spinal implantation of crystalline testosterone (Hart, 1967; Hart and Haugen, 1968). Male rats castrated at birth and spinalized as adults show impairment of genital responses following transection, suggesting that testicular neonatal androgens in some way organize neural tissues at the spinal level (Hart, 1968b). This might well contribute to the deficient male mating behavior in neonatally castrated rats. It is equally unlikely, however, that changes at this level can provide an explanation for the increased feminine mating behavior in neonatally castrated male rats.

The brain. The more frequent display of feminine sexual behavior in neonatally castrated male rats is assumed to be due to changes at the diencephalic level of the nervous system. There is some evidence for this assumption available (Nadler, 1968). Supportive evidence stems from preoptic self-stimulation experiments. Self-stimulation was shown to be androgen dependent in male rats, bearing chronic electrodes in the preoptic area (Van de Poll and Van Dis, 1971). Castration or cessation of androgen treatment, on the other hand, resulted in a decline of self-stimulation frequencies, and in some animals even in the total cessation of pedal-pressing. In neonatally castrated male rats, however, presence or absence of androgens does not affect preoptic self-stimulation. This differential effect between neonatally castrated and adult-castrated male rats was found only for self-stimulation with electrodes located in the preoptic area, but not when the electrodes were in the lateral hypothalamic area. These results do not provide conclusive support for the hypothesis that increased feminine sexual behavior in neonatally castrated males is due to changes at the level of the preoptic anterior hypothalamic area. Together with studies using electrical stimulation, lesioning or testosterone-implantation — all of which point towards a role of this area in the regulation of both male and female sexual behavior — the hypothesis nevertheless appears worthwile to explore (Vaughan and Fisher, 1962; Davidson, 1966; Heimer and Larsson, 1966/67; Van Dis and Larsson, 1971; Powers and Valenstein, 1972). Gonadal hormones influence the animals pheromonal production as well as the reactions to the pheromonal qualities of other animals. In rats, the presence of androgens is responsible for the preference that males show for the odor of estrous females. There is some evidence that androgens influence only the preference and not the olfactory acuity (Le Magnen, 1952; Carr and Caul, 1962). In studying heterotypical mating responses, this kind of stimulus-factors in the testing situation are of methodological importance. Sexually highly excited male rats will mount both male and female partners, but females are more readily followed and mounted, quite possibly due to her pheromonal qualities.

A final possible locus of action underlying sexual differentiation of behavior is the action of steroid metabolizing enzymes since sex differences and effects of neonatal castration have been shown for such enzymatic activities. The behavioral relevance is still unknown, however, some hormones may act differently upon male and female central nervous systems because of the modifying effects of such enzymes before or after reaching the brain.

SOME EXPERIMENTS CONCERNING DIFFERENTIATION OF SEXUAL BEHAVIOR

In the following some questions will be raised and some experimental results from our laboratory will be summarized which are of relevance to the concept of differentiation of sexual behavior.

(1) The problem rises of the precise changes in the levels of masculine and feminine behavior in the same animals following differential perinatal androgen manipulation in male and female rats.

(2) Some results will be reported from experiments in which the interdependency of masculine and feminine behavior systems is studied.

(3) Some prospect is given concerning experiments going on currently in our laboratory, which try to identify neuroanatomically and separate the masculine and feminine neural substrate in male rats.

(1) Most studies on sexual differentiation have investigated female mating patterns

References p. 328–329

TABLE I

MASCULINE AND FEMININE SEXUAL BEHAVIOR SCORES OF NEONATALLY CASTRATED MALE RATS, ADULT CASTRATED MALES, NEONATALLY TESTOSTERONE-TREATED ADULT OVARIECTOMIZED FEMALES, AND ADULT OVARIECTOMIZED CONTROL FEMALES FOLLOWING ESTROGEN TREATMENT IN ADULTHOOD

Groups	*N*	*Masculine sexual behavior*			*Feminine sexual behavior*		
		mounting %	*median*	*range*	*lordosis* %	*median*	*range*
Female neon. TP	11	81	10	0–18	72	20	0–80
Female control	12	8	0	0–1	100	100	80–100
Male neon. castr.	10	40	0	0–11	100	100	70–100
Male control	12	50	3	0–21	92	65	0–100

following treatment with estrogen alone or with estrogen and progesterone, and male patterns following androgen treatment. The effect of heterologous hormone treatment is less well studied. In the same rats the effect of 5-day treatment of 50 μg estradiol-benzoate was studied upon both masculine and feminine sexual responses in 4 groups of animals: neonatal and adult castrated males, females treated neonatally with 250 μg testosterone and control females, the latter two spayed in adulthood. Control females and neonatally castrated males exhibited high levels of feminine mating behavior, adult castrated males less, but still substantial, and only low lordosis-quotients were seen in neonatally androgenized females. For the male sexual behavior the mounting activity showed a completely reversed order of these 4 groups (Table I).

It is concluded that the way of reacting upon hormone administration is primarily dependent upon the disposition developed under influence of the perinatal presence or absence of gonadal hormones, and less upon the kind of hormone treatment given in adulthood.

(2) The female mating pattern, hopping, darting, earwiggling and lordosis is only seldom reported in male rats. This is probably due to the implicit assumption in most studies that female mating patterns can be best studied in castrated male rats. The effect of estrogen–progesterone, estrogen–oil, and oil–progesterone treatment was therefore studied in intact male rats. The doses used were 75 μg estradiol-benzoate, 36 h later followed by 1 mg progesterone, 8 h later followed by the behavioral testing. Estrogen–oil and estrogen–progesterone treatment resulted in the display of lordosis, respectively in 56 % and 86 % of the males. Comparison of the mean lordosis-quotients revealed a clear and significant facilitating effect of progesterone in these males.

The relation between masculine and feminine sexual behavior can be adequately studied in male rats in which, as shown above, both masculine and feminine behavior can be investigated. When the scores for female sexual behavior and for male sexual behavior of individual animals are correlated, there does not seem to be any systematic relation between the amount of feminine mating activity (lordosis-quotient) and the amount of masculine mating activity (number of ejaculations in a 30-min test, intromission-latency, ejaculation-latency).

Another way to investigate this question is to eliminate or inactivate the masculine sexual behavior system and observing the consequences for the activity of the feminine sexual behavior system. This can be done, for example, by giving male rats the opportunity to mate during several hours with different females. After reaching 10–15 ejaculations, males will be sexually exhausted and it takes several days before the sexual behavior system is functionally recovered. Using animals ejaculating only once as a control group, sexual exhaustion did not have any effect upon subsequent feminine sexual activity.

Finally, the most drastic way to inactivate the masculine sexual behavior system appeared to be castration. After a few weeks the masculine sexual behavior had disappeared. If these animals are then treated with estrogen and progesterone, they show less feminine sexual activity in comparison with intact males. Summarizing the different approaches, it has to be concluded that there is no particular strong negative or positive relation between the male and female sexual behavior system. Apparently, both systems are functioning to some extent independently in adulthood.

(3) From the evidence presented thusfar, it is clear, that the neural substrate regulating masculine sexual behavior and the neural substrate regulating feminine sexual behavior must be present in males as well as in females, either as a single inseparable system or as two systems. The few studies done in this field suggest the latter. Singer (1968) reported that preoptic lesions in the female rat disrupted testosterone-induced mounting behavior. The preoptic area appears to be essential for the execution of masculine sexual behavior in the male rat (Heimer and Larsson, 1966/67). Anterior hypothalamic lesions, on the other hand, resulted in Singer's study in cessation of the female mating patterns in female animals, while partly damaging the masculine pattern. More recently, Powers and Valenstein (1972) gathered evidence that the preoptic area exerts an inhibitory influence upon pathways mediating estrous behavior in the female rat. Lesions of the preoptic area produced increased feminine sexual behavior. Because of the generally assumed difficulty in inducing female sexual behavior patterns in the intact male, thusfar little attention has been paid to the neural regulation of feminine sexual behavior in male rats. Preliminary results of a study in which radio-frequency lesions are made in the medial preoptic area show that some male rats, not displaying any masculine mating any longer, exhibited unaffected or even increased levels of feminine mating behavior. Darting and hopping movements and even earwiggling were seen in some of the lesioned animals. This study is still in progress. Possibly these results will contribute to understanding of the neural aspects of the differentiation of sexual behavior.

Sex-linked differences in non-sexual behavior patterns are well known in many species and concern many different classes of behavior, *e.g.* eating and drinking behavior, general activity, emotional behaviors, avoidance conditioning, parental activities, aggressive behaviors and many others. The existence of these sex-linked differences, however, does not necessarily imply that they develop according to the same mechanism of differentiation as in the case of sexual behavior. Parallels have been found in the case of some other classes of behavior, *e.g.* in aggressive behavior, but other factors are probably also involved.

Extrapolation to primates, including man, has made some progress in the field of

References p. 328–329

sexual behavior by the work of Money (1971). A comparative approach concerning sex differences in non-sexual behavior has challenging prospects because of the existence of several psychiatric disorders that have incidence ratios strongly in favor of one or the other sex.

REFERENCES

ADLER, N. AND BERMANT, G. (1966) Sexual behavior of male rats; Effects of reduced sensory feedback. *J. physiol. Psychol.*, **61**, 240–243.

AREN-ENGELBREKTSSON, B., LARSSON, K., SÖDERSTEN, P. AND WILHELMSON, M. (1970) The female lordosis pattern induced in male rats by estrogen. *Horm. Behav.*, **1**, 181–188.

BEACH, F. A. (1938) Sex reversals in the mating pattern of the rat. *J. genet. Psychol.*, **52**, 329–344.

BEACH, F. A. (1945) Bisexual mating behavior in the male rat: effects of castration and hormone administration. *Physiol. Zool.*, **18**, 390–402.

BEACH, F. A. (1968) Factors involved in the control of mounting behavior by female mammals. In *Perspective in Reproduction and sexual Behavior*, M. DIAMOND (Ed.), Indiana Univ. Press, Bloomington, pp. 83–131.

BEACH, F. A. AND LEVINSON, G. (1950) Effects of androgen on the glans penis and mating behavior of castrated male rats. *J. exp. Zool.*, **114**, 159–171.

BEACH, F. A., NOBLE, R. G. AND ORNDOFF, R. K. (1969) Effects of perinatal androgen treatment on responses of male rats to gonadal hormones in adulthood. *J. comp. physiol. Psychol.*, **68**, 490–497.

BEACH, F. A. AND RASQUIN, P. (1942) Masculine copulatory behavior in intact and castrated female rats. *Endocrinology*, **31**, 393–409.

CARLSSON, S. G. AND LARSSON, K. (1964) Mating in male rats after local anesthetization of the glans penis. *Z. Tierpsychol.* **21**, 854–856.

CARR, W. J. AND CAUL, W. F. (1962) The effect of castration in rat upon the discrimination of sex odors. *Anim. Behav.*, **10**, 20–27.

DAVIDSON, J. M. (1966) Activation of the male rat's sexual behavior by intracranial implantation of androgen. *Endocrinology*, **79**, 783–794.

DAVIDSON, J. M. AND BLOCH, G. J. (1969) Neuroendocrine aspects of male reproduction. *Biol. Reprod.*, **1**, 67–92.

DÖRNER, G. (1974) Environmental dependent brain organization and neuroendocrine, neurovegetative and neuronal behavioral functions. In *Integrative Hypothalamic Activity*, *Progr. Brain Res.*, *Vol. 41*, Elsevier, Amsterdam, pp. 221–237.

GRADY, K. L., PHOENIX, CH. H. AND YOUNG, W. C. (1965) Role of the developing rat testis in differentiation of the neural tissues mediating behavior. *J. comp. physiol. Psychol.*, **59**, 176–182.

HART, B. L. (1967) Testosterone regulation of sexual reflexes in spinal male rats. *Science*, **155**, 1283–1284.

HART, B. L. (1968a) Sexual reflexes and mating behavior in the male rat. *J. comp. physiol. Psychol.*, **65**, 453–460.

HART, B. L. (1968b) Neonatal castration: Influence on neural organization of sexual reflexes in male rats. *Science*, **160**, 1135–1136.

HART, B. L. AND HAUGEN, C. M. (1968) Activation of sexual reflexes in male rats by spinal implantation of testosterone. *Physiol. Behav.*, **3**, 735–738.

HEIMER, L. AND LARSSON, K. (1966/1967) Impairment of mating behavior in male rats following lesions in the preoptic-anterior hypothalamic continuum. *Brain Res.*, **3**, 248–263.

JOST, A. (1970) Hormonal factors in the sex differentiation of the mammalian foetus. *Phil. Trans. B*, **259**, 119–130.

LE MAGNEN, J. (1952) Les phénomènes olfactosexuels chez le rat blanc. *Arch. Sci. physiol.*, **6**, 295–331.

MONEY, J. (1971) Clinical aspects of prenatal steroidal action on sexually dimorphic behavior. In *Steroid Hormones and Brain Function*, *UCLA Forum Med. Sci. No. 15*, C. N. SAWYER AND R. A. GORSKI (Eds.), Univ. California Press, Los Angeles, Calif., pp. 325–338.

NADLER, R. D. (1968) Masculinization of female rats by intracranial implantation of androgen in infancy. *J. comp. physiol. Psychol.*, **66**, 157–167.

NEUMANN, F., STEINBECK, H. AND HAHN, J. D. (1970) Hormones and brain differentiation. In *The*

Hypothalamus, L. MARTINI, M. MOTTA AND F. FRASCHINI (Eds.), Academic Press, New York, pp. 569–603.

POWERS, B. AND VALENSTEIN, E. S. (1972) Sexual receptivity: facilitation by medial preoptic lesions in female rats. *Science*, **175**, 1003–1005.

SINGER, J. J. (1968) Hypothalamic control of male and female sexual behavior in female rats. *J. comp. physiol. Psychol.*, **66**, 728–742.

SÖDERSTEN, P. (1972) Mounting behavior in the female rat during the estrous cycle, after ovariectomy, and after estrogen or testosterone administration. *Horm. Behav.*, **3**, 307–320.

SÖDERSTEN, P. (1973) Increased mounting behavior in the female rat following a single neonatal injection of testosterone propionate. *Horm. Behav.*, **4**, 1–17.

STONE, C. P. (1924) A note on 'feminine' behavior in adult male rats. *Amer. J. Physiol.*, **68**, 39–41.

VAUGHAN, E. AND FISHER, A. E. (1962) Male sexual behavior induced by intracranial electrical stimulation. *Science*, **137**, 758–760.

VAN DE POLL, N. E. AND VAN DIS, H. (1971) Sexual motivation and medial preoptic self-stimulation in male rats. *Psychon. Sci.*, **25**, 137–138.

VAN DIS, H. AND LARSSON, K. (1971) Induction of sexual arousal in the castrated male rat by intracranial stimulation. *Physiol. Behav.*, **6**, 85–86.

WHALEN, R. E. (1968) Differentiation of the neural mechanisms which control gonadotropin secretion and sexual behavior. In *Perspectives in Reproduction and Sexual Behavior*, M. DIAMOND (Ed.), Indiana Univ. Press, Bloomington, Ind., pp. 303–340.

WHALEN, R. E. AND EDWARDS, D. A. (1967) Hormonal determinants of the development of masculine and feminine behavior in male and female rats. *Anat. Rec.*, **157**, 173–180.

YOUNG, W. C. (1965) The organization of sexual behavior by hormonal action during prenatal and larval periods. In *Sex and Behavior*, F. A. BEACH (Ed.), Wiley, New York, pp. 89–107.

DISCUSSION

DREWETT: Concerning the experiments you did on neonatal androgenization I should like to know how much testosterone you gave to the newborn rats? And what age were the females you did the mating tests with, and in what stage were their ovaries?

VAN DE POLL: We administered 250 μg of testosterone to the newborn female rats. For the mating tests we used receptive females that were primed with estrogen and progesterone. We checked the behavioral receptivity of these females by stroking their backs and looking if ear-wiggling occurred, which is a normal sign of receptivity in female rats.

DONOVAN: I wonder if you can give me the details of your experiments with self-stimulation. It seems to me a very difficult experiment to do, because the testosterone-treated females will have different brain sizes. The stimulation electrodes may therefore not be localized in identical places.

VAN DE POLL: We only did the self-stimulation experiments in male rats and neonatally castrated males. We checked the placement of the electrodes, and they appeared all to be in the medial preoptic area. All adult castrated males showed a decline in self-stimulation frequency after castration or after cessation of testosterone treatment, whereas all neonatally castrated males did not react on cessation of the treatment; this latter group remained self-stimulating in frequencies as they did before.

DONOVAN: The animals you used have not only different sexually differentiated brains, but they have also different experiences of testosterone during their life. To push it to an extreme, we could say that you were referring to sexual differentiation of *e.g.* the liver, which may modify the steroids that were applied.

SALAMAN: I don't see how we can reconcile this idea that the liver is in fact the center that is differentiated, with several studies that have shown that very small, systemically ineffective doses of both estrogen or testosterone, neonatally implanted in the region of the basal hypothalamus can produce all the symptoms of an anovulatory syndrome induced by a larger systemic dose. These experiments point definitely to the brain as the differentiating center for sexual behavior.

Some Functions of Hormones and the Hypothalamus in the Sexual Activity of Primates

J. HERBERT

Department of Anatomy, University of Cambridge, Cambridge (Great Britain)

INTRODUCTION

This paper reviews some recent work on the neuro-humoral control of sexual activity in primates. It is an attempt to consider 3 problems basic to such studies: (1) the conditions necessary to assess the effects of hormones on sexual behavior in primates, (2) the behavioral and somatic "points of action" of each hormone studied, (3) whether hormones exert an appreciable degree of control on the behavior of primates, whether this differs in any fundamental way from non-primates and whether the results obtained from monkeys are applicable to man.

The importance of studying the effects of hormones on behavior in primates relates to intrinsic interest in these large-brained animals, in which, it has been said, hormones may have lost some of their influence (Ford and Beach, 1952) though these substances are well-known to exert a high degree of control on the behavior of non-primates (Young, 1961). If there is an appreciable degree of hormonal control in primates, how different is this in primates and non-primates? Knowledge of the physiological basis of human sexuality is still astonishingly slight, despite the pervasive role of sexuality in human affairs, and the devastating effects of disordered sexual activity on both a person and those around him. It is therefore imperative to develop a model for studying primate sexual behavior experimentally. How far can the results of studies on monkeys be applied to man?

EXPERIMENTAL CONDITIONS

Most of the work to be discussed in this paper concerns investigations on primates carried out in the laboratory. This is not to imply that field work has no contribution to make; indeed, it was the observation on wild and semi-wild groups that sexual interaction varied at different phases of the female's menstrual cycle that first suggested that hormones might play some role in this kind of behavior (Zuckerman, 1932). More recent studies in the field have gone on to show that factors such as the season of the year, composition of the group, an animal's kinship, its habitat and the re-

References p. 346–347

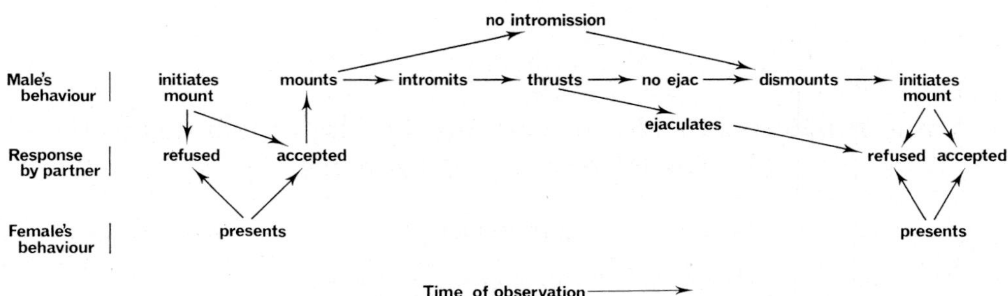

Fig. 1. Diagrammatic representation of part of the mounting sequence of the male rhesus monkey. The behaviors and responses resulting in a mount are shown; several such mounts normally occur before the male ejaculates. A mount can be initiated either by the male (above) or the female (below). Several points during the ensuing interaction are critical in determining the behavioral pattern (*e.g.* accepted/refused; intromits/no intromission; ejaculates/no ejaculation). The female's role is normally limited to determining whether a mount shall, or shall not, be initiated.

lationship between different groups can all have effects on an individual's sexual activity (De Vore, 1965).

But studying the action of hormones requires that all, or as many as possible, of these other variables be kept constant or otherwise accounted for. This is most easily accomplished in the laboratory, though it must also be recognized that removing observations to captivity may significantly alter behavior, as well as the effects of hormones on it. Experiments under laboratory conditions must be designed with prior knowledge of the animal's behavior and social organization in the wild.

Two categories of variable which might intervene between a hormone and its effects are particularly important. An animal's genetic constitution, or its previous humoral or behavioral experience can alter its response to hormones (Goy, 1970). The display of particular behaviors by other animals, or the way in which a social group of animals is organized, can also have powerful effects on the primary interaction between hormones and behavior (De Vore, 1965). In some experiments these factors have been reduced or eliminated as far as possible using particular experimental conditions. In others, they have been deliberately explored as important variables in their own right.

Recent developments in assays and associated procedures have led to great precision in the measurement of hormone levels and the way that these substances affect target tissues. In studies of behavior, the behavioral analysis needs to be of comparable sophistication and sensitivity if we are to understand a hormone's point of action on a behavioral system. For example, it would not be very useful to measure precisely changes in the level and turnover of serum testosterone in male monkeys and then to correlate this with a measurement as crude as, say, the "number of copulations" because so much potential information would have been lost.

In general terms, behavioral interaction between two animals can be subdivided into a number of components, which represent either one animal's action or the other animal's response to it (Fig. 1). Measurements of these two categories can give, of course, very different information. As an example of the first category, we can subdivide the male rhesus monkey's sexual activity into: number of mounts, interval between successive mounts (mounting rate), number of thrusts given during each mount, time taken to ejaculate etc. Each component is not, of course, necessarily independent of the others, though each may have peculiar neurohumoral control mechanisms (see below). Thus, they should be measured separately. As an example of the second, we also quantify the response by the female to the male's behavior; for example, the proportion of his attempts to mount that she allows or refuses. This parameter may be entirely independent of the number of times the male tries to mount; alternatively, in some cases, the female may perhaps allow only a certain number of attempts in a given period, or she may refuse a male who takes a long time to intromit. Similarly we measure the female's behavior (*e.g.* her presentations) as well as the male's response to it (*e.g.* the proportion that incites the male to mount). Complicated as they may seem, these interrelationships and subdivisions are essential if we are to dissect out the precise role of hormones (and other factors) on behavior (see Hinde, 1971 for further discussion). For example, a certain treatment may decrease the frequency with which a male monkey ejaculates. Suitable analysis could show whether this was the result of his mounting rate having decreased to levels providing insufficient stimulation to evoke ejaculation, or whether he was simply failing to intromit as frequently, or thrusting less during each mount, or whether some change had taken place in his nervous system so that an input which had been effective before treatment was now inoperative.

Finally, we have to take into account variables derived from the behavior pattern itself. For example, it would not be legitimate to compare directly the proportion of presentations inducing a male to mount in two sets of observations if we were to ignore possible coincident changes in the time at which the male ejaculates, since ejaculation itself alters his sexual responsiveness to the female (Dixson *et al.*, 1973).

In a behavioral experiment, therefore, a hormone has two points of action; a *somatic* one, which relates to the tissue on which it is acting to produce its effects on behavior, and a *behavioral* one, which relates to the particular component or components of the behavior of this or another animal that is altered in consequence. With this in mind, we can begin to assess the effects of humoral changes in either the female or male monkey upon the behavior of the animal itself, or that of its partner.

HORMONES IN THE FEMALE

Effects on the female's behavior

Observation on a number of primate species, either in the wild or captivity, in groups or in pairs, has shown that sexual interaction between females and males varies during the female's menstrual cycle. Whilst results have varied somewhat according to conditions and methods of measurement, interactions are usually greatest at mid-cycle, and at their lowest point during the luteal phase (Zuckerman, 1932; Ball and Hartman, 1935; Scruton and Herbert, 1970; and many others). It has been frequently pointed out that a female monkey may continue to mate, at least at basal levels,

References p. 346–347

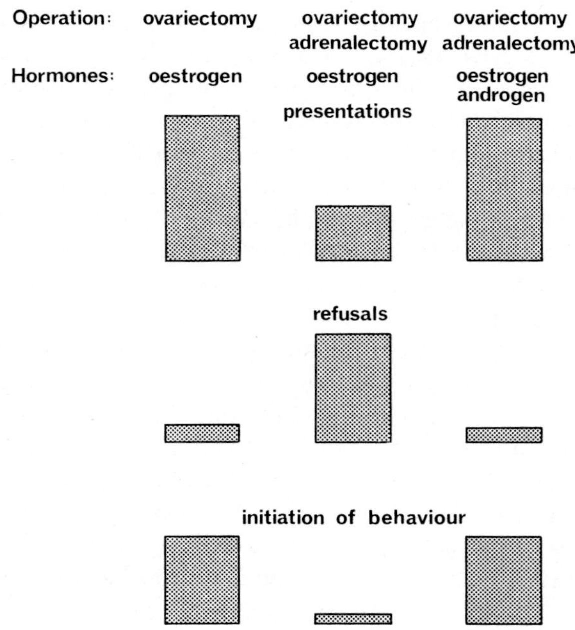

Fig. 2. Diagrammatic representation of the effect of adrenalectomy upon the female monkey's receptivity. Removal of the adrenals results in her presentations decreasing (above), she refuses the male more frequently (middle) and initiates much less sexual activity (below). Giving her androgens (either testosterone or androstenedione) reverses this effect. Estrogen, which is given throughout, is ineffective. Original data in Everitt *et al.*, 1971.

throughout the cycle and is thus distinguished from many sub-primate females, in whom the estrous period is usually sharply circumscribed and, often, rather brief.

More detailed studies on oppositely-sexed pairs of rhesus monkeys have now shown that these changes in behavior are secondary to alterations in two parameters of the female's behavior; her receptivity and her attractiveness.

Receptivity is defined as the willingness of the female to allow the male to mate. This is dependent upon qualities in the male (*e.g.* his sexual attractiveness, dominance etc.) as well as those in the female. It is also not necessarily a single entity. We should distinguish, for example, between behavior by which the female initiates sexual interaction (*e.g.* by "presenting to the male") from that by which she permits behavior initiated by the male (*e.g.* by accepting his attempts to mount her). These two categories of behavior may be controlled by somewhat different mechanisms, though evidence on this point has not yet been firmly established (Dixson *et al.*, 1973).

Females in which androgens have been withdrawn by removing both ovaries and adrenals are sexually unreceptive (Everitt *et al.*, 1971). They initiate very little sexual interaction, and may not allow the males to mount. Their behavior can be restored to normal by giving them either testosterone or androstenedione (Fig. 2). Both these androgens are secreted from the adrenals and, in lesser amounts, from the ovaries, though much of the testosterone present in the female's plasma is derived by conversion of secreted androstenedione by peripheral mechanisms (Osborn and Yannonne, 1971; Baird *et al.*, 1968). Estrogen, cortisol, progesterone, and dehydroepiandrosterone (the third major androgen secreted by the adrenals) are ineffective. Thus, androgen deprivation leads, specifically, to sexual unreceptivity in the female rhesus monkey. Conversely, giving extra testosterone to ovariectomized females may stimulate their presenting behavior and thus they seek to initiate more sexual interaction with the males (Trimble and Herbert, 1968).

The same mechanism probably operates in women. Giving androgen is said to stimulate "libido" which may correspond, at least in part, with sexual receptivity as measured in monkeys (Everitt and Herbert, 1972). Removal of a woman's adrenals and ovaries, operations sometimes carried out for advanced carcinoma, has been reported to depress libido promptly, whereas simply removing the ovaries does not seem to effect a woman's sexuality (Sopchak and Sutherland, 1960). Hence, it has been suggested, estrogen has only a minor role to play in controlling sexuality in women (Money, 1961), a conclusion well supported by the experimental work on monkeys described above. Taken together, these findings suggest that the female primate's sexual receptivity (or "drive") is regulated principally by androgen levels. These levels fluctuate during the cycle, being highest at its mid point, and are depressed by progesterone which is secreted maximally during the luteal phase (Hess and Resko, 1973). Thus, receptivity should also be maximal at mid cycle and lowest during the luteal phase, and be depressed in ovariectomized monkeys and in women by progesterone or related compounds. The available data suggest this to be so. Progesterone given to rhesus monkeys can make them unreceptive (Everitt and Herbert, 1972) and reduction in libido is quite a common complaint in women taking

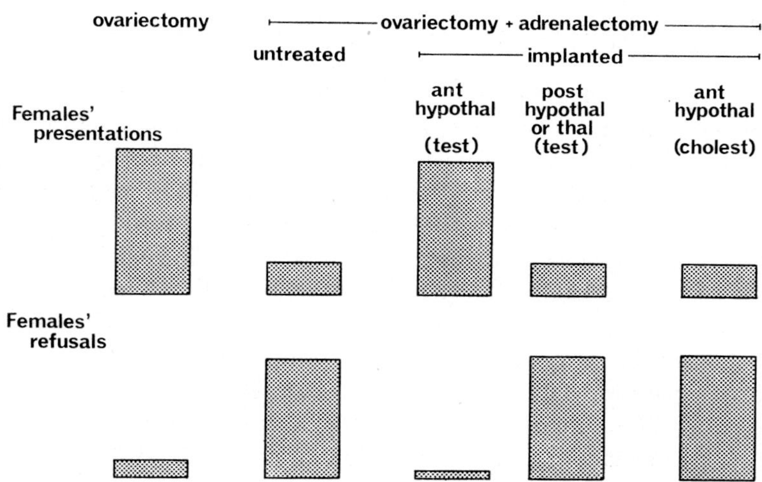

Fig. 3. The effect of intracranial implants of testosterone on sexual receptivity in adrenalectomized female rhesus monkeys. Only implants in the anterior hypothalamus are effective in restoring receptivity (test = testosterone; cholest = cholesterol). Compare with Fig. 2. Original data in Everitt *et al.*, 1974.

References p. 346–347

the contraceptive "pill" and has been correlated with the amount of "progestin" it contains (Grant and Mears, 1967).

These conclusions are clearly different from those in non-primate females, in which sexual receptivity is controlled by estrogen — with the addition of progesterone in the case of some rodents and ungulates (Young, 1961). The results of experiments in primates may explain why their receptivity varies so much less during the cycle than non-primates, since androgens are secreted throughout the cycle in quite large amounts.

Experimental work has established that estrogen affects sexual behavior in non-primates by acting on the hypothalamus. The anterior part of the hypothalamus is the site most frequently implicated, though there are other studies which try to point to more posterior areas (see Davidson, 1972). Does the difference in hormonal control of receptivity between primates and non-primates imply a different site of action?

Experiments in which testosterone-propionate has been implanted into the brain of ovariectomized and adrenalectomized (*i.e.* unreceptive) rhesus monkeys have shown that the anterior hypothalamus may be the site of action of this hormone in controlling receptivity. Extremely small unilateral implants (about 150 μg of hormone) into this area were able to restore these females' receptivity, whereas similar implants elsewhere (*e.g.* the posterior hypothalamus or cerebral cortex) were ineffective, as were cholesterol implants into the anterior hypothalamus (Fig. 3). Androgens therefore can activate receptivity by a direct action on the female's brain. Moreover, the site at which hormones regulate receptivity in primates and non-primates seems similar, even though the effective hormone itself differs (Everitt *et al.*, 1974).

The hypothalamus contains large concentrations of various monoamines. Recently these (particularly 5-hydroxytryptamine — 5HT) have been implicated in the control of sexual behavior in females. For example, giving estrogen-treated rats either reserpine or tetrabenazine — two drugs that deplete monoamines — can activate sexual receptivity by replacing the synergistic action of progesterone on estrogen (Meyerson, 1964). Further studies in which 5HT levels have been reduced by more specific methods have shown that such procedures potentiate estrous behavior (Meyerson and Lewander, 1970; Zemlan *et al.*, 1973). It is thus particularly interesting that the anterior hypothalamus contains high concentrations of 5HT, localized principally within terminals ending in the suprachiasmatic nucleus (Fuxe *et al.*, 1969). Does androgen alter receptivity in monkeys by modifying, in some way, the action of 5HT in this part of the brain? Information available at the moment suggests that it might. Giving adrenalectomized females 75–100 mg/kg *p*-chlorophenylalanine (PCPA), a drug which depletes 5HT, can restore their sexual receptivity, thus mimicking the effects of androgens. If 5HTP, the precursor of 5HT whose formation is prevented by PCPA, is administered together with the latter, then the effect is reversed and the animals become unreceptive again (Fig. 4). Monitoring the levels of 5-hydroxyindole acetic acid (5HIAA), the principal metabolite of 5HT in the female's CSF, suggests that these changes in behavior can be correlated with those in 5HIAA concentration (Fig. 4) (Everitt *et al.*, 1974). Further work in which implants of these and other drugs are made directly into the hypothalamus are necessary before firm conclusions can be

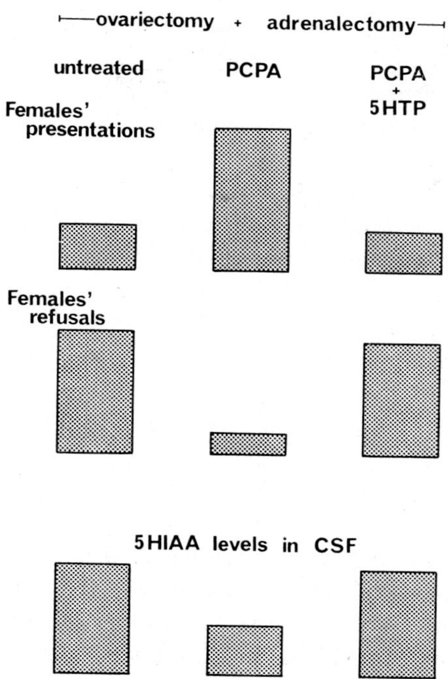

Fig. 4. Diagrammatic representation of the reversal of sexual unreceptivity in adrenalectomized female rhesus monkeys by PCPA. Giving PCPA restores receptivity in such monkeys in a manner comparable with that of androgen (see Fig. 2). Simultaneously administering 5HTP, whose formation is prevented by PCPA, reverses the latter's effect. Below is shown the changes in 5HIAA levels in the CSF after the same treatments. Original data in Everitt *et al.*, 1974.

drawn about the specificity of these results. But assuming for the moment that androgens do act by affecting 5HT in a specific part of the hypothalamus there are a number of further implications. The neurons from which 5HT-containing fibers arise lie in the midbrain raphe (Fuxe *et al.*, 1969). Therefore, androgens must be acting either on the hypothalamic presynaptic terminals (*e.g.* by preventing 5HT-release or storage) or by interfering in some way with postsynaptic mechanisms (*e.g.* by altering receptors, or enzymes that inactivate amines). Preliminary studies suggest that the turnover of 5HIAA in the CSF of female rhesus monkeys is lowered by testosterone (Everitt *et al.*, 1974).

It should not be forgotten that a second and quite different interpretation of these results is possible. Reducing 5HT levels can alter other kinds of behavior — for example aggressive behavior, eating, "emotionality" and sleep (Weissman and Harbert, 1972). It is thus possible that the common factor underlying these changes and those in sexual behavior is a generalized increase in sensitivity to incoming stimuli (*e.g.* the level of the animal's "arousal"). Changes in behavior would depend on the conditions under which behavior was observed. In the experiments described above the effects would be primarily on sexual behavior since a large part of the afferent signals passing between the animals are sexual in nature. Giving male rats electric shocks, which can be classed as a "non-specific" procedure, results in copulation if a female is present but fighting in the presence of a second male (Caggiula, 1972). It is necessary to decide by further experiments whether depleting 5HT in discrete parts of the brain can specifically alter sexual behavior; a preliminary report suggests that this may be so in the rat (Zemlan *et al.*, 1973).

It is also to be noted that estrogen, whose specific role in the regulation of sexual activity in non-primates is hardly in doubt, can also affect a wide variety of other behaviors such as eating, aggression, olfactory, auditory and tactile sensitivity (Vernikos-Danellis, 1972). Thus the fact that 5HT-containing neural systems may be implicated in other behaviors does not, *a priori*, rule out a specific role in controlling sexual receptivity in female primates.

Effect on the male's behavior

Changing androgen levels in the female can affect the behavior of the males. Males paired with unreceptive females commonly show decreased levels of mounting activity, probably because part of the sexual stimuli which they would otherwise receive (*i.e.* the female's presentations) are lacking. Furthermore, the fact that females may refuse their attempts to mount would also be expected to decrease the male's sexual activity (Herbert, 1970; Dixson *et al.*, 1973). In these cases, therefore, changing hormonal levels in the female alters the behavior of the male as a secondary consequence of the hormone's primary effect upon the female's behavior.

Changes in the levels of estrogen and progesterone in the female have a more direct effect on the male's behavior. This is because they alter the female's *attractiveness*, which is defined as her ability to stimulate mounting activity by the male. Attractiveness can be measured in two ways: the proportion of the female's presentations which, in the interval before he ejaculates, causes the male to mount, and by the proportion of

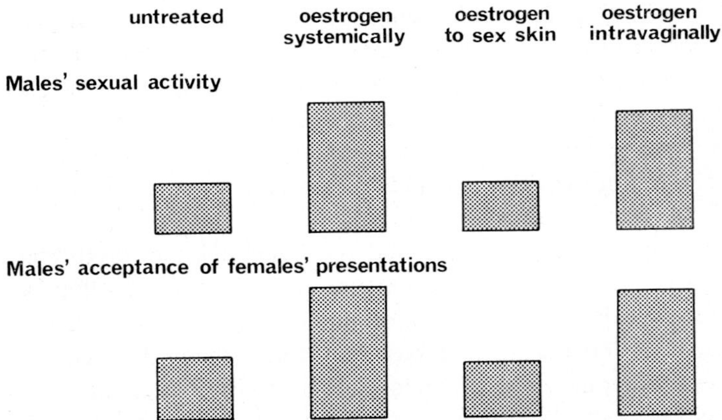

Fig. 5. Diagrammatic representation of the effects on the male rhesus monkey's sexual activity of giving estrogen to the ovariectomized female. Systemic administration stimulates his mounting activity, and this effect can be reproduced by intravaginal estrogen but not by estrogen applied to the female's sexual skin. Original data in Herbert, 1966, 1970.

mounts the male makes without first receiving presentation signals from the female (Herbert, 1970; Dixson *et al.*, 1973). Giving estrogen to the female rhesus monkey increases her sexual attractiveness (Fig. 5). The female's presentations become a more effective stimulus but the frequency with which the signal is given is not directly altered; neither is there a change, or so far as can be determined at the moment, in the way in which they are given. The second effect on the male is to increase markedly the proportion of mounts which he makes without waiting for a presentation signal. Thus, the site of action of estrogen in the female is on a non-behavioral mechanism, and some hormone-dependent stimulus is causing the male to initiate more sexual interaction.

In some species, *e.g.* baboons, chimpanzees and talapoin monkeys, estrogen causes the female's sexual skin (an area of specialized skin surrounding the perineum) to swell. The significance of this phenomenon has been extensively discussed without, however, there being any adequate experimental evidence on its function. However, male talapoin monkeys "look at" the ovariectomized female's sexual skin more often after she has been given estrogen (which causes the skin to swell) (Scruton and Herbert, 1970; Dixson *et al.*, 1973). Changes in color of the skin also occur in some species; for example, the skin in the female rhesus is normally bright red, fades to pale pink after ovariectomy and is restored to its original color by estrogen. Yet applying estrogen

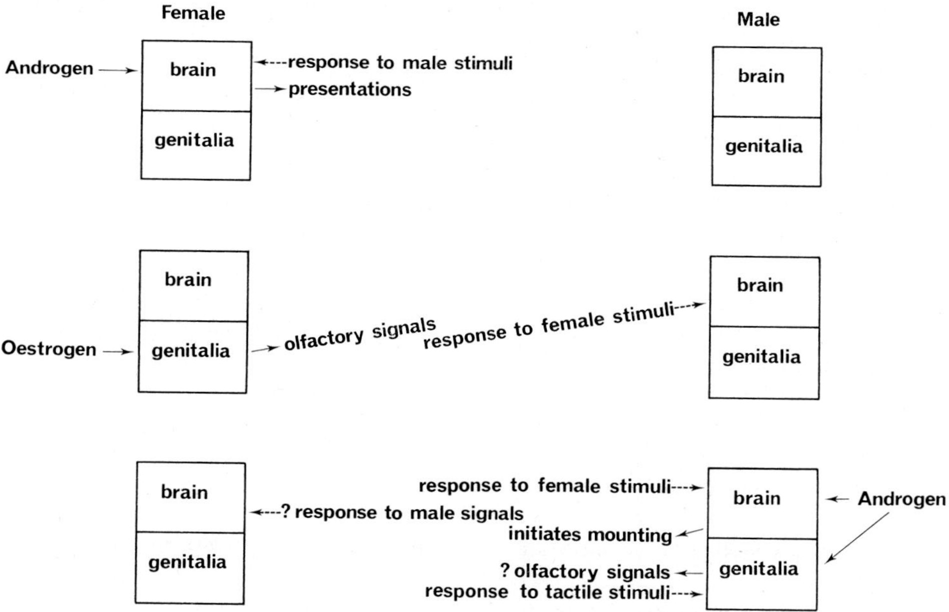

Fig. 6. Somatic and behavioral sites of action of hormones in the sexual behavior of female and male rhesus monkeys. In the female, androgen (above) acts on the female's CNS to cause changes in her behavior. Estrogen (middle) acts on the genitalia causing her to omit olfactory signals which change the behavior of the male. In the male (below) androgen has at least two sites of action, on the brain and on the genitalia, which directly cause changes in his behavior, and the hormone may also alter sensory signals transmitted to the female.

to the sexual skin restores the latter's color without activating the male's sexual behavior (Herbert, 1970; Dixson *et al.*, 1973). Estrogen introduced into the vagina, however, is much more effective, a finding suggesting that an olfactory rather than a visual stimulus is responsible (Herbert, 1966) (Fig. 5). Further work has supported this. Lissak (1962), investigating cats, concluded that valeric acid derived from the vagina acted as an important "pheromone" in sexual interaction. Studies on rhesus monkeys described in a series of preliminary reports indicate that valeric acid, together with other fatty acids such as acetic and butyric, may fulfil a similar role in these animals, and ovariectomized females can apparently be made attractive if these substances are applied to their genitalia (Curtis *et al.*, 1971); the presence of these substances in the vaginal secretions is accentuated by estrogen treatment.

The appropriate hormonal manipulation can thus produce an unattractive but receptive female (ovariectomized but given androgen) or an unreceptive but attractive one (adrenalectomized but given estrogen). The behavioral sites of action of the two sets of hormones are summarized in Fig. 6. Androgen acts directly on the *female's* brain causing both an increase in a particular behavior pattern (presentations) as well as changing the female's response to certain stimuli (not yet well defined) from the male. Estrogen, on the other hand, acts on the female's genitalia, its behavioral point of action being on the *male*, rendering him more responsive to signals (both visual and olfactory) derived from the female. Progesterone may antagonize the effects of both androgen and estrogen and the way that these various hormonal factors may interrelate as hormone levels fluctuate during the cycle is shown in Fig. 7 (see also Everitt and Herbert, 1972; Dixson *et al.*, 1973).

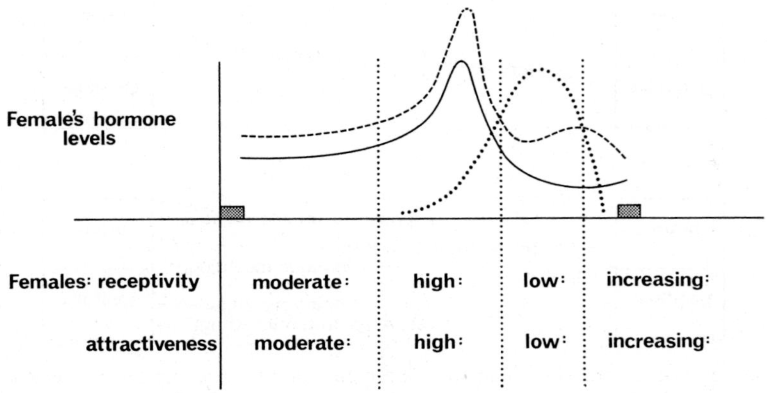

Fig. 7. Showing the fluctuations in hormone levels during the menstrual cycle of the female rhesus monkey (above) and the consequent effect on her sexual receptivity and attractiveness (below) (dashed line = estradiol; uninterrupted line = testosterone; dotted line = progesterone; cross-hatched block = menstruation) (see Everitt and Herbert, 1972; Dixson *et al.*, 1973).

HORMONES IN THE MALE

Effects upon the male's behavior

Until very recently, there was little reliable information on the effects of hormones on the male monkey's behavior. Now there is evidence that castration reduces sexual activity though its effects can be very variable. Some animals seldom ejaculate after operation whilst the behavior of others continues almost unaffected (Phoenix, 1973). As a rule, however, sexual behavior declines gradually after castration, though some degree of mounting activity persists. Erection, mounting and intromission seem to be particularly affected; should the male mount and intromit then the parameters of the ensuing mounting sequence (*e.g.* mounting rate, ejaculating time etc.) seem relatively unaltered (Phoenix, 1973). Treatment with testosterone propionate (1–2 mg/kg) stimulated mounting behavior and frequency of ejaculation in most males, though not always to preoperative levels (Wilson *et al.*, 1972). This may be the result of decreased sensitivity to androgens of neural structures which results from prolonged androgen deficit — a similar phenomenon has been described in birds (Hutchison, 1971). Testosterone is converted peripherally by some target tissues to dihydrotestosterone (DHT). There is considerable agreement that DHT does not restore behavior of castrated male rats (McDonald *et al.*, 1970) and it is interesting that DHT was able to stimulate mounting behavior and ejaculation to some degree in castrated male rhesus monkeys (Phoenix, 1973). It is thus possible that the nature of the central

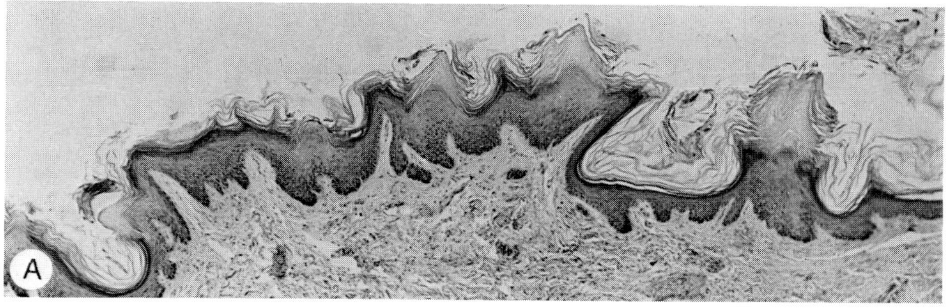

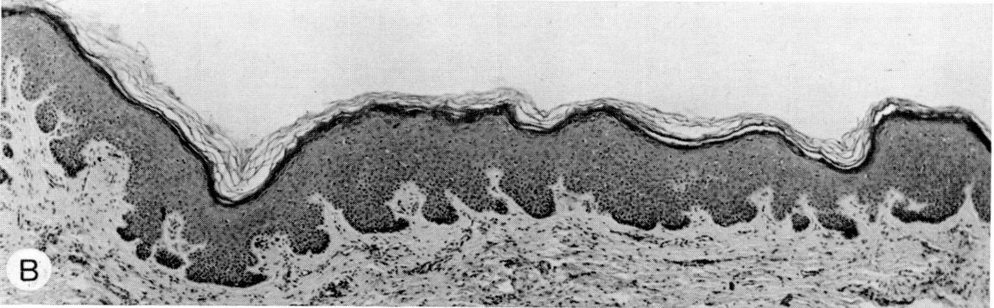

Fig. 8. The effect of castration on the glans penis of the male rhesus monkey. A: the glans from an intact male (above) shows numerous spines, whereas B: in the castrate (below) these are absent.

hormone-sensitive receptors mediating sexual behavior in male primates may differ somewhat from those found in rats.

Since estrogen restores, at least partially, sexual behavior in male rats (Pfaff, 1970), the suggestion has been made that testosterone is converted to estrogen in the rat brain (the enzyme necessary for this to occur has been isolated) and that the latter hormone stimulates behavior (McDonald *et al.*, 1970). Whether this is the case in primates (or even in non-primates) has not been established satisfactorily. The whole question of the behavioral site of action of hormones in male primates (Fig. 6) (as in non-primates) is complicated by the role of the penile spines, which are small outgrowths of the epidermis covering the glans penis, each spine being closely related to underlying receptors in the dermis. The penile spines degenerate after castration in rats, rhesus and talapoin monkeys (Fig. 8). The problem has been to separate the effect on behavior of these peripheral changes (which may affect sensory input from the penis) from those occurring in the male's brain. Thus, it has been reported that a combination of estrogen (presumably stimulating central mechanisms) and DHT (which restores the penile spines) can stimulate the castrated male rat's sexual behavior more effectively than either treatment alone (Larsson *et al.*, 1973).

Sensory inflow through the dorsal nerves of the penis is important for the male monkey's sexual behavior. If these nerves are sectioned then behavior is seriously affected. Certain components of the male's mounting activity are crucially dependent

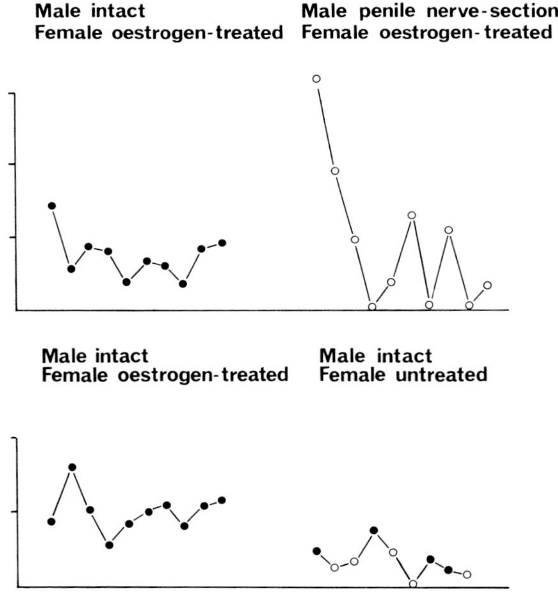

Fig. 9. The effect on the male's sexual activity of either (above) cutting the dorsal nerves of the penis (*i.e.* removing tactile input) or (below) rendering the females unattractive by withdrawing estrogen (*i.e.* removing olfactory input). Both result in a marked decrement in the male's behavior, though this occurs more gradually after nerve section than after estrogen withdrawal. Results from 10 consecutive day tests are shown in each block (● ejaculates; ○ no ejaculation).

upon this afferent input. For example, ejaculation was almost completely abolished after the nerves had been removed entirely, and thrusting became highly incoordinated, though mounting rate and the assumption of the normal mounting position were not directly affected (Herbert, 1973). Furthermore, completely nerve-sectioned males eventually mounted their female partners very little (Fig. 9) perhaps because this activity had become "unrewarding" in the absence of penile sensation, though their behavior could be stimulated for a while if they were paired with "unfamiliar" females. Presumably the latter were emitting novel visual or olfactory stimuli which were sufficiently arousing to overcome initially the lack of penile sensory input.

It is now possible to compare these effects with those of the other afferent inputs from the female monkey. If olfactory stimuli are changed (*e.g.* by giving progesterone to the female) mounting rates are markedly depressed; thrusting rates may also decline though the execution of thrusts is unaltered. Reducing visual stimuli (by preventing the females "presenting") has a rather similar effect, though in some cases the number of thrusts per mount actually increases — as if the male was attempting to concentrate his sexual activity to as few mounts as possible.

Fig. 9 compares the effects on the male's behavior of removing afferent input either through the dorsal nerves (by sectioning them), or through the olfactory pathways (by withdrawing estrogen from the female). In both cases, sexual activity declines and ejaculation becomes less frequent. In the first, however, mounting declines gradually and ejaculation ceases because of the loss of tactile stimuli during thrusting. In the second, the decline is more abrupt, and ejaculation is diminished because the overall level of the male's behavior has decreased to a degree where, as a secondary consequence, sufficient stimulation for ejaculation is not being received during mounting and thrusting.

It must be recognized that the pathways, taken by these 3 inputs within the male, are not yet understood. Visual inputs are probably processed at some stage in the cerebral cortex since this area is likely to be involved in the recognition of the critical features of a "presentation". What these features are has not yet been adequately defined. The problem is how a particular input to the cerebral cortex can activate the male's sexual arousal — possibly by a secondary effect upon a specific region of his hypothalamus.

The specificity of the olfactory input may depend upon the olfactory receptors; there may be certain receptors that are activated by particular odors and that directly excite neural structures (*e.g.* in the hypothalamus) concerned with sexual behavior. Pathways certainly exist between the primary olfactory centers and the hypothalamus (Powell *et al.*, 1965). Alternatively, the same kind of cortical processing may occur more centrally as in the case of visual input. Since it might be possible to train the male monkey to respond sexually to an olfactory (or visual) stimulus which was not, initially, sexually arousing, the latter mechanism may exist irrespective of the former.

Penile input through the dorsal nerve seems, at first sight, to be specific in the sense that this stimulus is chiefly, or exclusively, concerned with sexual arousal. The pathways taken by this source of stimulation are, therefore, of particular interest and must be further investigated.

References p. 346–347

Effect on the female's behavior

It is of the greatest interest to determine whether a female responds differently to a castrate or an intact male monkey; whether she tends to refuse more, or initiate less sexual interaction with an untreated castrate than one receiving testosterone — that is, whether a castrated male is less attractive. Preliminary results from talapoin monkeys suggest that this may be the case. Females present much more to castrated dominant male talapoins if the latter are given testosterone (Dixson *et al.*, 1973), though whether this is because the males are showing more sexual interest in the females or because the hormones induce some change (*e.g.* in the smell or appearance of the male) is not yet known.

The problem about interpreting effects such as these is that a monkey's behavior can produce profound changes in the sexual behavior of another. For example, two females, both receiving estrogen, were paired simultaneously with a male. One was consistently preferred by the male to the other, for reasons unknown. If this favored female was then made less attractive (by giving her progesterone) the second female (now more preferred) presented much more frequently. Since her treatment had remained the same, the change in her presentations could only have been the result of alterations in the behavior of either the male or the former favorite female (Everitt and Herbert, 1969; Herbert, 1970).

Interaction between dominance status and hormones

It is well-known that in some species of primate — particularly those living in large multi-male groups — competition between males has profound effects on mating activity (Crook, 1972). In general, the most dominant male (*i.e.* the male which by other criteria seems to take precedence over others) tends to mate most frequently with the most attractive females, though it is not necessarily the most dominant males that indulge in the most mating activity. Competition between females seems much less pronounced, though it may occur. Even so, different females seem to mate with very different frequencies, which may be correlated with other qualities (*e.g.* their relative attractiveness) rather than their respective dominance status (De Vore, 1965).

Work on talapoin monkeys has shown how powerfully these "social" factors can alter the effects of hormones on behavior. Groups of 8 monkeys, containing 5 females and 3 males, were studied. The males (as well as the females) formed a well-marked dominance hierarchy, the females often outranking the males. An intact male, or a castrate treated with testosterone, only mounted a female if he was the most dominant male of the group, even though measurements of his penile erections and the number of times he "looked at" the females genitalia indicated that his sexual interest seemed to have been stimulated by hormones (Dixson *et al.*, 1972, 1973). If the most dominant males were removed, the next male began mounting and so on; otherwise, despite high doses of testosterone, a subordinate male's mounting activity remained suppressed. The critical stimulus originating from dominant males and responsible for these findings and, in particular, why female talapoins (who frequently dominate males) do not suppress the male's sexual behavior remains to be investigated. It is obviously very important to take into account the interaction between this aspect of aggressive behavior and the effect of hormones. Had the study been limited to subordinate males, it might have been concluded that testosterone exercised little or no control over the mounting behavior of male talapoins: this would, quite clearly, have been incorrect.

EFFECT OF BEHAVIOR ON GONADAL HORMONES

Recent findings have shown that a primate's behavior (or the response to another animal's behavior) can alter his hormone levels. So far this work has been confined to males. In general, it seems that sexual stimuli (or activity) increases testosterone levels, presumably as a consequence of promoting the secretion of ICSH from the pituitary. On the other hand, testosterone is decreased by aggressive stimuli (or fearful responses). Thus, the testes of male rhesus monkeys were reactivated during the non-breeding season, if the females of the group were given estrogen (Vandenburgh, 1969). If a group of males was exposed to the sight and sound of neighboring females, then the males showed testicular reactivation as the females began to come into breeding condition. If the barrier between males and females was made more complete, so that they could not see each other, then the males' testes remained in the non-breeding condition (Gordon and Bernstein, 1973). Testosterone levels may be raised in man by sexual activity or anticipation of it (Anon, 1970).

Conversely, male monkeys in a situation in which they are dominated by others show lowered serum testosterone levels (Rose *et al.*, 1971); removing the more dominant males increases testosterone. Men exposed to fearful or stressful situations have also been found to have reduced serum testosterone (Kreuz *et al.*, 1972).

The best known of the ways in which behavior can affect hormones in non-primate females is the induction of ovulation by coitus in animals such as cats, rabbits and ferrets; it may occur in spontaneous ovulators, such as the rat under appropriate conditions (see Herbert, 1973). Whether or not mating behavior can influence the secretion of LH in the female primate has not yet been demonstrated, though its significance in the context of certain methods of birth control is clearly apparent.

The hypothalamus contains areas that are closely concerned with the secretion of gonadotropins, as well as those implicated in the expression of sexual and aggressive behavior. The evidence discussed briefly above suggests that there may be a closer and more direct functional link between these parts of the hypothalamus than has been suspected.

ACKNOWLEDGEMENTS

Much of the work described here was supported by grants from the Medical Research

Council. I am indebted to my colleagues Drs. A. F. Dixson, B. J. Everitt, P. B. Gradwell and D. M. Scruton who carried out many of these experiments.

REFERENCES

ANON (1970) Effects of sexual activity on beard growth in man. *Nature (Lond.)*, **226**, 869–870.

BALL, J. AND HARTMAN, C. G. (1935) Sexual excitability as related to the menstrual cycle in the monkey. *Amer. J. Obstet. Gynec.*, **29**, 117–119.

BAIRD, D. T., HORTON, R., LONGCOPE, C. AND TAIT, J. F. (1968) Steroid prehormones. *Perspect. Biol. Med.*, **2**, 384–421.

CAGGIULA, A. R. (1972) Shock-elicited copulation and aggression in male rats. *J. comp. physiol. Psychol.*, **80**, 393–397.

CROOK, J. H. (1972) Sexual selection, dimorphism and social organisation in the primates. In *Sexual Selection and the Descent of Man*, B. CAMPBELL (Ed.), Aldine, Chicago, Ill.

CURTIS, R. F., BALLANTINE, J. A., KEVERNE, E. B., BONSALL, R. W. AND MICHAEL, R. P. (1971) Identification of primate sexual pheromones and the proportion of synthetic attractants. *Nature (Lond.)*, **232**, 396–398.

DAVIDSON, J. M. (1972) Hormones and reproductive behavior. In *Hormones and Behavior*, S. LEVINE (Ed.), Academic Press, New York.

DE VORE, I. (Ed.) (1965) *Primate Behavior*, Holt, Rinehart and Winston, New York.

DIXSON, A. F., HERBERT, J. AND RUDD, B. T. (1972) Gonadal hormones and behaviour in captive groups of talapoin monkeys *(Miopithecus talapoin)*. *J. Endocr.*, **57**, xli.

DIXSON, A. F., EVERITT, B. J., HERBERT, J., RUGMAN, S. M. AND SCRUTON, D. M. (1973) Hormonal and other determinants of sexual attractiveness and receptivity in rhesus and talapoin monkeys. In *Primate Reproductive Behavior*, C. PHOENIX (Ed.), Karger, Basel.

EVERITT, B. J. AND HERBERT, J. (1969) The role of ovarian hormones in the sexual preference of rhesus monkeys. *Anim. Behav.*, **17**, 738–746.

EVERITT, B. J. AND HERBERT, J. (1972) Hormonal correlates of sexual behaviour in subhuman primates. *Dan. med. Bull.*, **19**, 246–258.

EVERITT, B. J., GRADWELL, P. AND HERBERT, J. (1974) In preparation.

EVERITT, B. J., HERBERT, J. AND HAMER, J. D. (1971) Sexual receptivity of bilaterally adrenalectomised female rhesus monkeys. *Physiol. Behav.*, **8**, 409–415.

FORD, C. S. AND BEACH, F. A. (1952) *Patterns of Sexual Behaviour*, Methuen, London.

FUXE, K., HÖKFELT, T. AND UNGERTEDT, V. (1969) Distribution of monoamines in the mammalian central nervous system by histochemical studies. In *Metabolism of Amines in the Brain*, G. HARPER (Ed.), Macmillan, London.

GORDON, T. P. AND BERNSTEIN, I. S. (1973) Seasonal variation in sexual behaviour of all male rhesus troops. *Amer. J. Physical Anthrop.*, **38**, 221–226.

GOY, R. W. (1970) Early hormonal influences on the development of sexual and sex-related behaviour. *The Neurosciences*, F. O. SCHMITT (Ed.), Rockefeller Univ. Press, New York.

GRANT, E. C. G. AND MEARS, E. (1967) Mental effects of oral contraceptives. *Lancet*, **ii**, 945.

HERBERT, J. (1966) The effect of oestrogen applied directly to the genitalia upon the sexual attractiveness of the female rhesus monkey. *Excerpta Med. int. cong. Ser.*, **3**, 212.

HERBERT, J. (1970) Hormones and reproductive behavior in rhesus and talapoin monkeys. *J. Reprod. Fertil.*, **Suppl. 11**, 119–140.

HERBERT, J. (1973) The role of the dorsal nerves of the penis in the sexual behavior of the male rhesus monkey. *Physiol. Behav.*, **10**, 293–300.

HESS, D. L. AND RESKO, J. A. (1973) The effects of progesterone on the patterns of testosterone and estradiol concentrations in the systemic plasma of the female rhesus monkey during the intermenstrual period. *Endocrinology*, **92**, 446–453.

HINDE, R. A. (1971) Development of social behaviour. In *Behavior of Non-Human Primates*, A. M. SCHRIER AND F. STOLLNITZ (Eds.), Academic Press, New York, pp. 1–68.

HUTCHISON, J. B. (1971) Effect of hypothalamic implants of gonadal steroids on courtship behaviour in Barbary doves *(Streptopelia risoria)*. *J. Endocr.*, **50**, 97–113.

KREUZ, L. E., ROSE, R. M. AND JENNINGS, J. R. (1972) Suppression of plasma testosterone levels and psychological stress. *Arch. gen. Psychiat.*, **26**, 479–512.

LARSSEN, K., SÖDERSREN, P. AND BERER, C. (1973) Induction of male sexual behaviour by oestradiol benzoate in combination with dehydrotestosterone. *J. Endocr.*, **57**, 563–564.

LISSAK, K. (1962) Olfactory-induced sexual behaviour in female cats. *Excerpta Med. Int. cong. Ser.*, **47**, 653–656.

MCDONALD, P., BEYER, C., NEWTON, F., BRIAN, B., BAKER, R., TAN, H. S., SAMPSON, C., HITCHING, P., GREENHILL, R. AND PRITCHARD, D. (1970) Failure of 5-dihydroxytestosterone to initiate sexual behaviour in the castrated male rat. *Nature (Lond.)*, **227**, 964–965.

MEYERSON, B. J. (1964) Central nervous monoamines and hormone-induced estrus behaviour in the spayed rat. *Acta physiol. scand.*, **62**, Suppl. **241**, 1–32.

MEYERSON, B. J. AND LEWANDER, T. (1970) Serotonin synthesis inhibition and oestrous behaviour in female rats. *Life Sci.*, **9**, 661–671.

MONEY, J. (1961) Sex hormones and human eroticism. In *Sex and Internal Secretions*, W. C. YOUNG (Ed.), Williams and Wilkins, Baltimore, Md.

OSBORN, R. H. AND YANNONNE, M. E. (1971) Plasma androgens in the normal and androgenic female. A review. *Obstet. Gynec. Survey*, **26**, 195–228.

PFAFF, D. (1970) Nature of sex hormone effects on rat sex behaviour; specificity of effects and individual patterns of response. *J. comp. physiol. Psychol.*, **73**, 349–358.

PHOENIX, C. (1973) The role of testosterone in the sexual behaviour of laboratory male rhesus. In *Primate Reproductive Behavior*, C. PHOENIX (Ed.), Karger, Basel.

POWELL, T. P. S., COWAN, W. L. AND RAISMAN, G. (1965) The central olfactory connections. *J. Anat. (Lond.)*, **99**, 791–813.

ROSE, R. M., HOLADAY, J. W. AND BERNSTEIN, I. S. (1971) Plasma testosterone, dominance rank and aggressive behaviour in male rhesus monkeys. *Nature (Lond.)*, **231**, 366–368.

SCRUTON, D. M. AND HERBERT, J. (1970) The menstrual cycle and its effect upon behaviour in the talapoin monkey *(Miopithicus talapoin)*. *J. Zool.*, **162**, 419–436.

SOPCHAK, A. L. AND SUTHERLAND, A. M. (1960) Physiological impact of cancer and its treatment. vii. Exogenous sex hormones and their relation to lifelong adaptations in women with metastatic cancer of the breast. *Cancer*, **13**, 528–531.

TRIMBLE, M. R. AND HERBERT, J. (1968) The effect of testosterone or oestradiol upon the sexual and associated behaviour of the adult female rhesus monkey. *J. Endocr.*, **42**, 171–185.

VERNIKOS-DANELLIS, J. (1972) Effect of hormones on the central nervous system. In *Hormones and Behavior*, S. LEVINE (Ed.), Academic Press, New York.

VANDENBURGH, J. G. (1969) Endocrine coordination in monkeys: male sexual responses to the female. *Physiol. Behav.*, **4**, 261–264.

WEISSMAN, A. AND HARBERT, C. A. (1972) Recent developments relating serotonin to behaviour. *Ann. Rep. Med. Chem.*, **7**, 47–58.

WILSON, M., PLANT, T. M. AND MICHAEL, R. P. (1972) Androgens and the sexual behaviour of male rhesus monkeys. *J. Endocr.*, **52**, ii.

YOUNG, W. C. (1961) The hormones and sexual behavior. In *Sex and Internal Secretions*, W. C. YOUNG (Ed.), Williams and Williams, Baltimore, Md.

ZEMLAN, F. P., WARD, I. L., CROWLEY, W. R. AND MARGULES, D. L. (1973) Activation of lordotic responding in female rats by suppression of serotonergic activity. *Science*, **179**, 1010–1011.

ZUCKERMAN, S. (1932) *The Social Life of Monkeys and Apes*, Kogan Paul, London.

DISCUSSION

VAN DELFT: I would like to ask whether you observe PCPA-effects also in adrenalectomized animals.

HERBERT: We can restore normal sexual behavior by PCPA in adrenalectomized females, which can otherwise only be done by giving androgenic hormones. The animals were adrenalectomized and ovariectomized, and given PCPA in a low dose of 75–100 mg/kg, given once every 4 days. The first observation was made 4 days after the first injection.

DONOVAN: Did you see any side-effects ?

HERBERT: In order to look for effects, other than on sexual behavior, you must set up the right experimental conditions. We have not done this. All we can say is that as far as differences between males and females are concerned, there is no change in aggressive interaction and the animals remain perfectly healthy.

DE WIED: At least from rat studies it is known that removing the adrenal gland will interfere with serotonine synthesis in the brain. I wonder whether some of your results in the adrenalectomized animals would be due to the fact that the metabolism of serotonin is different. Did you measure that ?

HERBERT: The animals received cortisol at a dose of 20 mg/day, which restores the urinary cortisol level to normal, and also a one-monthly injection of DOCA-pivalate. So the animals are essentially normal as regards their corticosteroids.

The Role of DNA, RNA and Protein Synthesis in Sexual Differentiation of the Brain

D. F. SALAMAN

Department of Anatomy, The Medical School, University of Bristol (Great Britain)

INTRODUCTION

The sexual differentiation of gonadotropin secretion and of mating behavior in the rat is known to occur during the first few days of post-natal life under the influence of testicular androgens (Harris, 1964: Barraclough, 1966). Several studies have localized the action of androgen in the brain and more particularly in the hypothalamus (Wagner *et al.*, 1966; Nadler, 1968, 1972; Sutherland and Gorski, 1972) but how the permanent, or almost permanent effects are produced is not yet known. There is increasing evidence that the initial action may involve aromatization of the androgen to estrogen in the hypothalamus (Naftolin *et al.*, 1971, 1972; McDonald and Doughty, 1972) but the prolonged effect remains unexplained. The process of androgenization can be at least partially prevented by the simultaneous administration of several quite different agents; *e.g.* tranquilizers, barbiturates, other steroids (Arai and Gorski, 1968), an antiandrogen (Wollman and Hamilton, 1967), an antiestrogen (McDonald and Doughty, 1972), a mixture of gonadotropins (Sheridan *et al.*, 1973) and certain antibiotics (Kobayashi and Gorski, 1970; Gorski and Shryne, 1972). The last observation suggests an analogy with other processes of hormone-induced differentiation, notably in the mammary gland (Turkington and Kadohama, 1972) and the exocrine pancreas (Wessells, 1964), in which an obligatory round of DNA synthesis and cell division is involved which itself requires some specific protein and RNA synthesis. In addition it is now generally agreed that both estrogens and androgens act on their peripheral target tissues at the level of the genome by an initial activation of RNA and protein synthesis which may be followed by DNA synthesis and cell division (Baulieu *et al.*, 1972).

In view of this analogy we decided to study the effects of selective inhibitors of DNA, RNA, and protein synthesis on the sexual differentiation of the brain induced by testosterone propionate (TP) (Salaman, 1973; Salaman and Birkett, 1974). In addition to this indirect approach, we have made some preliminary studies of RNA polymerase in the hypothalamus during TP treatment and have initiated an autoradiographic study of DNA synthesis by [^{3}H]thymidine incorporation in the brain of the 4-day-old rat.

References p. 360–361

MATERIAL AND METHODS

Female rats were treated on the fourth day of life (day of birth 0). They then weighed about 10 g (range 8–13 g). TP and the inhibitors were injected subcutaneously, the latter in divided doses. The progress of the anovulatory syndrome was followed by daily vaginal smearing for at least 3 weeks prior to death (or ovariectomy) at 80–90 days. Animals were assessed as acyclic if they showed at least 8 consecutive days of cornified smears. The ovaries were removed on the day of estrus from all cyclic rats and were weighed and examined for the presence of recent (functional) corpora lutea while the oviducts were examined for the presence of eggs.

Choice of inhibitors

The inhibitor of DNA synthesis chosen was hydroxyurea (OH-U). This has been shown to produce a total but reversible inhibition of DNA synthesis without interfering with overall RNA synthesis (Schwartz *et al.*, 1965). It may however block the synthesis of the specific messenger RNAs (mRNAs) for histones or histone-like proteins (Kedes and Gross, 1969) which is tightly coupled to DNA replication. Hydroxyurea is sufficiently non-toxic in rodents to be used in high doses, the highest used in this study being 8 mg/10 g pup. Two inhibitors of RNA synthesis were used with different selective actions: α-amanitin (AMAN) a fungal toxin and actinomycin D (ACT.D). AMAN has been shown, *in vitro*, to act selectively on the nucleoplasmic (extranucleolar) RNA polymerase which is responsible for the synthesis of all non-ribosomal RNA including mRNA (Roeder and Rutter, 1970; Tata *et al.*, 1972) *In vivo*, although it acts initially on nucleoplasmic RNA, the inhibition soon spreads to all forms of RNA, probably because of the dependence of ribosomal RNA synthesis on continued production of specific mRNA species (Tata *et al.*, 1972). ACT.D, at the low doses tolerated by the newborn rats in this study, has a selective action on ribosomal RNA (rRNA) synthesis but little effect on nucleoplasmic RNA synthesis (Perry and Kelley, 1970).

The inhibitor of protein synthesis used was puromycin (PURO) which blocks peptide bond formation with the liberation of unfinished polypeptide chains. Cycloheximide was also tested at several doses but was found to be too toxic for use with neonatal rats (all treated animals died within 48 h). In one group of rats 5-fluorouracil (FU) was used. This compound is thought to be incorporated into newly synthesized RNA and thereby to induce errors of translation leading to the synthesis of non-functional proteins. It has been shown to block androgenic stimulation of peripheral tissues (Dorfman, 1963).

RESULTS

Protection against 30 μg of TP

In a preliminary experiment a dose of TP of 30 μg given 4 days postnatally was chosen

TABLE I

EFFECT OF 30 μg OF TP COMBINED WITH INHIBITORS OF DNA, RNA AND PROTEIN SYNTHESIS ON SUBSEQUENT VAGINAL CYCLES AND OVARIAN FUNCTION

Treatment	*Dose of inhibitor*	*Vaginal cycles*		*Ovaries at 80 days*				
	μg/10 g b. w./inj.	*No. acyclic/total*	%	*No. with recent corpora/total*	%	*No. with ova in oviduct/total*	%	*Bilateral wt. mg/100 g ± S.E.M.*
Control	—	0/14**	0	14/14**	100	13/14**	93	42.3 ± 1.7§§§
TP 30 μg	—	7/10	70	2/10	20	1/10	10	24.5 ± 3.2
TP + OH-U	4000	5/10	50	5/10	50	4/10	40	32.4 ± 2.9
TP + OH-U	8000	2/7	29	6/7*	86	3/7	43	38.3 ± 3.1§§§
TP + AMAN	0.3	1/9*	11	8/9**	89	6/9*	67	36.2 ± 2.6§
TP + AMAN	1.5	0/6*	0	6/6**	100	4/6*	67	38.6 ± 1.6§§§
TP + ACT. D	0.3	6/9	67	3/9	33	1/9	11	24.3 ± 2.6
TP + ACT. D	0.6	12/17	71	5/17	29	4/17	25	27.8 ± 3.6
TP + PURO	80	3/8	38	5/8	63	4/8	50	34.6 ± 2.6§
TP + PURO	300†	3/8	38	5/8	63	5/8*	63	36.4 ± 3.9§
TP + FU	80	3/10	30	7/10*	70	6/10*	60	36.8 ± 3.2§

Inhibitors given at 0 and 6 h after TP, except † given at 0, 5 and 10 h after TP.
* $P < 0.05$; ** $P < 0.01$ *vs.* TP alone by Fisher Exact probability test.
§ $P < 0.05$; §§ $P < 0.01$; §§§ P 0.001 *vs.* TP alone by Mann–Whitney U test.

References p. 360–361

TABLE II

EFFECT OF 80 μg OF TP COMBINED WITH INHIBITORS OF DNA, RNA AND PROTEIN SYNTHESIS ON SUBSEQUENT VAGINAL CYCLES AND OVARIAN FUNCTION

Treatment	*Dose of inhibitor*	*Vaginal cycles*		*Ovaries at 80 days*				
	μg/10 g b. w./inj.	*No. acyclic/total*	%	*No. with recent corpora/total*	%	*No. with ova in oviduct/total*	%	*Bilateral wt. mg/100 g ± S.E.M.*
Control	—	0/14**	0	14/14**	100	13/14**	93	42.3 ± 1.7§§§
TP 80 μg	—	20/21	95	2/21	10	0/21	0	16.8 ± 1.1
TP + OH-U	5000	4/13**	31	9/13**	69	6/13**	46	32.7 ± 2.6§§§
TP + AMAN	0.25	2/8**	25	6/8*	75	4/8*	50	29.4 ± 3.4§§
TP + AMAN	1.5	1/9**	11	9/9**	100	4/9*	44	35.5 ± 2.6§§§
TP + ACT. D	0.6	12/13	92	2/13	15	1/13	8	19.1 ± 1.9
TP + PURO	300†	4/6	67	2/6	33	2/6	33	22.2 ± 3.1

Inhibitors given at 0 and 6 h after TP except † given at 0, 5, and 10 h after TP.
* $P < 0.05$; ** $P < 0.01$ *vs.* TP alone by Fisher Exact probability test.
§§ $P < 0.01$; §§§ $P < 0.001$ *vs.* TP alone by Mann–Whitney U test.

as this dose had been shown to be sufficient to produce sterility in 90% of the adults (Kobayashi and Gorski, 1970). Of the animals we treated with this dose, however, only 30% were acyclic at puberty; this proportion rose to 70% by 80 days (the delayed anovulatory syndrome of Gorski (1968)) but at this age 20% of the animals still had functional corpora lutea in their ovaries. This may be due to strain differences between the rats used (Sutherland and Gorski, 1972).

The results of combining this dose of TP with divided doses of the various inhibitors given 0 and 6 h later are shown in Table I. Ovarian weights appeared to be distributed not normally but bimodally and so were treated not by parametric methods but by the non-parametric Mann–Whitney test.

The RNA synthesis inhibitor AMAN at 0.3 μg/10 g and 1.5 μg/10 g provided almost complete protection against androgenization by 30 μg TP ($P < 0.001$). The other inhibitors were less effective but OH-U at 8 μg/10 g gave a significant protection with respect to ovarian morphology ($P < 0.05$) and weight ($P < 0.01$) while PURO and FU gave a similar degree of protection with respect to ovarian weight ($P < 0.05$). Only ACT.D was completely without effect at any dose tolerated by the 4-day-old rats.

All these inhibitors were tested alone in doses equal to the highest used in this study and were found to be without effect on vaginal cycles or ovarian morphology and weight.

Protection against 80 μg of TP

As 30 μg of TP failed to produce complete androgenization in our rats we repeated the experiment using 80 μg of TP, which should be adequate to achieve this (Barraclough, 1966). Only one dose level of the less effective inhibitors was used but AMAN was again tested at two dose levels (0.25 μg/10 g and 1.5 μg/10 g). The results of these treatments are shown in Table II. Acyclicity was seen in 95% of the TP treated animals while only 10% had functional corpora in their ovaries at 90 days.

AMAN was again highly effective in protecting against the androgenizing action of this higher dose of TP. Only 11 and 25% of the animals treated with the low and high dose respectively, became acyclic by 90 days ($P < 0.01$), while 75 and 100% respectively had corpora lutea in their ovaries ($P < 0.05$, $P < 0.01$). The ovaries were markedly heavier than in the group treated with TP alone ($P < 0.001$) and were not significantly lighter than those of oil treated controls. The animals treated with OH-U at 5 mg/10 g also showed clear protection as assessed by vaginal cycles ($P < 0.01$), ovarian function ($P < 0.01$) and ovarian weight ($P < 0.001$). However PURO, even though given in 3 divided doses of 300 μg/10 g at 0, 5 and 10 h after TP, failed to provide significant protection against 80 μg of TP ($P < 0.10$). Once again ACT.D was completely without effect on the development of the anovulatory syndrome.

Protection against 200 μg of TP

In view of the above results we wished to determine whether the differentiation of

References p. 360–361

TABLE III

EFFECT OF 200 μg OF TP COMBINED WITH INHIBITORS OF DNA AND RNA SYNTHESIS ON SUBSEQUENT VAGINAL CYCLES AND OVARIAN FUNCTION

Treatment	*Dose of inhibitor*	*Vaginal cycles*		*Ovaries at 80 days*				
	µg/10 g b. w./inj.	*No. acyclic/total*	%	*No. with recent corpora/total*	%	*No. with ova in oviduct/total*	%	*Bilateral wt. mg/100 g ± S.E.M.*
Control	—	1/16**	6	15/16**	94	13/16**	81	33.6 ± 1.5§§§
TP 200 µg	—	13/13	100	0/13	0	0/13	0	16.1 ± 1.5
TP + OH-U	4000	2/11**	18	8/11**	73	6/11**	55	35.0 ± 2.4§§§
TP + AMAN	1.5	11/15	73	5/15*	33	3/15	20	23.8 ± 2.0§§
OH-U Control	4000	0/8	0	8/8	100	6/8	75	36.6 ± 1.9
AMAN Control	1.5	0/13	0	13/13	100	11/13	85	35.4 ± 1.2

Inhibitors given at 0 and 6 h after TP.
* $P < 0.05$, ** $P < 0.01$ *vs.* TP alone; by Fisher's Exact probability test.
§§ $P < 0.01$, §§§ $P < 0.001$ *vs.* TP alone; by Mann–Whitney test.

sexual behavior, like that of gonadotropin secretion, was influenced by inhibition of DNA or RNA synthesis during the critical exposure to androgen. To test this possibility it was first necessary to establish some protective effect against a higher dose of TP as clear inhibition of lordosis behavior in the ovariectomized, estrogen/progesterone primed female is not obtained with doses of TP below 100 μg on day 4 (Clemens, *et al.*, 1970). We therefore gave groups of female rats 200 μg of TP on day 4 combined with two injections of AMAN or OH-U and followed their vaginal cycles prior to ovariectomy at 90 days. The effects on vaginal cycles and ovarian function are shown in Table III. Considerable protection against androgenization was still achieved with OH-U; the effect is highly significant for vaginal cycles ($P < 0.01$), ovarian function ($P < 0.01$) and ovarian weight ($P < 0.001$). The ovarian weights were indeed equal to those of control females. AMAN was also effective though not as effective as it was against lower doses of TP; it increased significantly the proportion of ovaries with corpora ($P < 0.05$) and the ovarian weight ($P < 0.01$), but the ovaries were nonetheless much smaller than those of the controls ($P < 0.001$). Again the inhibitors alone had no effect on any of these parameters. A study of the female sexual behavior of these animals after priming with estrogen/progesterone is at present under way (Salaman and Drewett, 1974, in preparation).

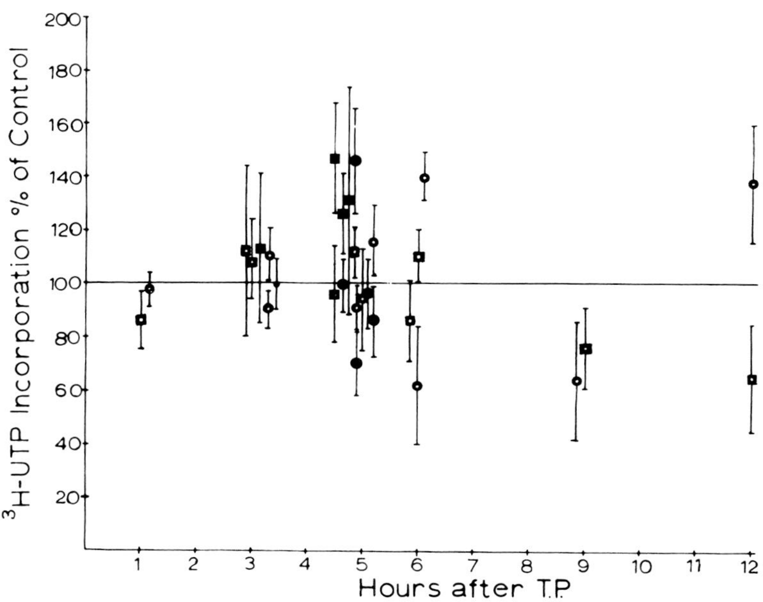

Fig. 1. RNA polymerase activity in hypothalamic cell nuclei after TP treatment to 4-day-old female rats. Activity was expressed as disint/min of [^{3}H]UTP incorporated/10 min/μg DNA and then as a percentage of control incorporation assessed in parallel. Each point is the mean of 4 or 5 determinations and the bars represent S. E. limits. The enzymes were assayed at high ionic strength in the presence of $(NH_4)_2SO_4$ (0.4 *M*) and $MnCl_2$ (3 m*M*) with or without the addition of 2 μg/ml of α-amanitin (Salaman *et al.*, 1972). □, ■ amanitin resistant activity (enzyme A, nucleolar); ○, ●, amanitin sensitive activity (enzyme B, nucleoplasmic). ○, □, after 80 μg TP; ●, ■ after 1 mg TP.

References p. 360–361

Effect of TP on RNA synthesis in the neonatal hypothalamus

In view of the results obtained with an inhibitor of RNA synthesis (AMAN) we decided to measure the rate of RNA synthesis directly by assay of RNA polymerase activity in isolated cell nuclei from the hypothalami of 4-day-old female rats at various times after treatment. Nuclei were isolated by centrifugation through 0.25 *M* sucrose at 800 × *g* and washed once in detergent (10% Triton X-100). The method of assay has been described previously and makes use of α-amanitin to distinguish between the nucleoplasmic (B) and nucleolar (A) polymerases (Salaman *et al.*, 1972). The activity of each enzyme was expressed as a percentage of the control activity in the same experiment and is shown graphically in Fig. 1. There was considerable variation between experiments as evidenced by the scatter of points both above and below the control level, but no reproducible changes were observed over a 12 h period after TP treatment even when a dose of 1 mg of TP was used.

DNA synthesis in the neonatal hypothalamus

Since OH-U was found to be as effective as or more effective than AMAN as an inhibitor of neonatal androgenization it was important to establish that DNA

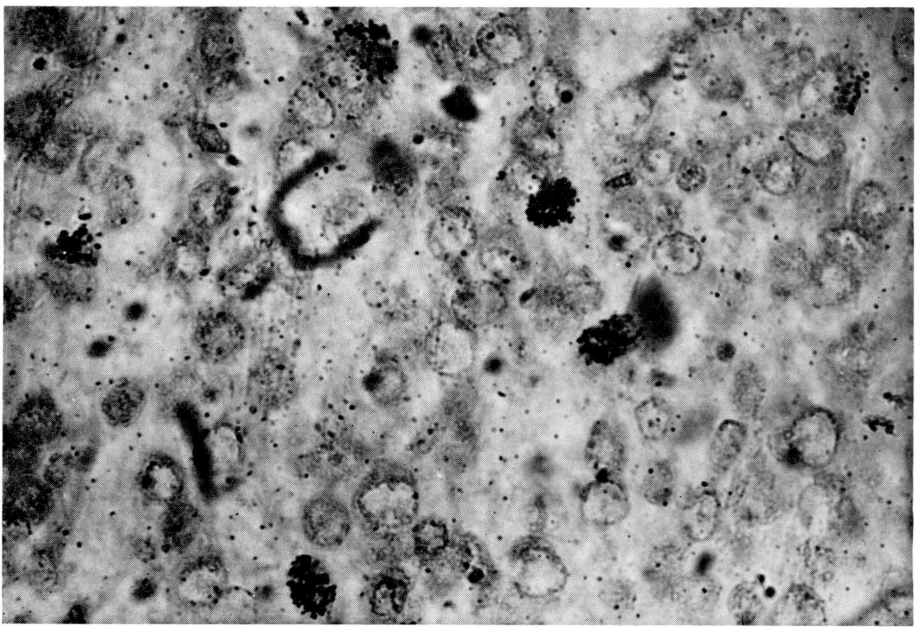

Fig. 2. Autoradiograph of a parasagittal 8 μm section from the medial preoptic area (between anterior commissure and optic chiasma, 0.2 mm from mid-line) of a 4-day-old female rat after [^{3}H]thymidine injection (50 μCi, s.c., 3 h before death). Counterstained with nuclear fast red, exposed for 1 week at 4 °C. Magnification × 800. The photomicrograph was made by Mr. P. F. Heap using a double exposure technique.

synthesis was in fact occurring in the hypothalamus of the 4-day-old rat. We approached this problem by the use of [^{3}H]thymidine autoradiography; [^{3}H]thymidine was injected subcutaneously into the neonatal females 3 h before death, and the brain was then rapidly removed and fixed for histology. The autoradiographs obtained showed clear evidence of thymidine incorporation into the nuclei of cells throughout the brain including all regions of the hypothalamus. A typical high power field in the preoptic area is illustrated in Fig. 2 which shows 6 labeled cells. It appeared that cells of all types were labeled to some extent though there were more examples of labeled glial (astrocyte and oligodendrocyte) nuclei than of neuronal nuclei. A similar distribution of thymidine incorporation into the nuclei of different cell types in the neonatal brain has recently been obtained by biochemical separation methods (Burdman, 1972; Mathias and Stambolova, 1973).

After 3 h of exposure to [^{3}H]thymidine the percentage of cells labeled in the anterior hypothalamus was 1.2%, in the ventromedial/paraventricular region it was 0.95% and in the medial preoptic area, 1.3%. The effect of TP on the number of cells incorporating [^{3}H]thymidine (*i.e.* undergoing DNA replication) in these areas is at present under investigation.

DISCUSSION

It is clear from the results with inhibitors that substances that produce a blockade of DNA synthesis or of nucleoplasmic RNA synthesis can protect the newborn female rat almost completely against the permanent neural organizing effect of neonatal TP. Inhibitors of protein synthesis appear to have a similar but much less marked effect and cannot protect effectively against higher doses of TP. This may have been due to our inability to produce a complete block to protein synthesis in the brain with the doses of puromycin used. With the high dose of puromycin we obtained only 28% inhibition of protein synthesis ([^{35}S]methionine incorporation into protein) in the hypothalamus despite an inhibition of 77% in the liver. Thus the lack of effect of puromycin is still compatible with an obligatory role of protein synthesis in neonatal androgenization which might involve only one or a few specific proteins at the initial stage.

The dependence of sexual differentiation of the brain on DNA synthesis, suggested by these results, is entirely compatible with what is known of the process of differentiation in the mammary gland after initial exposure to insulin and hydrocortisone, and in this tissue an early rise in RNA synthesis precedes the initiation of DNA synthesis which itself is followed by cell division into differentiated daughter cells (Turkingtor and Kadohama, 1972). Histone and histone-like protein synthesis has been shown to be tightly linked to DNA replication, and hydroxyurea is known to block both processes, the former by an action on histone mRNA synthesis (Kedes and Gross, 1969). Our combined data on hydroxyurea and α-amanitin action are thus compatible with an action of TP at the level of the genome, affecting the synthesis of certain histones or histone-like chromatin proteins via an induction of their specific mRNAs.

References p. 360–361

They are equally compatible however, with an obligatory requirement for ongoing DNA synthesis (and presumably cell division) for the activation of the differentiation process. The results of [^{3}H]thymidine incorporation clearly indicate that DNA synthesis is proceeding at a significant rate in the hypothalamus, as elsewhere in the brain, of the 4-day-old rat and that both neuronal and glial cells are involved; this confirms the recent report of Mathias and Stambolova (1973) which was based on biochemical evidence from whole rat brains at 10 days of age. Alternatively a stimulation of DNA synthesis and cell division in certain neurons of the hypothalamus may be implicated but as yet we have no direct evidence of this.

There are also parallels here with the action of estrogen in stimulating growth and cell division in the ovariectomized rat uterus. Here the earliest changes include induction of one or a few RNA species and of a "Key Intermediate Protein" both of which effects are blocked by amanitin (Baulieu *et al.*, 1972). Later nucleolar RNA (rRNA) synthesis is stimulated and only after 24 h does DNA synthesis and cell division commence. However one major difference between steroid action in the uterus and in the brain is apparent. Whereas in the uterus the effects are all reversible and no permanent changes remain, in the brain the changes following neonatal androgen are irreversible and permanent and so may rightly be termed a differentiation. The example of the mammary gland is more relevant to the brain as the differentiation of the alveolar cells is irreversible and cell division (asymmetric cell division according to Turkington and Kadohama (1972)) is an obligatory early step in the differentiation process. It would therefore be worth looking for such a mechanism in the brain by the use of mitotic inhibitors such as colcemid or mitomycin C.

The sexual differentiation of the newborn rat leaves some permanent traces in the brain of the adult. Differences have been observed in the distribution of synapses on the dendrites of neurons in the preoptic area between male and female rats (Raisman and Field, 1971). Differences were also seen in the rate of nuclear RNA synthesis in the anterior hypothalamus between androgenized and normal female rats (Salaman, 1970). Qualitative changes in the species of RNA synthesized by the brain of androgenized rats were reported by Shimada and Gorbman (1970) but this result could not be reproduced in a subsequent experiment from their laboratory when only quantitative differences were found (Namiki *et al.*, 1972).

The results we have obtained, particularly with hydroxyurea, are in apparent conflict with those obtained recently by Barnea and Lindner (1972). These authors were unable to achieve any significant protection against a dose of 100 μg TP by injecting hydroxyurea (1 mg), bromodeoxyuridine or cycloheximide intracerebrally. The failure of cycloheximide to protect against this dose of TP is entirely compatible with our findings with puromycin and 80 μg TP, but against a lower dose of TP (30 μg) some protection has been achieved with intracerebral administration of cycloheximide (Gorski and Shryne, 1972) and systemic administration of puromycin (Table I). Barnea and Lindner (1972) were able to show that hydroxyurea at doses of 1 mg intracerebrally caused almost complete cessation of DNA synthesis for 6 h, but despite this, repeated doses did not influence androgenization. These authors comment that very little is known of the time course of the action of TP and suggest that the effect may be much more delayed than hitherto supposed. This may possibly account for our success with hydroxyurea given systemically in two doses which might have had a longer lasting effect on the brain than smaller doses given intracranially.

It has been suggested that agents that protect against androgenization may act peripherally on the metabolism of the steroid, perhaps in the liver (Sutherland and Gorski, 1972; Barnea and Lindner, 1972), so as to lower the effective dose reaching the brain. This may be true of barbiturates which are known to effect steroid metabolizing enzymes in the liver (Kuntzman, 1969), but it is difficult to envisage a range of inhibitors of DNA, RNA and protein synthesis with very different modes of action and side effects all actively *stimulating* the degradation of TP peripherally. Any other form of metabolism such as aromatization (Naftolin *et al.*, 1971, 1972) which may be involved in the action of neonatal TP must take place in the brain in view of the clear evidence that minute doses, which are ineffective systemically, when administered into the region of the hypothalamus bring about effective androgenization (Wagner *et al.*, 1966; Nadler, 1968, 1972; Sutherland and Gorski, 1972). That aromatization to estrogen may well be an essential first step in the action of TP is indicated by the finding of McDonald and Doughty (1972), that the antiestrogen, MER 25, had a striking protective effect against 30 μg of TP, and is of course consistent with the reports that estrogens given systemically (Barraclough, 1966) or locally to the hypothalamus (Sutherland and Gorski, 1972) are as effective as or more effective than TP itself in producing androgenization. The possibility remains that the inhibitors may be interfering with the aromatization process in the brain rather than with the differentiation process which must follow it. This possibility could be tested by combining the same inhibitors with androgenizing doses of estradiol benzoate. The failure of the study of RNA polymerase activity in the hypothalamus to reveal any reproducible changes after TP treatment does not rule out the possibility (or probability) that selective effects on RNA synthesis are involved. The assay had a considerable intrinsic variability, as seen in Fig. 1, and this would have masked changes of less than $\pm$ 40%. There is a slight elevation of the nucleoplasmic polymerase at 12 h after TP and it is possible that at later times this might have become more marked. However the analogy with the uterus suggests that an early effect on selective RNA synthesis would be undetectable in the assay of total RNA polymerase activity (Baulieu *et al.*, 1972). In conclusion, our results suggest that sexual differentiation of the brain may be closely analogous to the process of differentiation of specific cell types in other adult organs, *e.g.* the mammary gland (Turkington and Kadohama, 1972), the exocrine pancreas (Wessells, 1964) or the erythropoietic system (Harrison *et al.*, 1973).

SUMMARY

Administration of inhibitors of DNA, RNA or protein synthesis systemically to newborn female rats treated with a single dose of testosterone propionate (TP) produced a significant protection against the permanent sterilizing effect of the androgen.

Inhibitors of protein synthesis, puromycin and 5-fluorouracil were effective only

References p. 360–361

against a low (threshold) dose of TP (30 μg). The inhibitor of nucleoplasmic RNA (including mRNA) synthesis, α-amanitin, gave almost complete protection against low and moderate (80 μg) doses of TP and considerable protection against a high dose (200 μg). The inhibitor of DNA synthesis, hydroxyurea, was equally effective against moderate and high doses of TP. None of these inhibitors when given alone had any significant effect on normal sexual development of the female. Direct assay of RNA polymerase in the hypothalamus of the neonatal rat failed to reveal any reproducible changes after TP, but the variability was large. [^{3}H]Thymidine autoradiography established that DNA synthesis persists in the hypothalamus, as elsewhere, in the brain of the 4-day-old rat, and that both neuronal and glial cells are involved.

These results are compatible with a mode of action of androgen at the level of gene transcription initiating a coordinated sequence of selective mRNA and protein synthesis which leads to a round of DNA synthesis and cell division. The hormone may alternatively be acting directly on histone or histone-like mRNA synthesis which would entail a requirement for ongoing DNA synthesis.

ACKNOWLEDGEMENTS

I am particularly grateful to Mrs. Sonya Birkett for her skilled assistance in the early part of this study and Mrs. Diana Roberts in the later part; also to Mrs. Margaret Plaster for the autoradiography and to Dr. R. F. Drewett for carrying out the behavioral study (which is still in progress) and for useful criticism of the manuscript.

This work was supported by a grant from the Nuffield Foundation (No. 10/61A/3).

REFERENCES

ARAI, Y. AND GORSKI, R. A. (1968) Protection against the neural organizing effect of exogenous androgen in the neonatal female rat. *Endocrinology*, **82**, 1005–1009.

BARNEA, A. AND LINDNER, H. R. (1972) Short-term inhibition of macromolecular synthesis and androgen-induced sexual differentiation of the rat brain. *Brain Res.*, **45**, 479–487.

BARRACLOUGH, C. A. (1966) Modifications in the CNS regulation of reproduction after exposure of prepubertal rats to steroid hormones. *Recent Progr. Hormone Res.*, **22**, 503–539.

BAULIEU, E. E., WIRA, C. R., MILGROM, E. AND RAYNAUD-JAMMET, C. (1972) Ribonucleic acid synthesis and oestradiol action in the uterus. *Acta Endocr. (Kbh.)*, Suppl. **168**, 396–415.

BURDMAN, J. (1972) The relationship between DNA synthesis and the synthesis of nuclear proteins in rat brain during development. *J. Neurochem.*, **19**, 1459–1469.

CLEMENS, L. G., SHRYNE, J. AND GORSKI, R. A. (1970) Androgen and development of progesterone responsiveness in male and female rats. *Physiol. Behav.*, **5**, 673–678.

DORFMAN, R. I. (1963) The antiandrogenic activity of 5-fluorouracil. *Steroids*, **2**, 555–561.

GORSKI, R. A. (1968) Influence of age on the response to paranatal administration of a low dose of androgen. *Endocrinology*, **82**, 1001–1004.

GORSKI, R. A. AND SHRYNE, J. (1972) Intracerebral antibiotics and androgenization of the neonatal female rat. *Neuroendocrinology*, **10**, 109–120.

HARRIS, G. W. (1964) Sex hormones, brain development and brain function. *Endocrinology*, **75**, 627–648.

HARRISON, P. R., CONKIE, D. AND PAUL, J. (1973) Role of cell division and nucleic acid synthesis in erythropoiesis-induced maturation of liver cells *in vitro*. In *The Cell Cycle in Development and Differentiation*, M. BALLS AND F. S. BILLETT (Eds.), Cambridge Univ. Press, London, pp. 341–364.

KEDES, L. H. AND GROSS, P. R. (1969) Identification in cleaving embryos of three RNA species serving as templates for the synthesis of nuclear proteins. *Nature (Lond.)*, **223**, 1335–1339.

KOBAYASHI, F. AND GORSKI, R. A. (1970) Effects of antibiotics on androgenization of the neonatal female rat. *Endocrinology*, **86**, 285–289.

KUNTZMAN, R. (1969) Drugs and enzyme action. *Ann. Rev. Pharmacol.*, **9**, 21–36.

McDONALD, P. G. AND DOUGHTY, C. (1972) Inhibition of androgen-sterilization in the female rat by administration of an antioestrogen. *J. Endocr.*, **55**, 455–456.

MATHIAS, A. P. AND STAMBOLOVA, M. (1973) The activities of deoxyribonucleic acid polymerase and other enzymes in the various types of nuclei found in liver and brain tissue. *Biochem. Soc. Trans. (Lond.)*, **1**, 624–626.

NADLER, R. D. (1968) Masculinization of female rats by intracranial implantation of androgen in infancy. *J. comp. physiol. Psychol.*, **66**, 157–167.

NADLER, R. D. (1972) Intrahypothalamic locus for induction of androgen sterilization in neonatal female rats. *Neuroendocrinology*, **9**, 349–357.

NAFTOLIN, F., RYAN, K. J. AND PETRO, Z. (1971) Aromatization of androstenedione by the diencephalon. *J. clin. Endocr.*, **33**, 368–370.

NAFTOLIN, F., RYAN, K. J. AND PETRO, Z. (1972) Aromatization of androstenedione by the anterior hypothalamus of adult male and female rats. *Endocrinology*, **90**, 295–298.

NAMIKI, H., RUCH, W. AND GORBMAN, A. (1972) Further studies of qualitative changes in RNA transcription in brains of female rats given testosterone soon after birth. *Comp. Biochem. Physiol.*, **42** B, 563–568.

PERRY, R. P. AND KELLEY, D. E. (1970) Inhibition of RNA synthesis by actinomycin D: characteristic dose response of different RNA species. *J. Cell Physiol.*, **76**, 127–139.

RAISMAN, G. AND FIELD, P. M. (1971) Sexual dimorphism in the preoptic area of the rat. *Science*, **173**, 731–733.

ROEDER, R. G. AND RUTTER, W. J. (1970) Specific nucleolar and nucleoplasmic RNA polymerases. *Proc. nat. Acad. Sci. (Wash.)*, **65**, 675–682.

SALAMAN, D. F. (1970) RNA synthesis in the rat anterior hypothalamus and pituitary: relation to neonatal androgen and the oestrous cycle. *J. Endocr.*, **48**, 125–137.

SALAMAN, D. F. (1973) The action of metabolic inhibitors on the sexual differentiation of the hypothalamus induced by neonatal androgen (abstract). *J. Anat. (Lond.)*, **114**, 305.

SALAMAN, D. F. AND BIRKETT, S. I. (1974) Androgen-induced sexual differentiation of the brain is blocked by inhibition of DNA and RNA synthesis. *Nature (Lond.)*, **247**, 109–112.

SALAMAN, D. F., BETTERIDGE, S. AND KORNER, A. (1972) Early effects of growth hormone on nucleolar and nucleoplasmic RNA synthesis and RNA polymerase activity in normal rat liver. *Biochim. biophys. Acta (Amst.)*, **272**, 382–395.

SCHWARTZ, H. S., GAROFALO, M., STERNBERG, S. S. AND PHILIPS, F. S. (1965) Inhibition of deoxyribonucleic acid synthesis in regenerating liver of rats. *Cancer Res.*, **25**, 1867–1870.

SHERIDAN, P. J., ZARROW, M. X. AND DENENBERG, V. H. (1973) The role of gonadotrophins in the development of cyclicity in the rat. *Endocrinology*, **92**, 500–508.

SHIMADA, H. AND GORBMAN, A. (1970) Long lasting changes in RNA synthesis in the fore-brains of female rats treated with testosterone soon after birth. *Biochem. biophys. Res. Commun.*, **38**, 423–430.

SUTHERLAND, S. D. AND GORSKI, R. A. (1972) An evaluation of the inhibition of androgenization of the neonatal female rat brain by barbiturate. *Neuroendocrinology*, **10**, 94–108.

TATA, J. R., HAMILTON, M. J. AND SHIELDS, D. (1972) Effects of α-amanitin *in vivo* on RNA polymerase and nuclear RNA synthesis. *Nature New Biol.*, **238**, 161–164.

TURKINGTON, R. W. AND KADOHAMA, N. (1972) Gene activation in mammary cells. *Acta Endocr. (Kbh.)*, Suppl. **168**, 346–363.

WAGNER, J. W., ERWIN, W. AND CRITCHLOW, V. (1966) Androgen sterilization produced by intracerebral implants of testosterone in neonatal female rats. *Endocrinology*, **79**, 1135–1142.

WESSELLS, N. K. (1964) DNA synthesis, mitosis, and differentiation in pancreatic acinar cells *in vitro*. *J. Cell Biol.*, **20**, 415–433.

WOLLMAN, A. L. AND HAMILTON, J. B. (1967) Prevention by cyproterone acetate of androgenic, but not of gonadotrophic, elicitation of persistent estrus in rats. *Endocrinology*, **81**, 350–356.

DISCUSSION

DONOVAN: Since you were giving DNA, RNA and protein synthesis-inhibiting drugs systematically, I should like to know whether there are systemic effects of these rather toxic drugs on sexual maturation. I noticed *e.g.* that you expressed your data per 100 g body weight. Does this mean that the injected animals were smaller than normal animals ?

SALAMAN: There were indeed significant differences in body weight between the various groups, particularly with the DNA synthesis inhibitor, hydroxyurea. However, all the major differences are equally apparent when the data are expressed as absolute figures or relative to body weight.

PILGRIM: I am much impressed by the high number of labeled cells you showed in the hypothalamus. I am, however, still a bit suspicious because the bulk of the hypothalamic neurons is probably formed in an earlier stage, whereas glial cell production will go on for a longer time. So I think it would be necessary not only to count all the labeled cells, but also to differentiate between glial cells and neurons.

SALAMAN: I quite agree with that. However, there is also some biochemical evidence for multiplication of neurons during this period. By means of gradient or zonal centrifugation of brain nuclei (Burdman, 1972, Mathias and Stambolova, 1973)*, separation has been obtained between nuclei of neurons and of small oligodendrocytes and astrocytes. It appeared that although there was a higher rate of DNA synthesis per cell nucleus in the oligodendrocyte fractions there was clear evidence of DNA synthesis in the neuronal fraction as well.

* BURDMAN, J. (1972) The relationship between DNA synthesis and the synthesis of nuclear proteins in rat brain during development. *J. Neurochem.*, **19**, 1459–1469.

MATHIAS, A. P. AND STAMBOLOVA, M. (1973) The activities of deoxyribonucleic acid polymerase and other enzymes in the various types of nuclei found in liver and brain tissue. *Biochem. Soc. Trans. (Lond.)*, **1**, 624–626.

An Evaluation of the Acute Effects of Electrochemical Stimulation of Limbic Structures on Ovulation in Cyclic Female Rats

P. VAN DER SCHOOT

Faculty of Medicine, Erasmus University, Rotterdam (The Netherlands)

INTRODUCTION

Data are accumulating which suggest that the amygdala has a stimulatory effect (Velasco and Taleisnik, 1969a), and the hippocampus an inhibitory effect (Velasco and Taleisnik, 1969b) on the release of an ovulatory amount of gonadotropic hormones in the rat. However in these and also in other studies on the same subject (Bunn and Everett, 1957; Arai, 1971) few normal, cyclic rats were used; instead mostly female rats were used in persistent estrus induced by continuous illumination (PE-♀♀) or ovariectomized rats pretreated with steroid hormones.

It seemed of interest, therefore, to investigate whether the results of these studies would also be applicable to the release of the ovulatory amount of gonadotropic hormones in normal cyclic female rats. The present study deals with the effect of electrochemical stimulation (ECS) (Everett and Radford, 1961) of the medial amygdala or ventral hippocampus in regular 5-day cyclic female rats.

MATERIALS AND METHODS

Inbred (R $\times$ U)F_1 hybrid female rats, bred in our laboratory, were used. The rats were maintained under standardized laboratory conditions. In the first experiment with PE-♀♀, rats were kept under continuous illumination for 3 months beginning at 2 months of age. Daily vaginal smears taken after this period showed that almost all animals were in persistent estrus. The PE-condition was verified at autopsy by examination of the ovaries. Experimental animals were smeared for at least 2 weeks prior to the experiments. For the other experiments, female rats were exposed to a 14:10 regimen of light and darkness, lights off from 19.00 to 05.00 h. Experimental animals were only used at 4–7 months of age and after at least two 5-day cycles immediately preceding the experiment.

Brain stimulation was performed through application of DC-current (1 mA for 20 sec) via monopolar steel electrodes (electrode = anode; tip diameter 0.2 mm; tip length 0.5 mm). The electrode was brought into position with a stereotaxic instrument

References p. 370

under light ether anesthesia, using De Groot's (1959) atlas. Such procedure probably results in a stimulatory effect for only a few hours following electrode placement (Hillarp *et al.*, 1954; Lincoln *et al.*, 1973). Therefore this procedure may give immediate stimulatory effects or long-term lesioning effects.

The localization of the stimulated focus and/or lesion was studied by histological examination of the brain. After autopsy the brains were fixed therefore in Bouin's solution and stained following the technique of Klüver and Barrera (Romeis, 1968); sometimes Perls (Romeis, 1968) staining method was also used for exact localization of the iron in the tissue, surrounding the electrode-tip.

The ovulatory response was studied by counting eggs in the oviducts and rupture points in the ovaries. In the case of absence of ovulation, oocytes, and sometimes also follicles, were studied microscopically. Both methods have been shown to be useful indicators of the exposure of the preovulatory follicles to a subovulatory amount of gonadotropic hormones (Vermeiden and Zeilmaker, 1973).

For statistical evaluation of the results the Fisher test (Siegel, 1956) was used.

RESULTS

Localization of the ECS-foci

The passage of DC-current through the brain tissue brought about a mass of coagulated tissue around the electrode-tip with a diameter of about 1 mm; outside this lesion there was an area, in which only necrotic neural tissue was found. The diameter of the inner and outer area together was about 2 mm.

In cases of ECS-POA these areas were present between the optic chiasm and anterior commissure, leaving these two structures intact, except for perforation of the anterior commissure by the electrode track. The wall of the third ventricle was mostly the medial border of the area. The center of the POA-foci was found around the level of the anterior + 7.4 plane of the De Groot's atlas. The foci in the medial amygdala (AME) were generally situated within the triangular area below the optic tract and medial to the stria terminalis at the level: anterior + 5.0 of De Groot's atlas. The damaged area never reached the diencephalon, lying immediately dorsal and medial of this part of the AME. The electrode tracks were mostly just antero-lateral of the upper part of the stria terminalis, although perforations of the stria occurred also. In case of perforation of the stria terminalis (diameter at that level about 1 mm) by the electrode (diameter 0.25 mm) only relatively few fibers of the stria were cut. The foci in the ventral hippocampus (VH) were present medial to the tapetum and lateral ventricle, in the ventral hippocampal mass at the level anterior + 3.0 of De Groot's atlas; the electrode tracks went caudal to the stria terminalis, mostly within the hippocampal tissue or just within the fimbria hippocampi.

TABLE I

ECS OF THE BRAIN AND OVULATION IN CONSTANT LIGHT-INDUCED PERSISTENTLY ESTROUS FEMALE RATS

	Full ovulation	*Partial ovulation**	*Delayed GTH-release***	*No effect*
ECS-POA***	4/5			1/5
ECS-AME	7/10	—		3/10
ECS-VH	4/10	2/10	1/10	3/10
ECS-FC			1/4	3/4
Pt-lesion AME	1/7			6/7

* Partial ovulation, *i.e.* 2–9 oocytes in both ampullae together and a number of follicles (2–10) with mature oocytes in the ovaries.
** Delayed GTH-release, *i.e.* no ampullary oocytes but early stages of oocyte-maturation in pre-ovulatory follicles.
*** ECS was given bilaterally in all cases. AME: medial amygdala; FC: frontal cortex; POA: preoptic area; VH: ventral hippocampus; Pt: platinum electrode.

ECS of the brain and ovulation in PE-♀♀ (Table I)

Stimulation was performed in the POA, in AME or in VH; in the latter two areas, stimulation was done bilaterally, since pilot experiments revealed, that unilateral ECS-AME only incidentally resulted in ovulation.

Animals in which the frontal cortex had been stimulated (FC), animals in which bilateral lesions had been placed into the AME using platinum electrodes, and animals subjected to ether anesthesia, combined with ACTH- or solvent-injection i.v. (12 animals, not included in the table), served as controls.

In accordance with the literature, bilateral ECS-AME appeared to be effective for induction of ovulation. Interesting in this respect is that the number of ova released per ovulating animal was about the same as the number of ova, released by cyclical animals per cycle (cyclic rats: 12.6 ova; PE-♀♀: 10.9 ova). In contrast to data in the literature (Velasco and Taleisnik, 1969a), ovulation also occurred, even when the electrode tracks had damaged the stria terminals bilaterally.

Ovulation occurred following ECS-VH in a significant number of cases, also in contrast to the literature (Velasco and Taleisnik, 1969b). Only 2 out of a total number of 23 animals, which served as controls, had ovulated.

It is concluded, that PE-♀♀ may ovulate following ECS of AME or VH, which structures have afferent connections with POA, AHA or MBH. Results with control animals suggest that this response is probably specifically due to stimulation of these structures and not merely due to aspecific side-effects of surgical manipulation.

Advancement of ovulation in 5-day cyclic female rats by ECS — experiment I (Table II, a + b)

Cyclic female rats were stimulated in the early afternoon of the third diestrus day

References p. 370

TABLE IIa

ECS OF THE BRAIN AND OVULATION IN CYCLIC FEMALE RATS

Experiment I: experimental procedure

Diestrus-3	1400–1700 h	ECS-POA (unilat.) or ECS-AME (bilat.) or ECS-VH (bilat.)
↓		
Proestrus	0900–1200 h	laparotomy of 50% of the animals: unilateral removal of fallopian tube and ovary
↓		
Estrus	0900–1200 h	autopsy of all animals: inspection of ovaries and fallopian tubes; fixation of brain and ovaries for histological study

TABLE IIb

ECS OF THE BRAIN AND OVULATION IN CYCLIC FEMALE RATS

Experiment I: results

	*Ovulation advanced**	*Ovulation at normal time*	*Ovulation inhibited*
ECS-POA	6/11	5/11	
ECS-AME	0/5	1/5	4/5
ECS-VH	0/8	0/8	8/8

* In all cases of ovulation at least 10 oocytes or oocytes + fresh CL were observed; partial ovulation was not observed.

(D-3). Unilateral ECS-POA was followed by ovulation at proestrus in 6 out of 11 animals. By contrast, advancement did not occur after bilateral ECS-AME or ECS-VH. Further study of these non-ovulating animals revealed that ovulation never occurred during the cycle under investigation: on the expected day of estrus the follicles that had been destined for ovulation were going into atresia.

Advancement of ovulation in 5-day cyclic female rats by ECS of the brain — experiment II (Table III, a + b)

Since data are available suggesting that an abnormally strong gonadotropic hormone stimulus is necessary in order to get advanced ovulation after stimulation at D-3 (Everett, 1964), the previous experiment was repeated with rats early at proestrus. Following this stimulation procedure, a number of the animals were anesthetized

TABLE IIIa

ECS OF THE BRAIN AND OVULATION IN CYCLIC FEMALE RATS

Experiment II: experimental procedure		
Proestrus	0900–1000 h	ECS-POA (unilat.) or ECS-AME (bilat.) or ECS-VH (bilat.)
↓		
Proestrus	1330 h	50% of the animals were given Nembutal
↓		
Estrus	0900–1200 h	autopsy: inspection ovaries and tubes; histology of ovaries and brain

TABLE IIIb

ECS OF THE BRAIN AND OVULATION IN CYCLIC FEMALE RATS

Experiment II: results

	Nembutal at 1330 h	*No Nembutal*
ECS-POA	5/5*	4/4
ECS-AME	0/5	5/9
ECS-VH	0/7	2/8

* Number of animals with ovulation/number tested.

with Nembutal (33 mg/kg b.w.) at 13.30 h, in order to block a possible release of ovulatory GTH during the normal "critical hours".

It appeared that neither bilateral ECS-AME nor ECS-VH provoked ovulation. In all animals that were injected with Nembutal, preovulatory follicles, containing immature oocytes, were present on the next morning. By contrast, ovulation occurred in a number of animals, that had not been given Nembutal, suggesting that ovulation occurred following an ovulatory GTH-release during the normal "critical hours".

Effect of unilateral ECS-AME or ECS-VH at D-3 or early at P (Table IV, a + b)

In order to investigate the possibility that the absence of ovulation following bilateral ECS-AME or ECS-VH was due to aspecific, traumatic side-effects of the surgical manipulation, the effect of unilateral ECS-AME and ECS-VH was studied. It appeared, that neither unilateral ECS-AME nor ECS-VH interfered to any extent with spontaneous ovulation.

References p. 370

TABLE IVa

ECS OF THE BRAIN AND OVULATION IN CYCLIC FEMALE RATS

Experiment III: experimental procedure		
Diestrus-3 or Proestrus	1400–1700 h 0600–1300 h	ECS-AME (unilat.) or ECS-VH (unilat.)
↓ Proestrus	1330 h	some animals were given Nembutal
↓ Estrus	0900–1200 h	autopsy

TABLE IVb

ECS OF THE BRAIN AND OVULATION IN CYCLIC FEMALE RATS

Experiment III: results	*Ovulation advanced**	*Ovulation at normal time*
ECS-AME (D-3)	1/9	6/7
ECS-AME (P)	1/10	3/3
ECS-VH (P)	—	4/4

* Advancement by 24 h after stimulation at D-3; advancement by some hours after stimulation early at P.

DISCUSSION

The present data confirm that ovulation may be induced in PE-♀♀ by ECS-AME (Bunn and Everett, 1957; Velasco and Taleisnik, 1969a). In contrast to other data (Velasco and Taleisnik, 1969b) ovulation also occurred following ECS-VH.

It may be assumed that ovulation in PE-♀♀ was an aspecific effect of exposure of these animals to stressful conditions, such as is noticed by Zolovick (1972) and reported by Brown-Grant *et al.* (1973). However, in the present experiments ovulation only occasionally occurred following exposure of PE-♀♀ to one of the control treatments.

In contrast with the results in PE-♀♀, ovulation was not provoked in cyclic female rats by ECS-AME or ECS-VH at D-3 or P. Results in the literature indicate that ECS-AME may overcome blockade of ovulation obtained with atropine and reserpine, but not with urethane, in proestrous female rats (Velasco and Taleisnik, 1969a). It thus appears that ECS-AME may be followed by ovulation under specific conditions which are present in both PE-♀♀ and in reserpine- or atropine-blocked animals, but not in normal cyclic animals.

Interference with the expected ovulation by both bilateral ECS-AME and bilateral ECS-VH might be explained as a result of the stimulatory effect of ECS or alternatively, from its lesion effect. There is some evidence that the stimulatory effect of ECS on the surrounding neural tissue lasts for only a few hours (Hillarp *et al.*, 1954; Lincoln *et al.*, 1973; Turgeon and Barraclough, 1973). Interference with ovulation after ECS at D-3 then could be explained as being due to stimulation of the AME or VH one day before. However several authors agree that non-stimulatory interruption of limbic system afferents to the hypothalamus interferes with ovulation in short-term experiments (Velasco and Taleisnik, 1971; Brown-Grant and Raisman, 1972; Velasco, 1972; Wildschut, 1972). These data suggest that the interfering effects of ECS at D-3 on ovulation are probably the result from neural damage, caused by ECS. If this interpretation is correct, it follows that the occurrence of normal ovulation in cyclic female rats is dependent on input from both AME and VH.

The difference in response to ECS of AME or VH between cyclic female rats and PE-♀♀ suggests that the functional relation between these structures and the hypothalamus changes following exposure of female rats to continuous illumination. Such a change could account for the well-known fact that ovulation in PE-♀♀ becomes dependent on external stimuli, such as copulation (Dempsey and Searles, 1943) or non-specific stress (Brown-Grant *et al.*, 1973).

SUMMARY

In order to investigate the possible role of the medial amygdala (AME) and ventral hippocampus (VH) in ovulation in normal cyclic female rats, effects of electrochemical stimulation (ECS) of these structures were studied during diestrus-3 (D-3) or proestrus (P).

In a preliminary study it was established that both ECS-AME and ECS-VH could provoke ovulation in female rats, being in persistent estrus following exposure to continuous illumination. Similar animals that were subjected to control procedures ovulated to a significantly less degree.

Experiments with cyclic female rats revealed that neither ECS-AME nor ECS-VH provoked ovulation after application of the stimulus at D-3 or early at P. By contrast both experimental procedures interfered with ovulation during the cycle under investigation. The data are interpreted as indicating that stimulation of AME or VH is not effective in inducing the release of ovulatory gonadotropic hormones and that destruction of large parts of AME or VH — due to the lesion effects of ECS — interferes with ovulation. It is concluded (1) that the effects of ECS-AME and ECS-VH in female rats are different under different hormonal conditions, (2) that both AME and VH are of critical importance in the normal process of ovulation in cyclic female rats, and (3) that in contrast with the situation in the preoptic area, the role of AME and VH in ovulation cannot simply be imitated by ECS of one of these structures.

References p. 370

REFERENCES

ARAI, Y. (1971) Effects of electrochemical stimulation of the amygdala on induction of ovulation in different types of persistent estrous rats and castrated male rats with an ovarian transplant. *Endocr. jap.*, **18**, 211–214.

BROWN-GRANT, K. AND RAISMAN, G. (1972) Reproductive function in the rat following selective destruction of afferent fibres to the hypothalamus from the limbic system. *Brain Res.*, **46**, 23–42.

BROWN-GRANT, K., DAVIDSON, J. M. AND GREIG, F. (1973) Induced ovulation in albino rats exposed to constant light. *J. Endocr.*, **57**, 7–22.

BUNN, J. P. AND EVERETT, J. W. (1957) Ovulation in persistent-estrous rats after electrical stimulation of the brain. *Proc. Soc. exp. Biol. (N.Y.)*, **96**, 369–371.

DEMPSEY, E. W. AND SEARLES, H. F. (1943) Environmental modification of certain endocrine phenomena. *Endocrinology*, **32**, 119–128.

EVERETT, J. W. (1964) Preoptic stimulative lesions and ovulation in the rat: 'thresholds' and LH-release time in late diestrus and proestrus. In *Major Problems in Neuroendocrinology*, E. BAJUSZ AND G. JASMIN (Eds.), Karger, Basel, pp. 346–366.

EVERETT, J. W. AND RADFORD, H. M. (1961) Irritative deposits from stainless steel electrodes in the preoptic rat brain causing release of pituitary gonadotropin. *Proc. Soc. exp. Biol. (N.Y.)*, **108**, 604–609.

GROOT, J. DE (1959) *The Rat Brain in Stereotaxic Coordinates*, North-Holland, Amsterdam.

HILLARP, N. A., OLIVECRONA, H. AND SILFVERSKIÖLD, W. (1954) Evidence for participation of the preoptic area in male mating behaviour. *Experientia (Basel)*, **10**, 224–225.

LINCOLN, D. W., SCHOOT, P. VAN DER AND ZEILMAKER, G. H. (1973) EEG arousal, sympathetic activation and the induction of ovulation by electrochemical stimulation of the rat hypothalamus. *J. Endocr.*, **57**, xix–xx.

ROMEIS, B. (1968) *Mikroskopische Technik*, Oldenbourg, München.

SIEGEL, S. (1956) *Non-parametric Statistics*, McGraw-Hill, New York, pp. 256–270.

TURGEON, J. AND BARRACLOUGH, C. A. (1973) Temporal patterns of LH-release following graded preoptic electrochemical stimulation in pro-estrus rats. *Endocrinology*, **92**, 755–761.

VELASCO, M. E. (1972) Opposite effects of platinum and stainless-steel lesions of the amygdala on gonadotropin secretion. *Neuroendocrinology*, **10**, 301–308.

VELASCO, M. E. AND TALEISNIK, S. (1969a) Release of gonadotropins induced by amygdaloid stimulation in the rat. *Endocrinology*, **84**, 132–139.

VELASCO, M. E. AND TALEISNIK, S. (1969b) Effect of hippocampal stimulation on the release of gonadotropin. *Endocrinology*, **85**, 1154–1159.

VELASCO, M. E. AND TALEISNIK, S. (1971) Effects of interruption of amygdaloid and hippocampal afferents to the medial hypothalamus on gonadotropin release. *J. Endocr.*, **51**, 41–55.

VERMEIDEN, J. P. W. AND ZEILMAKER, G. H. (1973) Relationship between maturation division, ovulation and luteinization induced by gonadotropic stimulation in pro-oestrous rats. *J. Endocr.*, **57**, lvii–lviii.

WILDSCHUT, J. (1972) *Oestradiol en de Ovulatie bij de Rat*. Thesis, University of Leiden, De Boer, Zouterwolde.

ZOLOVICK, A. J. (1972) Effects of lesions and electrical stimulation of the amygdala on hypothalamic-hypophyseal–regulation. In *The Neurobiology of the Amygdala*, B. E. ELEFTHERIOU (Ed.), Plenum Press, New York, pp. 643–683.

The Lateral Hypothalamus and Adjunctive Drinking

MATTHEW J. WAYNER

Brain Research Laboratory, Syracuse University, Syracuse, N.Y. (U.S.A.)

INTRODUCTION

The hypothalamus is one of the most interesting parts of the brain. Relatively simple experimental manipulations such as lesions, electrical stimulation, and application of various chemicals produce reliable and dramatic changes in the behavior of many different species of typical laboratory animals. The fact that the diencephalon — thalamus, subthalamus, and hypothalamus — has important integrative functions in control and coordination of skeletal and autonomic motor activity has not been emphasized frequently (Koella, 1969). The possibility that the lateral region of the hypothalamus (LH), a confluence of major pathways and reciprocating interconnections between the limbic-forebrain and limbic-midbrain (Morgane, 1964, 1969, 1974), has important motor control functions involved in the production of ingestive behavior has also not been emphasized frequently until recently (Wayner, 1970, 1974; Wayner and Carey, 1973). The purpose of the present report is to review some facts which demonstrate motor functions of the LH in the rat and to relate them to motor activity and adjunctive behavior.

EFFECTS OF LH ELECTRICAL STIMULATION AND ABLATION ON SPINAL REFLEX EXCITABILITY

In addition to the apparent adipsia and aphagia reported frequently following bilateral destruction of the LH (Epstein and Teitelbaum, 1964), a decrease in spontaneous motor activity occurs (Gladfelter and Brobeck, 1962) which might be related to a reduction in the ability to respond to certain types of sensory stimulation (Marshall *et al.*, 1971). Reflex discharges evoked in the third lumbar ventral root of rats by dorsal root electrical stimulation are attenuated in animals with LH lesions which produce hypoactivity (Miles and Gladfelter, 1969). The effect of decreased excitability is presented in Fig. 1 where a typical two component evoked discharge from a normal animal is illustrated in part A and a similar response in an animal with destruction of the LH is illustrated in part B. A statistical analysis of the data revealed a significant decrease in amplitude of both components. If electrolytic lesions are produced in the LH during the acute phase of the experiment when the ventral root reflex discharges are being obtained, there were no observable effects on spinal reflex excitability for as

References p. 392–394

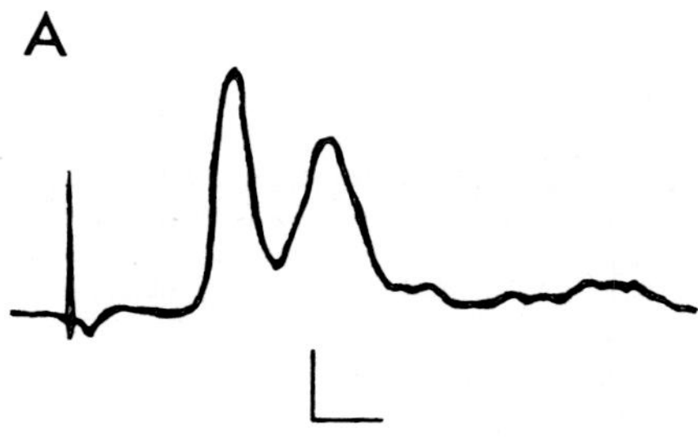

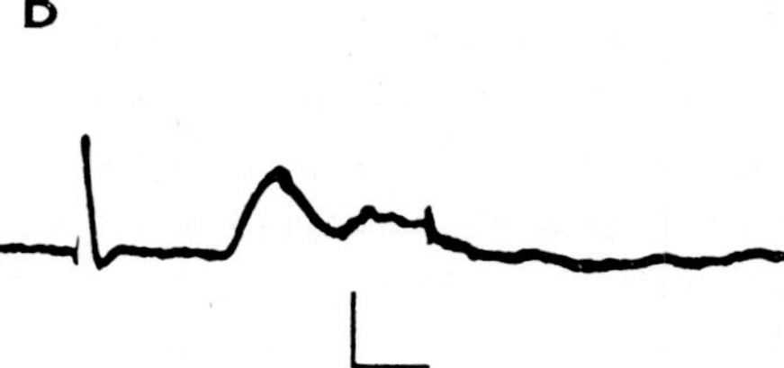

Fig. 1. Tracings of third lumbar ventral root reflex discharges in response to dorsal root stimulation from a control animal (A) and from a rat with lesions in the LH which affects locomotor activity (B). Vertical sensitivity is 100 μV; horizontal scale is 1 msec. Reproduced with permission from Miles and Gladfelter (1969).

long as 3 h. However, if the lesions are made chemically, a decrease in reflex discharge amplitude occurs within several minutes. Apparently, electrolytic lesions produce a local excitatory effect on remaining LH neurons which maintains a high level of excitability in the spinal reflex pathways. If under similar conditions the same region of the LH is stimulated electrically, 100 Hz train of square pulses 5–10 V in amplitude, reflex discharge facilitation occurs as illustrated in part B of Fig. 2. A tracing of an evoked ventral root reflex discharge before LH stimulation is presented in part A of Fig. 2. Ipsilateral or contralateral LH stimulation resulted in an increase in the total area of both components. The increase in the second component was more variable than in the monosynaptic discharge. The latency decreased in most animals studied.

Results of a similar experiment are illustrated in Fig. 3 where fifth lumbar ventral root reflex discharges are evoked by square wave stimulation applied to the peroneal nerve. The left tracing was obtained before LH stimulation. The next two tracings from left to right were obtained during LH stimulation of 60 Hz, 30 μA, indicated by the horizontal line. The last tracing to the right was obtained after LH stimulation was terminated. Reflex discharges were evoked once every 3 sec. A clear facilitation of both the monosynaptically excited component and multisynaptic responses can be observed during LH electrical stimulation.

These results indicate that the decrease in locomotor activity following destruction of the LH can be attributed to the maintenance of a higher level of excitability in

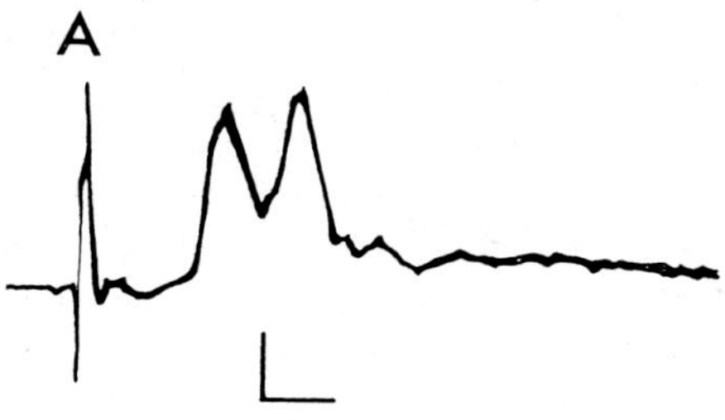

Fig. 2. Tracings of typical ventral root reflex discharges before (A) and during (B) LH stimulation. Vertical sensitivity is 100 μV; horizontal scale is 1 msec. Two stimulus artifacts due to LH stimulation are present in part B. Reproduced with permission from Miles and Gladfelter (1969).

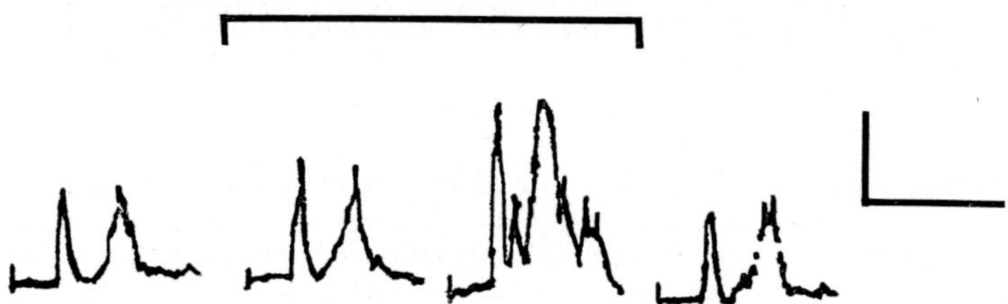

Fig. 3. Tracings of typical fifth lumbar ventral root reflex discharges evoked by peroneal nerve stimulation before, during and after LH stimulation by 60 Hz, 30 μA current. LH stimulation indicated by the horizontal line. Reflex discharges evoked 1/3 sec. Vertical calibration 1 mV; horizontal calibration is 10 msec.

spinal reflex pathways by cells and/or axons in this region under normal conditions. Also, the level of excitability in anesthetized acute preparations can be enhanced by electrical stimulation of the LH. The increased motor excitability and activity can be observed easily in the chronic preparation with permanently implanted LH electrodes where the effects occur very dramatically (Rosenquist and Hoebel, 1968; Valenstein *et al.*, 1969; Wayner, 1970).

EFFECTS OF SALT AROUSAL ON THE LH AND SPINAL REFLEX EXCITABILITY

As the rat with bilateral destruction of the LH is incapable of drinking in response to the administration of hypertonic saline (Epstein and Teitelbaum, 1964), cells critical to drinking under some conditions reside in this region of the brain. Drinking which follows the intravenous administration of hypertonic saline has been referred to as the salt arousal of drinking (Adolph, 1964). Since the extracellular content of sodium is regulated precisely, changes in body fluid volume might be detected as modulations in the extracellular concentration of sodium. Some recent evidence supports the existence of sodium sensitive cells in the LH and possibly in other parts of the hypothalamus close to the walls of the ventricles (Andersson, 1972; Myers and Veale, 1971). The existence of sodium and glucose sensitive cells of the LH in rat has been demonstrated by the direct application of these substances by means of microelectrophoresis and extracellular recordings of unit activity through multibarreled glass capillary electrode arrays (Oomura *et al.*, 1969). The change in discharge frequency in a sodium and glucose sensitive LH neuron is illustrated in Fig. 4 where the electrophoretic ejection

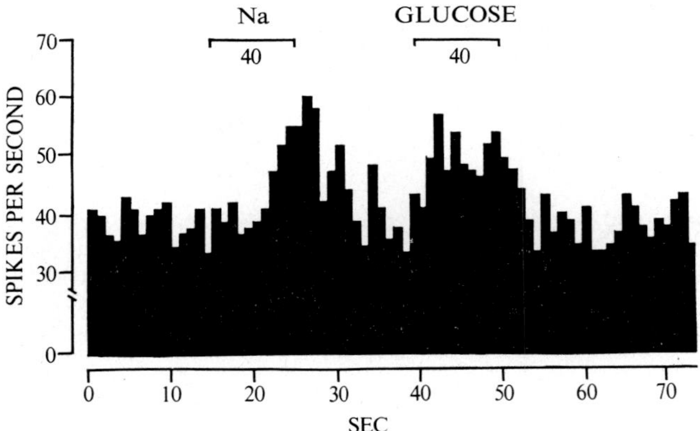

Fig. 4. The discharge frequency of an LH neuron sensitive to both sodium and glucose plotted as a function of time. The electrophoretic ejection of both substances by 40 nA indicated by the horizontal lines. The rat was anesthetized with urethane. This cell responded to both sodium and glucose with an increase in discharge frequency.

of sodium and glucose by 40 nA of current is indicated by the two short horizontal lines. Considerable evidence has existed for some time supporting the existence of osmosensitive cells in the lateral preoptic area, supraoptic nucleus and lateral hypothalamus involved in body fluid regulation (Andersson, 1972; Epstein, 1971; Novin and Durham, 1969; Wayner and Kahan, 1969) which respond to intracarotid or intravenous administration of hypertonic saline.

It has been known for some time that enhancement of certain types of simple motor acts can increase independently of consumption during the salt arousal of drinking (Wayner and Emmers, 1959). The increase in number of runs through an enclosed alley to obtain water following the subcutaneous injection of hypertonic saline is illustrated in Fig. 5. The increase in mean number of runs for both experiments was significant, whereas there were no significant changes in water consumption by any of the groups over the 4-h period following injection. The experiment was carried out

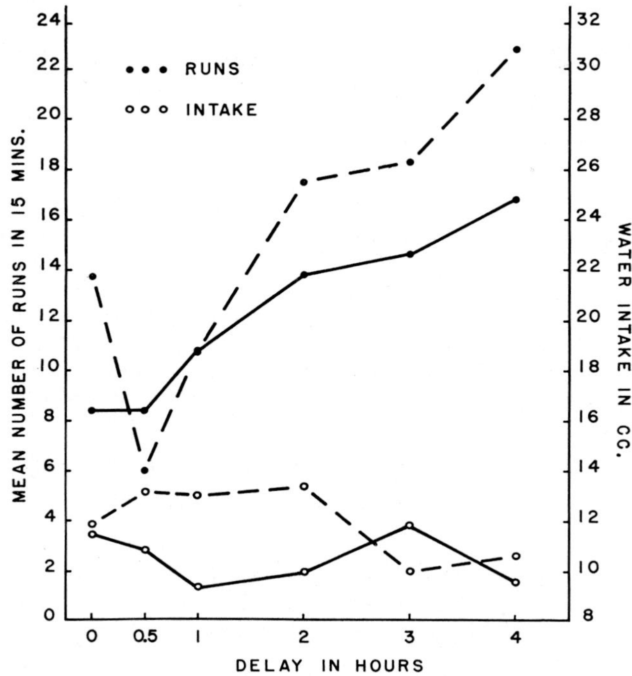

Fig. 5. Experiment 1: broken lines. Mean number of runs through an enclosed alley in 15 min and the cumulative 1-h intakes as a function of the delay in hours following a subcutaneous injection of 3.0 mEq of 15% NaCl. Experiment 2: solid lines. Same as experiment 1 except for 3-h intakes and that procaine hydrochloride was mixed with the hypertonic saline to eliminate pain associated with the injection. Reproduced with permission from Wayner and Emmers (1959).

the first time with subcutaneous injections of 3.0 mEq of 15% NaCl (solid lines), and the second time with 3.0 mEq of 15% NaCl plus procaine hydrochloride (broken lines), as a local anesthetic to eliminate the pain associated with the injections. The facilitation of running is obvious and cannot be attributed to different amounts of water imbibed.

Can the enhancement of running during salt arousal of drinking be attributed to an increase in discharge rate of LH neurons sensitive to sodium which in turn produce an increase in the excitability of spinal reflexes? Typical unit discharges are illustrated in Fig. 6, which were recorded from the ventromedial hypothalamic nucleus (VHM), in the upper tracing; and from the LH, in the lower tracing before, in the top part of the figure, and during, in the lower part of the figure, the intravenous infusion of 0.10 ml of 15% NaCl in a rat anesthetized with urethane. The two cells appear to be reciprocally related in time as illustrated in Fig. 7, where spikes/10 sec are plotted as a function of successive 10 sec intervals. Under these conditions the amount of NaCl administered is critical and relatively low doses, just above threshold, seem to be most effective. Large doses produce nonspecific variable activity in most parts of the brain. Consequently, with small intravenous doses the latencies are usually long, 350 sec as illustrated in Fig. 7. The increase in the LH neuron and the associated decrease in the

References p. 392–394

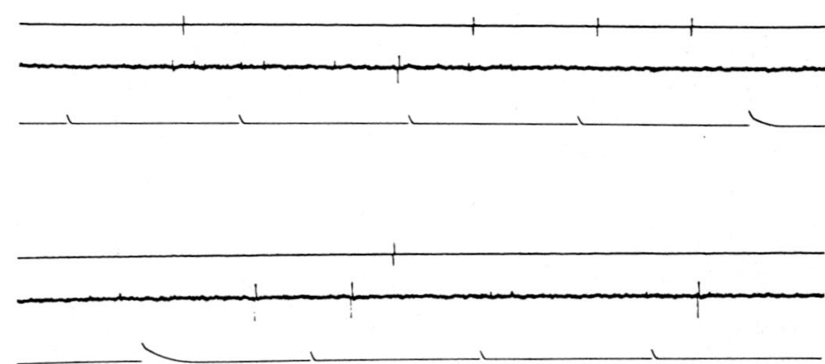

Fig. 6. Top: upper tracing VMH unit before salt arousal. Middle tracing LH unit before salt arousal. Bottom: upper tracing VMH unit during salt modulation. Middle tracing LH unit during salt modulation. The reciprocal change in activity due to salt arousal is clear. Lower tracing: time calibration 0.1 sec. Reproduced with permission from Wayner and Kahan (1969).

VMH neuron discharge frequencies are obvious and the change persisted for a relatively long time, approximately 12 min, although only 400 sec are illustrated in Fig. 7.

In similar acute preparations, laminectomies were performed and the effects of intracarotid administration of NaCl were assessed in terms of the amplitude of ventral root reflex discharges evoked by stimulation of appropriate afferents in the hind leg. An increase in spinal reflex excitability associated with salt arousal of drinking is illustrated in Fig. 8. Excitability was measured in terms of the relative change in the

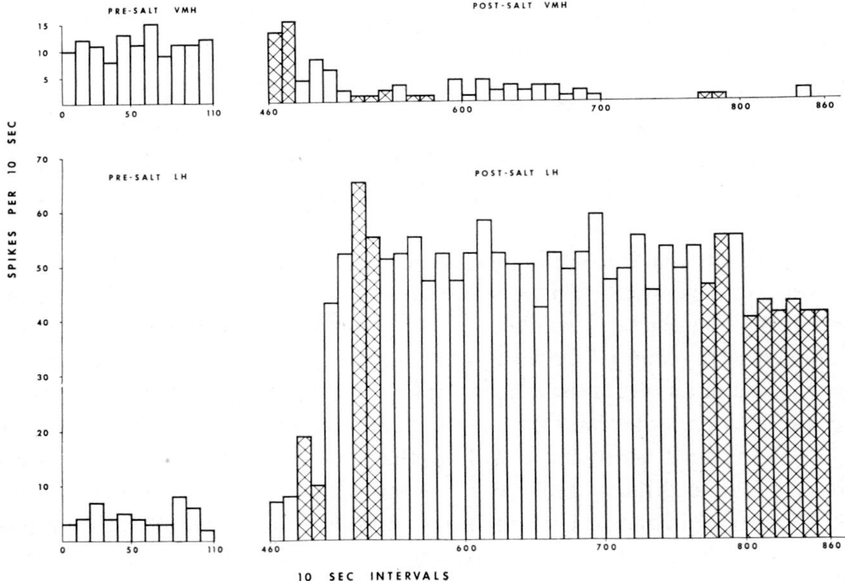

Fig. 7. Spikes per 10 sec plotted as a function of successive 10-sec intervals as indicated from the VMH and LH units before and following the intravenous administration of NaCl. Changes in stationarity indicated by the crosshatching. Reproduced with permission from Wayner and Kahan (1969).

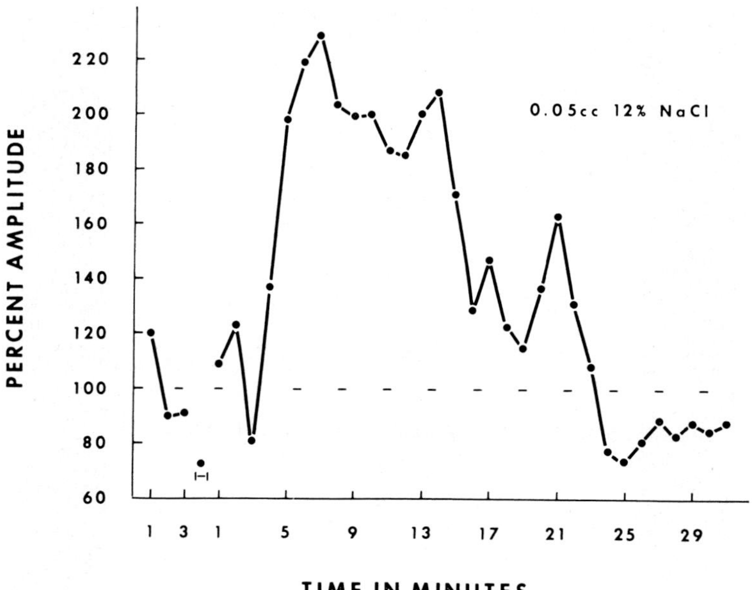

Fig. 8. Percent change in the fifth lumbar ventral root reflex discharges relative to the preinjection mean value following 0.05 ml of 12% NaCl. The duration of the infusion is indicated by the short horizontal line. Each point is the mean amplitude of 15 successive responses. Reproduced with permission from Wayner *et al.* (1966).

amplitude of successively evoked monosynaptic reflex discharges in the fifth lumbar ventral root by stimulation of the gastrocnemius nerve before, during, and after intracarotid injection of 0.05 ml of 12% NaCl. Ventral root reflex discharges were evoked every 4 sec. An increase in excitability is indicated by an increase in the amplitude of the ventral root compound action potential as illustrated in Fig. 9. As the afferent input remains relatively constant under these conditions, more spinal motoneurons respond to essentially the same input. The increase in spinal reflex excitability which accompanies salt arousal is reversed by distilled water, does not occur in the spinalized rat preparation and is not affected by adrenalectomy (Wayner *et al.*, 1966).

In another study, only animals with functional bilateral LH lesions and proven recovered lateral lesion symptoms were utilized in an effort to determine if the enhancement of spinal reflex excitability during salt arousal depends upon an intact LH. A rat with a typical recovered LH lesion syndrome is illustrated in Fig. 10. The animal began to recover from the aphagia on day 6 and continued to display a permanent adipsia with prandial drinking. Water intake decreased to 5.0 ml when food deprived for 24 h and the animal drank only 1.0 ml in 3 h following a subcutaneous injection of 8.0 mEq/kg of 15% NaCl. When this animal was subjected to a laminectomy and other experimental procedures similar to those for the rat in Fig. 8, there were no increases in spinal reflex excitability which could be attributed to salt arousal. The data are presented in the same way in Fig. 11. There were no significant changes in response to two injections of 12% NaCl or equiosmotic 18% mannitol. However,

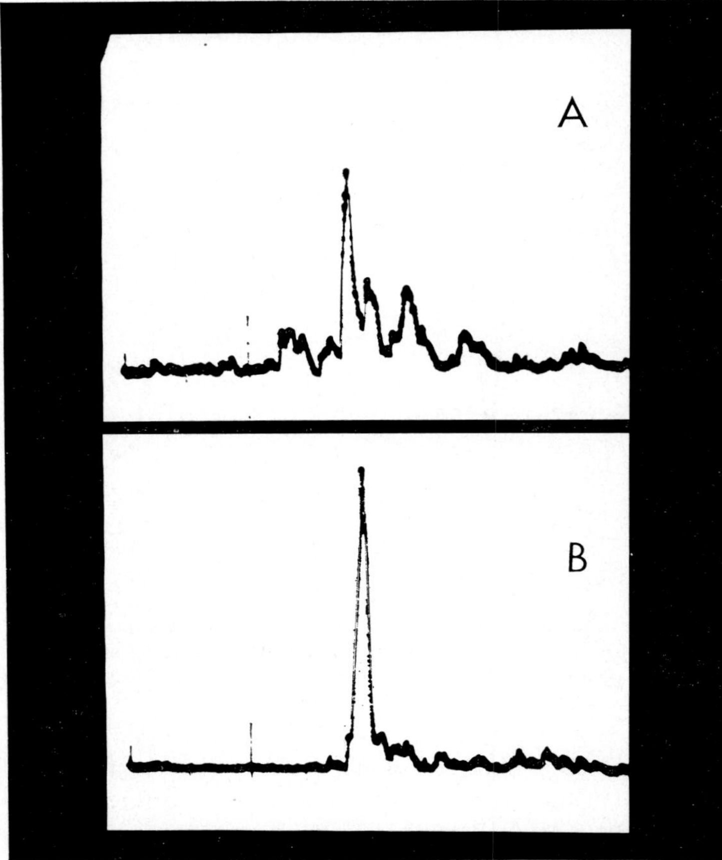

Fig. 9. Part A: fifth lumbar ventral root reflex discharge in response to simultaneous stimulation of both small nerves to the two heads of the gastrocnemius muscles. Part B: same as part A except recorded 5 min following the intracarotid injection of 0.15 ml of 9% NaCl.

strychnine sulfate did produce an immediate increase in the amplitude of the ventral root reflex discharge and indicated that the neurons were not depressed and the preparation was probably in good condition.

THE LH AND SPINAL REFLEX EXCITABILITY

These results demonstrate clearly that the LH is definitely involved in the determination of spinal reflex excitability. Sodium-sensitive neurons undoubtedly converge in the rostral portion of this system (Millhouse, 1969) upon pathways into and from the brain stem which are still a matter of speculation. Studies of fiber degeneration in the medial forebrain bundle from lesions in the LH (Wolf and Sutin, 1966) reveal that (a) ascending degeneration enters the preoptic region and the medial and fimbrial nuclei of the septum, (b) there is less degeneration in the lateral septal nucleus, and

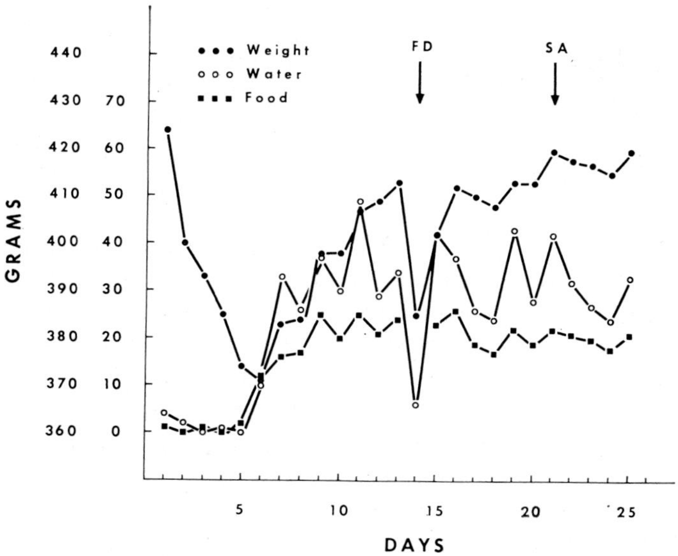

Fig. 10. Body weight and *ad lib.* food and water intake in grams for an animal with bilateral hypothalamic RF lesions for each of 25 days following surgery. FD: food deprived for 24 h. SA: the day on which the animal was injected subcutaneously with 8.0 mEq/kg of 15% NaCl. Reproduced with permission from Wayner *et al.* (1966).

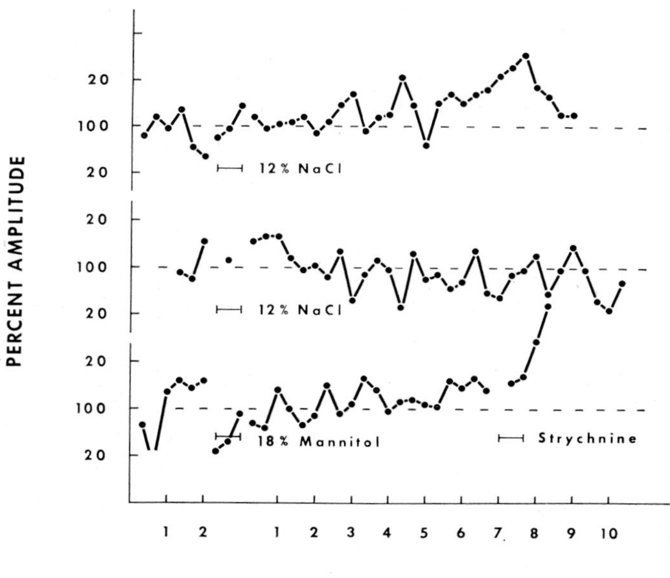

Fig. 11. Relative change, percent amplitude, in the monosynaptic component of the fifth lumbar ventral root reflex discharges following successive intracarotid injections of 0.05 ml 12% NaCl, 18% mannitol, and strychnine sulfate. Same animal as in Fig. 10. Reproduced with permission from Wayner *et al.* (1966).

References p. 392–394

(c) a few axons bypass the septum and terminate in the cingulate cortex; whereas (a) the descending degeneration enters the ventral tegmental area of Tsai with a few collateral fibers to the mammillary bodies, (b) in the midbrain fibers turn dorsomedially to enter the raphe nuclei and the central gray and dorsolaterally to enter the nucleus mesencephalicus profundus pars lateralis, (c) a small group of fibers leaves the anterior LH in a dorsomedial direction and passes through the inferior thalamic peduncle and stria medularis and terminates in the lateral habenular nucleus, and (d) another bundle leaves the posterior LH in a dorsomedial direction and passes through the posterior midline thalamus to enter the central gray. Very little is known about the brain stem reticular formation in rat except that facilitatory effects on hind limb movements are concentrated in the dorsomedial region whereas inhibitory sites are located in the ventrolateral portion (Emmers, 1959). Stimulation of the medial reticulospinal tract facilitated spinal reflexes and the ventral reticulospinal tract produced inhibition. Pyramidal tract stimulation usually inhibited leg flexion; however, contralateral stimulation sometimes produced facilitation. Although all postsynaptic inhibitory action on spinal motoneurons evoked by supraspinal descending impulses appears to be polysynaptic, a few cases of clear postsynaptic depolarizations in response to single hypothalamic stimuli have been reported in the cat (Shapovalov, 1966). Also, the highest rate of firing of lumbar motoneurons during supraspinal stimulation was recorded when hypothalamus and red nucleus were tetanized. The nature of these connections and the possibility that the effects on spinal reflexes are mediated primarily by descending pathways through the LH particularly in the rat is still uncertain. One important fact seems to be clear — the enhancement in the excitability of spinal motoneurons which comprise the final common motor path and produce a nonspecific facilitation of movements represents a nondiscriminatory increase in response probability at the behavioral level.

EFFECTS OF LH ELECTRICAL STIMULATION ON BEHAVIOR

The LH appears to be intimately involved in a brain stem motor control system which determines the normal level of spinal reflex excitability and motor activity. The mechanism by which these effects are carried out involves the supraspinal control of the alpha motoneuron and the dynamic and static fusimotor systems (Jansen, 1966; Shapovalov, 1966; Wayner and Carey, 1973). Osmosensitive cells of the lateral preoptic region (Blass and Epstein, 1971) and sodium and glucosensitive lateral hypothalamic neurons (Oomura *et al.*, 1969) apparently converge in the rostral LH. Therefore, electrical stimulation of the LH provides nonspecific partial stimulation of the motor control system which includes the caudal lateral hypothalamus, ventral tegmentum, mesencephalic and brain stem reticular formation. For example, thermosensitive cells of the preoptic area excite cells of the lower reticular formation which control the activity in gamma efferents and indirectly the excitability of the reflex pathways in which those afferents from the muscle spindles participate (Euler and Soderberg, 1957). Both synergistic inhibitory and facilitatory influences are enhanced and the reciprocal and coordinated nature of reflex activity and movements is preserved. Direct connections with spinal motoneurons might provide the basis for the initiation of rapid reflex activity and movements (Jansen, 1966; Shapovalov, 1966). Apparently these cells whose activity depends upon variations in blood temperature and composition above or below some preadapted range, set point, converge in the crucial middle portion of the LH onto the anterior path cells of the caudal region which also receive a rich supply of afferents from the ascending reticular formation. The entire extent of the lateral hypothalamic region appears to receive considerable excitation from taste, oropharyngeal and gastric, tactile, visual, auditory, olfactory and muscle receptors (Brooks *et al.*, 1962; Kawamura *et al.*, 1970; Nicolaidis, 1969a, b; Norgren, 1970). Evidence in a recent study (Emmers, 1973) demonstrates clearly in the cat that neurons of the supraoptic and paraventricular nuclei function as osmoreceptors whereas neurons of the LH and entopeduncular nucleus do not. Osmoreceptor neurons of the supraoptic nucleus inhibit cells of the LH which are normally excited by gustatory and splanchnic afferents via the accessory semilunar relay nucleus in the thalamus. Bilateral destruction of these nuclei in the rat produce a hypodipsia which for some animals was inadequate to maintain life (Emmers and Passamonte, 1972). Obviously, drinking will not occur without an appropriate afferent input into the LH from the mouth and viscera (Wayner, 1974). Apparently the motor control functions of the LH can be modulated easily by a variety of peripheral stimulation but primarily that from the mouth, tongue and viscera. Therefore, not only is the LH unique anatomically by providing a system with positive feedback but it is unusual in that the convergence of various inputs occurs at the confluence of the ascending and descending interconnections of the limbic-forebrain and limbic-midbrain regions (Morgane, 1964, 1969). Consequently, electrical stimulation of the LH should provide excitation of at least part of this system with an increase in reflex excitability and motor activity.

The effects of lateral hypothalamic electrical stimulation through chronic implanted electrodes in normal wakeful rats have been described in great detail and summarized recently (Valenstein, 1969; Valenstein *et al.*, 1969). In essence, an increase in general motor activity as indicated by discrete movements of the vibrissae, head, limbs and torso, chattering and grinding of the teeth, and sniffing; and coordinated simple acts such as locomotion, burrowing, object carrying, tail chasing, chewing and swallowing food, licking and swallowing water, head and mouth grooming and preening of the tail and fur occurs during the period of stimulation. During the first few periods with the current intensity at threshold, the initial effects are discrete movements and locomotion with tail chasing and grooming which become more pronounced and more frequent as the current is increased. If food is available, the animal might carry it in the mouth, roll it between the forepaws and on occasions lick it vigorously, begin to gnaw at it, or eat parts of it. At this point, just above the threshold, one of the simple acts such as gnawing or eating occurs more frequently, becomes stereotyped, and the discrete motor responses and other activities begin to diminish. Those activities which have orogastric sensory feedback such as object carrying in the mouth, chasing and holding the tail in the mouth, chewing and ingesting food, and licking and imbibition of water become stereotyped very readily. Eventually a particular act such as eating,

References p. 392–394

gnawing or drinking will be elicited reliably whenever the stimulus is turned on and will not occur during the interstimulus intervals.

An interesting feature of such behavior is that it can be switched to one of the other acts which occur readily and becomes stereotyped simply by changing the environmental stimuli without altering the pattern or intensity of LH stimulation through the same electrodes (Valenstein *et al.*, 1969). Sometimes the entire sequence of behavioral changes including spontaneous switching will occur within a 1-h test session. On other occasions, long periods of LH stimulation in the absence of previously eliciting stimuli are required before a new stereotyped behavior pattern will emerge in response to the remaining stimuli. The occurrence of ingestive behavior and switching can be facilitated by increasing oral ingestive positive sensory feedback and inhibited by mildly aversive orogastric sensory feedback (White *et al.*, 1970). Other more complex types of behavior such as fighting, stalking and killing (King and Hoebel, 1968; Panksepp, 1971) and copulation (Caggiula, 1970) require the presence of more complex stimuli such as a mouse or a receptive female. The readiness or ease with which a particular response pattern emerges during LH stimulation is partially determined by some indwelling prepotency behavior hierarchy (Valenstein *et al.*, 1969). Strain and individual differences suggest the importance of genetic factors and the previous history of the animal at the time of testing.

The fact that switching can occur through the same electrodes and at constant current intensities appears to be a serious challenge to a specific neural substrate explanation for the behavior which seems to be evoked by electrical stimulation of the LH. Such an explanation as implied by the concept of brain center can be very misleading. Electrical stimulation of the LH does not evoke eating, drinking or any behavior. It does, however, increase the excitability of spinal motoneurons, the final common motor pathway, which is associated with an increased probability that any response will occur to any above threshold stimulus (Wayner, 1970). In other words, the probabilities that simple movements and acts will occur in the presence of the appropriate environmental stimuli are increased during LH stimulation. Both proponents and critics of this point of view have failed to understand the nonspecificity implied by the results. The nonspecificity refers only to the behavior elicited during the electrical stimulation of the LH and not to the specific neural elements excited or the specific neural effect which is produced. What the animal will do during the first few periods of LH stimulation is predictable only within broad categories. Specific predictions would require knowledge of the animal's genetic predisposition and previous history or experience which is not available for current experimental situations.

The LH is unique because electrical stimulation here might very well potentiate or mimic orogastric sensory feedback normally provided via axons from the accessory semilunar nucleus of the thalamus. Oral related activities become stereotyped easily during LH stimulation because both the muscle movements and related sensory feedback involved in the control of those movements are facilitated by the electrical stimulation. Every somatic movement is under the same type of control from the muscle sensory receptors and associated with an increased frequency of occurrence in the development of a certain amount of stereotypy. There are very few wasted movements in a skilled performance. The common elements, specific neural pathways and motor units, persevere and become ingrained mainly because they are used more frequently (Wayner, 1970) as they tend to be elicited repetitively by the unchanging environmental stimuli. During the initial periods of stimulation simple motor responses and locomotion occur near the threshold. As the number of periods is increased with increasing stimulus current, the discrete motor responses begin to disappear and the animal begins to develop a definite postural attitude towards some prominent feature of the environment such as the drinking spout or food blocks. With subsequent stimulation the independent movements become more highly coordinated and integrated. The fact that the animal's activity in general decreases and certain postural adjustments frequently persist during the interstimulus intervals can easily be misinterpreted as a state of poststimulus inhibition. During the initial periods of LH stimulation the behavior tends to be adjunctive and only after the rat has made several contacts with some consumable object or continued self-directed activity such as tail preening or grooming does the stereotype begin to develop.

The important point in this context is that a general state of increased excitability and motor activity, similar to that which emerges during deprivation and on an intermittent schedule of reinforcement (Wayner, 1974), is being produced very rapidly and intensely by repetitive stimulation of the LH. The activity component of the effect is clearly not specific to a particular behavior and reflects a general increase in neural excitability within the motor system and the specific behavior elicited depends upon the environmental stimulation and the resultant sensory feedback. The similarity in the general motor activity which appears early in a deprivation schedule, during schedule induced polydipsia, and during the first few trials of electrical stimulation of the LH indicates that the same mechanism is involved and that deprivation also induces a nonspecific increase in excitability. Therefore, drinking or eating are not localized or encoded in the lateral hypothalamus or at any other site in the brain. Drinking continues to be elicited in a water deprived animal because the oropharyngeal, gastric and other visceral sensory inputs and the increase in plasma osmolality combine to increase the neural excitability of the motor system until the fluid activates gastric receptors which inhibit the LH or sufficient water is absorbed by the intestine to reduce the hypernatremia. The system displays positive feedback and operates so rapidly that it is almost impossible to observe in terms of the behavioral changes which occur during early deprivation in the presence of consumable objects. The adversive aspects of the system also operate efficiently and 4 licks at a taste discriminable fluid are sufficient to produce permanent bait shyness if the animal becomes ill (Halpern and Tapper, 1971). However, the development of the process can be studied in greater detail during electrical stimulation of the LH where the events appear to be a slow recapitulation of what happens following deprivation.

The specificity of drinking occurs as a result of licking which produces the contact and other sensory stimulation from the consumable object and the muscular contractions. The positive feedback maintains a high level of activity in the appropriate neural pathways. At the behavioral level the effect is associated with an increase in the probability that licking at the tube will occur in the presence of those environ-

References p. 392–394

mental stimuli. Whenever an increase in the excitability of the motor system occurs, the probability for licking as well as other potential activities will be increased and, as the probability for licking is already high, drinking will tend to occur in the same situation under the same conditions. When drinking becomes stereotyped it is more difficult to switch during electrical stimulation of the LH.

ADJUNCTIVE BEHAVIOR

Small volumes of hypertonic saline, when administered intracarotidly via chronic indwelling cannulae in normal wakeful rats, do not produce drinking but increases in locomotion and an increase in responsiveness to environmental stimuli which is similar to the behavior observed during LH electrical stimulation under the same conditions (Wayner, 1974; Wayner *et al.*, 1966). With larger doses and longer periods of infusion drinking eventually occurs. Because of the similarity of this behavior to that observed following brief periods of water deprivation and during LH electrical stimulation (Wayner, 1974), it seemed reasonable to examine another perplexing phenomenon which seems to involve an enhancement in the readiness to respond and an increase in motor activity. When rats are partially deprived of food and thereby reduced in body weight and the daily ration of food is delivered intermittently on a variety of reinforcement schedules, very strong and persistent post pellet drinking develops. This interesting phenomenon has been referred to as schedule-induced polydipsia (Falk, 1964). The copious drinking which occurs on a fixed interval 1-min schedule (FI-1) is illustrated in Fig. 12. The top tracing in each of the 4 pairs of tracings indicates licking, each downward deflection represents 5 licks. The other tracing is a typical cumulative record indicating number of lever presses and the delivery of the 45 mg Noyes pellet by the downward deflection. The top pair of tracings represents the results of a typical 1-h test session in animal 18 at 80% of initial body weight. The total number of licks (9048L), presses (968P) and water consumed (32.4 ml) are also presented. The bottom 3 pairs of tracings represent the results of a typical 3-h session in animal 15 at 80% of initial body weight. The post pellet pattern of licking and the differences between the two animals are pronounced. When animal 15 began training, the lick pattern for a 1-h session was similar to that in animal 18. The large volume of 98 ml of water consumed is typical of the longer sessions and is not germane in a physiological sense of water deficit or body fluid regulation and cannot be attributed to a food associated dry mouth (Falk, 1969). Schedule-induced polydipsia is only one of a general class of adjunctive behaviors which are induced under similar conditions of scheduled reinforcements (Falk, 1971). Schedule-induced polydipsia is also schedule-dependent (Falk, 1969) and the maximum effect occurs at approximately 3 min for a fixed interval food reinforcement schedule. As schedule-induced air licking (Mendelson and Chillag, 1970) and wheel turning (Levitsky and Collier, 1968) have been demonstrated, it is very improbable that these interesting adjunctive behaviors are relevant to any long term regulatory process. Consumption seems to be a fortuitous consequence of initial contact with a consumable object.

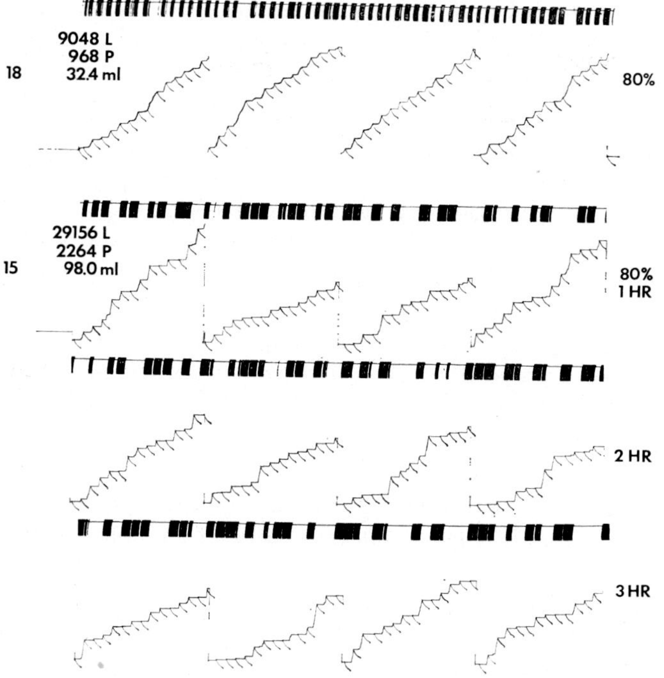

Fig. 12. Photographs of cumulative recordings. Each deflection in the top tracing of each pair represents 5 licks; each deflection in the bottom tracing indicates delivery of a pellet. The top two tracings for rat 18 on an FI-1 schedule for 1 h at 80% body weight. There were 9048 licks, 968 presses and 32.4 ml of water were consumed. The bottom 6 tracings for rat 15 on the same schedule except the test session is 3 h. There were 29,156 licks, 2264 presses and 98 ml of water were consumed.

Schedule-induced pica also occurs (Villarreal, 1972) and indicates that the sensory feedback need not be related to some special feature of natural foods.

The increased motor activity and other behavior emitted by a rat on an FI-1 food reinforced generator schedule for schedule-induced polydipsia is almost identical to that observed following water deprivation when the same test chamber is used for drinking (Wayner, 1974). There is one important difference. In the drinking test situation locomotion, turning movements, sniffing and some of the other behaviors begin to disappear at the end of the first 3 min and a stereotyped posture develops oriented towards the drinking tube with sustained licking for the next 4–5 min. At this point licking and the consumption of fluid tend to cease and extensive and vigorous grooming begins to occur. The delivery of the food pellet on an FI schedule is a very powerful complex stimulus which elicits this general activity associated with a state of increased motor excitability. The development of this change in excitability can be facilitated by the addition of signal lights to the test situation which indicate the delivery of the pellet. The schedule-dependency of adjunctive behavior provides a precise measure of the duration of the period of increased excitability and some indication of its relative intensity produced by the complex stimulation associated with the delivery of the pellet. It is interesting and significant that the period of apparent

References p. 392–394

enhanced excitability and increased general motor activity for both schedule-induced adjunctive behavior and following water deprivation is approximately 3–5 min.

One of the most important features of this phenomenon is the behavioral nonspecificity. In schedule-induced polydipsia, licking and drinking occur because it is the most probable activity associated with eating. Locomotion, grooming and other types of behavior also occur frequently. However, if the drinking spout is placed in a prominent position close to the food cup, licking and drinking is facilitated and the other behaviors begin to disappear and drinking becomes dominant as a stereotyped post-pellet phenomenon. Although drinking is the most probable response, it can be thwarted by making the spout inaccessible during the early test sessions and then other types of activity occur more frequently. Occasionally some other bizarre behavior becomes stereotyped which can be intensified by amphetamine (Wayner and Greenberg, 1973a). Displacement activities (Zeigler, 1964) are similar, if not identical phenomena, which might be explained in terms of the same mechanism. Once the licking and drinking become stereotyped, it can be used to sustain operant response schedules (Falk, 1969). Obviously the adjunctive activity is limited by the interreinforcement times of the generator schedule. If an animal is free to run in an activity wheel or to press a lever for food on a fixed interval schedule, running is suppressed when the schedule generates a high rate of pressing and increases when lever pressing decreases (Skinner and Morse, 1957). If an animal is on a variable interval 1-min food reinforcement schedule (VI-1) and given the choice of drinking or running, drinking occurs predominantly as a brief burst immediately following the reinforcement whereas running almost always follows drinking and continues to occur during the interreinforcement interval. The tendency to continue to run between reinforcements accounts for the decrease in lever pressing under these conditions and the absence of such an effect with drinking. Lever pressing is not a necessary condition for schedule-induced polydipsia and equivalent amounts of water are consumed when the pellets are dispensed noncontingently (Falk, 1969). When pellets are delivered noncontingently at various fixed time intervals, 1.0, 2.0, 3.0, 4.0 and 5.0 min, licking, water consumption, and lever pressing, if available, all display a schedule-dependency with a maximum effect at 4.0 min (Wayner and Greenberg, 1973b). Under these conditions, the lever pressing is definitely adjunctive because the partially deprived animals were naive and never trained to press for food.

Available evidence indicates that the relative intensity of schedule-induced adjunctive behavior is most closely related to the loss in body weight due to the partial food deprivation schedule (Falk, 1969). Apparently schedule-induced drinking decays very rapidly when body weight returns to normal. Adjunctive behavior also disappears rapidly when animals are switched from a VI or FI schedule to continuous reinforcement (Falk, 1969; Levitsky and Collier, 1968). These results suggest a causal relation between adjunctive behavior and some internal condition associated with food deprivation, loss in weight and the intermittent delivery of food which produces an increase in motor excitability. The fact that bilateral lesions in the LH result in a variable effect on lever pressing on a variable interval generator schedule with a consistent attenuation of the schedule-induced polydipsia (Falk, 1964) might be

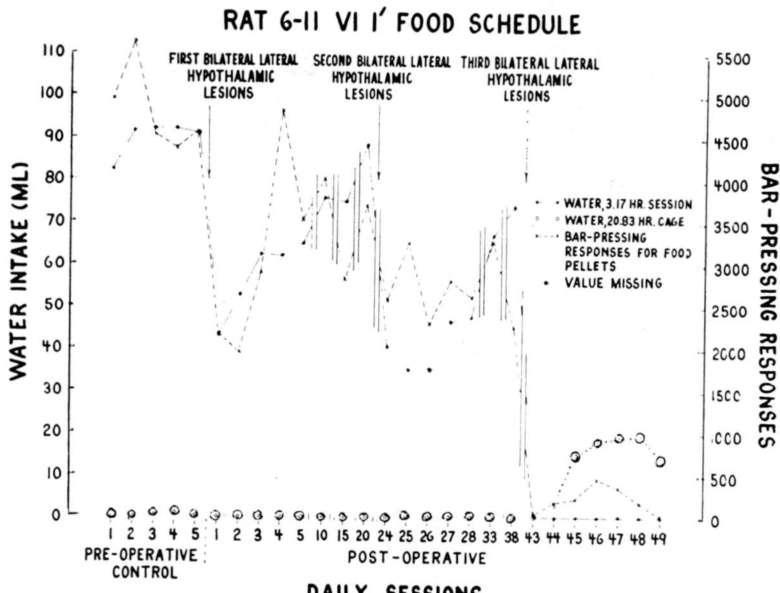

Fig. 13. Bar pressing and schedule-induced water intake in ml on a VI-1 food reinforcement schedule as a function of daily test sessions. Values are presented for daily sessions 5 days postoperatively and then only every fifth day is plotted. Reproduced with permission from Falk (1964).

attributed to differences in the size of lesion and suggest that an intact LH is required for the normal appearance of this phenomenon. These results are illustrated in Fig. 13 where bar pressing and schedule-induced water intake in ml are presented before and following 3 successive bilateral hypothalamic lesions. The first two lesions were produced by passing the electrode through the tissue without any current. The third lesion resulted from a lesion produced by 2 mA for 5 sec. The third lesion obliterated session drinking (solid circles connected by a solid line), although the animal was not adipsic as indicated by the increase in home cage drinking (open circles connected by a dotted line). Also, bar pressing displayed some post-lesion recovery.

If the increase in readiness to respond and motor activity which can be observed in a standard test chamber to accompany salt arousal of drinking, electrical stimulation of the LH, and schedule-induced adjunctive behavior can be attributed to a common mechanism in the LH involved in an enhancement of motor excitability, then it should be possible to substitute effective electrical stimulation of the LH for the delivery of the food pellet in a typical FI generator schedule and at least partially reproduce the phenomenon of schedule-induced polydipsia. Standard procedures were employed which have been described previously (Wayner *et al.*, 1973). Briefly, after stable schedule-induced polydipsia developed on an FI-30 sec schedule, animals were implanted bilaterally in the LH with bipolar stainless steel electrodes and tested until stable schedule-induced polydipsia returned. Animals were tested for the effects of brain stimulation in a different test chamber in another room. Only animals which ate during the stimulation through one pair of electrodes and displayed forward move-

References p. 392–394

ments and other locomotor activity, but no eating or other types of behavior, during stimulation through the other pair of electrodes were selected for further study. Constant current electrical stimulation for 3.0 sec duration through the pair of electrodes which resulted in the elicitation of eating was then substituted for the delivery of the pellet on the fixed interval 30 sec schedule. The same procedures were followed in testing the other pair of LH electrodes in each animal.

Of the 4 animals, only one displayed a clear and adequate substitution of LH electrical stimulation for the delivery of the food pellet as illustrated in Figs. 14–18.

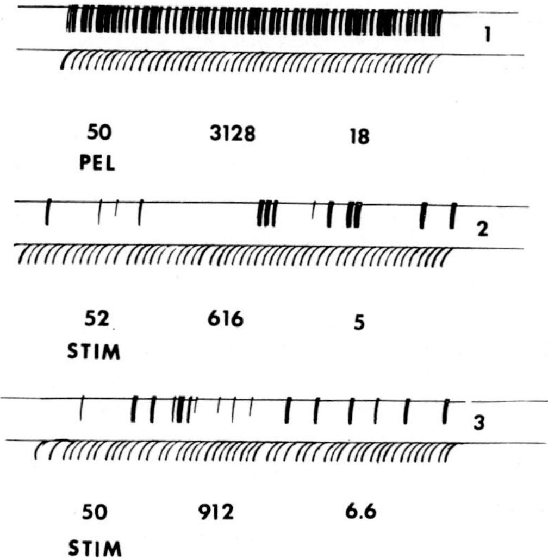

Fig. 14. Photographs of cumulative recordings. 1: typical post pellet schedule-induced polydipsia; each deflection in the top tracing represents 5 licks; each deflection in the bottom tracing indicates delivery of a pellet. 50 pellets (PEL) were delivered, 3128 licks occurred, and 18 ml of water were consumed. 2: same as 1 except the LH was stimulated electrically 52 times in place of the delivery of pellets. 3: same as 2. Reproduced with permission from Wayner *et al.* (1973).

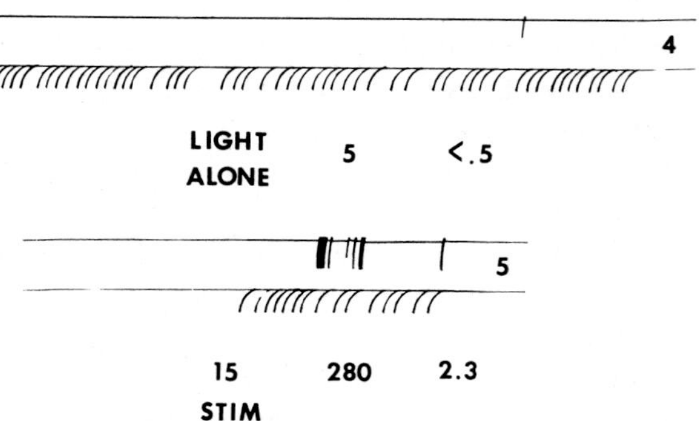

Fig. 15. Same as in Fig. 14 except in part 4 pellets and brain stimulation were not delivered. Pressing the bar after the 30 sec interval had elapsed turned on the signal lights for 2 sec (LIGHT ALONE). In part 5 the animal received 15 periods of LH electrical stimulation (STIM). Reproduced with permission from Wayner *et al.* (1973).

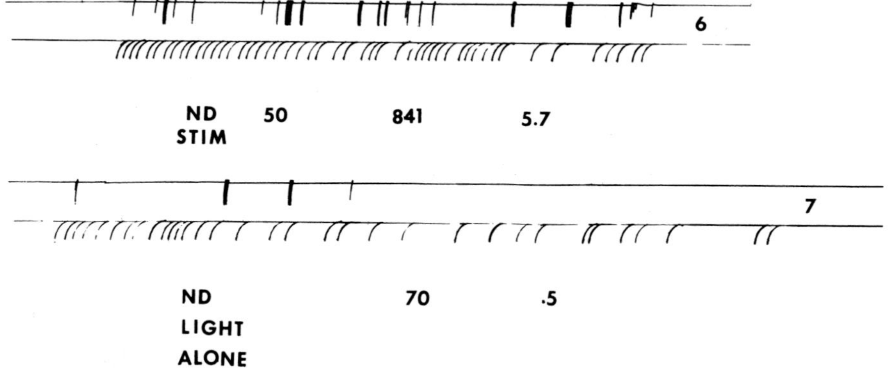

Fig. 16. Same as Fig. 14 except in part 6 the animal is not deprived (ND) and received LH electrical stimulation (STIM). In part 7 the animal is not deprived (ND) and does not receive any pellets or LH stimulation but the signal lights alone (LIGHT ALONE). Reproduced with permission from Wayner *et al.* (1973).

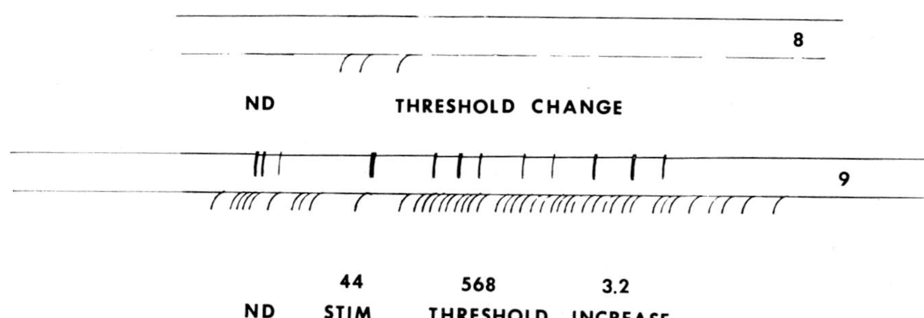

Fig. 17. Same as Fig. 14. The animal is not deprived (ND). The threshold for LH electrical stimulation increased in part 8. In part 9 the animal received a higher intensity LH electrical stimulation, otherwise same as part 6 in Fig. 16. Reproduced with permission from Wayner *et al.* (1973).

As the test conditions were different for the other 3 animals and because of variations in electrode position only the results on one animal will be presented.

In Fig. 14 there are 3 pairs of tracings labeled 1, 2, and 3 on the right side. The top tracing in each pair indicates every fifth lick by a downward deflection of the pen. The downward deflection of the pen in the lower tracing of each pair indicates the delivery of the pellet. The schedule-induced post-pellet licking and drinking in the top pair labeled 1 are obvious. The animal received 50 pellets (50 PEL), licked 3128 times and consumed 18 ml of water during the session. In this experiment the number of lever presses was not recorded. The animal was not removed from the test cage but the test session was continued, as indicated in part 2 of Fig. 14, and LH electrical stimulation (STIM) through the functional pair of electrodes which had previously resulted in eating was substituted for the delivery of the 45 mg pellet of food. The intensity of the electrical stimulation was the same as that employed earlier, 12–13 μA. Considerable post LH electrical stimulation licking and drinking occurred as indicated by the downward deflections of the upper tracing. Following 52 3-sec periods of LH electrical

References p. 392–394

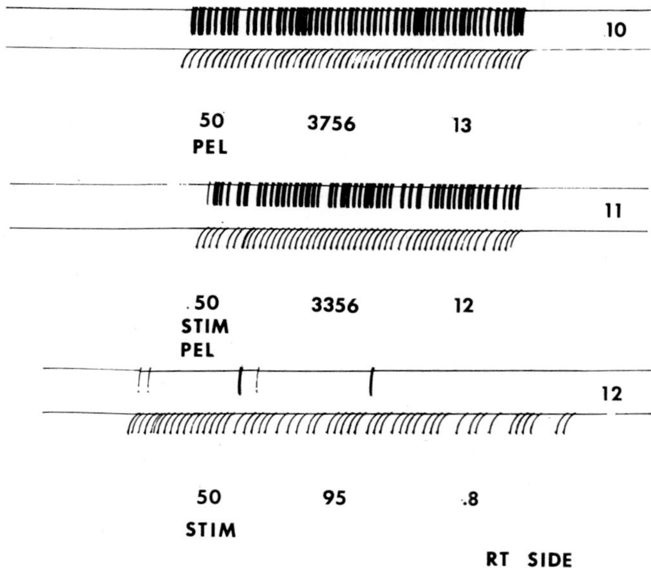

Fig. 18. Same as Fig. 14. Part 10 same as part 1 in Fig. 14. In part 11 the animal received both pellets (PEL) and LH stimulation (STIM) simultaneously. In part 12, the nonfunctional pair of LH electrodes on the right side (RT SIDE) were stimulated. Reproduced with permission from Wayner *et al.* (1973).

stimulation, the rat licked 616 times and drank 5 ml of water. The test session was then extended for another 50 periods of LH electrical stimulation as indicated in part 3 of Fig. 14. The animal licked 912 times and consumed 6.6 ml of water during this third consecutive test period. Following this period the animal was left in the test chamber to press the lever for the lights alone (LIGHT ALONE), as illustrated in part 4 of Fig. 15 and did not receive LH electrical stimulation or pellets of food. After the lights had been turned on 50 times, this session was terminated. The animal had licked 5 times and consumed less than 0.5 ml of water. Following this session the animal was tested again with the stimulus lights plus LH electrical stimulation as illustrated in part 5 of Fig. 15. Following 15 periods of LH electrical stimulation the animal licked 280 times and drank 2.3 ml of water.

For the next 7 days the animal was placed on *ad lib.* feeding until its body weight returned to normal. The animal was then returned to the test chamber where it was treated as before except that it was not deprived and received LH electrical stimulation in place of the food pellet on an FI-30 sec schedule. Results are illustrated in part 6 of Fig. 16 where ND indicates nondeprived. During the period of 50 LH stimulations the animal licked 841 times and drank 5.7 ml of water. Following this session the constant current device was turned off and the animal remained in the test chamber until it had depressed the lever a sufficient number of times to receive 33 light signals alone. As indicated in part 7 of Fig. 16 the animal licked the tube 70 times and consumed 0.5 ml of water during this period. On the following day an attempt to repeat the results of the last session failed and it appeared as if the threshold for LH electrical stimulation had increased. The failure is illustrated in part 8 of Fig. 17. The animal was then retested for eating during LH electrical stimulation and the stimulating current had to be increased from 12–13 to 15–16 μA before eating would occur reliably. Apparently the electrode assembly became loose. The animal was then tested as usual, employing the more intense electrical stimulation in place of the food pellets, as illustrated in part 9 of Fig. 17, and it licked 568 times and drank 3.2 ml of water following 44 LH stimulations.

For the next 7 days the animal was partially deprived of food and reduced to 80% of its free feeding body weight. On the next day it was tested on a standard FI-30 sec schedule with food pellet reinforcement and the schedule-induced polydipsia appeared normal, as illustrated in part 10 of Fig. 18, and the animal licked 3756 times and consumed 13 ml of water. Results were very similar to those of part 1 in Fig. 14. Because of the odd chance that these results were not due to the LH electrical stimulation and that the LH stimulation might even be disruptive in some respect, the animal was tested with LH stimulation applied simultaneously with the delivery of the pellet. These results are illustrated in part 11 of Fig. 18. The animal licked 3356 times and drank 12 ml of water, only slightly less on both measures as compared to the prior session. As it seemed obvious that the electrode assembly was becoming loose and would probably break free, the animal was tested again for eating and other behavioral changes during electrical stimulation through the nonfunctional pair of electrodes on the animal's right side (RT SIDE) and then returned to the test chamber. Electrical stimulation was delivered to the LH in place of the food pellets on an FI-30 sec schedule as before. Results are illustrated in part 12 of Fig. 18. The animal licked 95 times primarily in two bursts, and consumed 0.8 ml of water.

These results indicate that it is possible to substitute effectively electrical stimulation of the LH for the delivery of the food pellet in a typical schedule-induced polydipsia experiment. However, the placements of the electrodes are critical. This is particularly true for well placed bilateral LH implants which might destroy sufficient tissue to eliminate the schedule-induced polydipsia. Therefore, the bilateral implants must be asymmetrical but similar in placement to provide adequate control data. For the animal in Figs. 14–18, histological examination revealed that both pairs of stimulating electrodes were in the LH but the functional pair was more posterior. The signal lights in the test box are also important and, although the exact role is not known, do contribute to the effect.

DISCUSSION

These results demonstrate that the LH is involved in the control of spinal reflex excitability and motor activity. Electrical stimulation of the LH, salt arousal of drinking, schedule-induced polydipsia, and water deprivation all produce a common initial state of increased responsiveness from which specific behavioral sequences emerge. The development of stereotyped behavior elicited by a given set of environmental stimuli depends upon positive sensory feedback from the interacting muscles

References p. 392–394

into the LH. Muscular activity which produces sensory feedback coincidental with the internal sustaining stimulation will endure by maintaining a high level of excitability in these pathways, whereas sensory feedback which is not in phase will not produce coincidental positive feedback and these effects will habituate more rapidly and become increasingly less effective. A rat drinks not because of a need for water or thirst or because some drinking center in the brain is activated but initially due to a general increase in motor excitability mediated via the LH and the incidental chance occurrence of making contact with the drinking device and the first burst of licking. The sensory feedback from this interaction seems essential for drinking to continue and to occur again under similar circumstances in the future. With repeated exposures to the same external stimuli and internal stimulation, stereotyped drinking appears to follow the same general course of development under all 4 sets of experimental conditions and indicates a common neural mechanism. At present the specific mechanisms and pathways involved are relatively unknown and, because of the profuse and intricate interconnections of the lateral preoptic–lateral hypothalamic neuropil and brain stem reticulum, might remain obscure indefinitely.

ACKNOWLEDGEMENTS

This research was supported by NSF Grants GB-18414X and GB-35506, NIMH Grant 15473 and Training Grant MH-06969.

REFERENCES

Adolph, E. F. (1964) Regulation of body water content through water ingestion. In *Thirst in the Regulation of Body Water*, M. J. Wayner (Ed.), Pergamon Press, New York, pp. 5–17.

Andersson, B. (1972) Receptors subserving hunger and thirst. In *Handbook of Sensory Physiology, Vol. 3, Enteroceptors*, E. Neil (Ed.), Springer, New York, pp. 187–216.

Blass, E. M. and Epstein, A. N. (1971) A lateral preoptic osmosensitive zone for thirst in the rat. *J. comp. physiol. Psychol.*, **76**, 378–394.

Brooks, C. McC., Ushiyama, J. and Lange, G. (1962) Reactions of neurons in or near the supraoptic nuclei. *Amer. J. Physiol.*, **202**, 487–490.

Caggiula, A. R. (1970) Analysis of the copulation-reward properties of posterior hypothalamic stimulation in male rats. *J. comp. physiol. Psychol.*, **70**, 399–412.

Emmers, R. (1959) The effects of brain-stem stimulation on hind-leg movements in the hooded rat. *Exp. Neurol.*, **1**, 171–186.

Emmers, R. (1973) Interaction of neural systems which control body water. *Brain Res.*, **49**, 323–347.

Emmers, R. and Passamonte, P. (1972) Ineffectiveness of osmotic stimuli to induce water intake in rats with lesioned thalamic taste nucleus. *Physiologist*, **15**, 126.

Epstein, A. N. (1971) The lateral hypothalamic syndrome: its implications for the physiological psychology of hunger and thirst. In *Progress in Physiological Psychology, Vol. 4*, E. Stellar and M. Sprague (Eds.), Academic Press, New York, pp. 263–317.

Epstein, A. N. and Teitelbaum, P. (1964) Severe and persistent deficits in thirst produced by lateral hypothalamic damage. In *Thirst in the Regulation of Body Water*, M. J. Wayner (Ed.), Pergamon Press, New York, pp. 395–410.

Euler, C. von and Soderberg, U. (1957) The influence of hypothalamic thermoreceptive structures on the EEG and gamma motor activity. *Electroenceph. clin. Neurophysiol.*, **9**, 391–408.

Falk, J. L. (1964) Studies on schedule induced polydipsia. In *Thirst in the Regulation of Body Water*, M. J. Wayner (Ed.), Pergamon Press, New York, pp. 95–116.

FALK, J. L. (1969) Conditions producing psychogenic polydipsia in animals. *Ann. N.Y. Acad. Sci.*, **157**, 569–593.

FALK, J. L. (1971) The nature and determinants of adjunctive behavior. *Physiol. Behav.*, **6**, 577–588.

GLADFELTER, W. E. AND BROBECK, J. R. (1962) Decreased spontaneous locomotor activity in the rat induced by hypothalamic lesions. *Amer. J. Physiol.*, **203**, 811–817.

HALPERN, B. P. AND TAPPER, D. N. (1971) Taste stimuli: quality coding and time. *Science*, **171**, 1256–1258.

JANSEN, J. K. S. (1966) On fusimotor reflex activity. In *Muscular Afferents and Motor Control*, R. GRANIT (Ed.), Wiley, New York, pp. 91–105.

KAWAMURA, Y., KASAHARA, Y. AND FUNAKOSHI, M. (1970) A possible brain mechanism for rejection behavior to strong salt solution. *Physiol. Behav.*, **5**, 67–74.

KING, M. B. AND HOEBEL, B. G. (1968) Killing elicited by brain stimulation in rats. *Commun. behav. Biol., Part A*, **2**, 173–177.

KOELLA, W. P. (1969) Control of skeletal motor activity, with emphasis on the role of diencephalic mechanisms. In *The Hypothalamus*, W. HAYMAKER, E. ANDERSON AND W. J. H. NAUTA (Eds.), Thomas, Springfield, Ill., pp. 645–658.

LEVITSKY, D. AND COLLIER, G. (1968) Schedule-induced wheel running. *Physiol. Behav.*, **3**, 571–573.

MARSHALL, J. F., TURNER, B. H. AND TEITELBAUM, P. (1971) Sensory neglect produced by lateral hypothalamic damage. *Science*, **174**, 523–525.

MENDELSON, J. AND CHILLAG, D. (1970) Schedule-induced air licking in rats. *Physiol. Behav.*, **5**, 535–537.

MILES, P. R. AND GLADFELTER, W. E. (1969) Lateral hypothalamus and reflex discharges over ventral roots of rat spinal nerves. *Physiol. Behav.*, **4**, 671–675.

MILLHOUSE, O. E. (1969) A Golgi study of the descending medial forebrain bundle. *Brain Res.*, **15**, 341–363.

MORGANE, P. J. (1964) Limbic-hypothalamic-midbrain interaction in thirst and thirst motivated behavior. In *Thirst in the Regulation of Body Water*, M. J. WAYNER (Ed.), Pergamon Press, New York, pp. 429–455.

MORGANE, P. J. (1969) The function of the limbic and rhinic forebrain-limbic midbrain systems and reticular formation in the regulation of food and water intake. *Ann. N.Y. Acad. Sci.*, **157**, 806–848.

MORGANE, P. J. (1974) Anatomical basis of the central control of physiological regulation and behavior. In *Regulation of Food and Water Intake*, G. J. MOGENSON (Ed.), Western Ontario Press, London, Ontario, in press.

MYERS, R. D. AND VEALE, W. L. (1971) Spontaneous feeding in the satiated cat evoked by sodium or calcium ions perfused within the hypothalamus. *Physiol. Behav.*, **6**, 507–512.

NICOLAIDIS, S. (1969a) Discriminatory responses of hypothalamic osmosensitive units to gustatory stimulation in cats. In *Olfaction and Taste*, C. PFAFFMANN (Ed.), Rockefeller University Press, New York, pp. 569–573.

NICOLAIDIS, S. (1969b) Early systemic responses to orogastric stimulation in the regulation of food and water balance: functional and electrophysiological data. *Ann. N.Y. Acad. Sci.*, **157**, 1176–1203.

NORGREN, R. (1970) Gustatory responses in the hypothalamus. *Brain Res.*, **21**, 63–77.

NOVIN, D. AND DURHAM, R. (1969) Unit and D-C potential studies of the supraoptic nucleus. *Ann. N.Y. Acad. Sci.*, **157**, 740–754.

OOMURA, Y., ONO, T., OOYAMA, H. AND WAYNER, M. J. (1969) Glucose and osmosensitive neurons of the rat hypothalamus. *Nature (Lond.)*, **222**, 282–284.

PANKSEPP, J. (1971) Aggression elicited by electrical stimulation of the hypothalamus in albino rats. *Physiol. Behav.*, **6**, 321–329.

ROSENQUIST, A. C. AND HOEBEL, B. G. (1968) Wheel running elicited by electrical stimulation of the brain. *Physiol. Behav.*, **3**, 563–566.

SHAPOVALOV, A. I. (1966) Excitation and inhibition of spinal neurons during supraspinal stimulation. In *Muscular Afferents and Motor Control*, R. GRANIT (Ed.), Wiley, New York, pp. 331–348.

SKINNER, B. F. AND MORSE, W. H. (1957) Concurrent activity under fixed-interval reinforcement. *J. comp. physiol. Psychol.*, **50**, 279–281.

VALENSTEIN, E. S. (1969) Proceedings of a symposium: motivation, emotion and behavior elicited by activation of central nervous system sites. *Brain Behav. Evol.*, **2**, 289–376.

VALENSTEIN, E. S., COX, V. C. AND KAKOLEWSKI, J. W. (1969) The hypothalamus and motivated behavior. In *Reinforcement and Behavior*, J. T. TAPP (Ed.), Academic Press, New York, pp. 242–285.

VILLARREAL, J. (1972) *Schedule-induced pica.* Personal comm. Paper read at EPA Meetings held in Boston, Mass., April, 1972.
WAYNER, M. J. (1970) Motor control functions of the lateral hypothalamus and adjunctive behavior. *Physiol. Behav.*, **5**, 1319–1325.
WAYNER, M. J. (1974) Specificity of behavioral regulation. In *Regulation of Food and Water Intake*, G. J. MOGENSON (Ed.), Western Ontario Press, London, Ontario, in press.
WAYNER, M. J. AND CAREY, R. J. (1973) Basic drives. *Ann. Rev. Psychol.*, **24**, 53–80.
WAYNER, M. J. AND EMMERS, R. (1959) A test of the thirst deprivation trace hypothesis in the hooded rat. *J. comp. physiol. Psychol.*, **52**, 112–115.
WAYNER, M. J. AND GREENBERG, I. (1973a) Effects of D-amphetamine on schedule induced polydipsia. *Pharmacol. Biochem. Behav.*, **1**, 109–111.
WAYNER, M. J. AND GREENBERG, I. (1973b) Schedule dependence of schedule induced polydipsia and lever pressing. *Physiol. Behav.*, **10**, 965–966.
WAYNER, M. J. AND KAHAN, S. (1969) Central pathways involved during the salt arousal of drinking. *Ann. N.Y. Acad. Sci.*, **157**, 701–722.
WAYNER, M. J., KAHAN, S. AND STOLLER, W. (1966) Lateral hypothalamic mediation and spinal reflex facilitation during salt arousal of drinking. *Physiol. Behav.*, **1**, 341–350.
WAYNER, M. J., GREENBERG, I., FRALEY, S. AND FISHER, S. (1973) Effects of Δ^9-tetrahydrocannabinol and ethyl alcohol on adjunctive behavior and the lateral hypothalamus. *Physiol. Behav.*, **10**, 109–132.
WHITE, S. D., WAYNER, M. J. AND COTT, A. (1970) Effects of intensity, water deprivation, prior water ingestion and palatability on drinking evoked by lateral hypothalamic electric stimulation. *Physiol. Behav.*, **5**, 611–619.
WOLF, G. AND SUTIN, J. (1966) Fiber degeneration after lateral hypothalamic lesions in the rat. *J. comp. Neurol.*, **127**, 137–155.
ZEIGLER, H. P. (1964) Displacement activity and motivational theory: a case study in the history of ethology. *Psychol. Bull.*, **61**, 362–376.

DISCUSSION

DYER: During the presentation of your physiological data you showed various neurons, changing their firing rate following intracarotic injection of hypertonic saline. We know that such an injection often arouses the animal. Since there is good evidence that the activity of many neurons within the hypothalamus is related to changes in the EEG, I would like to know whether you recorded EEG-changes.

WAYNER: We did not record EEG during these experiments. We used small amounts of hypertonic saline, 0.05 ml of 6–9%. If larger doses are used, then cells begin to respond all over, particularly in the limbic system.

DYER: Do you think that the cells are responding to hypertonic saline are osmoreceptors?

WAYNER: My own personal thinking is that they are not osmoreceptors.

The Role of Hypothalamic Noradrenergic Neurons in Food Intake Regulation

J. L. SLANGEN

Rudolf Magnus Institute for Pharmacology, University of Utrecht, Utrecht (The Netherlands)

INTRODUCTION

Evidence has accumulated to support the existence of adrenergic receptor systems in the hypothalamus. Numerous biochemical studies have shown noradrenaline (NA) and dopamine (DA) to be present in this part of the brain (Holzbauer and Sharman, 1972). The Falck and Hillarp histochemical technique for visualizing monoamines provided evidence for the intraneuronal localization of noradrenaline (NA), dopamine (DA) and serotonin (5HT) to specific systems of cell bodies and nerve terminals (Falck *et al.*, 1962; Carlsson *et al.*, 1962). Also recently, a detailed mapping of the catecholamine pathways in the CNS has been completed (Ungerstedt, 1971a). A relatively small number of NA *cell bodies* in the pons and medulla give rise to large ascending and descending tracts. In the hypothalamus noradrenergic pathways are found mainly in the lateral part. Recently NA cell bodies were also found to be present in the caudal thalamus and posterior dorsal hypothalamus (Björklund and Nobin, 1973). Noradrenergic nerve *terminals* are found in a relatively small zone along the central part of the hypothalamus and in the periventricular area, the dorsomedial nucleus, nucleus arcuatus and perifornical area. Heavy innervation is evident in the supraoptic nucleus and the paraventricular nucleus. Dopaminergic *terminals* are mainly found in the median eminence. Several groups of DA *cell bodies* have been localized: (1) A8 and A9 to caudate nucleus, A10 to the accumbens nucleus (nigrostriatal system); (2) A12 (tubero-infundibular system); (3) a rostral periventricular cell group; (4) a group in the arcuate nuclei without projections to the median eminence; (5) a group in dorsal hypothalamus and caudal hypothalamus (Björklund and Nobin, 1973). An important DA *fiber system* (the nigrostriatal DA system) runs through the far lateral hypothalamus in the tip of the internal capsule.

Although the DA- and NA-systems represent only a small part of the total population of neurons in the hypothalamus, there is evidence that they form functionally different neuronal substrates for *e.g.*, neuroendocrine and temperature regulations (Hökfelt and Fuxe, 1972; Myers, 1969).

A specific technique used in investigating hypothalamic neural systems is the technique of injecting small amounts of chemicals, that are known to affect in a specific way a particular type of neuron, directly into the hypothalamus and then observing

References p. 405–406

their effects in unanesthetized and unrestrained subjects. A chronic cannula is implanted in the brain which allows for repeated drug injections into a specific site when the animal has recovered from the operation. Evidence for the physiological role of the lateral hypothalamus based only on electrical stimulation and ablation methods does not seem to establish conclusively whether the observed effects are due to destruction or stimulation of the numerous fiber tracks that pass through this area or due to destruction or stimulation of the relatively few cell concentrations. Thus direct pharmacological analysis of this inhomogenous region by application of drugs seems particularly interesting for investigating the physiological role of this area. This report will concentrate only on the role of the rat hypothalamic adrenergic system(s) involved in food intake.

Grossman first showed that the placement of small amounts (1–5 μg) of crystalline adrenaline (A) or NA in the area located between the fornix and mammillo-thalamic tract, lateral and dorsal to the ventromedial nuclei, induces vigorous and prolonged eating in satiated rats and enhances food intake in food-deprived rats (Grossman, 1960, 1962a). Control experiments revealed that strychnine sulfate, crystalline NaCl, (osmotic stimulation), barium chloride (vasoconstrictor), posterior pituitary extract, sodium nitrite (vasodilator), as well as substances with different pH values were not able to duplicate the effects of neuro-hormonal stimulation (Grossman, 1962a). Grossman found in addition that the *systemic* administration of 2.5 and 1 mg/kg of ethoxybutamoxane, a central adrenergic blocking agent, 1 h before the direct central adrenergic stimulation, prevented the previously observed eating effect. The food consumption of hungry rats was also reduced by previous central application of the adrenergic blocking agent phenoxybenzamine (Dibenzyline) (Grossman, 1962b). In 1964 Miller *et al.* showed in a dose-response study that about 11 μg of NA was an optimal dose for centrally eliciting eating of *liquid* food. When this dose of NA was injected into the jugular vein, via a permanently implanted catheter, of hungry rats they stopped eating or stopped pressing a bar for food delivery. Thus, the peripheral administration of NA reduced food intake (Miller, 1965). Further evidence for a hypothalamic adrenergic "feeding" system has been obtained by several investigators. It has been shown that inhibitors of the re-uptake of NA into the presynaptic terminal augment the eating response to exogenous NA in satiated rats (Booth, 1968; Slangen and Miller, 1969). Normal food intake in hungry rats is also enhanced when a re-uptake inhibitor like DMI (desmethylimipramine) is injected into the perifornical area (Montgomery *et al.*, 1969). It has been found that only a substance that specifically activates α-receptors consistently elicits eating in satiated rats. A specific activator of β-receptors has no such effect. Only central application of an adrenergic α-blocking agent and not a β-blocking agent can prevent eating after intrahypothalamic NA administration. Acetylcholine, DA, 5HT, histamine, NA precursors and catabolites as well as metabolically unrelated structural analogs of NA are unable to elicit an eating response when administered in the perifornical area (Booth, 1968; Slangen and Miller, 1969). Slangen and Miller (1969) have also found that the central administration of an MAO inhibitor followed by the local application of a substance which causes an intracellular increase of free NA induces eating in satiated rats. The local application of these drugs in the reverse sequence has no effect. NA has been injected into many different sites of the rat's diencephalon. Most of the evidence points to only one site in the hypothalamus that is specifically sensitive to the eating effect of adrenergic agonists, namely the perifornical area in the mediolateral hypothalamus just posterior to the anterior hypothalamus (Booth, 1967). The anatomical specificity of the effects of NA stimulation of the perifornical region is supported by the generally reported difficulty in consistently making implants into responsive loci even within the LH. The behavioral response elicited by NA when applied to this region seems to be restricted to food intake since other responses have not been observed. All of these results lend support to the hypothesis that in the hypothalamus of the rat a specifically α-adrenergic system is involved in mediating eating behavior in rat under normal circumstances.

When the β-adrenergic agonist isoproterenol was injected peripherally in *hungry* rats this substance was found to have an anorexic effect (Conte *et al.*, 1968) which could be partially blocked by a *peripherally* injected β-adrenergic blocker. Although these authors worked with peripherally administered drugs, they suggested that the effects of isoproterenol were mediated by a β-adrenergic system in the brain.

The hypothesis of a hypothalamic β-adrenergic system which functions antagonistically by suppressing eating behavior has been extensively tested by Leibowitz. Perifornical hypothalamic injections of 150 nmoles isoproterenol (about 37 μg) reduced food intake by about 60% in rats that were food-deprived for 18 h. Food intake returned to normal when this dose of isoproterenol was injected subsequent to propranolol. This β-adrenergic blocking agent was injected, also directly into the hypothalamus, in a dose of 140 nmoles. The α-adrenergic blocking agent phentolamine (70 nmoles) significantly enhanced the suppression of food intake after isoproterenol. The administration of these blockers alone was without effect on the food intake of 18 h food-deprived rats. In this respect the lack of effect of 70 nmoles phentolamine (in 4 μl) is somewhat surprising (Leibowitz, 1970a).

Similar results have been obtained by Goldman *et al.* (1971). These authors also observed a 65% decrease in food intake in 21 h food-deprived rats when 30 μg isoproterenol were administered in the perifornical area. This effect was reversed by pretreatment with 44 μg propranolol (and enhanced by the α-adrenergic antagonist tolazoline). When these drugs were injected *subcutaneously* in rats that were food-deprived for 21 h, both qualitatively and quantitatively similar results were obtained. These experiments have been replicated recently by Lehr and Goldman (1973). They confirmed that in 23 h food-deprived rats, peripherally administered isoproterenol (0.1 mg/kg s.c.) caused a more than 60% inhibition of the 1 h food intake which was obviated by propranolol (6.2 mg/kg) and enhanced by tolazoline (8.5 mg/kg).

The anorexic effect of peripheral D-amphetamine (0.5 mg/kg s.c.) is almost as strong as the aforementioned anorexic effect of s.c. administered isoproterenol (Lehr and Goldman, 1973). Anorexia induced by D-amphetamine cannot be inhibited however by pretreatment with propranolol, a drug that readily passes the blood–brain barrier. This amphetamine effect, therefore, does not seem to be dependent upon activation of a β-adrenergic satiety center as suggested by Leibowitz on the basis of experiments in which intrahypothalamically administered amphetamine (50–200

References p. 405–406

nmoles) suppressed the food intake of food-deprived rats and pretreatment with propranolol completely blocked amphetamine's anorexic effect (Leibowitz, 1970b).

Leibowitz, Lehr and their associates interpret their results as evidence for a central β-adrenergic satiety mechanism. This question, whether there exists such a β-adrenergic system in the hypothalamus of the rat, remains open, however, for the simple reason that in order to obtain an equally large suppression of food intake isoproterenol has to be injected centrally in a dose that is equal to, or even larger than, the dose injected subcutaneously.

There is, however, additional evidence in favor of a β-satiety system. Leibowitz has reported differences between medial and lateral parts of the hypothalamus with respect to their responsiveness to alpha and beta drugs. In food-deprived rats, isoproterenol (100 and 150 nmoles) reduced food intake when injected in the perifornical area and in far lateral hypothalamic areas (1.5 and 1.8 mm lateral) but had no effect on food intake when injected in VMA and DMA. This suggests that isoproterenol's satiety effect may after all be a central effect unless we assume that this drug, when injected centrally, on its route to the periphery, cannot leave the brain equally effectively from different sites. As mentioned before the ventromedial area of the hypothalamus was found to be insensitive to β-drugs. But it was sensitive in the same direction and to a similar extent as the perifornical area to α-drugs. These results are interpreted as evidence for two types of adrenergic receptors in the hypothalamus which play opposing roles in the regulation of food intake and are concentrated in anatomically different areas of the hypothalamus: the VMA containing primarily α-adrenergic "feeding" receptors, the LHA containing primarily β-adrenergic "satiety" receptors and the perifornical area containing both α "feeding" and β "satiety" receptors (Leibowitz, 1970c). On the basis of behavioral deficits (hyperphagia, overeating) observed following the destruction of the VMA this area has been assigned the role of a "satiety" center. Subsequently neuroanatomical and electrophysiological investigations have been performed in an attempt to provide experimental proof that the proposed scheme of a ventromedial satiety center inhibiting a lateral hypothalamic feeding center was correct (Stevenson, 1969). Following this tradition Leibowitz has proposed that the actions of both α- and β-receptors are inhibitory because the α-receptors which elicit eating are concentrated in a "satiety" area and the β-receptors which suppress eating are concentrated in a "feeding" area. It is clear, however, that this hypothesis of the inhibitory action of α-receptors rests entirely upon the concept of the VMA as a "satiety center", a concept which not only on the basis of anatomical and electrophysiological evidence is open to criticism (Rabin, 1972) but can also be criticized because it has been shown that destruction or manipulation of the VMH is not always followed by disturbances in the regulation of food intake (Reynolds, 1963; Voloschin *et al.*, 1968).

The hypothesis of an inhibitory action of the β-receptors rests upon the concept of the LH as a "feeding" center, a concept which has originated mainly because lesions in the lateral hypothalamus caused aphagia and adipsia, and because electrical stimulation of the LH caused stimulus bound eating. However, it has been reported that bilateral electrolytical and chemical destruction of the nigrostriatal dopamine system, inside as well as outside the hypothalamus, is associated with a syndrome of adipsia and aphagia (Ungerstedt, 1971b) that is identical to the selective deficit in food regulation produced by lateral hypothalamic lesions (Fibiger *et al.*, 1973; Marshall and Teitelbaum, 1973). Since it has not been shown that electrolytic lesions in the LH also cause aphagia when the nigrostriatal system is not damaged, the aphagia observed after lesioning the LH cannot be taken as evidence for a "hunger" system in the LH.

RECENT OBSERVATIONS

In order to investigate the differential involvement of adrenergic neurons in different parts of the hypothalamus we performed experiments in which we selectively destroyed the catecholaminergic systems in different loci by means of 6-hydroxydopamine (6-OHDA) (Ungerstedt, 1971c). If the adrenergic neurons in the LH predominantly have a "satiety" function then the destruction of these neurons will lead to some form of hyperphagia. Like-wise destruction of adrenergic neurons in the medial hypothalamus will interfere with the proposed "hunger" function of these neurons: permanent inhibition will lead to some form of hypophagia.

In the following experiments, male albino rats of an inbred Wistar strain (weighing 200–400 g) were anesthetized and placed in a stereotactic apparatus. Using standard operating and stereotactic techniques, 6-OHDA was injected bilaterally into the hypothalamus through a 0.35 mm (outer diameter) needle which was mounted on a SGE 5 μl syringe. Eight μg 6-OHDA and 0.8 μg ascorbic acid, dissolved in distilled water, were injected slowly at each site in a volume of 1 μl. The vehicle-injected controls received only 1 μl distilled water at each site. In 60 animals, divided over 6 groups of 7 experimental and 3 control animals each, 6-OHDA or the vehicle was injected at the rostrocaudal level between A3990 and A4890 according to the atlas of König and Klippel (1963). The groups differed in that the injections were made at different distances from the midline of the brain and at different depths. The results are summarized in 2 tables and 3 figures.

Table I summarizes the sites at which 6-OHDA was administered, expressed as coordinates of the König and Klippel atlas, and the number of animals per group in which, after histological examination, the injection sites were found to be symmetrical. Two injections were made on each side of the brain. The drug or the vehicle was first injected at the greatest depth. Then the syringe and the needle were slowly moved towards the skull and 1 mm above the first injection site a second injection was made.

After the operation the rats were housed in individual cages and provided with food pellets and tap water *ad libitum*. Body weight was measured daily. Of some groups, food and water intake were also measured daily. After 2 or 3 weeks, or when rats showed signs of severe malaise, the animals received an overdose of pentobarbital and were perfused with saline and 10% buffered formalin. The brains were removed, allowed to harden and frozen for sectioning. Photomicrographs were made of the sections in which microscopically the deepest penetration of the needle was found.

It can be seen in Fig. 1 that injections made at about 1.6 mm from the midline are

References p. 405–406

TABLE I

COORDINATES OF INJECTION SITES AND NUMBER OF ANIMALS PER GROUP

Group	*A–P**	*Distance lateral from midline (mm)*	*Depth below horizontal zero (mm)*	*6-OHDA (n)*	*Control (n)*
1	4380–4890	0.5	2.5–3.5	6	2
2	4110–4620	0.8	2.5–3.5	5	2
3	3990–4890	1.0	2.5–3.5	7	2
4	4110–4890	1.2	2.2–3.2	4	3
5	4110–4890	1.4	2.0–3.0	5	2
6	4110–4890	1.6	1.6–2.6	6	2

* According to the atlas of König and Klippel (1963).

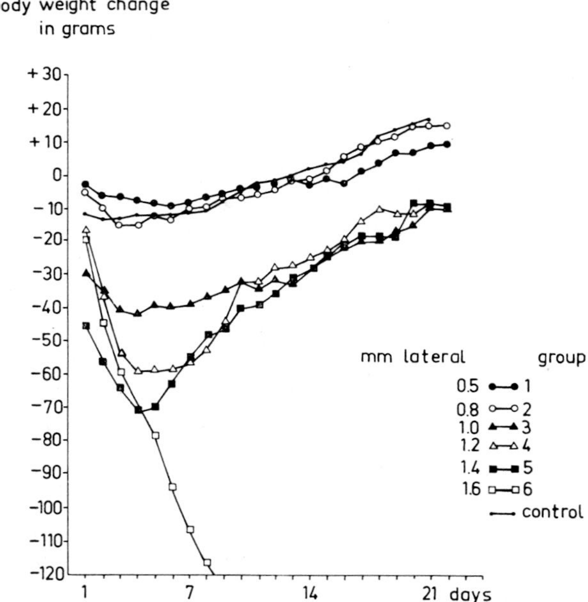

Fig. 1. Mean differences as compared to preoperative weight (= 0 gram) after bilateral administration of two doses of 8 μg 6-OHDA at different sites in rat hypothalamus (see text).

life threatening while injections made in the area from 0.5 to 1.4 mm laterally can cause a severe loss in body weight but do not prevent the animals recovering from the effects of 6-OHDA. About 10 days after the operation, the rate of increase in body weight tends to approach uniformity in all recovered animals and is not essentially different from the weight increase in control animals. Since no differences were found between body weight changes of the control animals of the different groups all control data are summarized in one curve in Fig. 1. All animals in the 1.6 mm lateral group were adipsic and aphagic immediately after the operation and never showed any interest again in water or food although in the first week after the operation they had a healthy appearance. In Fig. 2 (upper part) the 1.6 mm lateral injection site is superposed on the A4650 section taken from Ungerstedt's mapping study (Ungerstedt, 1971a). The black parallelogram in this A4650 drawing indicates the region in which injection sites were found that only transiently caused aphagia and adipsia. At all lateral levels there was a variation in anterior–posterior localization of about 1 mm (see Table I). Therefore it is not possible to correlate the loss of body weight of the recovered animals with the site of injection in an anterior–posterior direction. Probably the loss in body weight of the animals in the 1.4 mm lateral group was the greatest because the injections at that lateral level destroyed part of a system which is

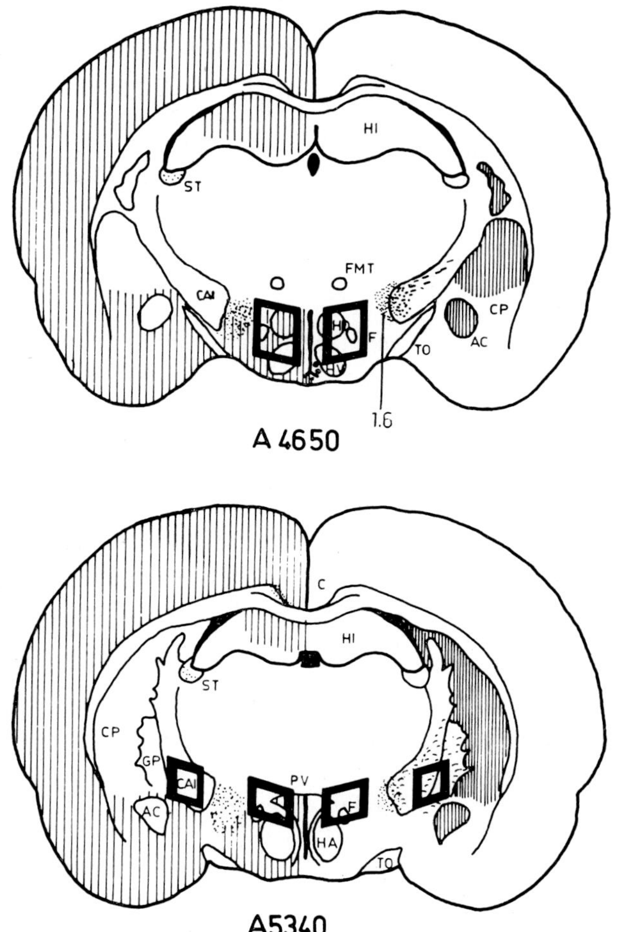

Fig. 2. Schematic drawing of areas in rat hypothalamus in which administration of 6-OHDA had *no* permanent effects on feeding. Noradrenergic fibers and terminals are indicated on the left side, dopaminergic fibers and terminals on the right side of the figure.

References p. 405–406

mainly located more laterally. And because only part of that system was destroyed recovery could take place.

The results of this experiment indicate that injecting a relatively small amount of 6-OHDA into the lateral part of the hypothalamus at 1.6 mm from the midline causes an absolute aphagia and adipsia, whereas no such effect can be seen after injecting the same amount of 6-OHDA into the region extending from the third ventricle to 1.4 mm lateral of it.

In an attempt to localize more precisely the area in the LH in which the administration of 6-OHDA causes aphagia and adipsia we used essentially the same methods. There were 10 groups of 6 experimental and 4 control animals each. The experimental animals received bilaterally *one single* injection of 8 μg 6-OHDA dissolved in 1 μl vehicle. The controls received likewise only the vehicle. Measurements of food intake, water intake and body weight were taken daily for 14 days, starting on the day of surgery. Each rat was given a weighed quantity of food pellets every day and after 24 h the remaining quantity, including any substantial spillage, was weighed to yield the amount of daily food intake. Fresh tap water was provided in inverted calibrated tubes and the water intake was read directly from the tubes.

All injection sites were aimed at the A–P level of 4620 μm and at positions 0.4, 0.8, 1.2, 1.6, 2.0, 2.4, 2.8 and 3.4 mm lateral to the midline. All coordinates are according to the atlas of König and Klippel (1963). Up to 1.4 mm lateral to the depth of the injecting site was kept constant at 3 mm below horizontal zero. From 1.4 mm lateral on, the depth was decreased in order to keep the injection sites parallel to the base of the brain and the optic tract.

Aphagic and adipsic animals were anesthetized and perfused when they were moribund. All other animals were sacrificed and perfused 2 or 3 weeks after surgery. Photographic records were made of the brain section showing the location of the needle tip.

Results are reported only for those animals which upon histological examination were found to have the needle tip perfectly symmetrical with respect to depth and lateral as well as A–P position. At this stage animals were assigned to different experimental groups depending on the site of injection. The specifications of the groups are summarized in Table II. In Fig. 3 the changes in body weight seen after the administration of 8 μg 6-OHDA are summarized. No statistical differences were found between data of control animals from different groups. Therefore these data are pooled and represented by only one curve. As can be seen in Fig. 3 the rate of body weight increase of all experimental animals in the groups 1, 2, 3, 4, 9 and 10 was roughly similar to the rate of body weight increase in control animals. In the 1.4 mm and 2.4 mm lateral groups the effect of 6-OHDA was severe but not life threatening. This is in sharp contrast with the effect of 6-OHDA in the groups 5, 6, 7 and 8 of which all animals became aphagic and adipsic. These animals showed a "hunched back" position and were hypoactive although they were not incapable of normal movements, including gnawing and licking movements as could incidentally be observed when these animals occasionally dragged food pellets or licked the water spontaneously from day 5 on. In groups 1, 2, 3 and 10 the mean daily food and water intake from the second

TABLE II

COORDINATES OF INJECTION SITES AND NUMBER OF ANIMALS PER GROUP

Group	*A–P**	*Distance lateral from midline (mm)*	*Depth below horizontal zero (mm)*	*6-OHDA (n)*	*Control (n)*
1	4230–4890	0.4–0.5	2.2–2.6	6	3
2	4230–4620	0.9–1.1	2.2–2.6	8	9
3	4380–4620	1.2–1.3	2.0–2.5	4	2
4	4230–4890	1.4	2.0–2.5	3	3
5	4380–4890	1.6	1.8–2.2	3	3
6	4380–4890	1.8	1.8–2.2	5	3
7	4380–5150	1.9–2.0	1.5–2.0	5	3
8	4380–5150	2.1	1.5–2.0	3	4
9	4380–4890	2.4–2.5	1.0–1.5	4	3
10	4380–4620	2.8–3.0	1.0–1.5	5	4

* According to the atlas of König and Klippel (1963).

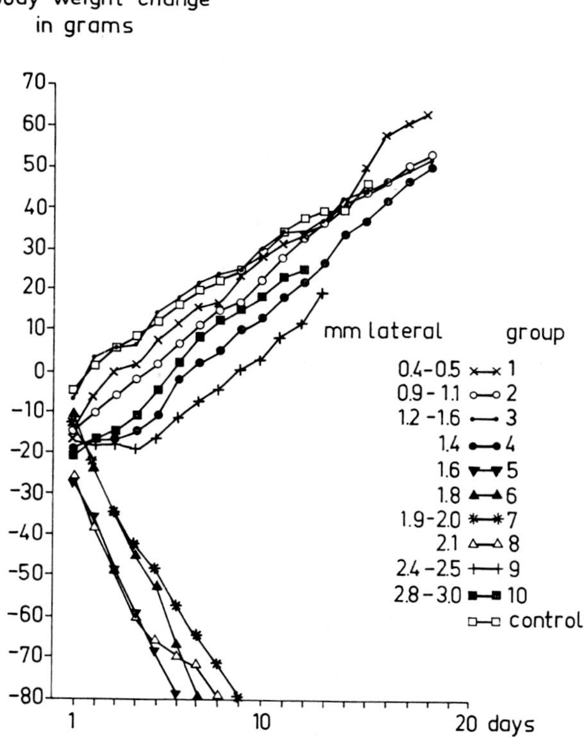

Fig. 3. Mean differences with preoperative weight (= 0 gram) after bilateral administration of a single dose of 8 μg 6-OHDA at different sites in rat hypothalamus.

References p. 405–406

postoperative day on was 23 g and 21 ml. This level of daily food and water intake was reached in group 4 on day 5 and in group 9 on day 8. The rate of body weight increase of non-aphagic animals in the first experiment is lower than the rate of increase of the non-aphagic animals in the second experiment. This effect can be ascribed to the double injection technique used in the first experiment.

Our results show clearly that 6-OHDA only affects food and water intake when injected in a very circumscribed region in the LH, extending from 1.6 to 2.2 mm lateral from the midline (see Fig. 2 lower part). This is precisely the area which contains the passing fibers from the nigrostriatal system as shown by the mapping studies of Ungerstedt (1971). Assuming that 6-OHDA affects intrahypothalamic aminergic interneurons and descending and ascending fibers alike, our findings strongly support Ungerstedt's suggestion that injecting 6-OHDA into the lateral hypothalamic area causes aphagia and adipsia because such a treatment results in an interruption of the nigrostriatal dopamine system.

The area from which other investigators have reported that eating can be elicited reliably by local application of sympathicomimetics does not extend further lateral than 1.4 mm. Therefore it is unlikely that the symptoms, seen after application of 6-OHDA more than 1.5 mm from the midline, are due to the interference with an intrahypothalamic aminergic neural network supposedly subserving feeding behavior.

In order to ascertain that 6-OHDA indeed destroys this intrahypothalamic noradrenergic system cannulas were implanted bilaterally in the perifornical region of 30 rats. Cannulated rats were tested 3 times. Injections were given every other day. In 13 rats the food intake after injection of 30 nmoles noradrenaline (NA) was 2.0 $\pm$ 1.1 g (mean $\pm$ S.D. in 1 h). Seven rats were weak eaters. Their mean eating response to 30 nmoles NA was 1.1 $\pm$ 0.7 g. After the NA-test, 8 μg 6-OHDA was injected bilaterally. It was found that 6-OHDA was still without effect on eating in these 20 rats. When 30 nmoles NA was injected 7 days after the 6-OHDA treatment an enhanced feeding response was obtained in two tests per cannula. The mean food intake of the strong eaters was 3.1 $\pm$ 1.3 g and of the weak eaters 3.2 $\pm$ 1.4 g. The difference between the mean response to NA before and after the 6-OHDA injection is statistically significant in both groups at the 0.1% level (Student's t-test). This enhanced eating response to NA suggests that a postsynaptic supersensitivity had developed as a result of the presynaptic 6-OHDA effect. In view of the fact that after 6-OHDA, NA had an enhanced effect it is highly unlikely that 6-OHDA had not reached the sites from which noradrenergic eating can be elicited. Nevertheless, when in the same animals a second intrahypothalamic injection of 16 μg 6-OHDA was given, no impairment of feeding was observed in 14 days.

In a series of experiments with 50 rats in which we implanted bilateral cannulas in the nigrostriatal system of the LH, we were unable to cause eating by injecting NA, but we also were unable to cause aphagia by injecting 6-OHDA via the cannula 2 weeks postoperatively. However, using the same coordinates, the same volume, and the same amount of 6-OHDA we always caused aphagia in rats when the 6-OHDA was injected via the needle during the stereotactic operation. A combination of factors may be responsible for this result. Perhaps the diffusion pattern at the tip of a permanently indwelling cannula is different from and less effective than the pattern that occurs at the tip of a needle which is withdrawn subsequent to the injection. If this is true, cannulas have to be placed very accurately on top of, or inside, a sensitive site in order to be effective. And in the case of the nigrostriatal system, we may have missed it altogether with our cannulas, or we may have correctly implanted only *one* cannula whereas a bilateral destruction of the nigrostriatal system is necessary for the aphagia effect. We have already presented evidence that the nigrostriatal system as it passes the LH is a relatively small structure of about 0.6 mm in a horizontal direction. Preliminary results from 20 rats indicate that the same seems to hold for the vertical direction. When 6-OHDA (8 μg) was injected bilaterally at A–P + 6.5, 2.0 mm lateral and at depths ranging from 6 to 9 mm below the skull (teethholder 5 mm above earbars) aphagia occurred only when the injections were made at a depth of 7.5 and 8.0 mm. The sites 6.0, 7.0, 8.5 and 9.0 mm deep were ineffective.

CONCLUSIONS

Our results do not support the localization of an adrenergic α-system in the VMA as proposed by Leibowitz because local destruction of this medial system did not lead to any form of hypophagia. The effect of 6-OHDA in the LH is highly dependent on the site of action. When the dopaminergic nigrostriatal system is interrupted, aphagia and adipsia occur. No effect on food and water intake is detected after injections of 6-OHDA in hypothalamic areas known to contain adrenergic terminals or fiber systems, including the perifornical area (PA) and the region around the ventromedial nucleus (VMA). Therefore, no *essential* role in the control of ingestive behavior should be assigned to the VMA or to the PA. The fact that structures or systems can mediate ingestive responses in the absence of the hypothalamus or in the absence of substructures of the hypothalamus does not necessarily mean that the hypothalamus plays no role in normal circumstances. The function of the nigrostriatal system in ingestive behavior remains to be elucidated.

REFERENCES

BJÖRKLUND, A. AND NOBIN, A. (1973) Fluorescence histochemical and microspectrofluorometric mapping of dopamine and noradrenaline cell groups in the rat diencephalon. *Brain Res.*, **51**, 193–205.

BOOTH, D. A. (1967) Localization of the adrenergic feeding system in the rat diencephalon. *Science*, **158**, 515–517.

BOOTH, D. A. (1968) Mechanism of action of norepinephrine in eliciting an eating response on injection into the rat hypothalamus. *J. Pharmacol. exp. Ther.*, **160**, 336–348.

CARLSSON, A., FALCK, B. AND HILLARP, N.-Å. (1962) Cellular localization of brain monoamines. *Acta physiol. scand.*, **56, Suppl. 196,** 1–28.

CONTE, M., LEHR, D., GOLDMAN, W. AND KRUKOWKI, M. (1968) Inhibition of food intake by beta-adrenergic stimulation. *Pharmacologist*, **10,** 180.

FALCK, B., HILLARP, N.-Å., THIEME, G. AND TORP, A. (1962) Fluorescence of catecholamines and related compounds condensed with formaldehyde. *J. Histochem. Cytochem.*, **10**, 348–354.

FIBIGER, H. C., ZIS, A. P. AND MCGEER, E. G. (1973) Feeding and drinking deficits after 6-hydroxydopamine administration in the rat: similarities to the lateral hypothalamic syndrome. *Brain Res.*, **55**, 135–148.

GOLDMAN, H. W., LEHR, D. AND FRIEDMAN, E. (1971) Antagonistic effects of alpha- and beta-adrenergically coded hypothalamic neurones on consummatory behaviour in the rat. *Nature (Lond.)*, **231**, 453–455.

GROSSMAN, S. P. (1960) Eating and drinking elicited by direct adrenergic or cholinergic stimulation of hypothalamus. *Science*, **132**, 301–302.

GROSSMAN, S. P. (1962a) Direct adrenergic and cholinergic stimulation of hypothalamic mechanisms. *Amer. J. Physiol.*, **202**, 872–882.

GROSSMAN, S. P. (1962b) Effects of adrenergic and cholinergic blocking agents on hypothalamic mechanisms. *Amer. J. Physiol.*, **202**, 1230–1236.

HÖKFELT, T. AND FUXE, K. (1972) On the morphology and the neuro-endocrine role of the hypothalamic catecholamine neurons. In *Brain–Endocrine Interaction*, K. M. KNIGGE, D. E. SCOTT AND A. WEINDL (Eds.), Karger, Basel, pp. 181–223.

HOLZBAUER, M. AND SHARMAN, D. F. (1972) The distribution of catecholamines in vertebrates. In *Catecholamines*, H. BLASCHKO AND E. MUSCHOLL (Eds.), Springer, Berlin, pp. 110–185.

KÖNIG, J. F. R. AND KLIPPEL, R. A. (1963) *The Rat Brain*, Williams and Wilkins, Baltimore, Md.

LEHR, D. AND GOLDMAN, W. (1973) Continued pharmacologic analysis of consummatory behaviour in the albino rat. *Europ. J. Pharmacol.*, **23**, 197–210.

LEIBOWITZ, S. F. (1970a) Hypothalamic beta adrenergic "satiety" system antagonizes an alpha adrenergic "hunger" system in the rat. *Nature (Lond.)*, **226**, 963–964.

LEIBOWITZ, S. F. (1970b) Amphetamine's anorexic *versus* hunger-inducing effects mediated respectively by hypothalamic beta- *versus* alpha-adrenergic receptors. *Proc. 78th Ann. Conv., APA*, 813–814.

LEIBOWITZ, S. F. (1970c) Reciprocal hunger regulating circuits involving alpha- and beta-adrenergic receptors located, respectively, in the ventromedial and lateral hypothalamus. *Proc. nat. Acad. Sci. (Wash.)*, **67**, 1063–1070.

MARSHALL, J. F. AND TEITELBAUM, PH. (1973) A comparison of the eating in response to hypothermic and glucoprivic challenges after nigral 6-hydroxydopamine and lateral hypothalamic electrolytic lesions in rats. *Brain Res.*, **55**, 229–233.

MILLER, N. E., GOTTESMAN, K. S. AND EMERY, N. (1964) Dose response to carbachol and norepinephrine in rat hypothalamus. *Amer. J. Physiol.*, **206**, 1384–1388.

MILLER, N. E. (1965) Chemical coding of behavior in the brain. *Science*, **148**, 328–338.

MONTGOMERY, R. B., SINGER, G., PURCELL, A. T., NARBETH, J. AND BOLT, A. G. (1969) Central control of hunger in the rat. *Nature (Lond.)*, **223**, 1278–1279.

MYERS, R. D. (1969) Temperature regulation: neurochemical systems in the hypothalamus. In *The Hypothalamus*, W. HAYMAKER, A. ANDERSON AND W. J. H. NAUTA (Eds.), Thomas, Springfield, Ill., pp. 506–520.

RABIN, B. M. (1972) Ventromedial hypothalamic control of food intake and satiety: a reappraisal. *Brain Res.*, **43**, 317–342.

REYNOLDS, R. W. (1963) Ventromedial hypothalamic lesions without hyperphagia. *Amer. J. Physiol.*, **204**, 60–62.

SLANGEN, J. L. AND MILLER, N. E. (1969) Pharmacological tests for the function of hypothalamic norepinephrine in eating behavior. *Physiol. Behav.*, **4**, 543–552.

STEVENSON, J. A. F. (1969) Neural control of food and water intake. In *The Hypothalamus*, W. HAYMAKER, E. ANDERSON AND W. J. H. NAUTA (Eds.), Thomas, Springfield, Ill., pp. 524–605.

UNGERSTEDT, U. (1971a) Stereotaxic mapping of the monoamine pathway in the rat brain. *Acta physiol. scand.*, **82, Suppl. 367**, 1–48.

UNGERSTEDT, U. (1971b) Adipsia and aphagia after 6-hydroxydopamine induced degeneration of the nigro-striatal dopamine system, *Acta physiol. scand.*, **82, Suppl. 367**, 95–122.

UNGERSTEDT, U. (1971c) Histochemical studies on the effects of intracerebral and intraventricular injections of 6-hydroxydopamine on monoamine neurons in the rat brain. In *6-Hydroxydopamine and Catecholamine Neurons*, T. MALMFORS AND H. THOENEN (Eds.), Elsevier, Amsterdam, pp. 101–127.

VOLOSCHIN, L., JOSEPH, S. A. AND KNIGGE, K. M. (1968) Endocrine function in male rats following complete and partial isolations of the hypothalamo–pituitary unit. *Neuroendocrinology*, **3**, 387–397.

DISCUSSION

DREWETT: I think you said that lesioning of the ventromedial hypothalamus does not cause hyperphagia. Are these your own data, or are you referring to the literature?

SLANGEN: Some investigators were able to cause hyperphagia by lesioning the ventromedial area (Cox *et al.*, 1969; Reynolds, 1963)*, but others failed to obtain hyperphagia. I have never been able to report positive results about hyperphagic animals, since I have always failed to obtain hyperphagia.

DREWETT: What type of lesions did you make in these experiments, and did you use male or female animals?

SLANGEN: We used male rats and have tried several types of lesioning techniques, *i.e.* DC-lesions and radiofrequency lesions.

DREWETT: I think that this is very interesting, because you measured food intake. One of the curious facts about male animals is that you can get an increase in food intake after the lesion, without an increase in body weight. Your data are the first that I heard of which definitely show that there is no increase in food intake after such a hypothalamic lesion. Most of the data in the literature only refer to body weight which does not allow a direct conclusion on food intake.

* COX, V. C., KAKOLEWSKI, J. AND VALENSTEIN, E. S. (1969) Ventromedial hypothalamic lesions and changes in body weight and food consumption in male and female rats. *J. comp. physiol. Psychol.*, **67**, 320–326.

REYNOLDS, R. W. (1963) Ventromedial hypothalamic lesions without hyperphagia. *Amer. J. Physiol.*, **204**, 60–62.

Intracranial Injection of Neurotransmitters and Hormones and Feeding in Sheep

J. MICHAEL FORBES AND CLIFTON A. BAILE

Department of Animal Physiology and Nutrition, University of Leeds, Leeds (Great Britain) and Smith Kline and French Laboratories, West Chester, Pa. and Monell Chemical Senses Center, University of Pennsylvania, Philadelphia, Pa. (U.S.A.)

INTRODUCTION

The control of feeding in ruminants differs in several important ways from that in the simple-stomached animal (Baile and Forbes, 1974). Because of this, and because of the economic importance of ruminants, work is in progress to study the central nervous control of feeding in sheep. This paper describes the effects of injections of putative neurotransmitters directly into hypothalamic tissue and of sex steroids into the cerebral ventricles.

METHODS

Castrate male sheep (40–60 kg) were prepared under general anesthesia with 6 guide tubes, 3 on each side of the midline, directed towards anterior, medial and lateral hypothalamus, using a modification of the X-ray technique of Tarttelin (1971). The guides were fixed to the skull with screws and acrylic resin and occluded with stylets. When the animals had reached a stable daily intake of the concentrate pelleted feed, offered *ad lib.*, injections were given by temporarily replacing a stylet with a fine tube which was advanced beyond the lower tip of the guide into hypothalamic tissue. The injection (1 μl) was given and the injection tube immediately withdrawn. The weight of food eaten at several time intervals after injection was recorded. Injections were at least 48 h apart. At the end of the experiments, the sheep were sacrificed, an injection of marker dye made into the hypothalamus, and the brain sectioned to identify injection loci according to the atlas of Richard (1967) after fixing in formol–saline.

In a second group of sheep access to lateral ventricles was obtained by implanting a tube in the parietal bone through which a needle could be advanced into the ventricle.

Each experiment was carried out with between 5 and 8 sheep.

References p. 414–415

RESULTS AND DISCUSSION

Intrahypothalamic injections

(1) Adrenergic agonists and antagonists. It was previously shown that noradrenaline injected into the lateral ventricle of satiated sheep caused increased feeding, with an optimum dose of 542 nmoles (Baile *et al.*, 1972). Noradrenaline injected into some loci in the anterior medial hypothalamus caused feeding with an optimum dose of 240 nmoles; this effect was blocked by pre-injection with the α-adrenergic antagonist, phenoxybenzamine, but not the β-adrenergic antagonist, LB46 (Fig. 1). Isoproterenol caused increased feeding at different loci with an optimum dose of 8 nmoles; this effect was blocked by LB46 but not by phenoxybenzamine (Fig. 1).

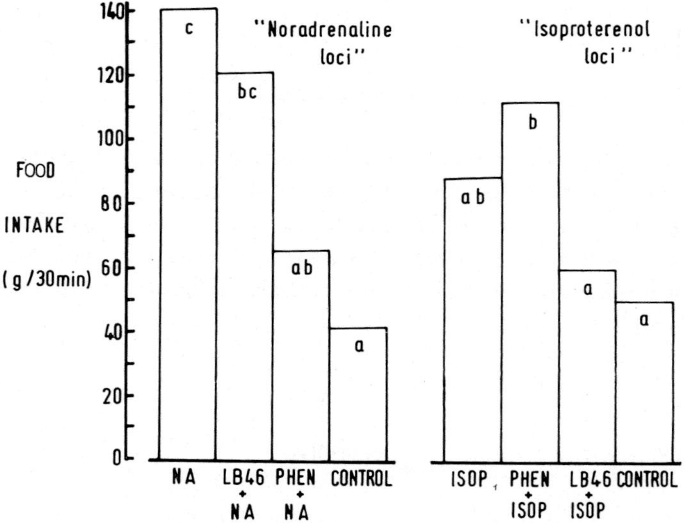

Fig. 1. Food intake of sheep following injections of α- and β-adrenergic agonists and antagonists into the anterior hypothalamus. Columns with different letters (a, b, c) differ significantly ($P < 0.05$). NA, noradrenaline; PHEN, phenoxybenzamine; ISOP, isoproterenol.

Lateral hypothalamic injection of isoproterenol had no effect on feeding whereas noradrenaline depressed food intake (Forbes and Baile, 1972). However, further more detailed studies failed to confirm this effect. Thus the responses to the α-adrenergic agonist introduced into the anterior hypothalamus gave results similar to those found in the rat; the β-agonist also stimulated intake in contrast to its effect in the rat (Leibowitz, 1970).

(2) Cholinergic agonists and antagonists. In the rat carbachol introduced into the anterior hypothalamus causes increased drinking but depresses feeding (Grossman, 1962). In the rabbit, however, low doses of carbachol into the lateral hypothalamus and preoptic area cause increased feeding (Sommer *et al.*, 1967).

Seventy-seven injections of 28 nmoles carbachol were given intrahypothalamically in 13 sheep. Twenty-eight of these induced a marked feeding response (175 ± 15 g in 60 min *versus* 44 ± 6 g for no response). Increased drinking was seen after 14 injections, only 3 of which also stimulated feeding. Eight of the feeding responsive loci were used for a dose-response study with doses of 1.75–112 nmoles; 3.5 nmoles gave a maximal response, with no further increase with greater doses. Eight more feeding responsive loci were used to study the chemical specificity of the carbachol effect. Atropine (28 nmoles) injected 3 min before carbachol (28 nmoles) blocked the response whereas phenoxybenzamine (120 nmoles) or LB46 (120 nmoles) did not. Atropine alone had no effect (Fig. 2).

The effects of carbachol on feeding are chemically and anatomically specific and in the opposite direction to that found in the rat.

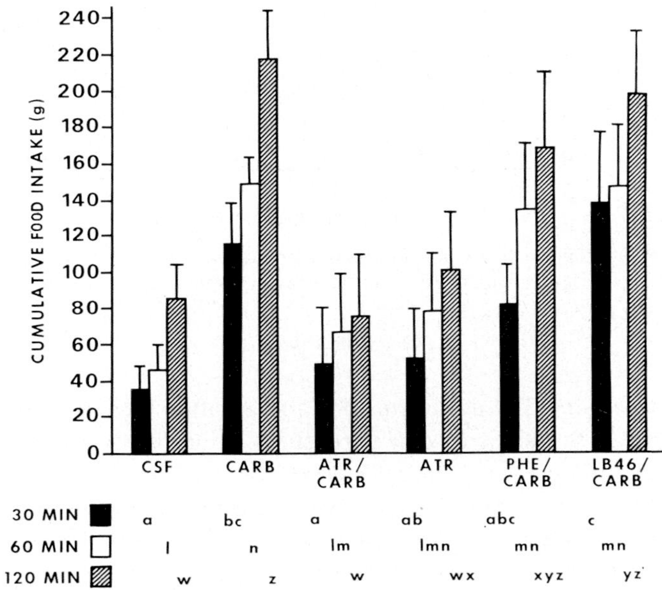

Fig. 2. Food intake of sheep following injections of carbachol and cholinergic and adrenergic antagonists into the hypothalamus. Columns with different subscripts differ significantly ($P < 0.05$). CSF, carrier; CARB, carbachol; ATR, atropine; PHE, phenoxybenzamine.

(3) Prostaglandins and antagonists. In view of the possibility that the effects of some hormones might be mediated by prostaglandins (PGs) and that PGs injected both peripherally (Scaramuzzi *et al.*, 1971) and hypothalamically (Baile *et al.*, 1971) depressed food intake in rats, PGs were injected into the hypothalamus of sheep.

PGE_1 (28 nmoles) was given into loci in the anterior hypothalamus where noradrenaline elicited feeding; a significant reduction in food intake followed, similar to the rat. PGE_1 given into loci where the β-adrenergic agonist, isoproterenol, elicited feeding gave a significant increase in feeding (Martin and Baile, 1973). PGE_2 had no effect at either class of locus. In the lateral hypothalamus, where neither α- nor β-

References p. 414–415

adrenergic agonists affected feeding, PGE_1 stimulated feeding while PGE_2 had no effect.

Of 3 PG antagonists tested, only polyphloretin phosphate (PPP) stimulated food intake, with an optimum dose of 6 nmoles. PPP given 30 min before PGE_1 blocked the PG induced depression of feeding (Fig. 3).

The effects of PGs on feeding are specific, with very different results with PGE_1 at α- or β-adrenergic loci and between PGE_1 and PGE_2 at the same loci.

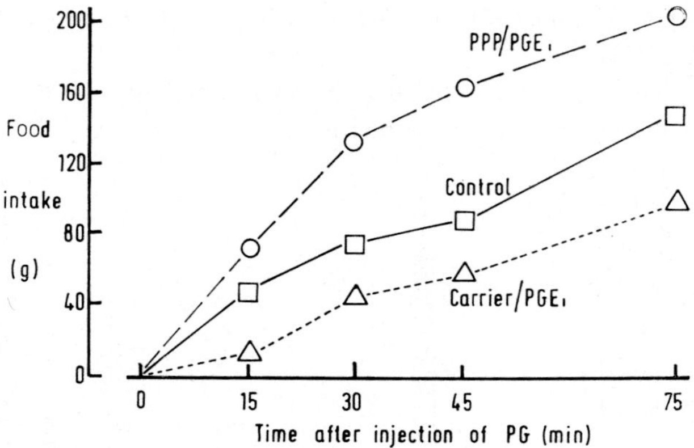

Fig. 3. Food intake of sheep following injections of prostaglandin E_1 (PGE_1) and polyphloretin phosphate (PPP) into the anterior hypothalamus.

Ventricular injections

(1) Estradiol benzoate. The decline in food intake often seen in late pregnancy in sheep might be due to the increase in estrogen secretion (Forbes, 1971). Intravenous infusion of estradiol into castrated male sheep depresses food intake (Forbes, 1972). Because estradiol receptors are known to exist in the hypothalamus and preoptic area (Kato and Villee, 1967) and because estrogen crystals placed in these areas depress food intake in rats (Wade, 1972), estrogen was injected into the lateral ventricle of sheep from where it passes into the third ventricle and has access to hypothalamic tissues.

Doses from 13 to 425 nmoles of estradiol benzoate were given with at least 24 h between injections. A biphasic response resulted, with significantly increased feeding for at least 2 h following 27 and 53 nmoles, no effect with 13, 106 or 160 nmoles and significant depression with 213, 319 and 425 nmoles (Fig. 4). Direct stimulation of intake by low levels of estrogen administration might contribute to the commercially exploited growth-stimulating effect of estrogen implants in ruminants.

(2) Estrogen and progesterone. Twenty-seven nmoles estradiol benzoate were given i.v. with or without 6 μmoles progesterone in a Latin Square experiment. Food intake in the first hour was increased 27% by estrogen alone but decreased 35% by progesterone alone, compared to control. Estrogen given simultaneously with progesterone

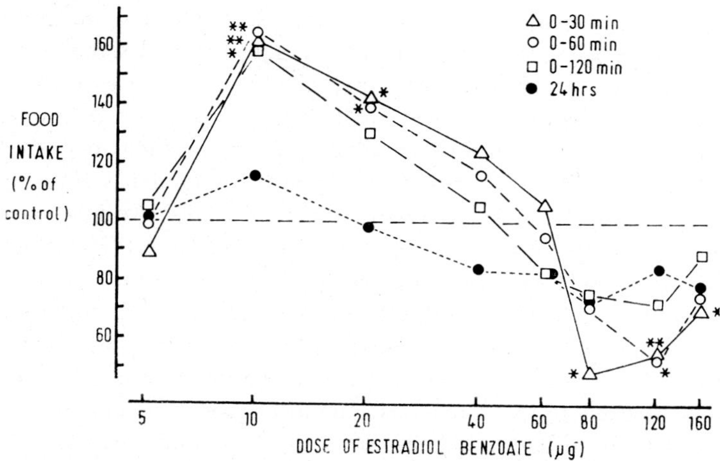

Fig. 4. Food intake of sheep following injections of estradiol benzoate into the lateral ventricle. Significant differences from control: * $P < 0.05$; ** $P < 0.001$.

resulted in a food intake which was only 8% greater than control. Thus progesterone counteracted the effect of estrogen. It is not known whether progesterone would counteract the intake depressing effect of higher doses of estradiol benzoate administered intraventricularly; progesterone has been shown to block the intake depressing effect of subcutaneously administered estrogen both in cows (Muir *et al.*, 1972) and rats (Paul and Duttagupta, 1973). This blocking effect of progesterone might be important in preventing excessive estrogen-induced changes in food intake during pregnancy.

CONCLUSIONS

These experiments show that there are several similarities but also several differences between the feeding responses of sheep and rats to centrally applied neural transmitters. The results obtained in various laboratories with rats have not always agreed and it is not yet possible to provide a definitive description of the neural circuit which control feeding in that species. We cannot say yet whether the situation in sheep is basically different from that in the non-ruminant, or whether the differences in the effects of transmitters are due to technical or superficial disparities.

The inhibition of feeding by high levels of centrally administered estrogen has previously been demonstrated in rats and is now shown to occur in sheep also. The stimulation of feeding which we have obtained in sheep with low levels of ventricular estrogen has not been previously reported.

SUMMARY

Sheep prepared with cannula guides directed towards the hypothalamus were used to

References p. 414–415

study the effects of intrahypothalamic injections of drugs on food intake. Each experiment involved 5–8 sheep. Noradrenaline (240 nmoles) caused significant short term increases at some loci in the anterior hypothalamus; isoproterenol (8 nmoles) had similar effects at other loci; both were blocked specifically by their respective antagonists. Carbachol (3.5–112 nmoles) stimulated feeding at some loci; the effect was blocked by atropine but not by α- or β-adrenergic antagonists. Prostaglandin E_1 (28 nmoles) depressed food intake at noradrenaline responsive loci, but stimulated feeding at isoproterenol responsive loci; the depressing effect was blocked by the prostaglandin antagonist, polyphloretin phosphate (6 nmoles).

In other sheep with lateral ventricular cannula guides, estradiol benzoate intraventricularly stimulated feeding at low doses (27 and 53 nmoles) but depressed intake at higher doses (213 nmoles and above). The stimulating effect of 27 nmoles estradiol benzoate was blocked by simultaneous administration of 6 μmoles progesterone.

These results show several differences from those of similar experiments with rats but the extent to which there exists a real dichotomy between the ruminant and the rodent in the central control of feeding is not yet known.

ACKNOWLEDGEMENTS

We thank Dr. F. H. Martin for permission to use some of his unpublished results from work performed in collaboration with C.A.B. The experiments involving hypothalamic injections were partially supported by the National Science Foundation and those involving ventricular injections by the Agricultural Research Council.

REFERENCES

Baile, C. A. and Forbes, J. M. (1974) Control of feed intake and regulation of energy balance in ruminants. *Physiol. Rev.*, **51**, 160–214.

Baile, C. A., Simpson, C. W., Bean, S. M. and Jacobs, H. J. (1971) Feeding effects of hypothalamic injections of prostaglandins. *Fed. Proc.*, **30**, 375.

Baile, C. A., Simpson, C. W., Krabill, L. F. and Martin, F. H. (1972) Adrenergic agonists and antagonists and feeding in sheep and cattle. *Life Sci.*, **11**, 661–668.

Forbes, J. M. (1971) Physiological changes affecting voluntary food intake in ruminants. *Proc. Nutr. Soc.*, **30**, 135–142.

Forbes, J. M. (1972) Effects of oestradiol-17β on voluntary food intake in sheep and goats. *J. Endocr.*, **52**, viii–ix.

Forbes, J. M. and Baile, C. A. (1972) Hypophagia following adrenergic agonists and antagonists injected into the lateral hypothalamus of sheep. *The Physiologist*, **15**, 136.

Grossman, S. P. (1962) Direct adrenergic and cholinergic stimulation of hypothalamic mechanisms. *Amer. J. Physiol.*, **202**, 872–882.

Kato, J. and Villee, C. A. (1967) Factors affecting uptake of oestradiol-6,7-^{3}H by the hypophysis and hypothalamus. *Endocrinology*, **80**, 1133–1138.

Leibowitz, S. F. (1970) Reciprocal hunger regulating circuits involving alpha- and beta-adrenergic receptors located, respectively, in the ventromedial and lateral hypothalamus. *Proc. nat. Acad. Sci. (Wash.)*, **67**, 1063–1070.

Martin, F. H. and Baile, C. A. (1973) Feeding elicited in sheep by intrahypothalamic injections of PGE_1. *Experientia (Basel)*, **29**, 306–307.

MUIR, L. A., HIBBS, J. W., CONRAD, H. R. AND SMITH, K. L. (1972) Effect of estrogen and progesterone on feed intake and hydroxyproline excretion following induced hypocalcaemia in cows. *J. Dairy Sci.*, **55**, 1613–1620.

PAUL, P. K. AND DUTTAGUPTA, P. N. (1973) Inhibition of oestrogen-induced increase in hepatic and uterine glycogen by progesterone in the rat. *Acta endocr. (Kbh.)*, **72**, 762–770.

RICHARD, P. (1967) *Atlas Stéréotaxique du Cerveau de Brébis.* Institut National de la Recherche Agronomique, Paris.

SCARAMUZZI, O., BAILE, C. A. AND MAYER, J. (1971) Prostaglandins and food intake in rats. *Experientia (Basel)*, **27**, 256–257.

SOMMER, S. R., NOVIN, D. AND LEVINE, M. (1967) Food and water intake after intrahypothalamic injections of carbachol in the rabbit. *Science*, **156**, 983–984.

TARTTELIN, M. F. (1971) A radiographic method for accurately locating deepseated structures in the brain stem of sheep. *Physiol. Behav.*, **7**, 789–792.

WADE, G. N. (1972) Gonadal hormones and behavioural regulation of body weight. *Physiol. Behav.*, **8**, 523–534.

DISCUSSION

DE WIED: In your presentation on feeding behavior of sheep you showed effects of carbachol on food and water intake. Was this an experiment in which the animals had access to both food and water?

FORBES: Yes. In all these tests the animals had access to food and water *ad libitum.* After about 40% of the injections they ate significantly more than they would normally eat. In about 10% of the injections they drank over 1 liter of water in one hour, which is much more than they would normally drink. In only 3 cases out of 77 injections were both significant feeding *and* drinking induced in the same animal in the same test. Since it may be a confusing factor to give both food and water, we should test carbachol in one situation with only food available, and in the other with only water available.

The Hypothalamo-Neurohypophyseal System and the Preservation of Conditioned Avoidance Behavior in Rats

D. DE WIED, B. BOHUS AND TJ. B. VAN WIMERSMA GREIDANUS

Rudolf Magnus Institute for Pharmacology, University of Utrecht, Utrecht (The Netherlands)

Removal of the posterior lobe of the pituitary in rats results in a disturbance of the release of ACTH in response to neurogenic and emotional stress, while that to somatic or systemic stress remains unaltered (Smelik, 1960; de Wied, 1961). Posterior lobectomized rats subjected to transfer into a strange environment, to sound or to pain respond significantly less to these stresses with their plasma corticosterone levels than sham-operated control animals. Treatment of posterior lobectomized rats with an extract of posterior pituitary tissue, Pitressin tannate in oil, was found to restore pituitary–adrenal activity almost completely in response to neurogenic stress. Although Arimura *et al.* (1965) similarly found a reduced pituitary–adrenal response to emotional stress in posterior lobectomized rats, they were unable to restore this deficiency by chronic treatment with Pitressin. These results nevertheless suggested that the absence of posterior pituitary principles is responsible for the defective release mechanism of ACTH. It would, however, also be possible that the deficient pituitary–adrenal response to neurogenic stimulation would have been the result of a defective translation of the neurogenic stimulus to the pituitary of posterior lobectomized rats and if this were true, the absence of posterior pituitary principles might have affected central nervous structures involved in the interpretation of environmental stimuli. It might be that this would be reflected in the behavior of these animals and therefore it was decided to study conditioned avoidance behavior in posterior lobectomized rats. This type of behavior was chosen because it is motivated by fear and anxiety.

Animals were trained in a shuttle box to avoid shock in response to a tone which was a buzzer presented for 5 sec which served as the conditioned stimulus (CS). If the animal did not cross the barrier within 5 sec, the unconditioned stimulus (US) was applied, *i.e.* electric shocks to the feet of the rat via the grid floor of the box (de Wied, 1965). Ten trials a day were given for 14 consecutive days with a reduced intertrial onset interval averaging 60 sec in the beginning and 30 sec at the end of the training. Trials were presented in a predetermined random sequence. After 14 days of acquisition, extinction was studied for 10–14 days. In this situation the US of shock does not follow the CS if the animal fails to make a conditioned avoidance response (CAR) within 5 sec. Intact rats, conditioned in this way however, are resistant to extinction. Although acquisition of the CAR of posterior lobectomized rats was indistinguishable from that of sham-operated controls, extinction of the CAR was markedly accelerated.

References p. 426–427

Treatment of posterior lobectomized rats with Pitressin tannate in oil (1 I.U. s.c./2 days) either during acquisition or during extinction restored the abnormal extinction behavior of these animals. Purified lysine vasopressin appeared to have an effect similar to that of Pitressin (de Wied, 1965; 1969).

Subsequent experiments were performed in intact rats. Animals were trained in the shuttle box till they achieved the criterion of learning *i.e.* 8 or more avoidances during a 10 trial session for 3 consecutive days. As soon as the criterion was reached, extinction was studied. Animals conditioned in this way are not resistant to extinction and the CAR extinguishes within a week or two. The subcutaneous injection of 1 I.U. Pitressin tannate in oil every 2 days either during acquisition or during extinction increased resistance to extinction (de Wied and Bohus, 1966). In contrast to arachis

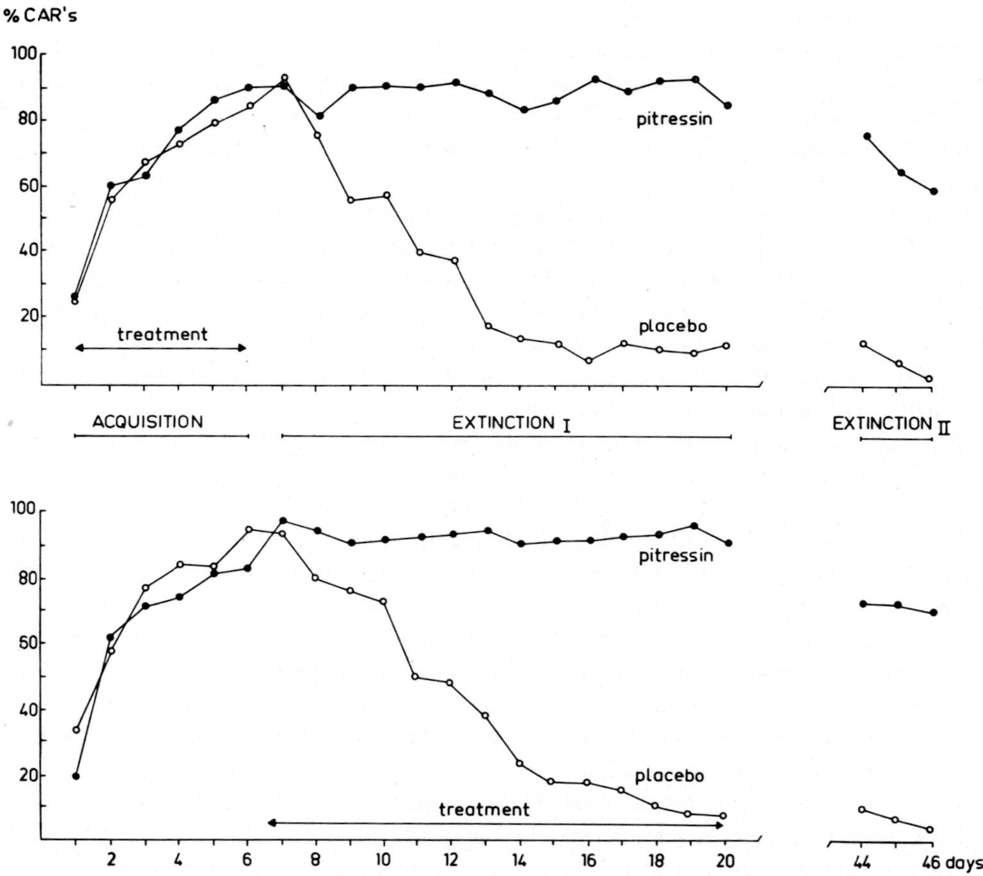

Fig. 1. Influence of long acting Pitressin tannate in oil (1 I.U. every other day, s.c.) and arachis oil (0.2 ml, s.c.), administered either during acquisition or during extinction to intact male rats on the maintenance of a shuttle box avoidance response. Following cessation of the first extinction period, a second extinction period of 3 consecutive sessions of 10 trials each was given 3 weeks later (9 animals per group).

oil treated rats, animals treated with Pitressin did not extinguish during the 14 days period of extinction. If these rats were studied again 3 weeks after termination of the first extinction period, the animals still made 60–70% positive responses (Fig. 1). From these experiments it was inferred that Pitressin has a "long term" effect on the preservation of a conditioned avoidance response irrespective of the time of treatment.

The principle present in the posterior pituitary responsible for this "long term" effect seemed to be vasopressin but definite proof using synthetic peptides was difficult to provide, because the preparation of long acting synthetic vasopressin or oxytocin is not easy. To circumvent this difficulty, experiments were designed in which the influence of various peptides not prepared for long lasting activity could be studied on the retention of avoidance behavior. For this, a pole jumping avoidance test was used (van Wimersma Greidanus and de Wied, 1971). Animals were conditioned to jump onto a pole within 5 sec after presentation of the CS, which is a light on top of the box. Rats which fail to jump, receive the US of shock to the feet via the grid floor. Ten conditioning trials were given each day for 3 days. Rats which made 10 or more avoidances during 3 acquisition sessions were used for extinction studies which started on the fourth day. Only rats that made 8 or more avoidances in the first extinction session were used for subsequent extinction sessions. A preparation of synthetic lysine vasopressin (LVP) containing 60 I.U./mg (Ferring batch no. 12865) was injected s.c. immediately after the first extinction session. It appeared that a single injection of 0.6 or 1.8 μg of LVP exhibits a long term effect on extinction of the avoidance response which is dose dependent. Other structurally and physiologically related peptides like oxytocin, angiotensin II, insulin or growth hormone at least in a dose of 1 μg failed to affect the rate of extinction of the pole jumping avoidance response (de Wied, 1971).

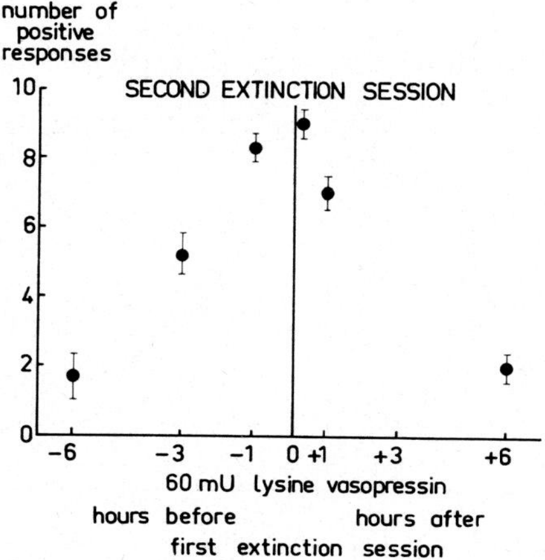

Fig. 2. Effect of LVP (1.0 μg) administered at various time intervals before the third acquisition trial of a pole jumping avoidance response on the rate of extinction of the avoidance response of the second extinction trial (4 animals per group).

References p. 426–427

Accordingly, the peptide in the Pitressin preparation responsible for the long term effect on extinction is lysine vasopressin. The effect is rather specific since the structurally related peptide oxytocin in a similar amount is without effect. Angiotensin II, which is an even more powerful pressor substance than vasopressin, similarly has no effect on extinction, indicating that circulatory changes are not primarily involved. This is also true with respect to the influence of LVP on carbohydrate metabolism, because neither insulin nor growth hormone are able to mimic the behavioral effect of the vasopressor molecule.

To determine the optimal effect of LVP, the peptide was administered at 6, 3 and 1 h before, or immediately, 1 or 6 h after the first extinction session. It appeared that the peptide had to be given within 1 h before or after this first extinction session in order to be completely effective as determined by the number of avoidances of the second extinction session of 10 trials 24 h after the treatment with LVP (Fig. 2). It is of interest to note, that the half-life of the antidiuretic and pressor activity of vasopressin has been estimated at around a few minutes (Czaczkes and Kleeman, 1964). The duration of the behavioral effect, however, seems to be more than 1 h since the administration of vasopressin 1 h before the first extinction session has approximately the same effect on extinction as an injection immediately after this session. One possibility is that some breakdown product exerts identical behavioral effects to those of vasopressin. This hypothesis was substantiated through the isolation of desglycinamide lysine vasopressin (DG-LVP) from hog pituitary material (Lande *et al.*, 1971). This peptide, which lacks the terminal amino acid glycine, has a behavioral effect

H.Cys.Tyr.Phe.Gln.Asn.Cys.Pro.Arg.Gly.NH2

H.Cys.Tyr.Phe.Gln.Asn.Cys.Pro.Lys.Gly.NH2

H.Cys.Tyr.Phe.Gln.Asn.Cys.Pro.Lys.OH

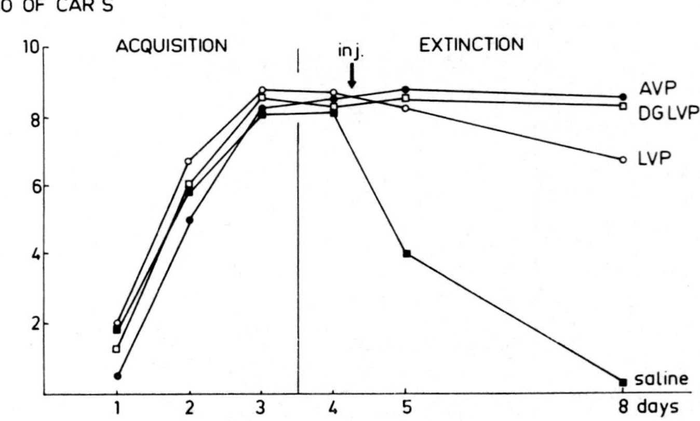

Fig. 3. Influence of a single s.c. injection of various vasopressin analogs on extinction of a pole jumping avoidance response. Peptides were injected in 0.5 ml saline immediately after the first extinction session on day 4. AVP (225 I.U./mg), 0.2 μg LVP (60 I.U./mg), 0.6 μg DG-LVP (0.3 μg/ml) (4 animals per group).

similar to that of the original lysine vasopressin molecule, but antidiuretic, pressor, oxytocic and CRF activities of DG-LVP are nearly completely destroyed by the removal of the C-terminal glycinamide (de Wied *et al.*, 1972). Fig. 3 depicts the effect of LVP, DG-LVP and arginine vasopressin (AVP) on extinction of a pole jumping avoidance response. All 3 peptides preserve the avoidance response in relatively low doses.

LVP not only affects active avoidance behavior, it also affects passive avoidance behavior, as studied in a simple "step through" type of passive avoidance situation. This situation uses the innate response of rats to prefer darkness to light. The apparatus consists of a dark box connected with an illuminated elevated platform (Ader *et al.*, 1972). Rats are placed on the platform and latency to enter the dark box is recorded. One trial is given on day 1; 3 such trials on day 2. At the end of the third trial, animals receive an electric shock in the dark box for 1 sec. The next day latency to enter is recorded again. LVP injected s.c. 1 h before the first retention trial considerably increases avoidance latency (Ader and de Wied, 1972). Avoidance latency is still augmented during the second retention trial 24 h later and long term preservation of the effect of LVP is also observed in passive avoidance behavior. A similar temporal relationship was found for the retention of passive avoidance as in active avoidance. Treatment administered as much as 6 h before the first retention trial is also ineffective in modifying passive avoidance behavior (Fig. 4).

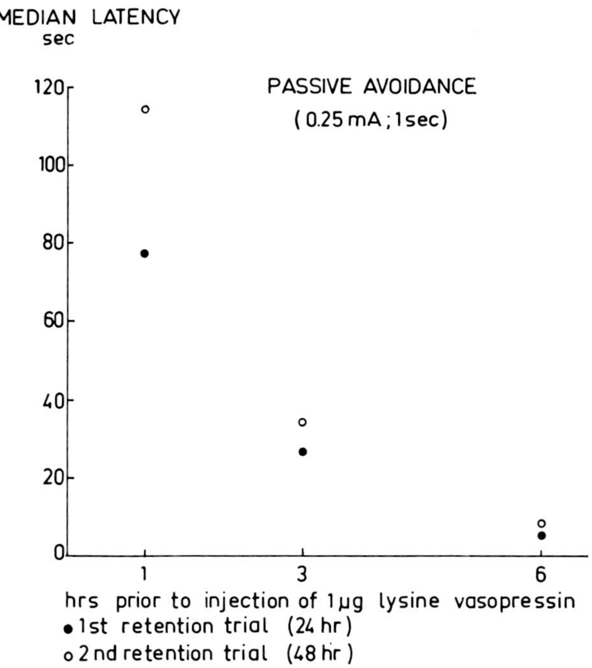

Fig. 4. Effect of LVP (1 μg) administered at various time intervals before the first retention trial of a passive avoidance response on avoidance latency in the second retention trial (5 animals per group).

References p. 426–427

Subsequently, the question was raised as to whether the long term behavioral effect of LVP is specific for the particular response acquired under the maximal influence of this peptide or whether generalization occurs to other aversively motivated responses. The experiments on passive avoidance behavior made it possible to study the interaction between two fear motivated responses when only one of these responses was tested under the maximal influences of LVP. Rats were trained in the active and in the passive avoidance situation. Half of the animals were trained in the pole jumping apparatus during the morning session and subjected to the step through procedure during the afternoon 6 h later. The other half were trained first in the passive avoidance test in the morning and subjected to the pole jumping avoidance test in the afternoon, 6 h later. Animals were injected on day 3 of the experiment with either saline or LVP 1 h prior to the morning session (Bohus *et al.*, 1972). LVP-treated rats made significantly more avoidances in the pole jumping test than saline-treated controls during extinction sessions on days 4 and 5. No such difference was found in median avoidance latencies in the passive avoidance test given 6 h later. Conversely, relative to controls, a marked increase in response latency was observed in the passive avoidance situation in rats treated with LVP 1 h prior to the first retention trial. Active avoidance behavior of LVP- and saline-treated rats, however, did not differ in the afternoon sessions. Thus, resistance to extinction of an active and a passive avoidance response is influenced by LVP only if treatment is provided shortly before the test and thus restricted to that behavior which is occurring during the period of optimal vasopressin influence. If the behavioral effect of vasopressin was due to a long lasting general arousal of some CNS mechanism, changes in performance in any number of other similar behavioral situations would have been likely. On the contrary, no evidence of generalization was observed. Specificity, therefore, of the effect of LVP for a particular behavior is evident even though both behavioral responses were aversively motivated presumably by the fear elicited by painful electric shocks.

These results clearly indicate the importance of a time dependent association between exogenous vasopressin and specific environmental signals. The effect of exogenous LVP in maintaining conditioned behavior then may be related to the relative ease with which a given behavioral response is elicited when situation specific cues reoccur. If this were true, it might be that an association between endogenous release of vasopressin or vasopressin analogs and specific environmental cues is of physiological significance in the maintenance of new behavioral patterns in spite of the non-specific nature of the release of this octapeptide.

In order to substantiate this hypothesis, vasopressor activity was determined during the first or second retention trial of rats trained in the passive avoidance test. Vasopressor activity was measured in the eye-plexus blood immediately following the first retention trial which was run 24 h after the animals had been exposed to electric shock in the dark box. The intensity of the shock was varied so as to vary avoidance latency. Eye-plexus blood was used because it contains much more vasopressin than peripheral blood and therefore can be measured directly in 0.1 ml of plasma. Vasopressor activity designated as antidiuretic hormone (ADH) activity was determined on urine flow in the alcohol anesthetized rat using a 2 + 1 point assay (de Wied, 1960). ADH activity

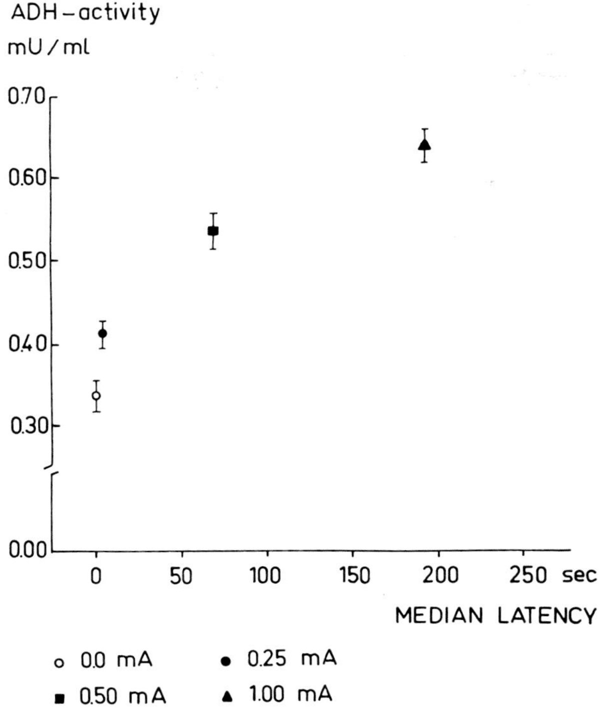

Fig. 5. ADH-activity in eye-plexus blood immediately following the first retention trial of a passive avoidance response. Blood was withdrawn from the eye-plexus of rats under ether anesthesia (6–7 animals per group).

in eye-plexus blood was higher in rats exposed to electric shock in the dark box 24 h previously (Fig. 5). The concentration in eye-plexus blood was greater the more the animal had experienced shock the day before. Latency to enter also increased with the intensity of the shock. Accordingly, vasopressin is released in high quantities under these conditions and it is possible, therefore, that the hormone contributes to subsequent behavioral performance when situation specific cues are presented (Thompson and de Wied, 1973).

The site of action of the behavioral effect of vasopressin has been explored by intracerebral microinjection of the octapeptide and in studies with lesions in the posterior thalamic area. For these studies, rats were trained in the pole jumping box. Subsequently, a stainless steel plate with 12 holes was placed on the skull and fixed with cement. Substances can thus be introduced into the brain of conscious rats through these 12 holes at various depths through a needle connected to a microsyringe via polyethylene tubing. LVP (0.1 μg/0.5 μl) or saline was injected intracerebrally (van Wimersma Greidanus *et al.*, 1973).

There appeared a difference in time over which LVP maintained the avoidance response. There were brain sites where LVP resulted in delay of extinction for approximately 1 day (parafascicular area, posterolateral and ventromedial thalamic nuclei

References p. 426–427

and cerebrospinal fluid) or for more than 1 week (nucl. parafascicularis, the parafascicular–posteromedial thalamic area, the transition from the reticular formation to the parafascicular nuclei, the cerebrospinal fluid). The difference in duration of the effect on extinction following intracerebral LVP might be explained by dose dependency, since not always the same amount of peptide might reach the effective site when applied close to the site of action. From these studies, the parafascicular area seemed to be a preferential locus of action of vasopressin.

In order to further investigate the significance of the parafascicular area in the behavioral effect of LVP, studies were performed in rats bearing lesions in the parafascicular nuclei. Lesions were placed bilaterally under ether anesthesia in these nuclei with the aid of a stereotaxic instrument using radiofrequency for destruction of the parafascicular area. Sham-operated animals underwent the same procedure except that no lesioning was performed. After recovery (5–7 days) animals were subjected to avoidance training in the pole jumping test. Ten trials a day were given for 4 days and on the fifth day extinction trials were run. Immediately following this last session, animals were injected s.c. with 0.5 ml of saline. On day 8 a second extinction session was run followed on day 9 by a retraining session of 10 trials, and an extinction session on day 10. Immediately after this session animals were injected s.c. with LVP or saline and extinction sessions were run on days 11, 12, and 15. The lesion of the parafascicular nuclei is illustrated in Fig. 6. In general, lesioned animals had difficulty

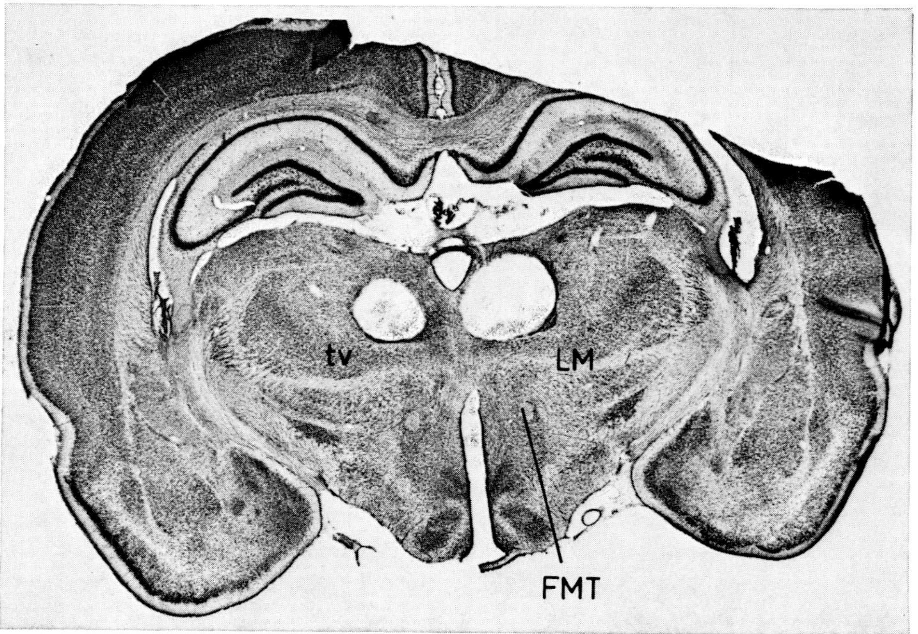

Fig. 6. Transverse section of rat brain showing destruction of the mediodorsal thalamus including the parafascicular area. FMT = fasciculus mammillothalamicus; LM = lemniscus medialis; tv = nucleus ventralis thalami.

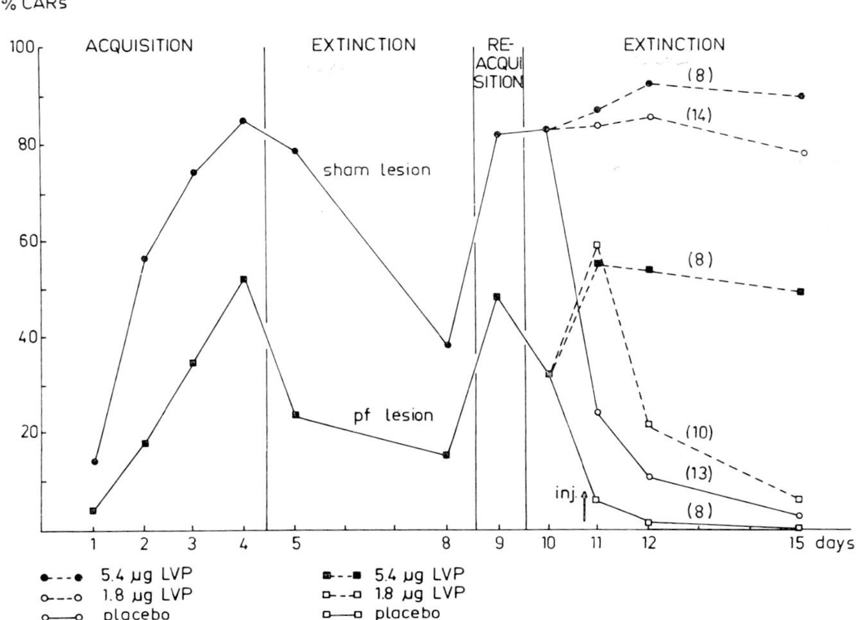

Fig. 7. Effect of LVP on extinction of a pole jumping avoidance response in rats bearing bilateral lesions in the nuclei parafascicularis. The number of animals per group is given between brackets.

in acquiring the pole jumping avoidance response and extinction was also rapid in these animals as compared to that of sham-operated rats. The administration of 1.8 μg LVP, which completely preserved the CAR in sham-operated animals, only slightly and for approximately 1 day delayed extinction of the CAR in lesioned rats. To induce resistance to extinction in these animals, three times the dose of LVP (5.4 μg) was necessary (Fig. 7). Thus an intact parafascicular area is not essential for the effect of LVP on extinction. The fact that more LVP was needed in lesioned rats might have to do with the poor learning of lesioned rats. Accordingly, other brain structures are involved in the consolidating effect of LVP on conditioned avoidance behavior. The fact that the parafascicular area is sensitive to LVP when implanted into these structures is difficult to understand at present. The posterior thalamic area is involved in the maintenance of conditioned avoidance behavior and it is considered a center of integration of external and internal information (Delacour, 1971). This locus does not seem to be specific for LVP since ACTH analogs and also corticosteroids affect avoidance behavior when implanted into this region of the brain (van Wimersma Greidanus and de Wied, 1969, 1971; Bohus, 1970; van Wimersma Greidanus *et al.*, 1973). It seems, therefore, that the non-specific thalamic area contains multiple receptors and thus serves as a sensor for information transmitted by endogenous factors released as a result of environmental changes.

The mode of action of LVP and analogs is unknown. The long term effect of these peptides suggests that rather permanent changes take place which facilitate the

References p. 426–427

consolidation of aversively motivated behavior, presumably by changing the properties of neuronal and synaptic membranes. In this respect it is relevant to mention the influence of DG-LVP on maze learning as affected by puromycin, an antibiotic which interferes with memory consolidation by an inhibitory effect on protein synthesis in the brain (Lande *et al.*, 1972). It was found that DG-LVP prevented puromycin-induced memory-blockade in mice. It is possible therefore that vasopressin facilitates avoidance learning by affecting protein synthesis in the nerve cell involved, which in turn may play a role in the regulation of neuronal connectivity.

In conclusion, the posterior pituitary and LVP profoundly affect acquisition and maintenance of aversely motivated active and passive avoidance behavior. The removal of the posterior pituitary interferes with the maintenance of an acquired response, the injection of LVP induces the reverse, it facilitates the preservation of such a response. The same is found with an endocrinologically active inert degradation product of LVP, *i.e.* DG-LVP. This suggests that LVP may act as a prohormone for behaviorally active peptide sequences in the same way as oxytocin seems to act as a prohormone for the MSH-release inhibiting factor, as recently shown by Celis and Taleisnik (1971) and Celis *et al.* (1971). A time-dependent association between exogenous vasopressin and specific environmental signals was found since LVP facilitated consolidation of that avoidance response which was acquired during optimal vasopressin influence. The effect of exogenous vasopressin in maintaining conditioned behavior for a long time may thus be related to the relative ease with which a given behavioral response is elicited when situation specific cues reoccur. If this were true, it might be that an association between endogenous release of vasopressin or related peptides and specific environmental cues is of physiological significance in the maintenance of new behavior patterns. The fact that the rate of vasopressin release appeared to be a function of the intensity of previously experienced aversive stimulation and the fact that the removal of the posterior pituitary in rats impairs the maintenance of an avoidance response is in keeping with this hypothesis.

REFERENCES

ADER, R. AND WIED, D., DE (1971) Effects of lysine vasopressin on passive avoidance learning. *Psychon. Sci.*, **29**, 46–48.

ADER, R., WEIJNEN, J. A. W. M. AND MOLEMAN, P. (1972) Retention of a passive avoidance response as a function of the intensity and duration of electric shock. *Psychon. Sci.*, **26**, 125–128.

ARIMURA, A., YAMAGUCHI, T., YOSHIMURA, K., IMAZEKI, T. AND ITOH, S. (1965) Role of the neurohypophysis in the release of adrenocorticotrophic hormone in the rat. *Jap. J. Physiol.*, **15**, 278–295.

BOHUS, B. (1970) Central nervous structures and the effect of ACTH and corticosteroids on avoidance behaviour: a study with intracerebral implantation of corticosteroids in the rat. In *Pituitary, Adrenal and the Brain, Progress in Brain Research, Vol. 32*, D. DE WIED AND J. A. W. M. WEIJNEN (Eds.), Elsevier, Amsterdam, pp. 171–184.

BOHUS, B., ADER, R. AND WIED, D., DE (1972) Effects of vasopressin on active and passive avoidance behavior. *Horm. Behav.*, **3**, 191–197.

CELIS, M. E. AND TALEISNIK, S. (1971) In vitro formation of a MSH-releasing agent by hypothalamic extracts. *Experientia (Basel)*, **27**, 1481–1482.

CELIS, M. E., TALEISNIK, S. AND WALTER, R. (1971) Regulation of formation and proposed structure of the factor inhibiting the release of melanocyte-stimulating hormone. *Proc. nat. Acad. Sci. (Wash.)*, **68**, 1428–1433.

CZACZKES, J. W. AND KLEEMAN, C. R. (1964) The effect of various states of hydration and the plasma concentration on the turnover of antidiuretic hormone in mammals. *J. clin. Invest.*, **43**, 1649–1658.

DELACOUR, J. (1971) Effects of medial thalamic lesions in the rat, a review and an interpretation. *Neuropsychologia*, **9**, 157–174.

LANDE, S., WITTER, A. AND WIED, D., DE (1971) Pituitary Peptides. An octapeptide that stimulates conditioned avoidance acquisition in hypophysectomized rats. *J. biol. Chem.*, **246**, 2058–2062.

LANDE, S., FLEXNER, J. B. AND FLEXNER, L. B. (1972) Effect of corticotropin and desglycinamide^9-lysine vasopressin on suppression of memory by puromycin. *Proc. nat. Acad. Sci. (Wash.)*, **69**, 558–560.

SMELIK, P. G. (1960) Mechanism of hypophysial response to psychic stress. *Acta endocr. (Kbh.)*, **33**, 437–443.

THOMPSON, E. A. AND WIED, D., DE (1973) The relationship between the antidiuretic activity of rat eyeplexus blood and passive avoidance behaviour. *Physiol. Behav.*, **11**, 377–380.

WIED, D., DE (1960) A simple automatic and sensitive method for the assay of antidiuretic hormone with notes on the antidiuretic potency of plasma under different experimental conditions. *Acta physiol. pharmacol. neerl.*, **9**, 69–81.

WIED, D., DE (1961) The significance of the antidiuretic hormone in the release mechanism of corticotropin. *Endocrinology*, **68**, 956–970.

WIED, D., DE (1965) The influence of the posterior and intermediate lobe of the pituitary and pituitary peptides on the maintenance of a conditioned avoidance response in rats. *Int. J. Neuropharmacol.*, **4**, 157–167.

WIED, D., DE (1969) Effects of peptide hormones on behavior. In *Frontiers in Neuroendocrinology*, W. F. GANONG AND L. MARTINI (Eds.), Oxford Univ. Press, New York, pp. 97–140.

WIED, D., DE (1971) Long term effect of vasopressin on the maintenance of a conditioned avoidance response in rats. *Nature (Lond.)*, **232**, 291–301.

WIED, D., DE AND BOHUS, B. (1966) Long term and short term effect on retention of a conditioned avoidance response in rats by treatment with long acting Pitressin and α-MSH. *Nature (Lond.)*, **212**, 1484–1486.

WIED, D., DE, GREVEN, H. M., LANDE, S. AND WITTER, A. (1972) Dissociation of the behavioral and endocrine effects of lysine vasopressin by tryptic digestion. *Brit. J. Pharmacol.*, **45**, 118–122.

WIMERSMA GREIDANUS, TJ. B. VAN AND WIED, D., DE (1969) Effects of intracerebral implantation of corticosteroids on extinction of an avoidance response in rats. *Physiol. Behav.*, **4**, 365–370.

WIMERSMA GREIDANUS, TJ. B. VAN AND WIED, D., DE (1971) Effects of systemic and intracerebral administration of two opposite acting ACTH-related peptides on extinction of conditioned avoidance behavior. *Neuroendocrinology*, **7**, 291–301.

WIMERSMA GREIDANUS, TJ. B. VAN, BOHUS, B. AND WIED, D., DE (1973) Effects of peptide hormones on behaviour. *In Endocrinology, Proc. 44th Int. Congr. Endocr., Int. Congr. Ser. no. 273*, R. SCOW (Ed.), Excerpta Medica, Amsterdam, pp. 197–201.

DISCUSSION

LEVINE: There has been recently emerging a number of remarkable publications in the field of peptides in general, like TRF, affecting some aspects of depression, and ACTH effects in relation to certain aspects of moods. Are you dealing with some generalized action of peptides on the brain, or with specific effects, or will it be both?

DE WIED: It is very difficult to answer this question. We are also a bit surprised about the fact that so many peptides do have effects on the brain. The specificity of these effects in terms of structures is however not as clear as one would like to see. TRF seems to potentiate the dopaminergic effects in the central nervous system. There have also been reports that it might be an antidepressant. Endröczi (1972)* found that after $ACTH_{1-10}$ and $ACTH_{4-10}$ patients were in a much better mood. This might point to some antidepressant effects of ACTH as well. I think that a lot of small peptides do have effects in the brain, however, on more or less specific structures.

---

* ENDRÖCZI, E. (1972) *Limbic System Learning and Pituitary–Adrenal Function*, Akademiai Kiadó, Budapest.

VAN SETERS: Are you able to dissociate your effects with LVP on the avoidance extinction entirely from effects on activation at the level of the anterior pituitary or the adrenal cortex?

DE WIED: In principle we have shown similar effects of vasopressin in hypophysectomized animals.

SWAAB: I should like to ask whether you have data on the rat with a hereditary hypothalamic diabetes insipidus, the so-called Brattleboro strain, in the avoidance situation.

DE WIED: Various members of our group work with this animal, but we find it very difficult, at this moment, to interpret our results. The behavior of the Long–Evans rat from which it is derived, is so different from that of our Wistar strain that we really need a similar animal which does not have any disturbance in the synthesis and the release of vasopressin, as a control animal. We expected that the heterozygous animal that produces vasopressin would be completely different from the homozygous one that is lacking vasopressin. We do find differences between homo- and heterozygous animals. Acquisition for example is disturbed in the homozygous animal, and also extinction is facilitated. But these differences are not as big as we had expected. However, the heterozygous animal has also a disturbance in the release of vasopressin, and is therefore not a reliable control animal. The data should be compared to the Long–Evans strain, from which they are derived.

VAN DER SCHOOT: I wonder whether you can get a more direct relationship with the physiological situation by stimulating the posterior lobe electrically like Dr. Lincoln did, instead of injecting vasopressin.

DE WIED: That may be another approach. However, there are so many peptides in the posterior lobe, that I would like to *know* exactly what is being released upon stimulation, we know exactly what we inject.

The Parafascicular Area as the Site of Action of ACTH Analogs on Avoidance Behavior

TJ. B. VAN WIMERSMA GREIDANUS, B. BOHUS AND D. DE WIED

Rudolf Magnus Institute for Pharmacology, University of Utrecht, Utrecht (The Netherlands)

The pituitary–adrenal system is involved in acquisition and extinction of conditioned avoidance behavior (de Wied, 1969; de Wied *et al.*, 1972). ACTH and peptides related to ACTH (MSH, $ACTH_{1-10}$ and $ACTH_{4-10}$) induce a short lasting inhibition of extinction of a shuttle box or pole jumping avoidance response (de Wied and Bohus, 1966). Not only peptides from anterior and intermediate pituitary origin are able to affect extinction of a conditioned avoidance response (CAR). Also peptides from posterior pituitary origin such as lysine–vasopressin (LVP) and desglycinamide-lysine–vasopressin induce inhibition of this extinction. A single s.c. injection of synthetic LVP markedly preserves a pole jumping avoidance response (de Wied, 1971; de Wied *et al.*, 1974) which lasts for several weeks following its administration (van Wimersma Greidanus *et al.*, 1973). The difference between the "long term" effect of vasopressin and the "short term" effect of ACTH analogs on extinction of a CAR might be explained if these peptides induce their effects via different sites in the brain. There is some evidence (van Wimersma Greidanus *et al.*, 1973) that the site of action of LVP on behavior is in the parafascicular area, since micro-injection of small amounts of this peptide in the posterior thalamic area, including the parafascicular nuclei, results in preservation of a pole jumping avoidance response. On the other hand, previous observations on the site of action of ACTH analogs in the brain (van Wimersma Greidanus and de Wied, 1971) revealed that the same area is important for the behavioral effect of these peptides since local application of crystalline $ACTH_{1-10}$ in this region results in a short term inhibition of extinction of the CAR. Also, micro-injections of $ACTH_{4-10}$, like the application of crystalline peptide into the parafascicular area, induced a short lasting inhibition of extinction of the CAR. Although these results suggest that both peptides, LVP and ACTH analogs act in the same brain area, further analysis of the site of action of these peptide hormones on behavior was made in rats bearing bilateral lesions in the parafascicular area (Fig. 1). Lesions in the parafascicular area interfered with acquisition and extinction of the avoidance response. Acquisition was poor and extinction facilitated. Nevertheless parafascicular lesioned animals were still able to preserve a pole jumping avoidance response following administration of LVP, although they needed higher amounts of this peptide to exhibit the same effect on extinction as sham operated rats (van Wimersma Greidanus

References p. 432

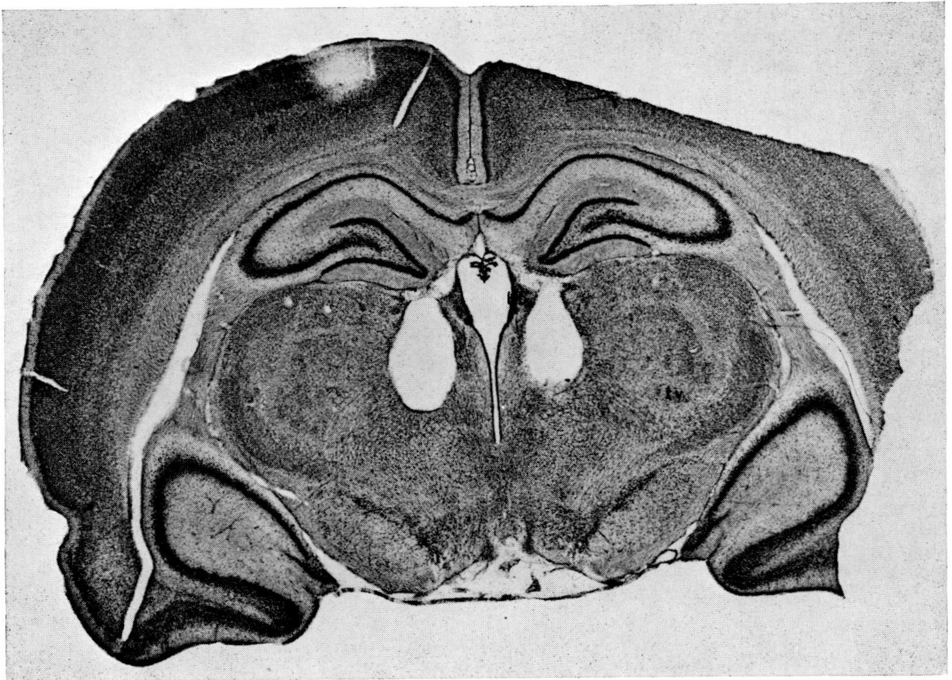

Fig. 1. Example of a bilateral flash-shaped lesion in the parafascicular area. Maximal median diameter in dorsoventral direction 1.5 mm and in mediolateral direction 0.7 mm. Hb: habenular complex; tv: nucleus ventralis thalami.

et al., 1973; de Wied *et al.*, 1974). The parafascicular area, therefore, is important but not essential for the behavioral effect of LVP.

Studies on the ACTH analog to be reported here had a completely different outcome. Lesioned and sham lesioned animals were trained to acquire a pole jump shock avoidance response during 4 days. Ten trials a day were given with a mean intertrial interval of 60 sec. On days 5 and 8, extinction sessions were run. Following a reacquisition session on day 9, extinction trials were run on days 10, 11, 12 and 15. $ACTH_{4-10}$ was administered daily 1 h before each extinction session following reacquisition.

It appeared that the treatment with $ACTH_{4-10}$ in sham operated rats resulted in a marked inhibition of extinction of these animals (Fig. 2). Administration of 1.0 μg $ACTH_{4-10}$ resulted in a delay of extinction of the CAR in all extinction sessions. Although the response rate at the end of the observation period was less than during the first extinction session the difference between the numbers of avoidance responses of the group treated with $ACTH_{4-10}$ still differed significantly from that of the saline treated rats. The dose of 3.0 μg $ACTH_{4-10}$ elicited a more marked inhibition of extinction of the avoidance response which was significantly different from that found after 1 μg of $ACTH_{4-10}$. Neither 1.0 μg nor 9.0 μg $ACTH_{4-10}$ were able to prevent extinction of the pole jumping avoidance response in rats bearing lesions in the parafascicular nuclei (Fig. 2).

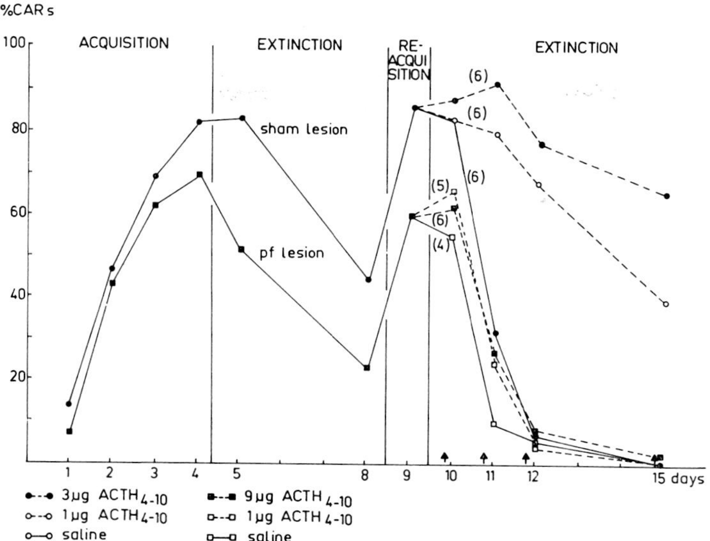

Fig. 2. Effect of graded doses of $ACTH_{4-10}$ on extinction of a pole jumping avoidance response in parafascicular lesioned rats and in sham operated animals. Number of animals shown in brackets.

The results reported in the present paper are in agreement with those obtained by Bohus and de Wied (1967b) who showed that bilateral lesions of the parafascicular nuclei block the inhibitory effects of long acting α-MSH preparations on extinction of a shuttle box avoidance response. The parafascicular region of the brain is thought to play a role in integrative functions (Cardo, 1965) and in processes concerned with extinction of conditioned avoidance behavior (Delacour, 1970, 1971). Bohus and de Wied (1967a) demonstrated that extensive lesions in the posterior thalamus interfere with both acquisition and extinction of a CAR, while smaller lesions destroying the parafascicular area only, mainly affect extinction of such a response. The results of the present experiments again point to the parafascicular area as an important structure in avoidance conditioning and in particular in extinction processes.

The present experiments indicate that an intact parafascicular area is essential for the effect of ACTH analogs on extinction of conditioned avoidance behavior in contrast to that of vasopressin and analogs. The influence, however, of ACTH analogs on the parafascicular nuclei is unknown and is, therefore, the subject of further experiments.

SUMMARY

Lesions in the parafascicular nuclei result in a poor avoidance performance in a pole jumping avoidance test. Acquisition is retarded and extinction is significantly facilitated. Whereas, daily s.c. injection of 1 μg or 3 μg of $ACTH_{4-10}$ prevents extinction of a pole jumping avoidance response in a dose dependent manner in sham operated

References p. 432

controls, injection of 1 μg or even 9 μg $ACTH_{4-10}$ fails to affect extinction in animals bearing bilateral lesions in the parafascicular area.

From these and previous results obtained after implantation of $ACTH_{1-10}$ in the parafascicular area (van Wimersma Greidanus and de Wied, 1971) it is concluded that the parafascicular nuclei are the site of action of the behavioral effect of ACTH analogs.

REFERENCES

Bohus, B. and De Wied, D. (1967a) Avoidance and escape behavior following medial thalamic lesions in rats. *J. comp. physiol. Psychol.*, **64**, 26–29.

Bohus, B. and De Wied, D. (1967b) Failure of α-MSH to delay extinction of conditioned avoidance behavior in rats with lesions in the parafascicular nuclei of the thalamus. *Physiol. Behav.*, **2**, 221–223.

Cardo, B. (1965) Rôle de certain noyaux thalamiques dans l'éboration et la conservation de divers conditionnements. *Psychol. franç.*, **10**, 344–351.

Delacour, J. (1970) Specific functions of a medial–thalamic structure in avoidance conditioning in the rat. In *Pituitary, Adrenal and the Brain, Progr. Brain Res., Volume 32*, D. de Wied and J. A. W. M. Weijnen (Eds.), Elsevier, Amsterdam, pp. 158–170.

Delacour, J. (1971) Effects of medial thalamic lesions in the rat, a review and an interpretation. *Neuropsychologia*, **9**, 157–174.

De Wied, D. (1969) Effects of peptide hormones on behavior. In *Frontiers in Neuroendocrinology*, W. F. Ganong and L. Martini (Eds.), Oxford Univ. Press, New York, pp. 97–140.

De Wied, D. (1971) Long term effect of vasopressin on the maintenance of a conditioned response in rats. *Nature (Lond.)*, **232**, 58–60.

De Wied, D. and Bohus, B. (1966) Long term and short term effects on retention of a conditioned avoidance response in rats by treatment with long acting Pitressin and α-MSH. *Nature (Lond.)*, **212**, 1484–1486.

De Wied, D., Delft, A. M. L. van, Gispen, W. H., Weijnen, J. A. W. M. and Wimersma Greidanus, Tj. B. van (1972) The role of pituitary–adrenal system hormones in active avoidance conditioning. In *Hormones and Behavior*, S. Levine (Ed.), Academic Press, New York, pp. 135–171.

De Wied, D., Bohus, B. and Wimersma Greidanus, Tj. B. van (1974) The hypothalamic neurohypophyseal system and the preservation of conditioned avoidance behavior in rats. In *Integrative Hypothalamic Activity, Progr. Brain Res., Vol. 41*, D. F. Swaab and J. P. Schadé (Eds.), Elsevier, Amsterdam, pp. 417–428.

Wimersma Greidanus, Tj. B. van and De Wied, D. (1971) Effects of systemic and intracerebral administration of two opposite acting ACTH-related peptides on extinction of conditioned avoidance behavior. *Neuroendocrinology*, **7**, 291–301.

Wimersma Greidanus, Tj. B. van, Bohus, B. and De Wied, D. (1973) Effects of peptide hormones on behaviour. In *Endocrinology, Proc. 4th. Int. Congr. Endocr., Int. Congr. Ser. no. 273*, R. Scow (Ed.), Excerpta Medica, Amsterdam, pp. 197–201.

Psychobiological Aspects of Lactation in Rats

JUDITH M. STERN* AND SEYMOUR LEVINE

Department of Psychiatry, Stanford University School of Medicine, Stanford, Calif. 94305 (U.S.A.)

INTRODUCTION

The psychobiology of the postpartum rat has been studied most extensively with respect to maternal behavior (*e.g.* Moltz, 1971; Rosenblatt and Lehrman, 1963; Wiesner and Sheard, 1933) and the neuroendocrine regulation of milk secretion (Cowie and Tindal, 1971; Cross, 1961). Care of the offspring and lactation are dramatic and demanding states entailing numerous behavioral and physiological changes, but only the principal changes have been studied in detail. As an outgrowth of our interest in the mother–litter interaction, we have been studying the responsiveness of lactating rats with respect to changes which have received little attention previously. These changes concern the lactating female rat's responsiveness to stress and diurnal rhythm in pituitary–adrenal activity and in behavior, functions which are regulated, in part, by hypothalamic mechanisms.

THE PITUITARY–ADRENAL SYSTEM DURING LACTATION

Several measures that we have examined suggest that the lactating rat may be "buffered" against disturbing stimuli. Pentobarbital anesthesia causes a temperature drop of 4–5 °C in 75 min in nonlactating females; in contrast, the lactating rats showed a much smaller temperature drop of 2 °C at the peak of lactation (Thoman *et al.*, 1968). The higher temperature of the lactator, due to the metabolic demands of milk secretion, may contribute to her resistance to disruption of temperature regulation by anesthesia. In response to the Ulrich and Azrin shock-induced fighting procedure (1962), lactating females showed an increased latency to fight and a reduced number of fights, compared to cycling females (Thoman *et al.*, 1970a). The reduction in fighting is surprising in light of the commonly held belief that the postpartum female increases her aggressive tendencies in an effort to defend her offspring. However, irritable (*i.e.* shock-induced) fighting, in the absence of the litter, may be more a reflection of general responsiveness to noxious stimuli than to postpartum aggression relevant to litter defense.

* Dr. Stern's new address: Department of Psychology, Rutgers College, Rutgers University, 88 College Avenue, New Brunswick, N.J. 08903, U.S.A.

References p. 442–444

One aspect of lactation that we have focused on is the pituitary–adrenal system. In the studies discussed above, it was found that the rise in plasma corticosterone in nonlactating females following pentobarbital (Thoman *et al.*, 1968), shock-induced fighting and ether (Thoman *et al.*, 1970a) was markedly reduced in the lactating rats. A brief review of the role and activity of the pituitary–adrenal system during lactation will provide perspective to our detailed studies of reduced adrenocortical responsiveness to noxious stimuli in lactating rats.

Glucocorticoids are known to play an important role in the initiation and maintenance of lactation in the rat as well as in other species (Cowie and Tindal, 1971). Lactation will proceed in the absence of the adrenals, but it certainly is not optimal (Cowie and Folley, 1947; Thoman *et al.*, 1970b). Glucocorticoids reverse the lactational deficiency produced by adrenalectomy (Cowie and Folley, 1947), via a direct action on the mammary gland as shown by effects on *in vitro* cultures (Elias and Rivera, 1959) and on biochemical pathways in the mammary gland (Greenbaum and Darby, 1964; Willmer and Foster, 1965). While administration of ACTH in lactating ruminants generally depresses milk secretion (Brush, 1960; Flux *et al.*, 1954), appropriate doses of cortisone acetate, hydrocortisone acetate and corticosterone enhance lactation in rats (Hahn and Turner, 1966; Johnson and Meites, 1958; Talwalker *et al.*, 1960). Thatcher and Tucker (1970) suggested that in late lactation, decreased adrenal secretions may be rate-limiting to maximal milk production in rats.

The suckling stimulus is a potent releaser of prolactin and oxytocin (Meites, 1966), the hormones necessary for milk secretion and release. Suckling has also been shown to stimulate the discharge of ACTH in the goat and sheep (Denamur *et al.*, 1965) and the rat (Grégoire, 1946; Voogt *et al.*, 1969). (See Cowie and Tindal, 1971, for a more extensive discussion of the suckling stimulus and its neuroendocrine consequences.) While it clearly is adaptive for the suckling stimulus to induce the release of a hormone (ACTH) which enhances milk secretion, the problem arises as to whether or not to consider the suckling stimulus in some way "stressful", since ACTH release usually is associated with stressful stimuli. Presumably, nursing, which postpartum animals engage in spontaneously and repeatedly, is not a noxious experience. However, the tactile stimulus to the teat must be of an intense and arousing nature to elicit oxytocin release (Tindal and Knaggs, 1970) and the metabolic demands of lactation, consequent to suckling, might be thought of as a stressor.

The finding of elevated morning basal levels of plasma corticosterone as a consequence of suckling (Kamoun, 1970a, b; Voogt *et al.*, 1969) is consistent with earlier measures of adrenal hyperfunction during lactation such as adrenal hypertrophy (Andersen and Kennedy, 1933), adrenal ascorbic acid depletion (Anderson and Turner, 1962), and increased adrenal cholesterol (Poulton and Reece, 1957). In light of these indices of adrenal hyperactivity during lactation, it is paradoxical that the pituitary–adrenal response to stress is reduced at this time. This is true not only in response to the potent physiological stressors of pentobarbital, shock, and ether, as discussed above, but also in response to the relatively mild stress of placement in a novel arena (open field) for 3 min (Stern *et al.*, 1973a). Perhaps the lactating rat is less aroused by these stimuli, with respect to corticosterone release, than is her nonlactating counterpart, but is particularly sensitive to the suckling stimulus. This does not appear to be the case, however: in 12-h litter-deprived rats, suckling evoked a smaller rise in plasma corticosterone than did ether (Stern and Voogt, 1974).

In order to further elucidate the pituitary–adrenal activity of the lactating rat, we have measured basal and stress secretions of plasma corticosterone during a 24-h period. We also measured levels of plasma ACTH in response to ether, as a function of suckling and the adrenal and plasma corticosterone response to exogenous ACTH (Stern *et al.*, 1973b). In these studies, Sprague–Dawley females were housed individually with their litters and were exposed to 12 h of light daily. Plasma corticosterone was measured by a fluorometric micro-method (Glick *et al.*, 1964) and plasma ACTH by a mouse bioassay modified after that of Hedner and Rerup (1962).

As shown in Fig. 1, the nonlactating, postparturient rat exhibits the typical female rat diurnal rhythm of plasma corticosterone (Critchlow *et al.*, 1963) with a 6-fold difference between the early morning trough and the late afternoon–early evening peak. In contrast, the elevated morning levels in the lactators almost obliterate the rhythm. Suckling-induced corticosterone release probably accounts for this increase since we have found that in females deprived of their litters for 24 h, morning basal levels of plasma corticosterone were not significantly different from nonlactating controls.

Fig. 1 also shows that 15 min after exposure to ether and jugular venipuncture, the plasma corticosterone response is reduced in lactating rats compared with postparturient nonlactating controls, and that this reduction occurs throughout the day and night. The suppression in the stress response of the lactators was less marked when

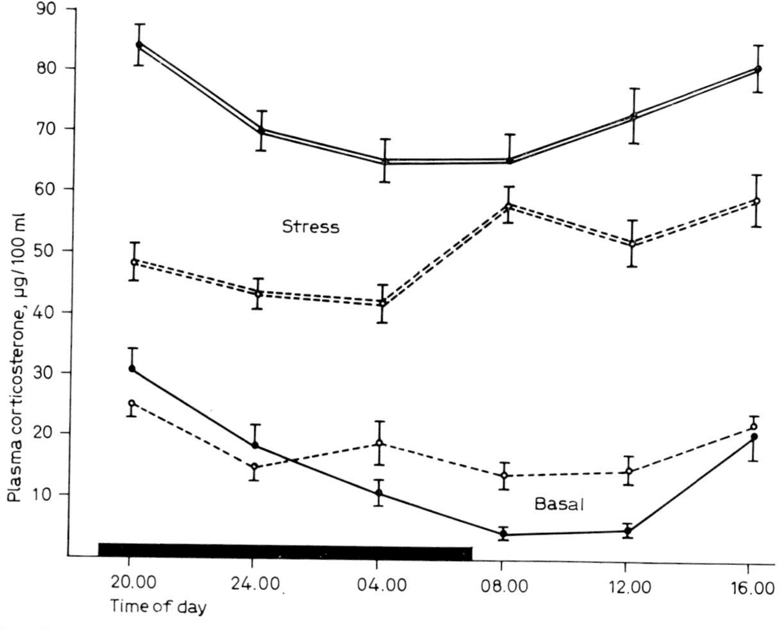

Fig. 1. Circadian variations in plasma corticosterone under basal and stress (ether) conditions. Each point (and vertical bar) is the mean (± S.E.M.) of 8 rats. ○ = lactating; ● = nonlactating.

References p. 442–444

the lights were on, possibly an indication of enhanced arousal during the day due to their maternal activities, as compared to nonlactators. A study of the acute time course of plasma corticosterone changes after ether (Zarrow *et al.*, 1972b) showed that the reduced stress response is not an artifact of the 15-min sampling procedure.

In lactating rats, stress levels of plasma ACTH were one-third that of nonlactating controls. It has been suggested that reduced pituitary–adrenal secretions after stress during lactation occur as a result of feedback inhibition by suckling-induced elevations in plasma corticosterone (Kamoun, 1970b). This explanation does not appear to be valid since we found reduced plasma ACTH and corticosterone in both continually lactating and 24-h litter-deprived lactating rats, indicating that recent suckling episodes and their consequences (*i.e.* ACTH and corticosterone increases) are not essential for the diminished pituitary–adrenal responsiveness to stress. In fact, as long as 4 days after weaning, whether weaning takes place on day 11 or day 21 postpartum, plasma corticosterone levels after pentobarbital anesthesia were reduced compared to controls (Thoman *et al.*, 1968).

The reduced plasma ACTH response to ether during lactation (Kamoun, 1970b; Stern *et al.*, 1973b) indicates that a pituitary or central nervous system component is at least in part responsible for the reduced plasma corticosterone levels. In addition, we found evidence that a peripheral component may also contribute to the reduced stress concentrations of plasma corticosterone (Stern *et al.*, 1973b). Secretion of ACTH can be blocked in rats by treatment with dexamethasone, a synthetic glucocorticoid. In rats so treated, exogenous ACTH administration resulted in higher adrenal and lower plasma corticosterone during lactation. Since the half-life and metabolic clearance rate of corticosterone do not change significantly during lactation (Kamoun, 1970a), this finding may reflect an inhibition of steroid release from the adrenal.

Exteroceptive stimuli can simulate suckling-induced hormonal secretions. After a period of nonsuckling in lactating rats, the sight, sound, and odor of pups was sufficient to result in prolactin discharge (Grosvenor, 1965) and corticosterone secretion (Zarrow *et al.*, 1972b). The conditioning of the oxytocin-induced milk-ejection reflex is well known in cows and in women (Tindal and Knaggs, 1970), as well as in rats (Deis, 1968). Since at least a recent (*i.e.* within 1–4 days) suckling episode is not necessary for altered pituitary–adrenal activity during lactation, we questioned whether maternal activities in the postpartum rat, in the absence of suckling stimulation, could bring about this change. Maternal behavior can be maintained in this manner by removal of the nipples (thelectomy) prior to conception and then twice-daily provision of pups nursed by donor mothers (Moltz *et al.*, 1967). Endocrine changes similar to lactation are induced by this procedure: resumption of cycling is delayed compared to rats completely deprived of pups postpartum and deciduomata are readily induced, changes suggestive of prolactin–progesterone secretion (Moltz *et al.*, 1969). We found, however, that the mother–pup interaction, in the absence of suckling, was not sufficient to diminish ACTH and corticosterone elevations following ether stress (Stern and Levine, 1972). Clearly then, physiological changes which occur in the suckled but not in the thelectomized or litter-deprived postpartum rat, must account for the difference in pituitary–adrenal responsiveness.

The possible endocrine differences that may be involved in these effects are the levels of prolactin, progesterone and estrogen. The raised levels of prolactin and (prolactin-induced) progesterone found in rat mothers with full litters are significantly reduced in lactators with small litters, and are reduced further in the complete absence of suckling postpartum (Amenomori *et al.*, 1970; Yoshinaga *et al.*, 1971). The reduced levels of gonadotropins (and therefore, estrogen) during lactation are directly attributable to the suckling stimulus, the degree of suppression being proportional to the intensity of suckling (Rothchild, 1960).

Our attempts to demonstrate a role of prolactin were unsuccessful. Injections of 1 or 2 I.U. of prolactin twice daily for 5 days in intact, ovariectomized and postpartum litter-deprived females did not consistently affect corticosterone levels after stress; when reductions were seen with this treatment, they were of a smaller magnitude than occurs during lactation (Stern and Levine, unpublished observations). Recently, Witorsch and Kitay (1972) reported that prolactin inhibits the activity of a 5α-reductase in the adrenal cortex, which ultimately has the effect of enhancing corticosterone production. In contrast, a recent brief report of Endröczi and Nyakas (1972) suggests that prolactin acts via the hypothalamus to exert an inhibitory effect on CNS–pituitary–adrenal function. Both effects may be operating simultaneously, since we observed a reduction in ACTH following stress and an increase in adrenal corticosterone in response to exogenous ACTH during lactation (Stern *et al.*, 1973b).

Because our own work with prolactin was not conclusive, studies on the possible roles of progesterone and estrogen were carried out. The inhibitory effects of progesterone on adrenal function have been demonstrated repeatedly (*e.g.* Steinetz *et al.*, 1965), while estrogen is known to augment the activity of the pituitary–adrenal system (Kitay, 1968). Consideration of the possible involvement of these hormones on the reduced stress response during lactation is bolstered by the findings that the reduced adrenocortical stress response is apparent on day 20 of pregnancy (Kamoun, 1970b), and probably as early as day 13 (Stern, in preparation); in rats, plasma progesterone is known to be high until about day 20 of gestation (Hashimoto *et al.*, 1968) and plasma estrogen is at diestrous levels until the periparturitional period (Yoshinaga *et al.*, 1969). Furthermore, in rats deprived of their litters shortly after delivery, the magnitude of the stress response increases between days 7 and 14 postpartum (Stern and Levine, 1972) and on day 3 postpartum, litter-deprived and suckled rats both have a reduced stress response as compared to cycling females (Stern, in preparation). Therefore, the effect of lactation should be termed a maintenance, rather than an induction, of reduced pituitary–adrenal activity after stress. While the mechanism may be different during pregnancy and lactation, the recrudescence of the response in the nonsuckled, postpartum rats shows the influence of a decreasing inhibitory and/or an increasing facilitatory factor after delivery in the absence of suckling.

Progesterone was ruled out as a factor by ovariectomizing rats on day 1 postpartum (Stern, in preparation). In females remaining with their litters, adrenocortical secretions after stress were reduced two weeks later whether or not the ovary was present. In females deprived of their litters, who resume estrus cycling in about a week, the presence of the ovary was necessary for the adrenocortical stress response to increase in magnitude. This latter finding pointed to estrogen as a possible factor.

Estrogen levels can be increased endogenously during lactation by reducing the intensity of suckling, accomplished with a litter of about two pups as opposed to a full-sized litter of 8–10 pups (Rothchild, 1960). We found that in mothers suckling only two pups the adrenocortical response to stress is equivalent to that of females completely deprived of pups postpartum (Stern *et al.*, 1973a) and ovariectomy diminishes the response (Stern, in preparation). Since prolactin levels are also reduced when the litter size is reduced (Amenomori *et al.*, 1970), the possible role of estrogen was examined further by daily subcutaneous injections of 0.5, 1.0 or 5.0 μg estradiol benzoate into intact and ovariectomized lactating females. Previously, we determined that 2 μg of estradiol benzoate daily for two weeks restores the corticosterone stress response of ovariectomized rats to that of intact, cycling females. Surprisingly, as much as 5 μg of estradiol benzoate per day for 14 or 21 days postpartum, had no effect on lactational levels of plasma corticosterone after ether stress (Stern, in preparation). It is possible that ACTH secretion was enhanced by estrogen during lactation, as it is in the cycling or ovariectomized female (Kitay, 1968), but that a change in corticosterone-binding globulin levels (Gala and Westphal, 1965) in the adrenal accounts for this observation. In any case, this experiment indicates that, with respect to circulating levels of plasma corticosterone after stress, the state of lactation protects the female from the effects of estrogen.

In light of our negative findings with estrogen treatment and ovariectomy (*i.e.* elimination of ovarian progesterone secretion) during lactation, and the recent report of Endröczi and Nyakas (1972) implicating prolactin in the reduced secretion of corticotropin-releasing factor after stress, we believe it likely that during lactation, prolactin plays a significant role in reduced ACTH and corticosterone secretion after stress. However, a definitive demonstration of such a role for prolactin must take into account the long-lasting nature of the effect, since, as discussed above, after 24 h (Stern *et al.*, 1973b) or 4 days (Thoman *et al.*, 1968) of litter deprivation, the adrenocortical stress response is still reduced. After as little as 12 h of suckling deprivation, prolactin levels are that of a cycling, diestrus female (about 10 ng/ml) but corticosterone levels after ether stress in these females are equivalent to those of lactating rats continuously with their litters, whose prolactin levels are very high (about 250 ng/ml) (Stern and Voogt, 1974).

The functional significance, if any, of lower corticosterone levels following stress in lactating rats remains to be elucidated. High titers of plasma corticosterone in the mother might reach the rat pups via the milk, resulting in a direct deleterious effect on the development of the young (Schapiro, 1968), although only trace amounts of corticoids seem to pass through the milk in rats (Zarrow *et al.*, 1970). Postnatal maternal stress has been found to alter the pituitary–adrenal physiology and behavior of the offspring (*e.g.* Levine and Thoman, 1969; Thoman and Levine, 1969). Reduced pituitary–adrenal secretions after stress may modify behavioral reactions, *per se*, and/or they may be a reflection of modifications in other behavioral and physiological changes after stress. Perhaps the mother's attentiveness to her litter is more stable when her degree of responsiveness to stress is reduced, thereby minimizing the consequences to the pups of disturbances in the maternal environment.

DIURNAL RHYTHMS IN BEHAVIOR DURING LACTATION

Another aspect of the psychobiology of lactation which our laboratory has studied recently concerns the diurnal rhythms of activity and the patterning of maternal behavior, including nursing. A striking example of this relationship is provided by the rabbit mother, who nurses her young once a day for a few minutes (Cross and Harris, 1952; Zarrow *et al.*, 1965). Clearly, a variety of factors are integrated with this pattern, including a well-insulated hair-lined maternal nest (Zarrow *et al.*, 1972a), rapid milk release, vigorous sucking and a large stomach capacity in the young. What determines the timing of the once-daily visit of the rabbit mother to her nest is a fascinating question which has yet to be answered.

The rat, unlike the rabbit, nurses her young at frequent intervals throughout the day (Bolles and Woods, 1964; Rosenblatt and Lehrman, 1963; Wiesner and Sheard, 1933). The continual presence of the rat mother in her nest is less mysterious than the behavior of the rabbit, since most mammals show frequent suckling, but no less remarkable considering the amount of time she devotes to her offspring. Grota and Ader (1969) monitored the activity of lactating rats in a dual-chambered cage, one chamber containing the litter and nest and the other containing food and water. Immediately after delivery, dams spent 85% of their time in the litter chamber (a measure of maternal behavior); there was a progressive decline in this measure so that by day 17 postpartum, the mothers spent only 30% of their time in the litter chamber. On a 24-h basis, mothers spent almost twice as much time in the litter chamber when the lights were on than when the lights were off in mid-lactation. Although the proportion was not indicated for the different periods of time postpartum, it was stated that the mothers spent more time with their litters during lights on than lights off throughout lactation. Further, the greatest number of bursts of activity (moving from cage to cage or eating) occurred when the mothers spent the least amount of time with the litter.

The data of Grota and Ader (1969) suggest that there is a diurnal rhythm of maternal behavior which is the inverse of the rat's nocturnal activity and eating rhythm, but they do not specify in detail what the mother is doing in the two chambers of her living cage or how the proportion of her time spent with the litter or engaged in other activities changes over time. For example, although mothers spent more time in the vicinity of their litters during the day, Bolles and Woods (1964) observed that mothers and pups often slept during nipple attachment and they reported their impression that actual nursing time (*i.e.* milk release) was similar during the day and the night. To obtain more information about maternal activity, we measured maternal food and water intake and litter weights every 12 h, at light onset and offset (Stern and Levin, in preparation). As can be seen from Fig. 2A, the lactating rat increases her daily food intake from 26 g on day 2 postpartum to over 60 g 2 weeks later. Normal cycling

References p. 442–444

females in our laboratory ingest about 18–20 g of rat chow pellets per day. At the beginning of lactation, the typical laboratory nocturnal pattern of food intake is evident, 75% of the daily intake occurring during lights off. As lactation proceeds, the nocturnal intake doubles and the diurnal intake triples, so that during mid-lactation, the nocturnal intake is 66.7% of the daily intake. Thus, the circadian rhythm and food intake is modulated during lactation, but not as much as that of plasma corticosterone.

Consonant with Grota and Ader's observations (1969) on time spent in the litter chamber, Fig. 2B shows that from day 4 to day 17 postpartum, litters gain more weight during lights on than during lights off. However, the proportional difference is not as dramatic as that seen for maternal food intake. The maximum difference was on day 4, when 66.7% of the daily weight gain occurred during lights on. Therefore, the lights on weight gain varied between 55.9% and 61.8% of the total. Time near the litter (Grota and Ader, 1969), therefore, is a good estimate of nursing time, but it underestimates the contribution of nocturnal nursing. On day 19, when the pups have had access to the mother's food supply for several days, a nocturnal pattern of intake was evident in the pups.

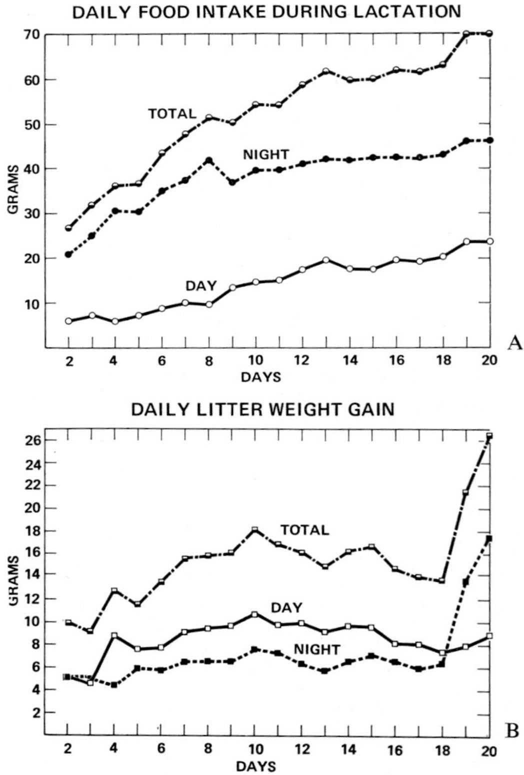

Fig. 2. A: daily food intake in the lactating rat; each point is the mean of 16 determinations. B: daily pup weight gain during lactation; each point is the mean of 16 determinations.

We wondered if the diurnal rhythm of maternal behavior, as measured by litter weight gains in our study, would be altered by imposing a restriction on the mothers' food access to lights on only or lights off only. It is known that when the rat's feeding time is restricted to less than 1 h each day, the daily food intake is sufficient to maintain body weight (Routtenberg and Kuznesof, 1967). Therefore, the lactating female should be able to meet her food needs in 3–4 h, leaving plenty of time for other activities, such as nursing. The data indicate that when the females are adapted to the 12-h restricted food schedule prior to conception, their food intake is 70–90% that of the lactating controls, a decrement reflected in reduced litter weight gains which were equivalent for the litters of day-fed only and night-fed only mothers. In the night-fed rats, the diurnal pattern of litter weight gain was very similar to the 24-h unrestricted feeding controls. In the day-fed group, however, the pattern was reversed, litters gaining more weight during lights off than lights on. In a recently completed experiment, preliminary analysis indicates that the diurnal rhythm of maternal behavior was reversed in the day-fed rats, and, in addition, the rate of milk secretion in these animals may be higher at night than during the day. Thus, rather than changing her meal-intake pattern of several meals spread out over time, there was an adjustment in when, during a 24-h period, the mother would spend time with, and nurse, her litter. As in the unrestricted feeding situation, the result was a maternal behavior rhythm the inverse of the activity (or feeding and drinking) pattern.

The influence of diurnal rhythms of behavior and physiology and of reduced pituitary–adrenal responsiveness to stress during lactation on the development of the offspring has yet to be explored. However, our present findings on the psychobiology of the lactating rat suggest that the mother is capable of meeting the many behavioral and physiological demands on her in a stable, yet flexible, manner.

SUMMARY

Care of the offspring and lactation are dramatic and demanding states entailing numerous behavioral and physiological changes. The present paper focuses on the lactating female rat's responsiveness to stress and diurnal rhythm in pituitary–adrenal activity and in behavior, functions which are regulated in part by hypothalamic mechanisms.

Alterations in the neuroendocrine function of the pituitary–adrenal system are observed during lactation. The pronounced circadian rhythm in plasma corticosterone level in rats is almost obliterated by lactation. Furthermore, following a variety of stressful stimuli, reduced levels of plasma ACTH and corticosterone are observed in lactating compared to nonlactating females. The possible endocrine involvement in these changes was studied. It was found that ovarian progesterone does not play a role in these changes during lactation and that the state of lactation protects the female from the effects of estrogen on pituitary–adrenal activity. It is possible that prolactin plays a significant role in the observed alterations in neuroendocrine function.

References p. 442–444

Diurnal rhythms of feeding activity and nursing were also studied. Mothers continued to show a pronounced nocturnal rhythm in food intake, although this rhythm is modulated compared to nonlactators. Mothers spend more time with their litters during the day than at night, as judged by the pattern of litter weight gain. The diurnal pattern of nursing is probably the inverse of the activity cycle: when mothers were permitted to feed only during the day, the maternal behavior and litter weight gain rhythms became nocturnal.

In conclusion, our present findings on the psychobiology of the lactating rat suggest that the mother is capable of meeting the many behavioral and physiological demands on her in a stable, yet flexible manner.

ACKNOWLEDGEMENTS

This study was supported by Research Grant NICH and HD 02881 from the National Institutes of Health and by the Leslie Fund of Chicago.

Dr. Seymour Levine was supported by Research Scientist Award MH-19936 from the National Institute of Mental Health.

REFERENCES

Amenomori, Y., Chen, C. L. and Meites, J. (1970) Serum prolactin levels in rats during different reproductive states. *Endocrinology*, **86**, 506–510.

Andersen, D. H. and Kennedy, H. S. (1933) Studies on the physiology of reproduction. V. The adrenal cortex in pregnancy and lactation. *J. Physiol. (Lond.)*, **77**, 159–173.

Anderson, R. R. and Turner, C. W. (1962) Adrenal ascorbic acid levels during pregnancy and lactation in rats and mice. *Endocrinology*, **70**, 796–800.

Bolles, R. C. and Woods, P. J. (1964) The ontogeny of behavior in the albino rat. *Anim. Behav.*, **XII**, 427–441.

Brush, M. G. (1960) The effect of ACTH injections on plasma corticosteroid levels and milk yield in the cow. *J. Endocrinol.*, **21**, 155–160.

Cowie, A. T. and Folley, S. J. (1947) Adrenalectomy and replacement therapy in lactating rats. 2. Effects of deoxycorticosterone acetate on lactation in adrenalectomized rats. *J. Endocrinol.*, **5**, 14–23.

Cowie, A. T. and Tindal, J. S. (1971) *The Physiology of Lactation*, Edward Arnold, London.

Critchlow, V., Liebelt, R. A., Bar-Sela, M., Mountcastle, W. and Lipscomb, H. S.(1963) Sex difference in resting pituitary–adrenal function in the rat. *Amer. J. Physiol.*, **205**, 807–815.

Cross, B. A. (1961) Neural control of lactation. In *Milk: The Mammary Gland and its Secretion, (Vol. 1)*, S. K. Kon and A. T. Cowie (Eds.), Academic Press, New York, pp. 229–280.

Cross, B. A. and Harris, G. W. (1952) The role of the neurohypophysis in the milk-ejection reflex. *J. Endocrinol.*, **8**, 148–161.

Deis, R. P. (1968) The effect of an exteroceptive stimulus on milk ejection in lactating rats. *J. Physiol. (Lond.)*, **197**, 37–46.

Denamur, R., Stoliaroff, M. et Desclin, J. (1965) Effets de la traite sur l'activité corticotrope hypophysaire des petits ruminants en lactation. *C. R. Acad. Sci. (Paris)*, **260**, 3175–3178.

Elias, J. J. and Rivera, E. (1959) Comparison of the responses of normal, precancerous, and neoplastic mouse mammary tissues to hormones *in vitro*. *Cancer Res.*, **19**, 505–511.

Endröczi, E. and Nyakas, Cs. (1972) Pituitary–adrenal response of lactating rats and the effect of prolactin administration. *Hormones*, **3**, 267.

FLUX, D. S., FOLLEY, S. J. AND ROWLAND, S. J. (1954) Effect of adrenocorticotrophic hormone on yield and composition of milk of cow. *J. Endocrinol.*, **10**, 333–339.

GALA, R. R. AND WESTPHAL, U. (1965) Corticosterone-binding globulin in the rat: possible role in the initiation of lactation. *Endocrinology*, **76**, 1079–1088.

GLICK, D., REDLICH, D. VON AND LEVINE, S. (1964) Fluorometric determination of corticosterone and cortisol in 0.02–0.05 milliliters of plasma or submilligram samples of adrenal tissue. *Endocrinology*, **74**, 653–655.

GREENBAUM, A. L. AND DARBY, F. J. (1964) The effect of adrenalectomy on the metabolism of the mammary glands of lactating rats. *Biochem. J.*, **91**, 307–317.

GRÉGOIRE, C. (1946) Factors involved in maintaining involution of the thymus during suckling. *J. Endocrinol.*, **5**, 68–87.

GROSVENOR, C. E. (1965) Evidence that exteroceptive stimuli can release prolactin from the pituitary gland of the lactating rat. *Endocrinology*, **76**, 340–342.

GROTA, L. J. AND ADER, R. (1969) Continuous recording of maternal behaviour in *Rattus norvegicus*. *Anim. Behav.*, **17**, 722–729.

HAHN, D. W. AND TURNER, C. W. (1966) Effect of corticosterone and aldosterone upon milk yield in the rat. *Proc. Soc. exp. Biol. (N.Y.)*, **121**, 1056–1058.

HASHIMOTO, I., HENRICKS, D. M., ANDERSON, L. L. AND MELAMPY, R. M. (1968) Progesterone and pregn-4-en-20α-ol-3-one in ovarian venous blood during various reproductive states in the rat. *Endocrinology*, **82**, 333–341.

HEDNER, P. AND RERUP, C. (1962) Plasma corticosteroid levels and adrenal ascorbic acid after intravenous corticotrophin injections and "stressful" stimuli in the rat. *Acta endocr. (Kbh.)*, **39**, 527–538.

JOHNSON, R. M. AND MEITES, J. (1958) Effects of cortisone acetate on milk production and mammary involution in parturient rats. *Endocrinology*, **63**, 290–294.

KAMOUN, A. (1970a) Activité cortico-surrénale au cours de la gestation, de la lactation et du développement pré et post-natal chez le rat. I. Concentration et cinétique de disparition de la corticostérone. *J. Physiol. (Paris)*, **62**, 5–32.

KAMOUN, A. (1970b) Activité cortico-surrénale au cours de la gestation, de la lactation et du développement pré et post-natal chez le rat. II. Régulation de l'activité corticotrope hypophysaire dans les conditions normales et au cours de l'agression. *J. Physiol. (Paris)*, **62**, 33–50.

KITAY, J. I. (1968) Effects of estrogen and androgen on the adrenal cortex of the rat. In *Functions of the Adrenal Cortex Vol. 2*, K. W. MCKERNS (Ed.), Appleton-Century-Crofts, New York, pp. 775–811.

LEVINE, S. AND THOMAN, E. B. (1969) Physiological and behavioral consequences of postnatal maternal stress in rats. *Physiol. Behav.*, **4**, 139–142.

MEITES, J. (1966) Control of mammary growth and lactation. In *Neuroendocrinology, Vol. 1*, L. MARTINI AND W. F. GANONG (Eds)., Academic Press, New York, pp. 669–707.

MOLTZ, H. (1971) The ontogeny of maternal behavior in some selected mammalian species. In *The Ontogeny of Vertebrate Behavior*, H. MOLTZ (Ed.), Academic Press, New York, pp. 263–313.

MOLTZ, H., GELLER, D. AND LEVIN, R. (1967) Maternal behavior in the totally mammectomized rat. *J. comp. physiol. Psychol.*, **64**, 225–229.

MOLTZ, H., LEVIN, R. AND LEON, M. (1969) Prolactin in the postpartum rat: synthesis and release in the absence of suckling stimulation. *Science*, **163**, 1083–1084.

POULTON, B. R. AND REECE, R. P. (1957) The activity of the pituitary–adrenal cortex axis during pregnancy and lactation. *Endocrinology*, **61**, 217–225.

ROSENBLATT, J. S. AND LEHRMAN, D S. (1963) Maternal behavior of the laboratory rat. In *Maternal Behavior in Mammals*, H. L. RHEINGOLD (Ed.), Wiley, New York, pp. 8–57.

ROTHCHILD, I. (1960) The corpus luteum–pituitary relationship. The association between the cause of luteotrophin secretion and the cause of follicular quiescence during lactation; the basis for a tentative theory of the corpus luteum–pituitary relationship in the rat. *Endocrinology*, **67**, 9–41.

ROUTTENBERG, A. AND KUZNESOF, A. W. (1967) Self-starvation of rats living in activity wheels on a restricted feeding schedule. *J. comp. physiol. Psychol.*, **64**, 414–421.

SCHAPIRO, S. (1968) Some physiological, biochemical, and behavioral consequences of neonatal hormone administration: cortisol and thyroxine. *Gen. comp. Endocrinol.*, **10**, 214–228.

STEINETZ, B. G., BEACH, V. L., DIPASQUALE, G. AND BATTISTA, J. V., JR. (1965) Effects of different gestagenic steroid types on plasma-free corticosteroid levels in ACTH-treated rats. *Steroids*, **5**, 93–108.

STERN, J. M. AND LEVINE, S. (1972) Pituitary–adrenal activity in the postpartum rat in the absence of suckling stimulation. *Horm. Behav.*, **3**, 237–246.

STERN, J. M. AND VOOGT, J. L. (1974) Comparison of plasma corticosterone and prolactin levels in cycling and lactating rats. *Neuroendocrinology*, **13**, 173–181.

STERN, J. M., ERSKINE, M. S. AND LEVINE, S. (1973a) Dissociation of open-field behavior and pituitary–adrenal function. *Horm. Behav.*, **4**, 149–162.

STERN, J. M., GOLDMAN, L. AND LEVINE, S. (1973b) Pituitary–adrenal responsiveness during lactation in rats. *Neuroendocrinology*, **12**, 179–191.

TALWALKER, P. K., MEITES, J. AND NICOLL, C. S. (1960) Effects of hydrocortisone, prolactin and oxytocin on lactational performance of rats. *Amer. J. Physiol.*, **199**, 1070–1072.

THATCHER, W. W. AND TUCKER, H. A. (1970) Adrenal function during prolonged lactation. *Proc. Soc. exp. Biol. (N.Y.)*, **134**, 915–918.

THOMAN, E. B. AND LEVINE, S. (1969) Role of maternal disturbance and temperature change in early experience studies. *Physiol. Behav.*, **4**, 143–145.

THOMAN, E. B., CONNER, R. L. AND LEVINE, S. (1970a) Lactation suppresses adrenal corticosteroid activity and aggressiveness in rats. *J. comp. physiol. Psychol.*, **70**, 364–369.

THOMAN, E. B., SPROUL, M., SEELER, B. AND LEVINE, S. (1970b) Influence of adrenalectomy in female rats on reproductive processes including effects on the foetus and offspring. *J. Endocr.*, **46**, 297–303.

THOMAN, E. B., WETZEL, A. AND LEVINE, S. (1968) Lactation prevents disruption of temperature regulation and suppresses adrenocortical activity in rats. *Commun. Behav. Biol.*, **2**, 165–171.

TINDAL, J. S. AND KNAGGS, G. S. (1970) Environmental stimuli and the mammary gland. *Mem. Soc. Endocr.*, **18**, 239–258.

ULRICH, R. E. AND AZRIN, N. H. (1962) Reflexive fighting in response to aversive stimulation. *J. exp. Anal. Behav.*, **5**, 511–520.

VOOGT, J. L., SAR, M. AND MEITES, J. (1969) Influence of cycling, pregnancy, labor, and suckling on corticosterone–ACTH levels. *Amer. J. Physiol.*, **216**, 655–658.

WIESNER, B. P. AND SHEARD, N. J. (1933) *Maternal Behaviour in the Rat*, Oliver and Boyd, London.

WILLMER, J. S. AND FOSTER, T. S. (1965) Restoration of hepatic and mammary gland hexosemonophosphate shunt activity in the adrenalectomized lactating rat by adrenal corticoids. *Canad. J. Physiol. Pharmacol.*, **43**, 905–913.

WITORSCH, R. J. AND KITAY, J. I. (1972) Pituitary hormones affecting adrenal 5α-reductase activity: ACTH, growth hormone and prolactin. *Endocrinology*, **91**, 764–769.

YOSHINAGA, K., HAWKINS, R. A. AND STOCKER, J. F. (1969) Estrogen secretion by the rat ovary *in vivo* during the estrous cycle and pregnancy. *Endocrinology*, **85**, 103–112.

YOSHINAGA, K., MOUDGAL, N. R. AND GREEP, R. O. (1971) Progestin secretion by the ovary in lactating rats: effect of LH-antiserum, LH and prolactin. *Endocrinology*, **88**, 1126–1130.

ZARROW, M. X., DENENBERG, V. H. AND ANDERSON, C. O. (1965) Rabbit: frequency of suckling in the pup. *Science*, **150**, 1835–1836.

ZARROW, M. X., DENENBERG, V. H. AND SACHS, B. D. (1972a) Hormones and maternal behavior in mammals. In *Hormones and Behavior*, S. LEVINE (Ed.), Academic Press, New York, pp. 105–134.

ZARROW, M. X., PHILPOTT, J. E. AND DENENBERG, V. H. (1970) Passage of ^{14}C-4-corticosterone from the rat mother to the foetus and neonate. *Nature (Lond.)*, **226**, 1058–1059.

ZARROW, M. X., SCHLEIN, P. A., DENENBERG, V. H. AND COHEN, H. A. (1972b) Sustained corticosterone release in lactating rats following olfactory stimulation from the pups. *Endocrinology*, **91**, 191–196.

The Hypothalamus and Behavioral Patterns

UWE JÜRGENS

Max-Planck-Institut für Psychiatrie, Munich (G.F.R.)

INTRODUCTION

The present paper will deal with the functional significance of the hypothalamus for the directly observable behavioral actions. There are essentially two approaches which make up the main bulk of our present knowledge concerning this problem: the first one is stimulation, using either electrical, chemical or thermic stimuli applied to discrete hypothalamic areas; the second one elimination, using either destruction or local anesthesia of the whole hypothalamus or of parts of it. In the present review, we will follow these two approaches and see how far they will carry us.

STIMULATION EXPERIMENTS

Principally, these experiments can be considered in 3 different aspects: firstly, the types of behavioral patterns elicitable from the hypothalamus; secondly, the localization of the stimulus responses within the hypothalamus; and thirdly, the interpretation of these responses in terms of the level of behavioral integration they occupy. As only the combination of all 3 aspects will give us a complete picture, we will discuss them all 3 in the order just mentioned.

(a) Behavioral patterns elicitable

The most consistently evoked stimulus responses from the hypothalamus are agonistic ones. Behavioral patterns of this type have been elicited hypothalamically not only in mammals, like the opossum (Roberts *et al.*, 1967), rat (Vergnes and Karli, 1970), cat (Hess, 1949), and monkey (Jürgens and Ploog, 1970), but also in fish (Demski and Knigge, 1971), reptiles (Distel, 1973) and birds (Åkerman, 1965; Putkonen, 1966; Maley, 1969). Agonistic behavior, principally, can be subdivided into 3 categories: (a) dominance gestures and pure attack behavior; (b) defence reactions, and (c) flight behavior, including certain types of submissive gestures. As there are no distinct boundaries between these categories it is understandable that several authors avoided such a classification of their stimulus responses and preferred a division into only two categories, one being aggressive behavior, the other flight behavior. Nevertheless, in

References p. 460–462

the more thoroughly studied species, like the rat (Woodworth, 1970, cited from Roberts, 1969) or the cat (Wasman and Flynn, 1962) it became evident that all 3 categories can be distinguished and elicited reproducibly from the hypothalamus. The special form that these behavioral patterns take depends of course upon the species studied and ranges, for instance in the case of defensive reactions, from opening the bill, erection of the neck feathers, tail spreading and rabrab calls in the duck, to pupil dilatation, piloerection, growling, hissing, and arching the back in the cat. On the other hand, even in the same animal sometimes a variety of different behavioral patterns from one and the same agonistic category occur. For example in the duck, when rushing around in panic, freezing as well as sneaking under cover may indicate a flight reaction.

Another stimulus response elicitable from the hypothalamus and its rostral continuation, the preoptic region, is sexual behavior. This stimulus response has also been recorded from a wide variety of species belonging to different vertebrate classes. However, with the exception of the rat, male sexual behavior has only been obtained as courtship behavior, but not as complete copulation behavior. So, in fish, for instance, it was possible to evoke circling around a conspecific with spread fins, but not spawning (Demski and Knigge, 1971); in frogs stimulation yielded mating calls, but not clasping of the female (Schmidt, 1968); and in pigeons the typical bowing ceremony which consists of walking in circles with feathers ruffled and uttering cooing calls accompanied by repeated bowing movements but lacking the copulatory components could be obtained (Åkerman 1965). There are also reports on evoked female sexual behavior — induced either by electrical stimulation (Meyer and Hess 1957) or by local implantation of sex hormones into discrete hypothalamic areas (review by Beach, 1970) — but these reports are, probably due to most authors' bias to concentrate on its male counterpart, still rather rare.

A third type of behavior often encountered as a hypothalamic stimulus response is feeding and drinking. These two alimentary reactions are the most thoroughly studied hypothalamically induced behavioral patterns and, perhaps as a consequence of this, among the most disputed ones — as we shall see later.

Another type of hypothalamically elicitable behavioral pattern consists of grooming and preening behavior, respectively. This has been described in the cat (MacDonnell and Flynn, 1968), rat (Cox and Valenstein, 1969), opossum (Roberts *et al.*, 1967), and pigeon (Åkerman, 1965). Furthermore, there is a group of behavioral patterns, like hoarding, gnawing, or prey-catching which, because of their limited occurrence in only a few species, are accordingly limited in their occurrence as stimulus responses.

Finally, two very basic behavior functions should be mentioned, one the contrary of the other, but both elicitable from the hypothalamic–preoptic region: exploratory behavior and sleep. The former consists of, in its fully developed form, alerting, orienting movements, sniffing, and locomotion; the latter, if elicited electrically (Clemente and Sterman, 1967) or thermically (Roberts and Robinson, 1969) represents a state of drowsy inactivity that, however, can be easily interrupted by external sensory stimuli.

This list of hypothalamically elicited behavioral patterns is of course far from being complete; it could be extended by adding a large number of behavioral patterns, especially those not directly related to the environmental situation and those forming transitions between behavioral and autonomic reactions; as for example shivering, panting, sneezing, vomiting, urination, defecation, and many others. But as the aim of this list is just to give an impression of the diversity of hypothalamic stimulus responses instead of being exhaustive we will finish here our listing and summarize the aforementioned.

The hypothalamic–preoptic region is an area from which a great variety of behavioral patterns belonging to the most diverse behavioral functions can be evoked. In fact, there is almost no behavioral pattern which has been recorded as a stimulus response from other parts of the brain that has not also been obtained from the hypothalamus. Nevertheless, one has to realize that the complete behavioral repertoire of a species has as yet not been attained by brain stimulation. A second characteristic of many hypothalamic stimulus responses is their complexity. The typical stimulus response from this region, apart from autonomic reactions, is not an elementary movement, like raising of the forepaw or rotating the head, but a highly integrated behavioral pattern which as a rule is well adjusted to the environmental situation.

(b) Localization of stimulus responses

Let us now consider the specific loci within the hypothalamus from which the different behavioral patterns are elicitable. The most thoroughly explored species in this respect are rat, cat, and opossum. Fig. 1 shows the hypothalamic distribution of 5 behavioral functions, namely attack, flight, defence, copulation, and feeding in the rat. It can be seen in the first column that there is a clear separation between those loci producing pure attack without defence and those producing flight behavior — represented by crouching. The latter lie in the most ventral part of the anterior hypothalamus and the medial part of the central hypothalamus. All of the former lie ventrolaterally from the fornix. In the second column defence reactions, as indicated by squealing calls, have been plotted. Their distribution is much less circumscribed and covers areas from which attack as well as those from which flight can be elicited. Nevertheless, again it can be seen that all defence reactions evoked either medial to the fornix or from the most basal part of the anterior hypothalamus are accompanied by flight behavior; whereas all defence reactions evoked lateral to the fornix are either accompanied by attack behavior or do occur without other agonistic behavioral components. In the third column, electrode positions are shown which are able to induce copulation in males. These electrode positions are all lying lateral to the fornix and somewhat dorsal to the area from which attack can be produced. Finally, eating is represented in the last column where its distribution shows a clear limitation to the dorsolateral and far-lateral ventral parts of the hypothalamus. Even though there is some overlap in the border zones, the focal points of these different behavioral functions are lying in different hypothalamic areas. Only the area for defence reactions extensively merges medially with flight reactions and laterally with attack. Thus, it appears that the more related two behavioral functions are, the greater is the overlapping of their hypothalamic representation. Therefore, in Fig. 2, 4 more closely related behavioral

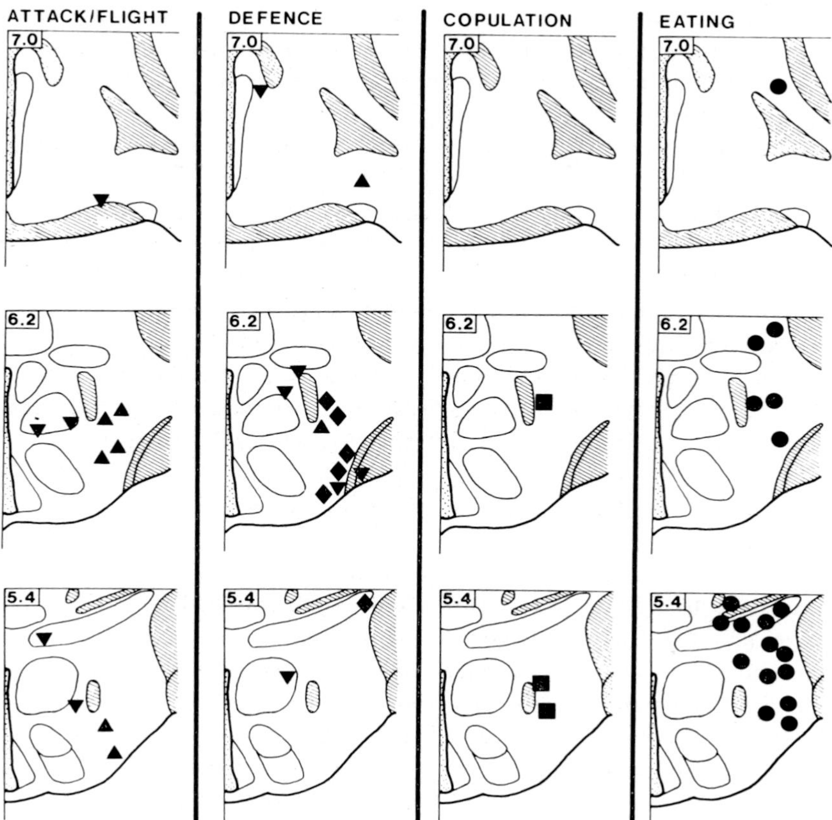

Fig. 1. Electrode positions in the hypothalamus of the rat from which the following behavior could be evoked: first column pure attack (without defence reaction) ▲ and flight (crouching) ▼; second column defence (squealing) either alone ◆, together with flight behavior ▼ or with attack ▲; third column copulation ■; fourth column eating ●. (Compiled from Caggiula, 1970; Cox and Valenstein, 1969; Woodworth, 1970, cited by Roberts, 1969.)

patterns are compared, namely eating, drinking, gnawing, and object carrying. Here it becomes evident that a distinction into different zones according to their behavioral patterns elicitable is no longer possible. As the diagrams show, this does not mean that each electrode position in the dorsolateral hypothalamus yields all 4 behavioral patterns — even though the number of loci evoking only one behavioral pattern are in the minority. These loci, however, cannot be clustered according to their types of stimulus responses.

If we turn now to another animal with a totally different feeding habit, what will be the consequences in the hypothalamic distribution of feeding loci? Fig. 3 shows the electrode positions from which eating, biting attack on a rat, and defence — rep-

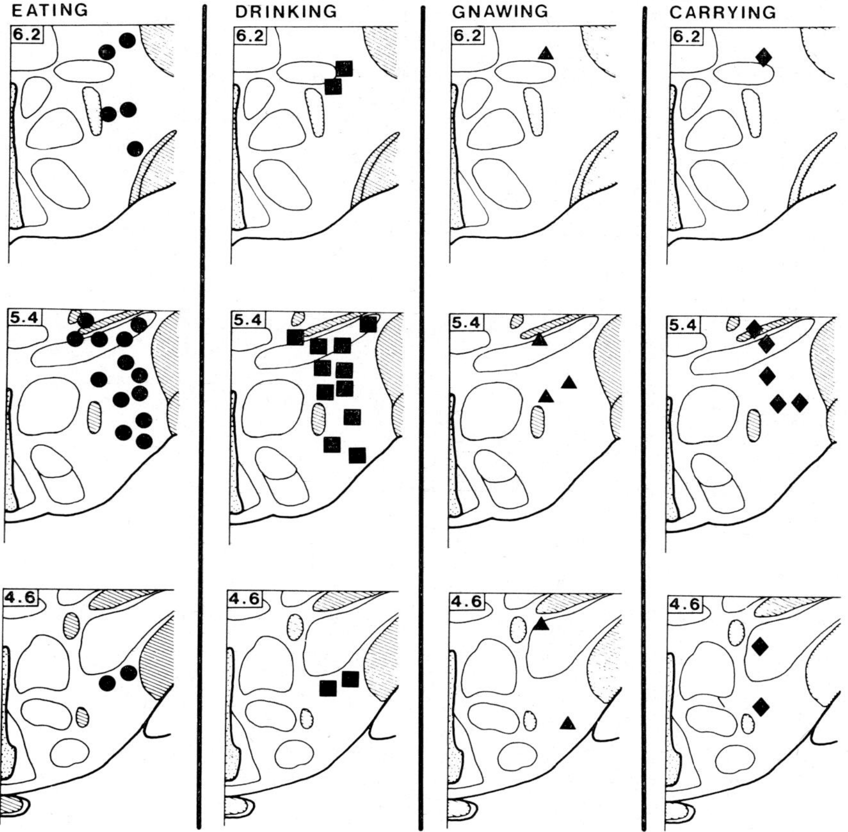

Fig. 2. Electrode positions in the hypothalamus of the rat from which eating, drinking, gnawing, and object carrying could be elicited by electrical stimulation. (Compiled from Cox and Valenstein, 1969; Valenstein *et al.*, 1970.)

resented by hissing — can be evoked in the cat. Again, we find an accumulation of feeding loci in the dorsolateral hypothalamus, but in contrast to the rat, the effective area reaches somewhat more ventrally. On the other hand, the area yielding the so-called stalking attack (Wasman and Flynn, 1962; later called "quiet biting attack": Flynn, 1967), where the cat approaches a rat and bites vigorously into its neck or head without displaying defence behavior, is not only obtained from a region ventrolateral to the fornix, but also from the dorsolateral hypothalamus. So, attacking and eating loci which show minimal overlap in the rat, almost completely intermingle in the cat. The explanation of this apparent discrepancy is that attack on a conspecific is another thing than attack on a prey, the former representing a pure agonistic reaction, the latter an alimentary reaction (with agonistic components only to the extent that

References p. 460–462

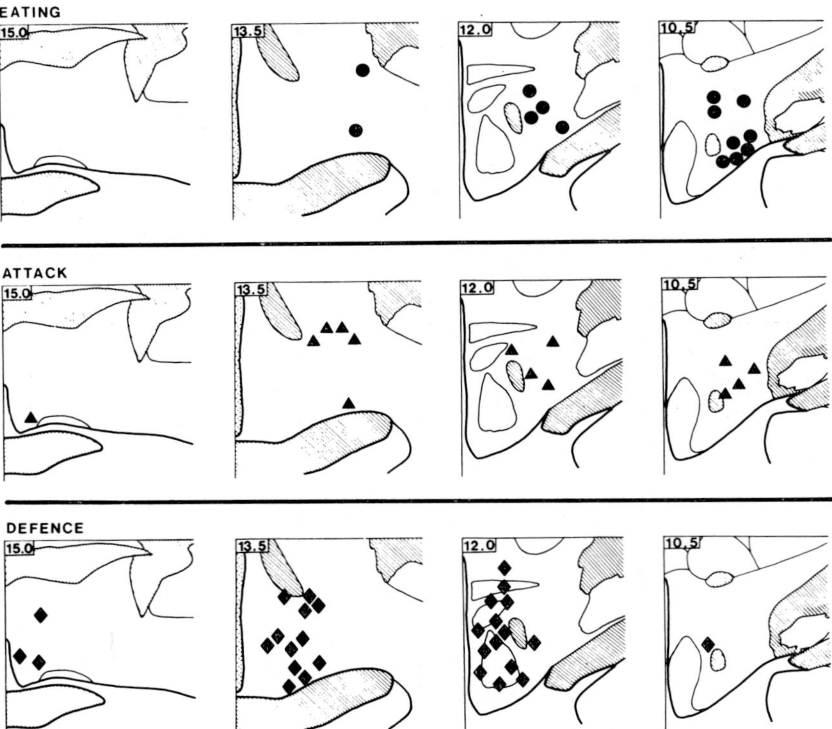

Fig. 3. Electrode positions in the hypothalamus and preoptic area of the cat yielding eating, biting attack on a rat without defence components, and defence characterized by hissing. (Compiled from Flynn, 1967; Nakao, 1971; Roberts and Kiess, 1964.)

the prey's counterattacks seriously jeopardize the predator). The overlap of feeding and prey-catching loci in the cat should be compared therefore with the overlap of feeding and object carrying in the rat; for, under natural circumstances both object carrying in the form of hoarding and prey-catching in the form of mouse killing serve to provide the animal with food.

That laboratory cats often do not kill rats spontaneously is not a proof against this, as experiments of Roberts and Bergquist (1968) with cats raised under total isolation showed that prey-catching is an innate reaction and therefore is executed under adequate conditions by all cats.

The third row of Fig. 3, finally, shows the electrode positions from which defence reactions (represented by hissing) could be elicited. A comparison of the third row with the first and second rows reveals that there is a clear separation between the eating and prey-catching area on the one side and the defence area on the other side. When compared with the defence, attack, and flight areas of the rat, it can be seen that all defence loci of the cat are lying within the flight and pure defence zone of the rat. There is only one single point lying within the pure attack or attack with defence zone of the rat. This leads one to assume that the cat's hissing has a higher probability to be followed by flight than the rat's squealing. This suggestion is corroborated firstly by the fact that the electrical elicitation of most of the hissing reactions is negatively reinforcing, that is to say aversive (Adams and Flynn, 1966; Nakao, 1971); secondly, ethological observations show that during intraspecific fights squealing in rats is often heard from the clearly dominant animal, which seems not to be the case with the cat's hissing reaction (Eibl-Eibesfeldt, 1967).

The last animal which shall be considered in this context is the opossum. Fig. 4 shows the hypothalamic and preoptic distribution of electrode positions yielding diverse male mating patterns, defensive threat patterns, and grooming. The male mating responses consisted of either mounting, rubbing the head in the female's (or dummy female's) fur, gentle biting on the neck of the female, penile erection, clicking calls, or of various combinations of these behavioral patterns. Intromission and

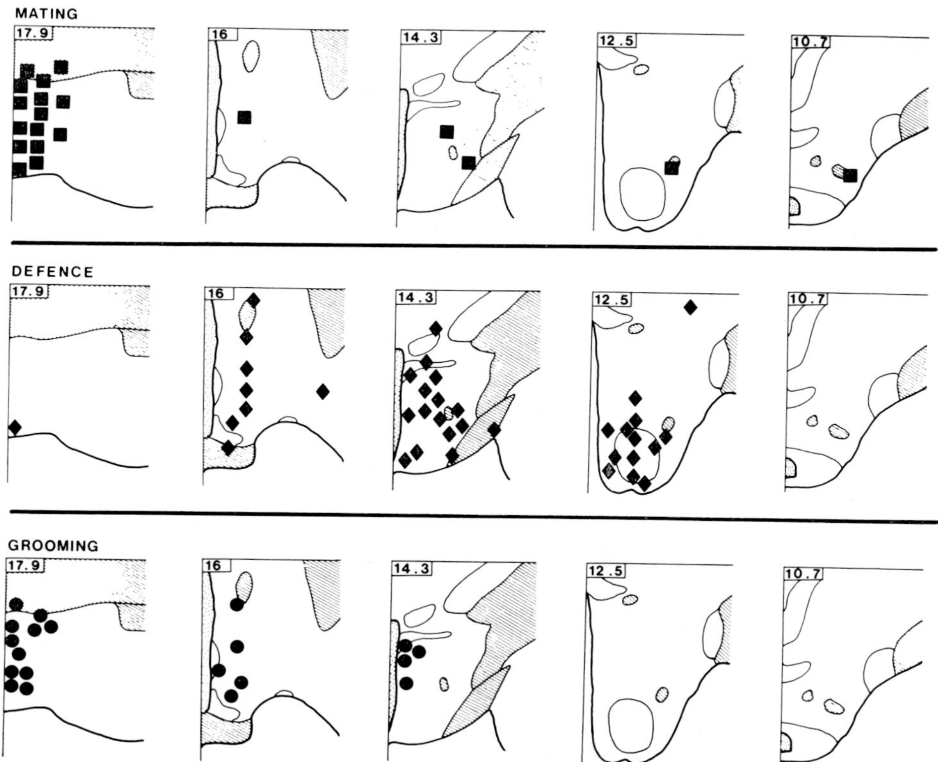

Fig. 4. Electrode positions in the hypothalamus and preoptic area of the opossum yielding male mating behavior characterized by the appearance of at least 2 of the following 5 behavioral patterns: mounting, rubbing of the head in the female's or dummy female's fur, gentle biting on the neck of the female, penile erection, clicking calls; defensive threat characterized by the appearance of at least 2 of the following 4 behavioral patterns: opening of the mouth, backing, head swinging, growling; grooming. (Modified after Roberts *et al.*, 1957.)

ejaculation, however, were not obtained. As the first row shows, there is a conspicuous accumulation of sexual stimulus responses in the medial preoptic area. From here a narrow band of positive loci continues caudalward following the course of the fornix. The relationship between the preoptic area and male sexual behavior seems to be a phylogenetically very old one. Even in fish, amphibians, and birds (see p. 446) male mating patterns have been elicited from this region. Concerning the band of mating loci along the fornix, there is a quite clear correspondence with results found in rats.

The next behavior to be considered is the defensive threat. A comparison with Figs. 1 and 3 reveals that its distribution is very similar to that of its counterparts in the rat and cat, resembling more closely the cat in the ventromedial part of the hypothalamus and the rat in the ventrolateral part of the hypothalamus. As the ventrolateral part of the central hypothalamus (except for the area immediately adjacent to the optic tract) has been presumed to be related to attack behavior, it might be inferred that the defensive threat of the opossum expresses a somewhat lower probability to flee than the hissing of the cat. Assuming this is the case, this would provide us with an explanation why there is such a complete overlapping of defensive threat responses and male mating responses in the central and anterior hypothalamus of the opossum — as the ethological literature provides us with hundreds of behavioral examples (called in German "imponieren") which clearly have this double function: namely to threaten an opponent and to attract a female.

The last behavioral pattern represented in Fig. 4 is grooming. Grooming can be elicited by electrical or thermic stimulation of the medial preoptic–anterior hypothalamic region. So, the effective area is rather limited; nevertheless, it shows a complete overlap with mating responses in the preoptic region and with defence threatening in the anterior hypothalamus. As there is no overlap of grooming and mating in the caudal part of the anterior hypothalamus and also no overlap of grooming and threatening in the preoptic area, this means that the grooming loci form an independent system. The fact that grooming can be elicited simultaneously with either threatening or mating might be explained in two ways: either the electrical stimulation because of its artificial character leads to an unselective excitation of different neuron populations which under natural conditions do not fire in that combination; or, despite the artificial character of the stimulation, grooming, when evoked together with mating or threatening, bears a true relationship to these behavior patterns. It could represent, for example, a preparatory mating behavior in the case of simultaneous occurrence with sexual responses and a displacement behavior in the case of simultaneous occurrence with threatening (for displacement activities are typical for conflict situations and threatening often represents a conflict, namely that between attack and flight).

So, what are the conclusions to be drawn from the stimulation experiments concerning the localization of behavioral patterns within the hypothalamus?

(1) It can be stated that the diverse hypothalamically elicitable behavioral patterns are not distributed diffusely throughout the hypothalamus, but are limited to specific areas. The only exception to this statement is the exploring reaction which indeed can, in some animals, be obtained anywhere within the hypothalamus, especially with intensities too low to evoke a more specific behavior or in the absence of an adequate

goal object. It therefore should be regarded as an unspecific expression of a more or less specific motivation.

(2) The area producing a distinct behavioral pattern when stimulated is relatively large, often reaching from the preoptic region down into the midbrain or from the base of the brain up to the ventral thalamus.

(3) The different behavioral patterns are not represented in a mosaic-like fashion; instead, there is frequent and extensive overlapping between them. The simultaneous occurrence of different behavioral patterns sometimes may be circumvented by changing from one type of stimulation to another. For example, eating and drinking, which can be simultaneously obtained in rats from the same electrode using electrical stimulation, can be elicited selectively using chemical stimulation (Grossman, 1960). Nevertheless, the fact remains that in these cases an anatomical distinction into separate areas according to their stimulus responses is not possible.

(4) It must be kept in mind that not all electrodes placed into an area related to a specific behavior do indeed evoke that behavior. Furthermore, electrodes placed into an area yielding various behavioral patterns do evoke in the one animal all of these, in another animal only behavioral pattern A, and in still another animal pattern B. So, individual differences clearly exist; the physiological basis of these differences is, however, totally unknown.

(c) Interpretation of the elicited behavior

This leads us to the question how to interpret the hypothalamically elicitable behavioral patterns in terms of functional integrity, compared with their spontaneous occurrence. Even a short look at the previously described behavioral patterns reveals that these are not performed as totally stereotyped motor coordinations, but instead represent well-adjusted complex behavior systems. So, eating as well as attacking, mating, or carrying behavior all need for their successful performance an adequate orientation with respect to their goal object. This suggests that hypothalamic stimulation in these cases does not elicit behavioral patterns directly but rather induces changes in the motivational state of the animal. Changes in the motivational state mean objectively that the statistical probability to react to a given stimulus in a specific way is altered. Furthermore, in the absence of adequate goal objects a high level of specific motivation may lead to a searching for this goal object; the latter is called in the language of ethology "appetitive behavior". Numerous experiments have proved that both criteria of motivational changes, namely the altered threshold for a specific reaction and the occurrence of appetitive behavior — if motivation is high and the goal object lacking — are fulfilled by many hypothalamically elicitable behavioral patterns. So, apart from the fact that the stimulus-bound elicitability itself demonstrates the lowered threshold of these behavioral patterns, it also is possible to induce eating in satiated animals (Morgane, 1961) or rat killing in cats which spontaneously do not kill rats under laboratory conditions (Wasman and Flynn, 1962). Furthermore, conditioning experiments have shown that for almost all hypothalamic behavioral patterns tested in this respect the adequate appetitive behavior was also produced. Rats, cats, opossums as well as goats when stimulated learned a maze, a lever-pressing response, or a visual discrimination task in order to obtain that goal object necessary to perform the specific behavioral pattern that would have occurred immediately if the adequate goal object had been present (review by Roberts, 1970).

Nevertheless, some authors doubted that these electrically elicited motivational changes resemble the spontaneously occurring ones (Valenstein *et al.*, 1970). According to this view, the stimulation primarily has an unspecific arousing effect. This would mean that the stimulus does not only raise the probability of occurrence of one single behavioral pattern, but of several simultaneously. Thus, the actually occurring behavior would be determined by the environmental situation and individual factors rather than by the anatomical locus of stimulation. We do not want to discuss here all the arguments supporting this interpretation, as they have been worked out explicitly by Valenstein (1969). Let us rather try to mediate between this position and another which tends to overemphasize the specificity of hypothalamic responses.

Firstly, as has already been mentioned, it is a fact that stimulation of the same area may yield different stimulus responses in different animals, even if the respectively responding area in the same animal may be relatively large. An explanation of this fact is still lacking. It should be kept in mind, however, that this apparent interchangeability of behavioral patterns is quite limited and neither comprehends all behavioral patterns nor is the same from one hypothalamic area to the other.

Secondly, electrocorticographic recordings in lightly anesthetized animals have shown that electrical stimulation of practically the whole hypothalamus leads to cortical desynchronization, that is to say arousal. Also it may be added that loci inducing sleep, if stimulated with low stimulus frequencies, do produce arousal when stimulated with the more commonly used higher frequencies. Furthermore, the same electrodes yielding cortical desynchronization showed a clear facilitating effect on all movements electrically elicitable from the cortex (Murphy and Gellhorn, 1945). So, hypothalamic stimulation does have an arousal effect. On the other hand, we know that a strong arousal has the tendency to be reacted out. Male rats, for instance, which are electrically shocked without having the possibility to escape, can thus be induced either to copulate or to fight against another rat, depending on the sex of the partner (Caggiula and Eibergen, 1969; Ulrich *et al.*, 1965). This example also demonstrates the influence the environmental situation can have on the behavioral response in an aroused animal. As hypothalamic stimulation, in fact, produces such an aroused animal, it becomes understandable why changes in the experimental setting may cause changes in the stimulus response. This variability of the stimulus response, however, does not necessarily mean that the evoked motivational change is not a specific one; it might merely indicate that the environmental situation was not adequate for the performance of the evoked behavioral tendency.

Thirdly, several experiments clearly indicate that there is indeed a difference between electrically elicited motivational changes and spontaneously occurring ones. For instance, hunger is commonly thought to be an aversive state; nevertheless, the stimulation of electrode positions resulting in eating is pleasant, even in the absence of food (Hoebel and Teitelbaum, 1962). Thirst leads to a preference for pure water to sugar water in rats; nevertheless, the stimulation of electrode positions yielding drinking reveals a preference for sugar water in a free choice situation (Valenstein *et al.*, 1968). Hoarding in rats is not bound to specific spatial relationships except that the hoarding objects must lie outside the hoarding place which is usually the nest; nevertheless, hoarding has only been obtained during hypothalamic stimulation if the experimental setting fulfilled very specific spatial conditions (Phillips *et al.*, 1969). Mating, finally, which under natural circumstances is concluded by copulation or spawning, respectively, is lacking these elements in most animals under hypothalamic stimulation. On the other hand, it can be shown that the rewarding effect of stimulating "feeding electrodes" is to some degree dependent on the state of food deprivation. So, satiation, i.v. glucose injection, or distension of the stomach with a balloon, decrease the self-stimulation rate; whereas deprivation or insulin injection increase it (Hoebel, 1968). A corresponding observation has been made using electrodes yielding male sexual responses: these showed, in contrast to feeding electrodes, a clear correlation between their self-stimulation rate and the testosterone level (Caggiula, 1970). Furthermore, it has been shown that electrical stimulation of feeding loci increases the quinine tolerance; this increase is directly related to the stimulus intensity applied and summates with the deprivation-induced quinine tolerance (Tenen and Miller, 1964). Another type of evidence concerning the specificity of hypothalamically induced motivational changes comes from transfer experiments. Rats, for instance, which have learned that the onset of a specific tone signals a food reward (which can be obtained by lever-pressing) show a much more pronounced reaction to that tone during stimulation of a feeding electrode than during presentation of that tone alone or during stimulation without the tone (Fantl and Schuckman, 1967). Finally, the fact that by switching from one electrode to another and back, it is possible to make one and the same animal feed, interrupt feeding in order to copulate, and then force it again to feed, is a strong argument for the specificity of the induced motivational changes (Caggiula, 1970).

So, what is the synthesis of both these controversial view-points? An observation that might indicate the direction wheie to look for this synthesis is the following: rats with electrodes which yield simultaneously eating and drinking decrease their stimulus-bound drinking, but not their stimulus-bound eating after preceding water intake; they increase their stimulus-bound drinking, but not their stimulus-bound eating after water deprivation; and the corresponding holds true also for eating (Devor *et al.*, 1970). This means that the electrical stimulus does not by-pass the homeostatic control of the organism; it just shifts the set point to a somewhat higher level. Complex stimulus responses, thus, always seem to represent a synthesis between the motivational state of the animal in the moment just before stimulus application and the more or less specific motivational effect induced by the stimulus. So, if a satiated animal is induced to eat, the stimulus surely does not evoke hunger: a synthesis between satiety and desire to eat is not hunger, it rather resembles the appetite aroused by delicious pastry, fruits, or similar things in man after having completed a not too opulent meal. The same holds true for drinking. Again the stim-

References p. 460–462

ulus surely is not able to put out of circuit the whole homeostatic control, thus producing thirst; instead, the shifting of the set point to a higher level may rather manifest itself by the desire of just drinking for pleasure. This would explain why the more palatable sugar solution is preferred to pure water in stimulus-bound drinking. This interpretation would also explain why stimulus-induced mating includes copulation in the rat, but not in the opossum or cat. Male laboratory rats do have even spontaneously a high probability to copulate with receptive females. Therefore, the amount of motivation-specific energy necessary to release this behavioral pattern is low. Cats and opossums are much more susceptible to the environmental situation and it is therefore difficult to impel them to copulate under laboratory conditions. So, the motivation-specific energy necessary to overcome this reluctance is high and obviously cannot be attained by electrical stimulation. Also, the fact that stimulation of feeding loci is pleasant instead of aversive, becomes understandable: the arousal of appetites in contrast to more severe urges is at least not unpleasant, otherwise the attraction of sex films would not be understandable. (There is some evidence that the self-stimulation effect obtained from the lateral hypothalamus does not totally depend on the motivational effect induced by that same stimulation (Huston, 1972). Nevertheless, one would expect that the aversive effect of the urge component would reduce the self-stimulation effect of the rewarding component. As there is an increase in self-stimulation with increasing deprivation state as well as stimulus intensity, this seems not to be the case.)

If we suggest that the effect of hypothalamic stimulation does not mimic the urges arising from severe deprivation or highly inciting environmental situations, this does not imply that the stimulation has an unspecific effect on motivation. The experimental results cited previously rather make it probable that the motivational change induced is in at least some cases a quite specific one. This change is, however, at the same time a very restricted one — restricted, namely, by the counteracting of certain control instances that check the actual state of the organism against the stimulus effect and tries to balance them out.

So far we have considered only the motivational effects of hypothalamic stimulation. We may ask now, if there are also direct motor effects. Several observations seem to indicate this. Defence reactions, for instance, which were totally unrelated to the environmental situation have been reported repeatedly from cats. These animals could be induced to hiss and growl together with piloerection and pupil dilatation without defending themselves against another attacking cat (Delgado, 1964) and without reacting to a rat running around in the same cage (Wasman and Flynn, 1962). Less drastic expressions of defence reactions, like piloerection and pupillary dilatation without vocalization, could even be evoked during milk-lapping or grooming (Masserman, 1941). Another example has been reported by Wayner and co-workers (White *et al.*, 1970). They showed that in the rat, licking rate in stimulus-bound drinking increases significantly with higher stimulus intensities; whereas in normal animals the licking rate is quite constant, irrespective of the deprivation state. It has also been observed that stimulus-bound drinkers may continue to display lapping at a tube connected to an empty water bottle during stimulation (Valenstein *et al.*, 1968).

Furthermore, even licking into the empty space and chewing without food intake are readily elicitable in several mammals and birds from the most lateral parts of the anterior hypothalamus (Hess, 1949; Putkonen, 1967; own observations).

These results clearly indicate that, concerning the level of behavioral integration, hypothalamic stimulus responses cannot be interpreted uniformly. There are behavioral patterns driven quite directly by the stimulus (even if they form the minority); there are others, which are initiated indirectly by a stimulus-induced motivational change (these are considerably influenced by the environmental situation); and there are still others, representing intermediates between these extremes. So, the level of behavioral integration of a hypothalamic stimulus response has to be determined for each anatomical locus as well as for each behavioral pattern anew.

LESION EXPERIMENTS

We will now have a short look at the results of lesion experiments. Let us begin with the most drastic treatment: the elimination of the whole hypothalamus. If one isolates the hypothalamus of the cat by inserting a specially shaped knife along the midline into the brain and rotating it so as to cut the sides and top of a cylinder, within which the hypothalamus rests, these animals are still able to display complete defence and attack behavior (Ellison and Flynn, 1968). The defence reaction includes lowering of the head, flattening of the ears, pupillary dilatation, piloerection, growling, and unsheathing of the claws; it can simply be evoked by pinching the tail or flank of the cat. Well-directed biting attack on a rat is also still elicitable. These attacks are visually guided and performed as in normal cats; they have to be initiated (or supported, respectively), however, by tail pinching, which is not necessary, of course, in normal rat-killing cats. The latter is a typical characteristic of hypothalamus-deprived animals: only very strong stimuli are able to make them react. Their spontaneous activity is almost zero. If left alone they either sit or lie motionless in the cage, resembling statues. When placed on their feet, they walk forward into some obstacle and then remain motionless in contact with the obstacle. So, what is lost in the hypothalamus-deprived animal is not the motor coordination itself, nor some basic behavioral pattern, but instead the initiative — or as one could also say, the motivation.

In order to produce losses in motivation it is not necessary to eliminate the whole hypothalamus. Circumscribed deficits have also been produced by lesioning of discrete parts of the hypothalamus. Destruction of the preoptic–anterior hypothalamic area, for instance, produces loss of sexual motivation in male rats and lesions somewhat more caudally also in female rats (Heimer and Larsson, 1966; Law and Meagher, 1958). This unresponsiveness to the sexual attractivity of the partner cannot be overcome by hormone treatment. It can be overcome, however, as in the case of the cat's attack reaction, by extraordinarily strong stimuli, like manipulation of the genital region (Dempsey and Rioch, 1939; Hart, 1967). This again indicates that the motor programs for the basic behavioral patterns are not stored in the hypothalamus. It is of interest in this context that the loss of male sexual motivation after preoptic–

References p. 460–462

anterior hypothalamic lesions is not restricted to mammals, for it also has been found in birds and even amphibians (Meyer and Salzen, 1970; Schmidt, 1968).

Another well-known, relatively specific motivational deficit is that of aphagia and adipsia; these occur after lesions in the lateral hypothalamus at the level of the ventromedial nucleus. Again, there is no deficit in the motor pattern of swallowing or mouthing, but just a lack of feeding motivation. And here also, highly palatable food placed directly in front of the animal may in certain stages of the aphagia overcome this deficit and thus induce the animal to eat (Teitelbaum and Epstein, 1962).

One has to be aware, on the other hand, that lesions in the hypothalamus can also have an enhancing effect on specific motivations. This is the case with ventromedial lesions in respect to food intake (Brobeck, 1946). Such an effect does not come, however, totally unexpected: that destruction of a structure supposed to regulate a specific motivation may cause loss of balance in both directions — thus resulting in a decrease as well as an increase of motivation — seems rather self-evident. In the case of feeding motivation it is substantiated by a mutual inhibitory influence between ventromedial and lateral hypothalamic neurons, so that activity of the latter increases and activity of the former decreases food motivation (Oomura *et al.*, 1964).

CONCLUSIONS

If we try to combine now the results of the stimulation studies with those of the lesion studies, what are the conclusions to be drawn from this, concerning the role of the hypothalamus for the elaboration of behavioral patterns? The lesion studies have revealed that the motor programs of those basic behavioral patterns electrically, chemically, or thermally elicitable from the hypothalamus are not stored in the hypothalamus itself. Transections at different levels of the brain stem indicate that at least some of these basic behavioral patterns, or so-called fixed action patterns in the language of ethology, are represented in the midbrain, lower brain stem, and even spinal cord (Bard and Macht, 1958; Bazett and Penfield, 1922; Maes, 1939). On the other hand, also for sensory perception, the hypothalamus is dispensable; cats with an isolated hypothalamus only attack rats, but never stuffed toys, and their attack is well-directed to the neck region of the rat. Furthermore, the identification of the rat can be carried out both by visual and olfactory cues.

If, then, the function of the hypothalamus is neither motor coordination nor sensory integration, the only possibility left is that it plays a role in the *mediation* between (external) stimulus and (behavioral) response. This mediation could manifest itself in one of two different ways: either the hypothalamus could represent a link between stimulus and response, or it could serve to modulate a stimulus–response relationship which itself is maintained elsewhere in the brain. The lesion experiments favor the latter proposition. This is in contrast to the role the hypothalamus plays in many autonomic reactions; for we know that its influence in this respect is a very direct one. This difference in its influence on autonomic reactions on the one side and behavioral reactions on the other becomes especially clear in thermoregulation. So, rats with lesions in the anterior hypothalamus are no longer able to maintain their body temperature in the cold by shivering, vasoconstriction, piloerection, and other autonomic reactions. They are, however, able to maintain their body temperature by pressing a bar to turn on a heat lamp (which is switched off automatically after a predetermined time interval and therefore can be maintained for a longer period only if repeated bar pressing is performed (Satinoff and Rutstein, 1970)). In this case the behavioral thermoregulation survives the autonomic thermoregulation, which means that there still exists a control mechanism that is able to perceive the temperature drop. The fact that hypothalamic lesions may severely interfere with the autonomic reaction to an external stimulus, but leave intact the behavioral reaction, also points to a more indirect influence of the hypothalamus on the actually ongoing behavior. This interpretation is further supported by the close relationship between hypothalamus and hypophysis; for the hypothalamus acts as a link between external stimuli and all those adjustments of the internal milieu, controlled by hypophyseal hormones, that serve to prepare the organism for efficient behavioral activity.

Finally, the stimulation studies also point in the same direction; as the adaptability of most hypothalamically elicitable behavioral patterns clearly indicates that we are dealing in these cases with motivational effects and not with pure motor effects. Those exceptions where behavioral patterns could be elicited which were not adapted to the environmental situation, like chewing or licking movements and the hissing reaction of the cat, do not refute this, for all 3 behavioral patterns could also be evoked as well-adjusted forms such as eating, drinking, and defence reactions from the hypothalamus, and furthermore survived the elimination of the hypothalamus in the cat (Ellison and Flynn, 1968). Their automaton-like display rather indicates that the hypothalamus may exert its modulating influence in a very direct manner on the motor coordinating structures.

To condense once more our knowledge concerning the role of the hypothalamus on the control of behavioral patterns, we should say: the hypothalamus on the one hand seems to transform internal needs and drive stimuli into behavior by changing the probability of occurrence of specific behavioral patterns; on the other hand, it seems to transform external stimuli into a more general behavioral readiness — thus influencing several motivations simultaneously — by means of diverse autonomic effects as well as a general arousal effect.

SUMMARY

The role of the hypothalamus for the elaboration of behavioral patterns is examined by reviewing some of the most relevant stimulation and lesion experiments.

(1) It is shown that the patterns elicitable from the hypothalamus belong to quite different behavioral functions, like mating, attack, flight, eating, or grooming.

(2) They usually are not performed as stereotyped motor coordinations totally unrelated to the environmental situation; on the contrary, they often possess moti-

References p. 460–462

vational properties, in the sense that in the absence of an adequate goal object stimulation is able to induce the respective appetitive (searching) behavior.

(3) Nevertheless, it has become clear that the stimulation-induced motivational changes are not identical with those spontaneously occurring motivational changes that arise from severe deprivation or highly inciting natural stimuli.

(4) The anatomical distributions of the different stimulus responses in most cases show extensive, although not complete overlapping.

(5) Lesion experiments indicate that for motor coordination of hypothalamically elicitable behavioral patterns as well as for sensory perception of external stimuli the hypothalamus only plays a minor role.

(6) It is therefore concluded that the hypothalamus mainly serves to modulate stimulus–response relationships — which themselves are maintained elsewhere in the brain — by changing the probability of occurrence of specific behavioral patterns, that is to say by changing the motivation.

REFERENCES

ADAMS, D. AND FLYNN, J. P. (1966) Transfer of an escape response from tail shock to brain-stimulated attack behavior. *J. exp. Anal. Behav.*, **9**, 401–408.

ÅKERMAN, B. (1965) Behavioural effects of electrical stimulation in the forebrain of the pigeon. *Behaviour*, **26**, 323–350.

BARD, P. AND MACHT, M. B. (1958) The behaviour of chronically decerebrate cats. In *Neurological Basis of Behaviour*, G. E. W. WOLSTENHOLME AND C. M. O'CONNOR (Eds.), Churchill, London, pp. 55–71.

BAZETT, H. C. AND PENFIELD, W. G. (1922) A study of the Sherrington decerebrate animal in the chronic as well as the acute condition. *Brain*, **45**, 185–265.

BEACH, F. A. (1970) Some effects of gonadal hormones on sexual behavior. In *The Hypothalamus*, L. MARTINI, M. MOTTA AND F. FRASCHINI (Eds.), Academic Press, New York, pp. 617–639.

BROBECK, J. R. (1946) Mechanism of the development of obesity in animals with hypothalamic lesions. *Physiol. Rev.*, **26**, 541–559.

CAGGIULA, A. R. (1970) Analysis of the copulation-reward properties of posterior hypothalamic stimulation in male rats. *J. comp. physiol. Psychol.*, **70**, 399–412.

CAGGIULA, A. R. AND EIBERGEN, R. (1969) Copulation of virgin male rats evoked by painful peripheral stimulation. *J. comp. physiol. Psychol.*, **69**, 414–419.

CLEMENTE, C. D. AND STERMAN, M. B. (1967) Limbic and other forebrain mechanisms in sleep induction and behavioral inhibition. In *Structure and Function of the Limbic System*, *Progr. Brain Res.*, *Vol. 27*, W. R. ADEY AND T. TOKIZANE (Eds.), Elsevier, Amsterdam, pp. 34–47.

COX, V. C. AND VALENSTEIN, E. S. (1969) Distribution of hypothalamic sites yielding stimulus-bound behavior. *Brain Behav. Evol.*, **2**, 359–376.

DELGADO, J. M. R. (1964) Free behavior and brain stimulation. *Int. Rev. Neurobiol.*, **6**, 349–449.

DEMPSEY, E. W. AND RIOCH, D. M. (1939) The localization in the brainstem of the oestrous responses of the female guinea pig. *J. Neurophysiol.*, **2**, 9–18.

DEMSKI, L. S. AND KNIGGE, K. M. (1971) The telencephalon and hypothalamus of the blue-gill *(Lepomis macrochirus)*: evoked feeding, aggressive and reproductive behavior with representative frontal sections. *J. comp. Neurol.*, **143**, 1–16.

DEVOR, M. G., WISE, R. A., MILGRAM, N. W. AND HOEBEL, B. G. (1970) Physiological control of hypothalamically elicited feeding and drinking. *J. comp. physiol. Psychol.*, **73**, 226–232.

DISTEL, H. (1973) *Die Auslösbarkeit von Imponieren, Abwehr- und Fluchtverhalten durch elektrische Hirnreizung bei Iguana iguana* (Reptilia), Doctoral Dissert., Univ. Munich.

EIBL-EIBESFELDT, I. (1967) *Grundriß der vergleichenden Verhaltensforschung*, Piper, Munich.

ELLISON, G. D. AND FLYNN, J. P. (1968) Organized aggressive behavior in cats after surgical isolation of the hypothalamus. *Arch. ital. Biol.*, **106**, 1–20.

FANTL, L. AND SCHUCKMAN, H. (1967) Lateral hypothalamus and hunger: responses to a secondary reinforcer with and without electrical stimulation. *Physiol. Behav.*, **2**, 355–357.

FLYNN, J. P. (1967) The neural basis of aggression in cats. In *Neurophysiology and Emotion*, D. C. GLASS (Ed.), Rockefeller Univ. Press, New York, pp. 40–60.

GROSSMAN, S. P. (1960) Eating or drinking elicited by direct adrenergic or cholinergic stimulation of hypothalamus. *Science*, **132**, 301–302.

HART, B. L. (1967) Sexual reflexes and mating behavior in the male dog. *J. comp. physiol. Psychol.*, **64**, 388–399.

HEIMER, L. AND LARSSON, K. (1966) Impairment of mating behavior in male rats following lesions in the preoptic–anterior hypothalamic continuum. *Brain Res.*, **3**, 248–263.

HESS, W. R. (1949) *Das Zwischenhirn. Syndrome, Lokalisationen, Funktionen*, Schwabe, Basel.

HOEBEL, B. G. (1968) Inhibition and disinhibition of self-stimulation and feeding: hypothalamic control and post-ingestional factors. *J. comp. physiol. Psychol.*, **66**, 89–100.

HOEBEL, B. G. AND TEITELBAUM, P. (1962) Hypothalamic control of feeding and self-stimulation. *Science*, **135**, 375–377.

HUSTON, J. P. (1972) Inhibition of hypothalamically motivated eating by rewarding stimulation through the same electrode. *Physiol. Behav.*, **8**, 1121–1125.

JÜRGENS, U. AND PLOOG, D. (1970) Cerebral representation of vocalization in the squirrel monkey. *Exp. Brain Res.*, **10**, 532–554.

LAW, O. T. AND MEAGHER, W. (1958) Hypothalamic lesions and sexual behavior in the female rat. *Science*, **128**, 1626–1627.

MACDONNELL, M. F. AND FLYNN, J. P. (1968) Attack elicited by stimulation of the thalamus and adjacent structures of cats. *Behaviour*, **31**, 185–202.

MAES, J. (1939) Neural mechanism of sexual behaviour in the female cat. *Nature (Lond.)*, **144**, 598.

MALEY, M. J. (1969) Electrical stimulation of agonistic behavior in the mallard. *Behaviour*, **34**, 138–160.

MASSERMAN, J. H. (1941) Is the hypothalamus a center of emotion? *Psychosom. Med.*, **3**, 3–25.

MEYER, A. E. AND HESS, W. R. (1957) Diencephal ausgelöstes Sexualverhalten und Schmeicheln bei der Katze. *Helv. physiol. pharmacol. Acta*, **15**, 401–407.

MEYER, C. C. AND SALZEN, E. A. (1970) Hypothalamic lesions and sexual behavior in the domestic chick. *J. comp. physiol. Psychol.*, **73**, 365–376.

MORGANE, P. J. (1961) Distinct "feeding" and "hunger motivating" systems in the lateral hypothalamus of the rat. *Science*, **133**, 887–888.

MURPHY, J. P. AND GELLHORN, E. (1945) The influence of hypothalamic stimulation on cortically induced movements and on action potentials of the cortex. *J. Neurophysiol.*, **8**, 339–364.

NAKAO, H. (1971) *Brain Stimulation and Learning*, Fischer, Jena.

OOMURA, Y., KIMURA, K., OOYAMA, H., MAENO, T., IKI, M. AND KUNIYOSHI, M. (1964) Reciprocal activities of the ventromedial and lateral hypothalamic areas of cats. *Science*, **143**, 484–485.

PHILLIPS, A. G., COX, V. C., KAKOLEWSKI, J. W. AND VALENSTEIN, E. S. (1969) Object carrying by rats: an approach to the behavior produced by stimulation. *Science*, **166**, 903–905.

PUTKONEN, P. T. S. (1966) Attack elicited by forebrain and hypothalamic stimulation in the chicken. *Experientia (Basel)*, **22**, 405–407.

PUTKONEN, P. T. S. (1967) Electrical stimulation of the avian brain. *Ann. Acad. Sci. fenn.*, A.V., **130**, 95.

ROBERTS, W. W. (1969) Are hypothalamic motivational mechanisms functionally and anatomically specific? *Brain Behav. Evol.*, **2**, 317–342.

ROBERTS, W. W. (1970) Hypothalamic mechanisms for motivational and species-typical behavior. In *The Neural Control of Behavior*, R. E. WHALEN, R. F. THOMPSON, M. VERZEANO AND N. M. WEINBERGER (Eds.), Academic Press, New York, pp. 175–206.

ROBERTS, W. W. AND BERGQUIST, E. H. (1968) Attack elicited by hypothalamic stimulation in cats raised in social isolation. *J. comp. physiol. Psychol.*, **66**, 590–595.

ROBERTS, W. W. AND ROBINSON, T. C. L. (1969) Relaxation and sleep induced by warming of preoptic region and anterior hypothalamus in cats. *Exp. Neurol.*, **25**, 282–294.

ROBERTS, W. W., STEINBERG, M. L. AND MEANS, L. W. (1967) Hypothalamic mechanisms for sexual, aggressive, and other motivational behaviors in the opossum, *Didelphis virginiana*. *J. comp. physiol. Psychol.*, **64**, 1–15.

SATINOFF, E. AND RUTSTEIN, J. (1970) Behavioral thermoregulation in rats with anterior hypothalamic lesions. *J. comp. physiol. Psychol.*, **71**, 77–82.

SCHMIDT, R. S. (1968) Preoptic activation of frog mating behavior. *Behaviour*, **30**, 239–257.

TEITELBAUM, P. AND EPSTEIN, A. N. (1962) The lateral hypothalamic syndrome: recovery of feeding and drinking after lateral hypothalamic lesions. *Psychol. Rev.*, **69**, 74–90.

TENEN, S. S. AND MILLER, N. E. (1964) Strength of electrical stimulation of lateral hypothalamus, food deprivation, and tolerance for quinine in food. *J. comp. physiol. Psychol.*, **58**, 55–62.

ULRICH, R. E., HUTCHINSON, R. R. AND AZRIN, H. H. (1965) Pain-elicited aggression. *Psychol. Rec.*, **15**, 111–126.

VALENSTEIN, E. S. (1969) Behavior elicited by hypothalamic stimulation. *Brain Behav. Evol.*, **2**, 295–316.

VALENSTEIN, E. S., COX, V. C. AND KAKOLEWSKI, J. W. (1970) Reexamination of the role of the hypothalamus in motivation. *Psychol. Rev.*, **77**, 16–31.

VALENSTEIN, E. S., KAKOLEWSKI, J. W. AND COX, V. C. (1968) A comparison of stimulus-bound drinking and drinking induced by water deprivation. *Comm. behav. Biol.*, **2**, 227–233.

VERGNES, M. ET KARLI, P. (1970) Déclenchement d'un comportement d'aggression par stimulation électrique de l'hypothalamus médian chez le rat. *Physiol. Behav.*, **5**, 1427–1430.

WASMAN, M. AND FLYNN, J. P. (1962) Directed attack elicited from hypothalamus. *Arch. Neurol. (Chic.)*, **6**, 220–227.

WHITE, S. D., WAYNER, M. J. AND COTT, A. (1970) Effects of intensity, water deprivation, prior water ingestion and palatability on drinking evoked by lateral hypothalamic electric stimulation. *Physiol. Behav.*, **5**, 611–619.

WOODWORTH, C. H. (1970) *Attack Elicited in Rats by Electrical Stimulation of the Lateral Hypothalamus*, Ph.D. Dissert., Univ. Minnesota.

DISCUSSION

SLANGEN: Since you reviewed the behavioral research in this area so well, you made it clear that we are perhaps at the end of our possibilities for research in the hypothalamus, using lesioning and stimulation techniques. I wonder, therefore, what your advice would be for the very near future.

JÜRGENS: What should be done in the future is to introduce much more differentiated ethological methods into these studies. Most lesioning and stimulation studies have been performed under rather unnatural conditions, where the animal has had no possibility of showing its whole behavioral repertoire. In my opinion, they should be performed again in a much more sophisticated experimental situation, with an improved knowledge of the ethology of the species. I, myself, am working with monkeys; with these animals our lack of knowledge about their behavior is the main handicap in interpreting stimulation and lesion results. In the case of a laboratory rat, however — which to me, as a zoologist, is a rather degenerated animal, which means an animal with a reduced behavior repertoire — it might be true that further stimulation and lesion experiments will not add many new aspects to our present knowledge concerning the hypothalamic control of behavioral patterns.

ROMIJN: You mentioned that when you stimulated the same spot in different animals you could not always evoke the same pattern. Are you sure that you are always exactly in the same place of the hypothalamus?

JÜRGENS: The commonly-used electrodes stimulate a brain area of about 1 cu. mm. Because this area is so large, it is no problem to always be absolutely at the same place.

WAYNER: There is another interesting thing about electrical stimulation. If you try to become more elegant in your methods, and you make small electrodes, and try to stimulate discrete areas, you don't get any effects anymore.

DE WIED: I have a question to ask to Dr. Jürgens about the preference for water and sugar water. You said that in a normal, thirsty animal there is a preference for pure water? What we see in our laboratory is that there is a preference to sugar water. We use sugar water as the preferential drinking fluid for animals in which we study 'poison avoidance behavior'. This means that we poison the animals at the moment they have touched for the first time sugar water. You must have a preference for this kind of water to elicit that kind of behavior. So I was a little bit surprised to hear that you said that a thirsty animal prefers pure water, but it may be the concentration of sugar that one is using.

JÜRGENS: Yes, the concentration has to be rather high. The authors to which I was referring in my paper (Valenstein *et al.*, 1968) used a 40% sucrose solution.

Effects of Adrenalectomy and Treatments with ACTH and Glucocorticoids on Isolation-Induced Aggressive Behavior in Male Albino Mice*

ANGELA E. POOLE AND PAUL BRAIN

Department of Zoology, University College of Swansea, Swansea (Great Britain)

INTRODUCTION

It is well-known that isolation induces fighting behavior in laboratory mice and that the longer the period of isolation, the more intense the behavior (*e.g.* Welch and Welch, 1971). Whilst some workers have related such fighting behavior to the deprivation aspect of isolation (*e.g.* Valzelli, 1969), Brain (1971) has stressed the importance of the comparison between the condition of isolation and territoriality in male mice.

It is apparent that, as well as being influenced by the "stress" of behavioral interactions, the hormones of the pituitary–adrenocortical axis (ACTH and glucocorticoids) can have a marked and rather specific influence on some social behaviors in rodents (see review by Brain, 1972). Sigg *et al.* (1966) showed that adrenalectomy slows up the rate at which isolation-induced aggressiveness is acquired in mice and Kostowski *et al.* (1970) have produced evidence that injection of glucocorticoids can increase isolation-induced aggressiveness in mice. Brain *et al.* (1971) investigated the influence of a number of manipulations of the pituitary–adrenocortical axis on this behavior in mice. They found that adrenalectomy before isolation reduced the amount of fighting shown in a "standard opponent test" (as described by Brain and Poole, 1973; see also Fig. 1),that dexamethasone injection into intact animals throughout the isolation period increased it and that long-acting ACTH injection into similar animals throughout the isolation period, significantly reduced fighting using this type of test. Many of these effects have been duplicated (Leshner, 1972; Candland and Leshner, 1973; Harding and Leshner, 1972) and a direct action of the adrenal axis more strongly indicated. This group found, that under their testing conditions (which differed from those in the Brain *et al.*, 1971 study), and in CFW mice, that the influence of the adrenal on isolation-induced fighting behavior appeared to be independent of any action that adrenal hormones might have on androgen titers (a suppressive action of glucocorticoids on testicular function in mammals is a well-established phenomenon). They demonstrated that glucocorticoid injections could restore fighting behavior in

* Supported by M.R.C. Project Grant G972/189 B.

References p. 471–472

Fig. 1. Some typical agonistic postures seen in encounters between two isolated male albino mice. 1: preliminary 'sniff'; 2: mutual attack; 3: attack on subordinate; 4: defeated mouse.

adrenalectomized mice, an effect which could not be obtained by androgen treatment and that a combination of androgens and glucocorticoids was necessary to restore the behavior in isolated males which had been both adrenalectomized and castrated.

These interesting findings would seem to demonstrate conclusively the importance of "stress" hormones in isolation-induced fighting behavior in mice but, in view of some conflicting results, *e.g.* the inability of Burge and Edwards (1971) to demonstrate an effect of adrenalectomy on fighting behavior in their tests with mice, and the negative result obtained by Bronson and Desjardins (1971) in their studies on the effects of ACTH administration on fighting behavior in mice, it was decided to study the effects of adrenalectomy, treatments with ACTH and dexamethasone in both adrenalectomized and intact animals using a variety of testing procedures, to attempt to determine whether such different results could be the result of methodological differences in treatments, testing etc.

METHODS

Experiment 1

Adult male TT strain mice were bilaterally adrenalectomized or mock operated under ether anesthesia. Ten mock operated animals were maintained on tap water and 10 on saline. All adrenalectomized animals were maintained on physiological saline and there were 10 animals in each treatment category. Categories consisted of animals injected i.p. with 0.1 ml placebo, mice injected with 0.2, 0.4 or 0.8 mg of dexamethasone solution (Organon) and mice receiving 1 mg s.c. implants of corticosterone (kindly donated by Organon) on the day of operation. Injected categories were treated on alternate days. The 7 categories were isolated for 27 days, after which they were subjected to a 5 min "standard opponent" test (described fully in Brain and Poole, 1973). The results are expressed as percentage aggression scores in Fig. 2. Percentage aggression score being the number of first attacks, tail rattles and defeats inflicted on the "opponent" by the mouse expressed as a percentage of the total possible.

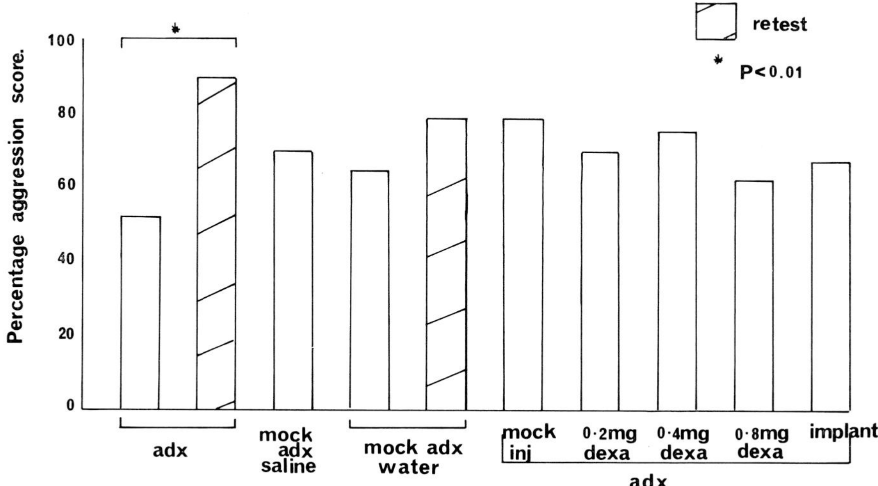

Fig. 2. Effects of adrenalectomy (adx) and steroid replacement on percentage aggression scores in isolated male mice given a standard opponent test.

Experiment 2

Adult male mice were adrenalectomized or mock adrenalectomized following different intervals of isolation (0, 1 or 3 weeks). Such mice were then isolated for a further period before receiving a "standard opponent" test (see Fig. 3), followed by tests between adrenalectomized and mock operated animals from the same schedules (see Fig. 4).

References p. 471–472

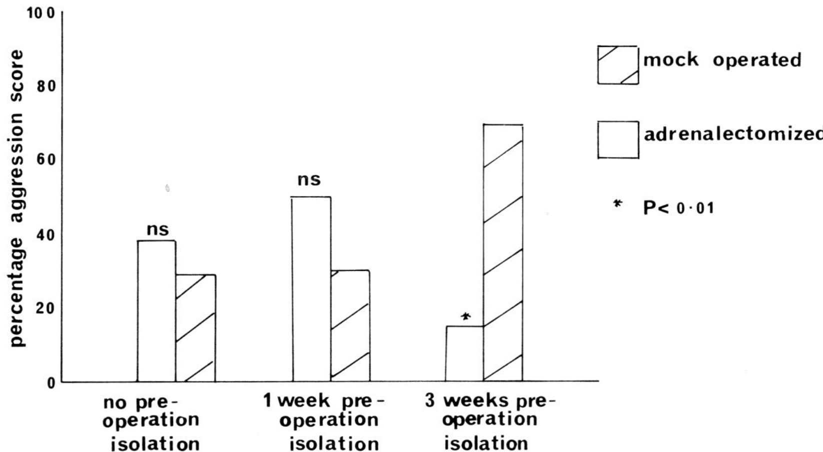

Fig. 3. Effect of pre-operational isolation period on percentage aggression score in adrenalectomized (adx) and mock adrenalectomized (mock adx) male albino mice given a 'standard opponent' test. ns = not significant. Each group consists of 12 animals.

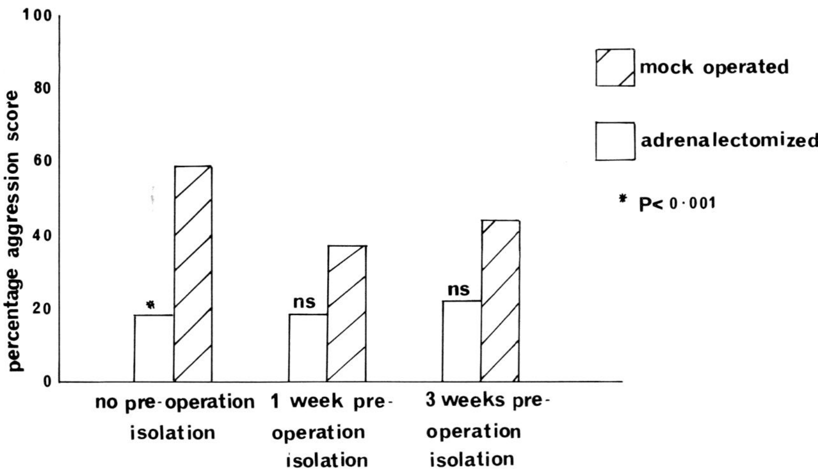

Fig. 4. Effect of pre-operational isolation period on percentage aggression score in adrenalectomy *vs.* mock adrenalectomy pairings in male albino mice. Each group consists of 12 animals.

Experiment 3

Utilizing animals which had had fighting experience and had been adrenalectomized or mock operated for over 5 weeks, an attempt was made to influence the aggressiveness of the adrenalectomized animals by giving them a single i.p. injection of 0.2 mg dexamethasone 24 h before pairing in a test with mock operated animals (see Fig. 5).

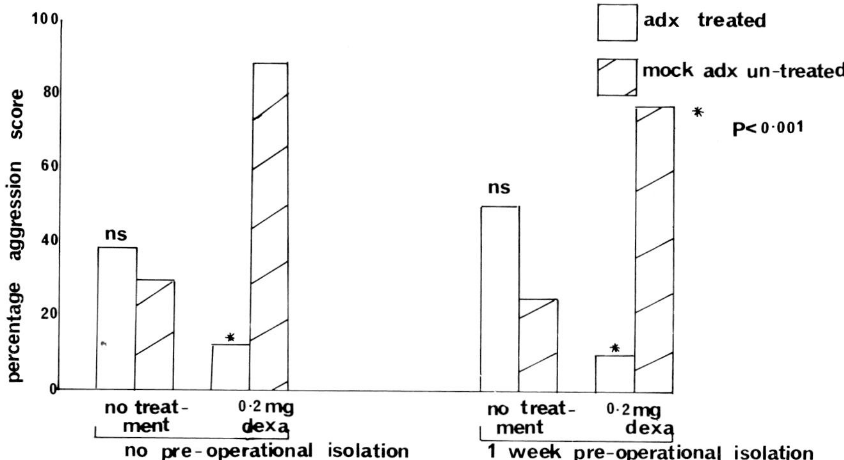

Fig. 5. Effects of treatment with dexamethasone on adrenalectomized (adx) mice in encounters with untreated intact animals in terms of percentage aggression score. Each group consists of at least 8 animals.

Experiment 4

A further series of tests employed fighting these animals within their operational category following hormone treatment or injection of placebo. Single injections of 4 I.U. ACTH (Cortrophin/Zn, Organon) i.p. were given over 3 days, 24 h before a test against a mock-injected animal in both adrenalectomized and mock operated categories (see Fig. 6). A similar series of tests were carried out utilizing 0.2 mg dexamethasone i.p. as the active factor (see Fig. 7).

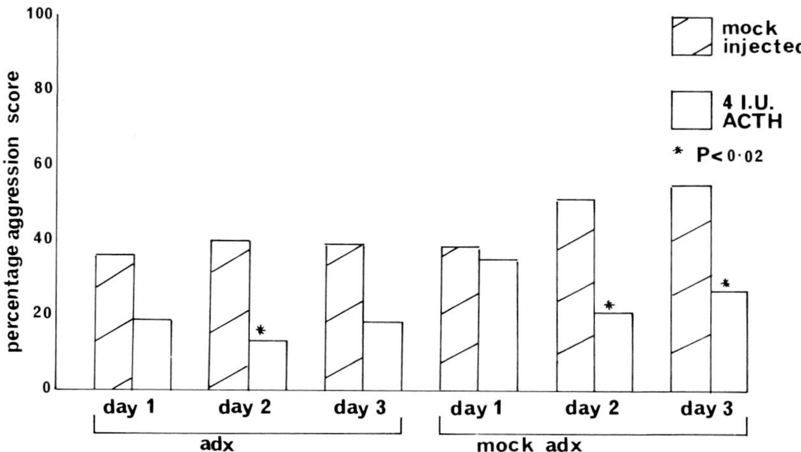

Fig. 6. Effect of ACTH on percentage aggression scores in experienced male mice following adrenalectomy (adx) or mock adrenalectomy (mock adx). Each group consists of at least 9 animals.

References p. 471–472

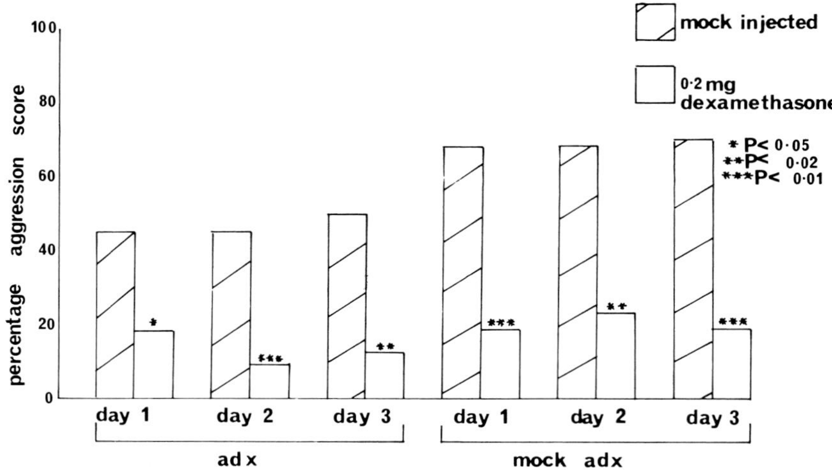

Fig. 7. Effect of dexamethasone or mock injection on percentage aggression scores in experienced male mice following adrenalectomy (adx) or mock adrenalectomy (mock adx). Each group consists of at least 9 animals.

RESULTS

Under the conditions employed in experiment 1 there appeared to be no effects of giving saline as opposed to tap water to drink, no effects of adrenalectomy as opposed to mock adrenalectomy (differing from the original finding, Brain *et al.*, 1971) and no consistent effects of steroid replacements (unlike Candland and Leshner, 1973). However, retesting mice led to a marked increase in fighting behavior.

In experiment 2 it was shown that using the "standard opponent" test, a difference between treatment and mock animals could only be demonstrated for the category with a pre-operational isolation period of 0 weeks (adrenalectomy < mock). A similar difference was found following adrenalectomy/mock encounters in the case of mice with a 3 week pre-operational isolation period.

Experiment 3 demonstrated that the injection of dexamethasone caused the adrenalectomized mice to be vigorously attacked in encounters with mock operated animals and that the treatment did nothing to restore aggressiveness in these mice as it appears to do in the study of Candland and Leshner (1973).

The results of experiment 4 demonstrated that single injections of 4 I.U. ACTH i.p. (Cortrophin/Zn, Organon) caused a marked suppression of fighting behavior as compared to mock injected mice 24 h later in both adrenalectomized and mock operated categories, and that similar treatment with 0.2 mg dexamethasone i.p. resulted in these animals continuing to show a condition in which mock injected animals were dominant to hormone treated mice in both categories the effect being apparent at least 10 days after the last injection.

DISCUSSION

The results confirm that modulation of isolation-induced fighting by the pituitary–adrenocortical axis is a complex phenomenon, strongly influenced by the animal's prior experience and the type of test employed. The fact that ACTH causes both adrenalectomized and mock operated animals to be subordinate with respect to mock injected animals, suggests a direct action rather than an effect mediated via the production of adrenal steroids. Under these conditions, dexamethasone did not have the effect of restoring aggressive behavior in adrenalectomized animals reported elsewhere but it should be noted that this study involved animals which had already been subjected to defeat. It may be that both ACTH and the glucocorticoids have part of their action on aggressive behavior by influencing pheromone production and hence the "attackability" of the treated animal.

SUMMARY

In view of some of the conflicting reports concerning the effects of adrenalectomy and/or glucocorticoid and ACTH treatment on isolation-induced aggression in laboratory mice, a number of experiments investigating some of the possible variables were carried out. It was found that, in the strain of mice used, bilateral adrenalectomy, with or without steroid replacement, did not markedly influence fighting behavior in a "standard opponent" test 27 days later. Varying the period of isolation before the operation occurred resulted in a significant suppression of fighting behavior in a "standard opponent" test in animals with the shortest pre-operational period of isolation (0 days). Matching adrenalectomized animals with mock-operated controls, however, led to the same result with animals from the longest pre-operational isolation category (3 weeks). Utilizing experienced animals, it was found that 0.2 mg of dexamethasone administered to adrenalectomized animals resulted in their being vigorously attacked by mock operated animals. A further series of tests employed fighting adrenalectomized *vs.* adrenalectomized, and mock operated *vs.* mock operated animals. Following administration of either 0.2 mg of dexamethasone or 4 I.U. of ACTH, treated animals in both adrenalectomized and mock operated categories were found to be subordinate to mock-treatment categories. It was concluded that both ACTH and glucocorticoids may have a direct action on aggressive behavior and that part of this influence may be by affecting the "attackability" of the treated animal.

REFERENCES

BRAIN, P. F. (1971) The physiology of population limitation in rodents — a review. *Commun. behav. Biol.*, **6**, 115–123.

BRAIN, P. F. (1972) Mammalian behavior and the adrenal cortex — a review. *Behav. Biol.*, **7**, 453–477.

BRAIN, P. F. NOWELL, N. W. AND WOUTERS, A. (1971) Some relationships between adrenal function and isolation-induced intermale aggression in albino mice. *Physiol. Behav.*, **6**, 27–29.

Brain, P. F. and Poole, A. E. (1973) Some studies on the use of 'standard opponents' in intermale aggression testing in TT albino mice. *Behaviour*, in press.

Bronson, F. H. and Desjardins, C. (1971) Steroid hormones and aggressive behavior in mammals. In *The Physiology of Aggression and Defeat*, B. E. Eleftheriou and J. P. Scott (Eds.), Plenum Press, London, pp. 43–64.

Burge, K. G. and Edwards, D. A. (1971) The adrenal gland and the pre and post castrational aggressive behavior of the male mouse. *Physiol. Behav.*, **7**, 885–888.

Candland, D. K. and Leshner, A. I. (1973) A model of agonistic behavior: endocrine and autonomic correlates. In *Advances in Limbic and Autonomic Nervous System Research*, L. DiCara (Ed.), Plenum Press, London, in press.

Harding, C. F. and Leshner, A. I. (1972) The effects of adrenalectomy on the aggressiveness of differently housed mice. *Physiol. Behav.*, **8**, 437–440.

Kostowski, W., Rewerski, W. and Piechocki, T. (1970) Effects of some steroids on aggressive behavior in mice and rats. *Neuroendocrinology*, **6**, 311–318.

Leshner, A. I. (1972) The adrenals and testes: two separate systems affecting aggressiveness. *Hormones*, **3**, 272–273.

Sigg, E. B., Day, C. and Colombo, C. (1966) Endocrine factors in isolation-induced aggressiveness in rodents. *Endocrinology*, **78**, 679–684.

Valzelli, L. (1969) Aggressive behaviour induced by isolation. In *Aggressive Behaviour*, S. Garattini and E. B. Sigg (Eds.), Excerpta Medica, Amsterdam, pp. 70–76.

Welch, A. S. and Welch, B. L. (1971) Isolation, reactivity and aggression: evidence for an involvement of brain catecholamines and serotonin. In *The Physiology of Aggression and Defeat*, B. E. Eleftheriou and J. P. Scott (Eds.), Plenum Press, London, pp. 91–142.

Some Studies on Endocrine Influences on Aggressive Behavior in the Golden Hamster (*Mesocricetus auratus* Waterhouse)

C. M. EVANS* AND P. F. BRAIN

Department of Zoology, University College of Swansea, Wales (Great Britain)

INTRODUCTION

The golden hamster is unusual with respect to other common laboratory rodents in that both the male and the non-estrus female show pronounced fighting behavior, a response which in the female is more intense than in the male (*e.g.* Swanson, 1967; Lerwill, 1968; Payne and Swanson, 1970). Hamsters also differ from rats and mice with respect to a number of endocrine functions. Whilst in these latter animals the adrenal gland is usually heavier and more active in the female than in the male of the species, the reverse is found to be true in the hamster (Chester Jones, 1955). Gaskin and Kitay (1970) found that testosterone appeared to increase adrenal cortical size in the hamster in a similar manner to the effect exerted by estrogens in rats and mice. The same authors (1971) produced evidence that, in the male hamster, testosterone enhances ACTH production leading to increased glucocorticoid secretion. They state that comparison with the rat reveals numerous divergences, indicating different mechanisms of action of the gonads on the hypothalamic–pituitary–adrenal axis in these species. From the above discussion it would consequently appear that the hamster provides an interesting subject for relating behavior to endocrine function.

One such area is that of the hormonal influences on isolation-induced fighting behavior, an interest of many students of mouse and rat behavior (reviewed by Brain, 1971). An isolation *versus* grouping comparison has been carried out in male and female golden hamsters (Brain, 1972) and it was found that, using one type of test, aggressiveness appears to be markedly increased by isolation in both the male and the female. Concomitant studies on endocrine function, however, while producing evidence for isolation-induced increases in androgen production in the male, provided little indication of any differential housing effects on adrenocortical activity in either sex or on gonadal function in the female.

Payne (1973) has also demonstrated that castration reduced the amount of agonistic behavior directed by isolated male hamsters towards "standard opponents". Such behavior can be restored by replacement therapy with androgens. However, several excellent studies using relatively large doses of hormone by Payne and Swanson (1971a, b; 1972a, b), have implicated sex hormones in the control of agonistic behavior in hamsters. These workers found that spaying caused females to be beaten in agonistic encounters with males, whereas progesterone treatment of such animals caused them

* Supported by a S.R.C. Studentship.

References p. 479–480

to evidence dominance in these encounters (an effect not obtained by testosterone treatment). So far as the male is concerned, castration causes the animal to show less fighting behavior than his opponents in encounters with animals of either sex. They also found that aggression towards intact males was increased in these animals by testosterone propionate or estradiol benzoate treatment, whereas progesterone increased such behavior directed towards intact females. Progesterone administered to castrated males caused intact males to show a decrease in aggressiveness, the treated animal becoming dominant. It is interesting to note that aggressiveness of castrated male hamsters can also be restored by ovarian implantation (Payne and Swanson, 1971c).

These workers conclude that it is likely that these hormone treatments can have a direct neural action on the behavior and an indirect effect possibly by influencing the production of olfactory cues.

The studies described usually involved social interactions between two individuals both of whom were likely to exhibit fighting behavior (except where fighting in the home cage is employed), a fact which makes it difficult to differentiate between these two suggested actions.

It is thought that it is likely that olfactory cues are involved in this behavior (as well as visual cues *e.g.* Grant *et al.*, 1970) because: (1) "pheromones" seem to play a decisive role in agonistic behavior in many rodent species (*e.g.* Mugford and Nowell, 1970, 1971, 1972); (2) olfaction has been shown to play a very important role in the social responses involved in the mating pattern of the male hamster (Murphy and Schneider, 1970); (3) hamsters have flank glands which are said to produce olfactory substances utilized in territorial marking (Giegel *et al.*, 1971).

Because of this suspected importance of the role of sex steroids in affecting olfactory cue production relevant to agonistic encounters and because a number of different types of "aggression tests" have been employed in mice, it was thought to be useful to study the actions of spaying and treatment with sex steroids on the intensity and duration of agonistic behavior directed by "trained fighters" towards such animals. It was felt that such a test which has been used with some success in studies concerned with the influence of olfactory cues on aggressiveness in mice (*e.g.* Mugford and Nowell, 1970) and in studies on the physiological responses to defeat in mice (Bronson and Eleftheriou, 1964; 1965) would possibly enable one to distinguish more readily between direct and indirect effects of sex steroid treatment on fighting behavior in male and female hamsters.

Consequently, an account of the type of behavioral test employed as well as some preliminary data on the effects of orchiectomy and/or steroid treatments in both male and female hamsters is presented.

EXPERIMENTS

Experiment 1: Methods

Twenty 6–7 week old male hamsters were isolated for a period of 47 days after which they were either castrated or mock operated. After a further 10 days, these hamsters were subjected to a trained fighter test. This test involves the exposure of experimental animals to highly aggressive opponents of known aggressiveness. Levels of fighting behavior in such a test are conveniently high and can be quantified. Behavioral sequences in such a test differ in a number of respects to those described by Grant and Mackintosh (1963) and Payne and Swanson (1970) who utilized rather different situations. The sequence can, however, be conveniently divided into 4 stages comparable to those presented in the Payne and Swanson study and here presented in Fig. 1.

Measures selected in order to quantify this behavior included: (1) the number of "aggressive" postures; (2) the number of attacks; (3) the latency of attack by one animal on the other; (4) the threat latency *i.e.* time from removal of the partition separating the animals to one animal (usually the trained fighter) initiating the first aggressive posture or attack; (5) the total investigation time *i.e.* time spent by the trained fighter following and sniffing the opponent; (6) the incidence and duration of ambivalent behaviors including grooming, rubbing, digging etc.

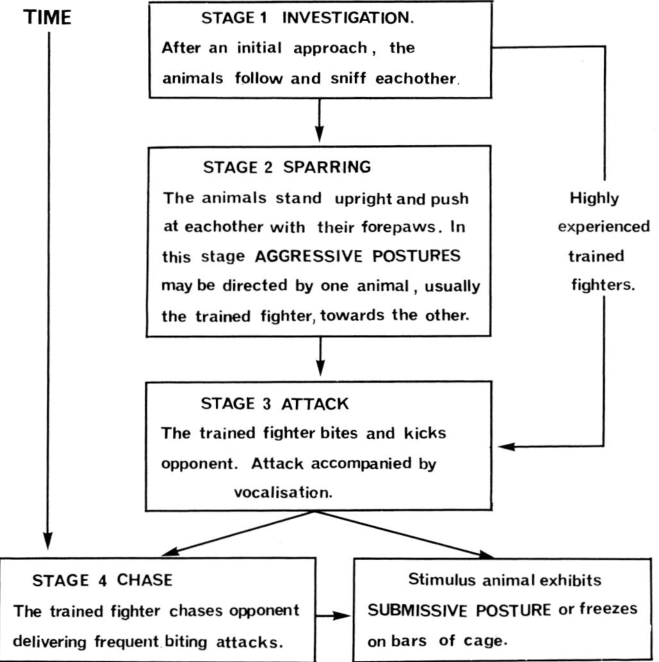

Fig. 1. Block-diagram illustrating the most common sequence of behaviors involved in a trained fighter test between hamsters of the same sex in a neutral arena.

References p. 479–480

All statistical comparisons were on the basis of the Mann–Whitney "U" test. (Some social postures of hamsters are shown in Fig. 2.) Experimental hamsters were tested in experiment 1 against mature male hamsters, which had been isolated for 50 days before being given a series of 5 min training tests against grouped male opponents in a neutral arena. Ten such animals were selected for the experiment which had shown spontaneous fighting in 4 out of the 5 previous training sessions.

Each trained fighter was fought against one animal from each treatment category, tests in a neutral arena being of 8 min duration after 1 min familiarization period.

Results and discussion

It was found that the number of attacks directed towards castrated animals were fewer than those directed towards intact opponents by trained fighters of the same sex ($P < 0.05$). The latency of the first attack directed by one animal towards its opponent was significantly longer in the case of castrated opponents *versus* trained fighters encounters ($P < 0.05$) and these animals were also investigated for a significantly longer period than in encounters between trained male fighters and intact animals ($P < 0.05$).

These results suggest that castration may inhibit attack by males, presumably by a reduction in androgen-dependent olfactory cues. It was interesting to note that a proportion of the castrates fought back when attacked and defeated their "trained fighter".

Experiment 2: Methods

Two categories of castrated animals consisting of 12 and 9 animals respectively were assembled and matched for previous experience. The first category was given 1 mg of testosterone propionate in 80% arachis oil and 20% ethyl oleate on each day over a 7-day period whilst the other category was given placebo. At the end of this period, the animals were subjected to a trained fighter test as described previously for experiment 1.

Results and discussion

A significantly shorter latency to both threat and fight was found in encounters between trained fighters and the testosterone-treated category compared to the latencies obtained between trained fighters and placebo-treated hamsters ($P < 0.05$). Trained fighters also directed more attacks and aggressive postures and spent more time investigating testosterone-treated animals than was evident in the response to placebo-treated hamsters. None of these differences, however, reached significance.

The results would appear to indicate that testosterone influences, to a small degree, the "attackability" of test hamsters by trained fighters.

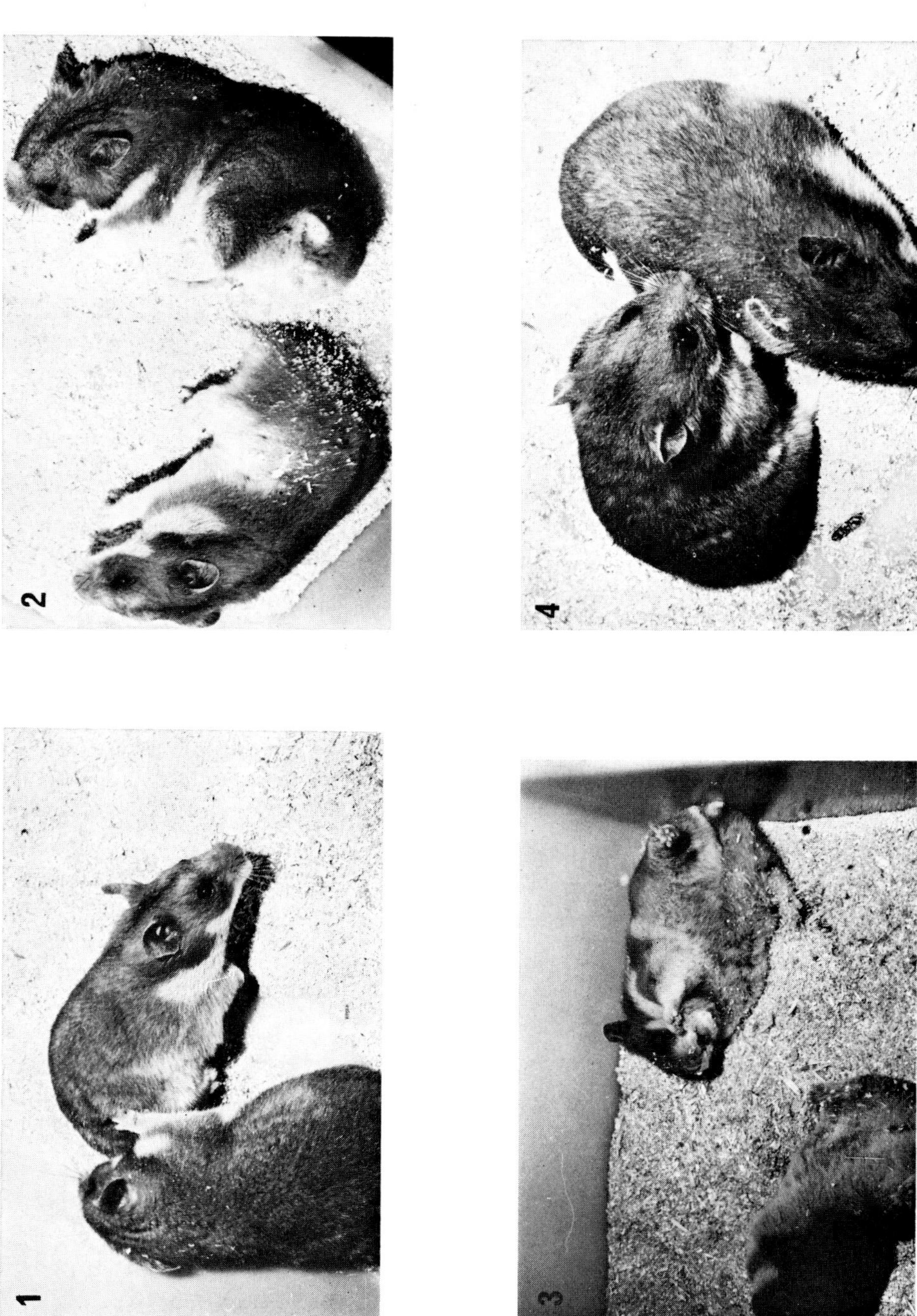

Fig. 2. Some agonistic postures in encounters between two female golden hamsters. 1: sniff; 2: animal on rihgt attacks; 3: animal on right attacks; 4: animal on left submits.

References p. 479–480

Experiment 3: Methods

Twenty-two 7–8 week old female hamsters maintained in groups of 3–4 were either bilaterally ovariectomized or mock operated. Seventeen days after operation they were subjected to trained fighter tests as described in experiment 1 against female opponents which had been isolated as adults and had been given a series of 5 min training sessions against grouped adult females. At the time of testing each trained fighter had shown spontaneous attack in at least 3 out of the 4 previous training sessions and had not exhibited lordotic behavior. All experimental tests and training sessions took place in the home cage of the trained fighter as this was found to facilitate aggressive behavior. Trained fighters were tested once against an ovariectomized and once against a mock operated animal.

Results and discussion

No significant differences on the basis of any of the measures selected could be demonstrated between agonistic encounters between the trained fighters and ovariectomized or mock operated animals. It is interesting, however, that a high incidence (in 48% of the encounters) of marking behavior was exhibited in these tests by the trained fighters.

CONCLUSIONS

Utilization of trained fighter techniques in studies on the effects of sex steroids on the "attackability" of hamsters seems initially promising. It appears that castration causes a reduction in the ease with which the animal is attacked by a more aggressive male (as has been demonstrated in the mouse, *e.g.* Brain and Poole, 1974) and that testosterone treatment of such animals increases to some extent the intensity of this behavior directed towards test animals by trained fighters. No effects of ovariectomy could be detected utilizing this type of test but much work remains to be done.

SUMMARY

Some endocrine and behavioral differences between golden hamsters and more common laboratory rodents are described. Recent studies on interactions between isolation-induced agonistic behavior in animals of both sexes in this species are described and the difficulty of assessing whether the hormonal effects so described have a direct neural action on the behavior, or have indirect effects possibly by influencing the production of olfactory cues, discussed. The application of a trained fighter technique to this type of study is reported. Some of the behavioral differences found in this type of encounter, in which it is hoped one will be able to distinguish more easily between direct and indirect effects of hormones on agonistic behavior, and more commonly applied tests are enumerated and some preliminary experiments, involving studying the responses of trained fighters to male and female opponents modified by spaying and/or treatments with sex steroids, are described.

It was found that utilization of the trained fighter technique in the study on the effects of sex steroids on the "attackability" of the hamster seemed initially promising. It appeared that castration caused a reduction in the ease with which the animal is attacked by a more aggressive opponent of the same sex (as has been demonstrated in the mouse *e.g.* Brain and Poole, 1974) and that testosterone propionate treatment of such animals increases, to some extent, the intensity with which this behavior is directed towards test animals by trained fighter males. Initial experiments failed to reveal any obvious effects of ovariectomy of hamsters on their being attacked by female trained fighters but much work remains to be done.

REFERENCES

BRAIN, P. F. (1971) The physiology of population limitation in rodents — a review. *Commun. behav. Biol.*, **6**, 115–123.

BRAIN, P. F. (1972) Effects of isolation/grouping on endocrine function and fighting behaviour in male and female Golden hamsters (*Mesocricetus auratus* Waterhouse). *Behav. Biol.*, **7**, 349–357.

BRAIN, P. F. AND POOLE, A. E. (1974) Some studies on the use of "standard opponents" in intermale aggression testing in TT albino mice. *Behaviour*, in press.

BRONSON, F. H. AND ELEFTHERIOU, B. E. (1964) Chronic physiological effects of fighting in mice. *Gen. comp. Endocr.*, **4**, 9–14.

BRONSON, F. H. AND ELEFTHERIOU, B. E. (1965) Relative effects of fighting on bound and unbound corticosterone in mice. *Proc. Soc. exp. Biol. (N.Y.)*, **118**, 146–149.

CHESTER JONES, I. (1955) Role of the adrenal cortex in reproduction. *Brit. Med. Bull.*, **11**, 156–160.

GASKIN, J. H. AND KITAY, J. I. (1970) Adrenocortical function in the hamster: sex differences and effects of gonadal hormones. *Endocrinology*, **87**, 779–786.

GASKIN, J. H. AND KITAY, J. I. (1971) Hypothalamic and pituitary regulation of adrenocortical function in the hamster: effects of gonadectomy and gonadal hormone replacement. *Endocrinology*, **89**, 1047–1053.

GIEGEL, J. L., STOLFI, L. M., WEINSTEIN, G. D. AND FROST, P. (1971) Androgenic regulation of nucleic acid and protein synthesis in the hamster flank organ and other tissues. *Endocrinology*, **89**, 904–909.

GRANT, E. C. AND MACKINTOSH, J. H. (1963) A comparison of the social postures of some common laboratory rodents. *Behaviour*, **21**, 246–259.

GRANT, E. C., MACKINTOSH, J. H. AND LERWILL, C. J. (1970) The effect of a visual stimulus on the agonistic behaviour of the golden hamster. *Z. Tierpsychol.*, **27**, 73–77.

LERWILL, C. J. (1968) *Agonistic Behaviour and Social Dominance in the Golden Hamster, Mesocricetus auratus* (Waterhouse). M.Sc. Thesis, University College of Swansea.

MUGFORD, R. A. AND NOWELL, N. W. (1970) The aggression of male mice against androgenized females. *Psychonom. Sci.*, **20**, 191–192.

MUGFORD, R. A. AND NOWELL, N. W. (1971) The preputial glands as a source of aggression-promoting odors in mice. *Physiol. Behav.*, **6**, 247–249.

MUGFORD, R. A. AND NOWELL, N. W. (1972) Paternal stimulation during infancy: effects upon aggression and open field performance of mice. *J. comp. physiol. Psychol.*, **79**, 30–36.

MURPHY, M. R. AND SCHNEIDER, G. E. (1970) Olfactory bulb removal eliminates mating behavior in the male golden hamster. *Science*, **167**, 302–304.

PAYNE, A. P. AND SWANSON, H. H. (1970) Agonistic behaviour between pairs of hamsters of the same and opposite sex in a neutral observation area. *Behaviour*, **36**, 259–269.

PAYNE, A. P. AND SWANSON, H. H. (1971a) Hormonal control of aggressive dominance in the female hamster. *Physiol. Behav.*, **6**, 355–357.

PAYNE, A. P. AND SWANSON, H. H. (1971b) Hormonal modification of aggressive behaviour between female golden hamsters. *J. Endocr.*, **51**, xvii–xviii.

PAYNE, A. P. AND SWANSON, H. H. (1971c) The effect of castration and ovarian implantation on aggressive behaviour of male hamsters. *J. Endocr.*, **51**, 217–218.

PAYNE, A. P. AND SWANSON, H. H. (1972a) The effect of sex hormones on the agonistic behavior of the male golden hamster (*Mesocricetus auratus*, Waterhouse). *Physiol. Behav.*, **8**, 687–691.

PAYNE, A. P. AND SWANSON, H. H. (1972b) The effect of sex hormones on aggression in the male golden hamster. *J. Endocr.*, **53**, lxi.

PAYNE, A. P. (1973) The effect of androgens on isolation-induced aggression in the male golden hamster. *J. Endocr.*, **57**, xxxvi.

SWANSON, H. H. (1967) Effects of pre- and post-pubertal gonadectomy on sex differences in growth, adrenal and pituitary weights of hamsters. *J. Endocr.*, **39**, 555–564.

DISCUSSION

HERBERT: In view of your research on aggressive behavior with trained fighters I am wondering, in general terms, how much a factor like training makes a difference. Since you are trying to compare untrained and trained fighters, you have to account for the fact that this condition in itself produces changes in specific mechanisms. Do you think this is an important factor?

EVANS: Certainly it may be the case that one would expect animals to show less discrimination with respect to opponents following training. However, in this series of experiments, experience did not seem to abolish the influence of olfactory cues on the aggressiveness of hamsters. I may also, in this respect, refer to similar work carried out by Payne (1973)*, investigating the same problem but using hamsters with apparently little or no prior experience of fighting. He obtained essentially similar results to mine with respect to effects of castration and androgen treatment on opponent "attack-ability". I feel that training may be useful under certain circumstances, in that, as the animals are screened for aggressiveness, a more uniform population, with respect to this behavior, than by random selection, is obtained.

* PAYNE, A. P. (1973) *Acta endocrinol. (Kbh.)*, **Suppl. 177**, 287.

Models of Behavior and the Hypothalamus

L. DE RUITER, P. R. WIEPKEMA AND J. G. VEENING*

Department of Zoology, University of Groningen, Groningen (The Netherlands)

INTRODUCTION

For our present purpose, a model of behavior can be defined as a set of statements from which truthful predictions of the future behavior of an individual can be derived. All such models rest on the assumption that, given sufficient information on present internal state and external input of the individual, its behavior in a subsequent time interval is more or less predictable. There are, however, two different ways in which the state of the individual can be specified. On the one hand, there is the purely behavioral approach, which attempts to define the internal state in terms of currently observed relations between sensory input and behavioral output of the individual. In this approach the individual is regarded as a black box. On the other hand, there is the physiological approach, which attempts to open the box and to specify the internal state in terms of relevant parameters of the mechanisms subserving the input–output relations of the individual. Since in this paper we are concerned with the behavioral functions of the hypothalamus, our ultimate aim must be a model of the latter kind. However, as we shall see, development of both kinds of model requires much the same approach at least in the initial phase. The present essay discusses some concepts useful for this approach, and their limitations.

Either kind of model primarily requires specification of the input received and of the output performed by the individual. As regards input, we shall simply take it for granted that an adequate description of the individual's environment is feasible at any moment. As regards output, we shall base ourselves on the customary ethological viewpoint (Hinde, 1970) that behavior is a time sequence of various recurrent elements, or unit activities, each of which can be precisely described in terms of patterns of effector activities, and their orientation to the environment. We shall not go into the problems involved in the definition of behavioral elements, or into the evolutionary, maturational or learning processes that determine what elements the behavioral repertoire as a whole will comprise (Hinde, 1970).

We merely state that the concept of element of behavior as here utilized embraces the entire overt output at a given moment, in other words, an individual does only one thing at a time. However, it never does "nothing" as long as it is alive. Sleeping and

* Present address: Department of Anatomy, Faculty of Medicine, Roman Catholic University, Nijmegen, The Netherlands.

References p. 505–507

resting are behaviors just as well as fighting or mating. According to this view the individual must willy nilly perform a lifelong sequence of choices among the unit activities that compose its behavioral repertoire. We will focus here on the investigation of these choices, which we term response selection.

From a functional point of view survival of the individual, and of the species to which it belongs, depends on its selecting the right responses. This has two aspects. Once the animal has engaged in a particular mode of behavior, it must continue that behavior up to its functional endpoint, if at all possible. It is a waste of energy to overtake but not to catch a prey. On the other hand, the individual must be able at all times to drop whatever it is doing, if its survival suddenly requires some other activity. This means that the mechanisms of response selection must strike a balance between stability and flexibility of behavior. We shall return to this point below.

BEHAVIOR SYSTEMS

If one could unravel the structure and the physiological properties of the information processing networks of the brain, one could predict its behavioral output without ever observing any actual behavior at all. However, the complexity of the central nervous system precludes this direct approach to the construction of models of behavior. The only possibility, then, is to study overt behavior first in order to obtain a starting point for the analysis of its neural mechanisms. We find that the concepts and methods of ethology are suitable tools for such an approach. Therefore, we shall briefly illustrate some traits of the ethologist's approach, taking vertebrate feeding behavior as our main example.

Leaving aside many complications we can regard feeding behavior as composed of 4 different elements: (1) food searching, (2) approaching a detected food object, (3) grasping the approached object, and (4) ingesting the grasped object (de Ruiter, 1967). Clearly, each element is a response to a particular set of external stimuli. Only perception of the general feeding environment is involved in searching; for the other 3 elements specific stimuli from the food object are required.

Now it is an important point that, given the appropriate eliciting stimuli for any one response, the latter may or may not ensue. Therefore, we must discuss responsiveness in terms of probability of occurrence (P_o) of a given response. This implies that even when feeding responses are strongly dominant, there is always a finite chance that the animal will do something quite different. Further, on repeated presentation of the appropriate external situation, P_o of any feeding response is found not to be constant but to fluctuate widely. Finally, these fluctuations prove to be positively correlated for the various elements of feeding behavior. For all 4 of them, P_o increases in the course of food deprivation, and when food is returned after deprivation, P_o decreases again in inverse relation to the amount ingested.

Familiar as they are, these facts lead to the important conclusion that P_o for all elements of feeding behavior is governed by the same set of internal factors. For this reason these elements can be classed together in a higher category, the feeding behavior system or feeding system for brevity. We shall term the complex of factors governing responsiveness of the feeding system the feeding motivation.

More generally, one may in principle apply this kind of analysis to the entire behavioral repertoire of an individual. That is to say, first the occurrence of the various elements is recorded in a suitably diversified range of internal and external conditions. Next, it is determined which elements manifest positively correlated fluctuations in P_o. On this basis the repertoire is divided into a number of behavior systems, each with its own "specific" motivation (see also p. 494 and 499). As one might expect, elements grouped together according to this criterion usually also prove to subserve the same biological function, so that it is convenient to speak of *e.g.* a sexual system, a parental system, and so on.

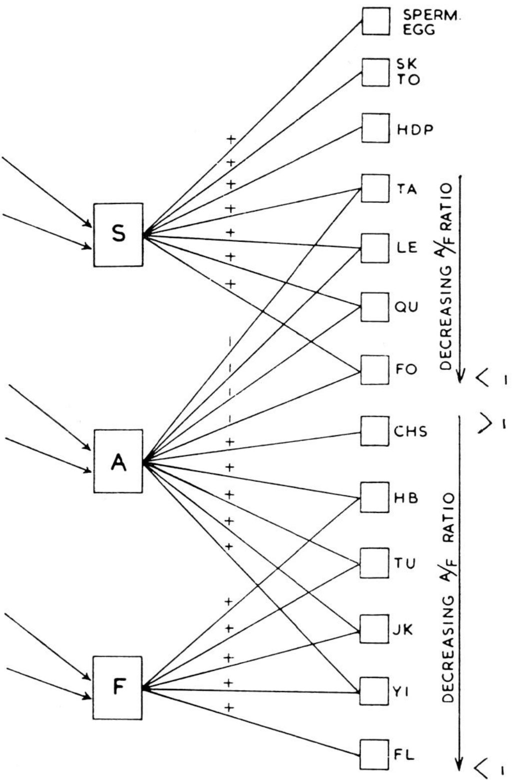

Fig. 1. Diagram of the reproductive behavior of the bitterling. S, A, F, motivations for sexual, aggressive, and fleeing behavior, respectively. Motivational signals (arrows from the left) determine the level of each of these motivations. Right hand column: array of behavior elements performed in the reproductive situation. Both male (*e.g.* ejaculation) and female elements (*e.g.* oviposition) are included; many elements are performed by both sexes. Note that certain "agonistic" elements are furthered by simultaneous activation of A and F motivations in a certain ratio. Certain sexual elements tend to be inhibited by simultaneous activation of A. From Wiepkema (1967). Abbreviations: CHS, chasing; FL, fleeing; FO, following; HB, head butting; HDB, head down posture; JK, jerking; LE, leading; QU, quivering; SK, skimming; TA, tail bend; TU, turning beat; YI, yielding.

One example of such a study is the analysis of reproductive behavior in the bitterling by one of us (Wiepkema, 1961). The results are summarized in Fig. 1. This shows that the behavior of the fish in the presence of its partner is governed by 3 different motivations, which may be termed sexual, aggressive, and fleeing motivations. Three points deserve attention here. Firstly the level *e.g.* of aggressiveness is much influenced by the environment (perception of a rival male). This shows that external stimuli may have two effects: not only response-eliciting, but also motivational. An operational criterion for the latter, of course, is whether temporary presentation of the stimulus affects subsequent responsiveness to this or other stimuli. Secondly, all 3 motivations mentioned may be activated at the same time, in degrees depending on the vicissitudes of the male over a certain foregoing time interval. Thirdly, several elements prove to be governed by more than one motivation, indicating that there is considerable overlap of behavior systems.

Similar studies on mammals, including primates (van Hooff, 1971), indicate that the method is fruitful here as well, but more work is needed to assess to what extent it can be applied to the entire repertoire, particularly in this highly evolved group. It should also be mentioned that there are methodological problems that need further consideration. One of these, to which we shall return below, lies in the measurement of P_o (see p. 493).

Yet, although these analyses are still in mid-stream, their relevance to our topic is unmistakable. First, they provide a basis for "black box" models of behavior, for if elements A and B prove to be governed by the same motivation, P_o^B can be predicted from measurement of P_o^A. Secondly, more than any other kind of behavior study known to us, this approach gives detailed information on the motivational state of the individual, *i.e.* on the functional states of the control mechanisms of behavior, and thereby provides a starting point for the physiological study of these mechanisms, as we shall see in more detail below.

PHYSIOLOGICAL STUDIES OF A BEHAVIOR SYSTEM AND THEIR LIMITATIONS

Granted that the concepts of behavior element, system, and specific motivation have aspects incompletely analyzed as yet, it nevertheless seems sensible if one wants to develop a physiological model of behavior to concentrate on a single behavior system first. This is indeed what many physiologists and physiological psychologists are doing, though sometimes perhaps on grounds of intuitive functional considerations rather than on the basis of a systematic analysis of the causal structure of behavior. For several systems these studies have already provided valuable insight into the processes underlying the respective motivations. We shall again take vertebrate feeding behavior as an illustrative case.

Much effort has been spent on analysis of the feeding motivation, particularly concerning the uptake of fuel for energy metabolism ("caloric regulation"). Several reviews are available (Code, 1967; Morgane, 1969; Hoebel, 1971), and a brief summary will suffice of the commonly accepted overall picture, which is one of a negative feedback control system. This summary will involve some statements that we shall have to qualify later on.

Feeding responses are facilitated by an extensive neural network in which the lateral hypothalamus (LH) appears to play a key role. Electrical or chemical stimulation of the LH in food satiated subjects may elicit feeding activities resembling those normally performed by food deprived subjects closely enough to convince many workers "that such stimulation has many of the properties of normal hunger" (Miller, 1961). LH lesions, depending on their size and site, produce more or less complete and prolonged suppression of food intake.

When activity of this motivational network causes the individual to ingest food, nutrients enter first into the digestive tract, subsequently into the blood stream, and finally — as long as the rates of ingestion and intestinal absorption exceed the immediate needs of catabolism — into the reserves. A variety of experiments indicate that the nutrient contents of the first two, and possibly also of the third of these compartments are monitored. These messages ("satiety signals") are fed back into the motivational network, where they exert an inhibitory effect.

In this negative feedback loop a major function is attributed to the ventromedial hypothalamus (VMH), which is assumed to pool the various satiety signals and to modulate the activity of the "feeding center" in the LH in accordance with the total satiety level. This conclusion is based on the well known fact that lesion of this ventromedial "satiety center" causes excessive hyperphagia (and obesity), whereas artificial stimulation of the VMH suppresses feeding. It may be added that in the case of the VMH, as well as in that of the LH, there is also some electrophysiological evidence that supports the views here presented (Hoebel, 1971).

If one wants to assess the explanatory value of this model (Fig. 2), one should realize that it is certainly far from complete. It may be mentioned in passing that it is not clear which of the major systems passing through the critical region of the LH (pallidofugal, nigrostriatal, medial forebrain bundle) is involved, or how the LH and VMH networks are functionally related to other brain areas involved in feeding behavior. A further serious shortcoming is that we have only rough ideas what positive motivational signals the LH feeding network receives. Presumably exteroceptor messages reporting the environmental food situation in the broadest sense are received at this level (Panksepp, 1971; Wyrwicka, 1969), so that P_o for feeding responses is set by the balance of these positive messages and the counteracting satiety signals (the greater the response-eliciting value (palatability, etc.) of the food objects encountered, the higher the satiety level needed to terminate feeding).

In view of such uncertainties any quantitative assessment of the predictive powers of the model is out of the question. True, it appears to have certain interesting qualitative potentialities, *e.g.* as regards the apparently paradoxical finickiness in subjects made hyperphagic by VMH lesions (de Ruiter *et al.*, 1969). However, in order to predict at what level the nutrient content of the body will stabilize under given conditions, or what the time pattern of feeding behavior will be in an *ad libitum* situation, we must know much more than we do at present about intensity and time course of the

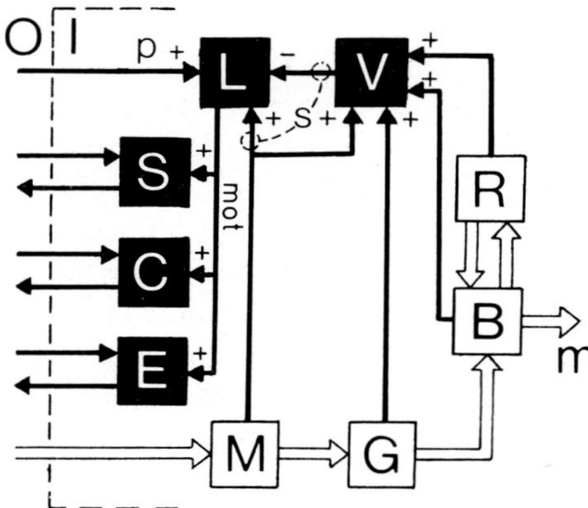

Fig. 2. Diagram of a physiological model of the feeding behavior system subserving caloric regulation. Black arrows symbolize messages (neural or humoral), open arrows fuel transport in the body. O, outside world; I, individual; S, C, E, central nervous mechanisms for performance of responses of searching, capturing, eating food objects; L, neural network facilitating these elements (situated in the LH, and elsewhere); V, network processing satiety messages (in the VMH and elsewhere); M, mouth cavity; G, gastrointestinal tract; B, blood stream; R, reserve tissues; m, energy metabolism; mot, specific motivation for feeding behavior; p, "attractiveness" (palatability, etc.) of the general external food situation (evaluated on basis of previous experiences); s, net satiety, *i.e.* resultant of positive and negative feedback signals. Performance of S, C, E, modifies sensory input from O, and on the other hand causes food to enter M. Interaction effects are disregarded in this model. After de Ruiter (1971).

various messages postulated in the model. To illustrate this, let us consider the best studied class of satiety signals, those arising from the blood.

There are a number of indications that in the rat, and probably in many other mammals, availability of blood glucose to the peripheral tissues is one of the determinants of P_0 for feeding responses (Anand, 1967; Steffens, 1970; Strubbe, in press). Glucose availability, as measured in terms of rate of uptake of glucose from the circulation, depends of course on presence in the blood not only of glucose but also of certain hormones, notably insulin. In the rat, under normal *ad libitum* conditions, blood sugar always remains at levels so high that it is unlikely that glucose itself can become the limiting factor (Steffens, 1969a, b). Fluctuations in glucose availability accompanying meals in the rat, as manifested in RQ measurements (Nicolaidis, 1969), no doubt are mediated by endocrine changes. Now in the rat feeding behavior remains in abeyance as long as insulin remains above a fairly sharply defined critical level, but is reinstated at once when the gradual decrease of insulin attains that level (Fig. 3). Admittedly, several pieces of the puzzle are still lacking: it remains to be shown that the insulin level critical for starting a meal is that at which energy metabolism begins to switch over from carbohydrate to other fuel, and above all proof is needed that the timing of *ad libitum* meals can be altered by manipulation of parameters of glucose availability, if this manipulation stays within the normal physiological range of parameters.

If glucose availability provides a signal controlling intake, as here envisaged, there must be "glucoreceptors" monitoring this variable. Indeed there is evidence that this

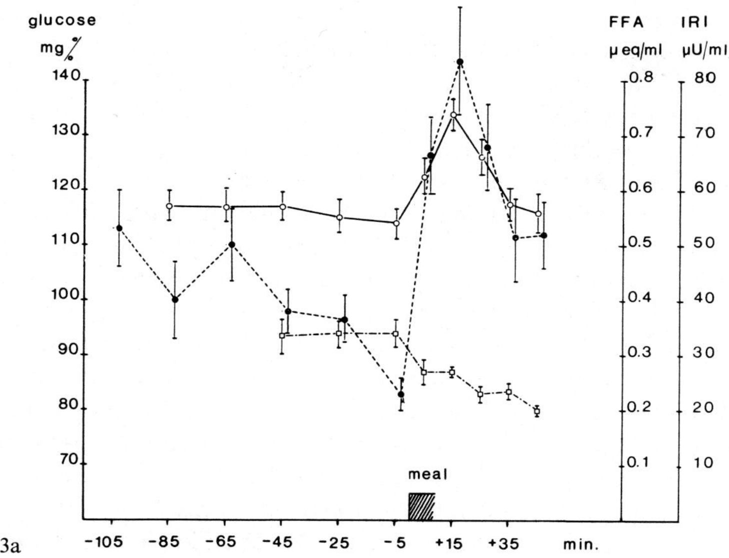

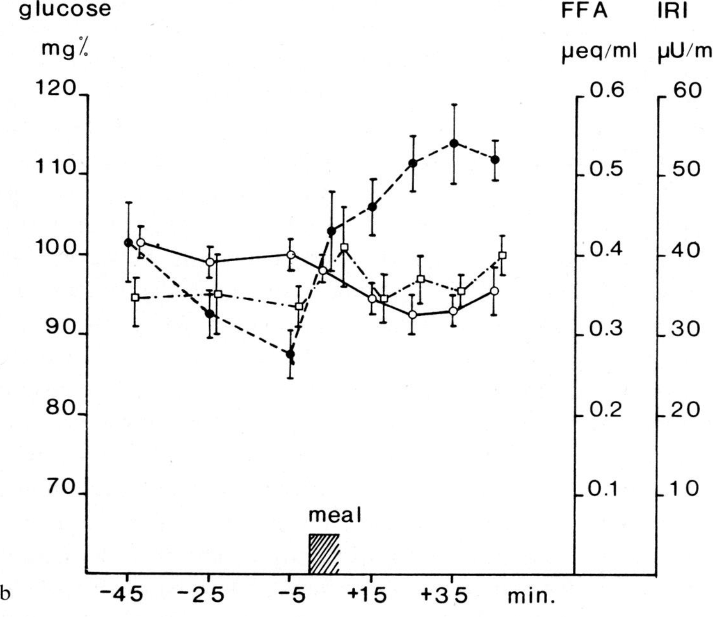

Fig. 3. Blood levels of glucose (○——○), insulin (IRI, ●----●), and free fatty acids (FFA, □-·-·-·-□) in relation to the time pattern of meals in the rat (daytime observations). a: normal diet (containing free glucose and starch); b: carbohydrate free diet. Abscissa: time in minutes; start of meal (shaded block, open ended because of variable duration) is zero time. Note that glucose and FFA remain constant up to beginning of meal. Feeding begins when insulin drops to a critical level; this is even the case with carbohydrate free diet. From Strubbe (in press).

References p. 505–507

function is performed by cells in the VMH itself (Anand, 1967; Oomura *et al.*, 1964) which report the combined presence of glucose and insulin in the blood (Debons *et al.*, 1970). In line with this, and with the above hypothesis of "glucostatic" triggering of feeding, it was recently found by Strubbe (in press) that local administration to the VMH in the rat of minute amounts of insulin antibody produces a marked, dose-dependent increase in food intake lasting about half an hour.

Now, regardless of whether waning of glucose availability starts the meal, can it be that its recovery causes feeding to stop? This seems doubtful, because duration of meals varies from about 4 to 20 min (meal size has a comparable range), whereas blood glucose rises sharply in the third minute (Steffens, 1969a) and blood insulin (Strubbe, unpublished) and RQ (Nicolaidis, 1969) do so already in the very first minute. One might therefore expect glucose availability signals to act more promptly than is compatible with observed duration of meals. Here we must face the fact that we can measure insulin and glucose concentrations but have no firm evidence as regards time course either of the message derived from these substances, or of any other one of the multiple signals that together determine the satiety level. For the time being we must leave it an open question whether variation in glucoreceptor messages or fluctuation of one or more other satiety signals is the critical event that starts and stops meals (we shall return to the termination of meals below, p. 493).

To sum up the state of these lines of work, it appears that for a long time to come we shall have only scattered bits of direct physiological information on the various parts of the complex control system. In this situation it is difficult to evaluate what may be concluded (and what not) from the available facts, or to draw up a priority program for further physiological study. Indeed a special system-theoretical analysis is needed here. Results of a preliminary study of this kind, so far restricted to deterministic models, are available (Geertsema, 1973; Reddingius and Geertsema, in press), from which we will take one illustrative example. It is commonly assumed that the constancy of the fat reserves (and hence of the body weight) is conclusive proof that negative feedback signals from the reserves are operating. However, "in spite of twenty years of searching, no one has yet managed to specify the stimulus which is correlated with body weight and serves such an important role in the control of food intake" (Hoebel, 1971). In this situation it is of interest that theoretical analysis shows that there are several different possible conditions under which constancy of the reserves is ensured. For instance, constancy will ensue simply if the rate of fat mobilization is proportional to the fat content of the reserves — perhaps not a grossly implausible possibility, and certainly one that is open to direct physiological investigation. On the other hand monitoring of the reserves by a separate receptor system could also be responsible for the constancy of body weight. For the present we must keep two alternative models in mind as concerns this question. Somewhat similar situations are found in respect of several other problems. Eventually the theoretical analysis shows that a large number of different models of the feeding system appear compatible with the data at hand. Now as a point of particular interest, the results also point to possible ways to narrow the field through fairly simple measurements of such output parameters as rates of ingestion and catabolism, and durations of meals and of the intervals separating one meal from the next. However, rather than elaborate these theoretical conclusions, the generality of which needs further investigation, we will terminate this paragraph with a concrete example how study of behavioral output of the system may shed light on its structure.

The feeding responses of many mammals are not scattered evenly over time, but tend to be aggregated in "meals" often lasting many minutes, separated from one another by intervals of several hours virtually devoid of feeding activities. This suggests that P_0 for feeding switches back and forth between two discrete values, one close to zero and the other to unity, with only brief transitional phases (Fig. 4). On the other hand, there is no clear reason to ascribe to the metabolic decay of satiety the steplike character needed to explain the abrupt change from low to high P_0 for feeding. As regards the subsequent return from high to low values, we may perhaps assume that the relatively rapid process of ingestion does give recovery of satiety a somewhat steplike nature, but then the problem arises why a meal, once initiated, can go on for prolonged and rather variable periods, since in an *ad libitum* situation the satiety deficit at the start presumably is small.

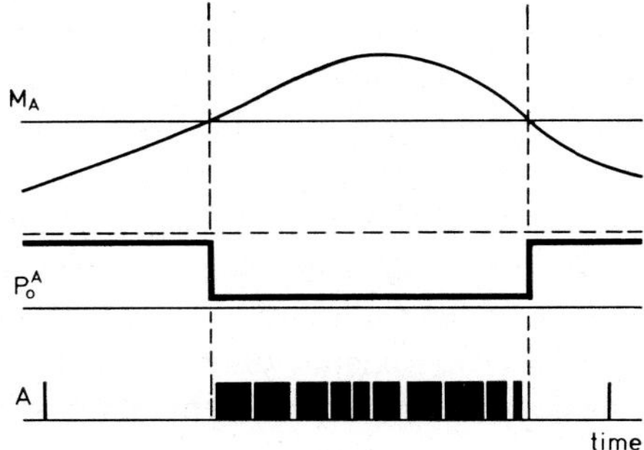

Fig. 4. Hypothetical diagram to illustrate the concept of abrupt "switching on and off" of feeding behavior at beginning and end of meals. When gradually changing specific motivation for feeding behavior (M_A; *cf.* mot. in Fig. 2) rises above a critical level, $P_0{}^A$ drops from values near zero (horizontal broken line) to values near unity. When M_A crosses the threshold level in the reverse direction, $P_0{}^A$ jumps back. As long as $P_0{}^A$ is high, many A responses (bottom line of diagram) are performed in bouts of long mean duration, with only brief bouts of other behavior interspersed. When $P_0{}^A$ is low, only isolated randomly scattered A responses occur. No effects of interaction are included in this diagram (*cf.* pp. 491–495).

The following observations on feeding activities of male mice (Wiepkema, 1971b) are pertinent to these questions. During a meal, in this as in other species, P_0 for feeding is high, but less than unity. Therefore, brief spells of other behavior alternate with bouts of uninterrupted feeding during the meal. The mean length of the feeding bouts, of course, is determined by P_0 for this behavior. Now if over a large number of meals, the averages are calculated of the lengths of the 1st, 2nd, 3rd, ... nth feeding bouts, it is found that bout length rises over approximately the first 6 bouts of the meal. In other words, the feeding motivation increases during the early stages of the meal. This may be due to either or both of two causes: (1) due to time lags in the system, satiety signals may continue to decrease during this time, and (2) contacts with food may set up positive motivational feedback signals in addition to the satiety messages. This point was settled by the following observation. When the food is given a bitter flavor, leaving its nutritional quality unchanged, the rise of motivation early in the meal is abolished. This rise therefore appears to be due to positive motivational feedback messages from ingestion of palatable food, presumably mediated by gustatory receptors.

Clearly such effects will promote a stepwise transition from non-feeding to feeding behavior, and will aid the animal to go on eating until a sizeable meal has been taken. Particularly in a natural environment, where encounters with food may be few and far between, such persistence seems useful.

However, what about the reverse change at the end of the meal? No doubt, rising satiety facilitates the latter transition, but is it not hampered by the positive feedback mechanism? Two points are of interest here. Firstly, the positive feedback effects decay rapidly (Wiepkema, 1971a). If the interruptions of meals by non-feeding behavior are lengthened from about 5 to about 15 sec by making the mouse feed in an unfamiliar environment, no increase in P_o for feeding is seen early in the meal. Apparently there is now a complete decay of the positive effects set up in one feeding bout during the interval preceding the next bout. Secondly, the satiety level determines both the sign and the magnitude of the motivational effect of contact with a food object (Wyrwicka, 1969). This effect was recently analyzed by Thomas (in press) in a study of food searching in the 3-spined stickleback. After finding and eating a prey, the fish makes strongly increased searching efforts near the site of the find for about 30 sec. After rejecting a discovered prey, the fish shows temporary cessation of feeding behavior, rapidly removes itself from the vicinity of that prey, and continues to avoid that site for some time. In these experiments hunger level of the fish and prey quality are kept as constant as possible. It is not certain therefore whether the rejections are due to stochastic fluctuations in motivation or to slight variations in the prey, but in any case a previously rejected prey may be eaten at a subsequent encounter. It is concluded that exteroceptor feedback mechanisms from responses to a food object will raise the feeding motivation, provided that hunger level and prey quality combine to trigger ingestion. Conversely, these feedback mechanisms lower this motivation, if the encounter results in rejection. Moreover, in the latter case avoidance is elicited, which ensures that stimuli from the rejected prey will no longer distract the predator's attention. Clearly such effects, if of general occurrence, will promote the stability of feeding behavior once it is initiated, and on the other hand they will aid the animal to move resolutely from this to the next activity at the stop of the meal.

These points illustrate how behavior observation may supplement physiological studies on the organization of a particular behavioral system. However, they also contain a warning that for proper understanding of any particular activity, we must consider the place this activity takes in the context of the animal's behavior as a whole. This is really an obvious point, because an increase of P_o for one behavior must entail a decrease for at least one other activity. Disregarding this complication, to which we shall return below, we can state that as concerns feeding behavior graded motivational effects of exteroceptor feedback mechanisms appear important for switching this activity on and off. The available ethological evidence being still very limited, it seems worth considering whether other lines of work lead to similar views.

Studies on intracranial self-stimulation (ICSS) by physiological psychologists provide a case in point. As is well known, animals will perform ICSS over electrodes in the LH through which stimulus bound feeding can also be elicited. Several workers have pointed out that this ICSS has a certain resemblance to feeding responses to palatable food (Fonberg, 1967; Hoebel, 1968; de Ruiter, 1968). Indeed, the apparently paradoxical fact that a rat will compel itself to excessive food intake by ICSS becomes explicable if we assume that in normal feeding, positive feedback mechanisms from palatable food are an important component of the feeding motivation, and that the electrical stimulus mimics these feedbacks. To this extent ICSS studies are consistent with our above conclusion on the role of positive feedback mechanisms in the control of feeding. Moreover, the tendency to self-stimulate in these sites is positively correlated with the hunger level. The more satiated the animal, the less rewarding ICSS becomes. In overfed subjects the stimulus even becomes aversive, so that the animal will work to escape from the stimulus (Hoebel and Thompson, 1969). If it is accepted that ICSS in these sites essentially mimics exteroceptor feedback messages normally set up by feeding behavior, the latter facts confirm that sign and magnitude of the motivational effects of these signals depend on the satiety level of the subject.

To conclude this section, let us repeat that there is still a great lack of data for quantitative specification of the model of feeding behavior here developed. As it stands, this model does explain roughly why the total amount of feeding behavior performed over a prolonged period is such that it will just compensate metabolic expenditure, but it does not enable us to predict at what points in time feeding will be switched on or off. When we turn to physiological, or combined physiological–behavioral models of other systems (Milner, 1970), we encounter similar situations, but little would be gained for our present discussion by reviewing these various models.

The general impression we are left with at this point is that each system has built-in provisions (positive feedback mechanisms) that ensure continued stable performance of the behavior in question once it is switched on, until negative feedback mechanisms finally gain the upper hand, and switch it off again. For the present we cannot accurately measure all these signals with physiological techniques. The best specification of the motivation of any system therefore is in terms of overt behavior. Now if we accept the picture that seems to emerge in this way, the following problem arises. In many different systems the respective specific motivations may be building up simultaneously. Yet the individual can do only one thing at a time. How does it decide what to do first? In order to deal with this question we must have a closer look at the phenomenon of interaction of behavior systems.

INTERACTION OF BEHAVIOR SYSTEMS

A typical example of interaction is provided by the interrelations of feeding and drinking behavior. The specific motivation for the latter is known to depend on

References p. 505–507

volume and concentration of the extracellular fluid (Morgane, 1969). Abnormal values of either parameter are reported by various messages to the neural network for drinking behavior, which extends through large parts of the brain, including the LH. Actually, chemical stimulation of one and the same site in the LH of the rat may affect both feeding and drinking (Grossman, 1967). Cholinergic stimulation of such sites increases drinking, and at the same time depresses feeding; adrenergic stimulation has the opposite results. The inhibitory effects cannot be attributed to mere competition for time between the two behaviors, for feeding is depressed even in the absence of water, and drinking in the absence of food. Rather it appears that a high level of the specific motivation for drinking lowers P_o for feeding responses (and *vice versa*), as evidenced also by the decrease of food intake in subjects undergoing water deprivation (Collier, 1969).

Ethological studies (Hinde, 1970) have revealed a great many examples of similar mutual inhibitory relations between all sorts of behavioral systems. Yet in other cases a different kind of interaction is found in which high motivation for one system also facilitates responses pertaining to some other system. The relation between feeding and drinking again provides an example. In rats, with food and water *ad libitum*, about 70% of all water intake is associated with the meals (Fitzsimons and le Magnen, 1969). Now food intake of course has physiological effects that cause real thirst, but this does not explain the association, for this persists even when water is given continually by slow continuous infusion through a gastric fistula, so that real thirst cannot arise during the meal (Fitzsimons, 1957). The effect must therefore be ascribed to direct facilitatory influence of the feeding motivation on drinking behavior.

As can be seen from these examples, inhibitory and facilitatory interactions between a given pair of systems may exist side by side; which of the two will manifest itself depends on the state of the individual and on the environmental situation. The effects of interaction on time patterns of overt behavior also vary widely. At one end of the scale, we find short interruptions of one activity by another, as when a rat briefly grooms its whiskers during a meal. In somewhat farther reaching cases the time pattern of one behavior is more seriously distorted by the exigencies of another system. Thus, if rats are kept on a feeding schedule that compels them to eat at abnormal hours, a large part of their drinking is also shifted to these hours (Fitzsimons and le Magnen, 1969). Finally, at the other extreme there are cases where not only the time pattern but the set point itself of one system is lastingly changed by the dictates of another. For instance, in mammals receiving water rations smaller than their *ad libitum* intake, feeding is also depressed so that, with food *ad libitum*, the nutrient reserves of the body shift to a lower set point which is maintained as long as water is rationed (Collier, 1969). The interactions are often more or less symmetrical, mutual inhibitory relationships between systems being particularly common. However, profound asymmetry may also occur, *e.g.* in the case of the demands of thermoregulation which receive great priority over those of feeding and drinking behavior (Hamilton, 1967).

The bearing of interaction on response selection and stability of behavior is obvious. In particular, it opens the possibility for a strongly motivated system to suppress competing systems, so that it gains undisputed control over the motor apparatus, and may even maintain this control for a while when its specific motivation is just beginning to subside, although, as pointed out by Geertsema (1973), the conditions under which such speculations hold, need further theoretical analysis.

It is worth to reconsider in this light the concept of "meals" in feeding behavior, which as we have seen (p. 489) rests on the assumption that P_o for feeding tends to switch abruptly from a low to a high value, and back again (Fig. 4). This implies that P_o for other behavior is low during meals, so that "interruptions" of feeding will be brief on the average. This point needs further verification, as appears from the rather wide range of maximal durations of interruptions within a meal allowed by different authors (Teitelbaum and Campbell, 1958; le Magnen and Tallon, 1966; Wiepkema, 1968; Kissileff, 1970). Yet several facts appear to confirm the validity of the meal concept. Bouts of other behavior may fall into two rather distinct though overlapping classes, short ones during meals, and long ones in the intervals between meals (Wiepkema, 1968). Further, if P_o for feeding is high throughout, we may expect mean duration of feeding bouts to remain long up to the end of the meal. What data we have (Evenhuis, unpublished; de Ruiter, unpublished) appear to confirm this expectation in the case of the mouse. Now, if generally true this point has an interesting implication. In view of physiological data it seems certain that satiety signals accumulate during the meal, and reduce the specific motivation for feeding. Hence, if P_o for feeding remains constantly high, this must be due to the feeding system maintaining its dominance through interaction. If so, termination of the meal (*cf.* p. 488) must also be brought about by interaction processes, the course of which is not governed exclusively by the specific feeding motivation. If due to this the meal happens to be small in comparison to metabolic needs, satiety will wane rapidly, and the specific feeding motivation will be reinstated sooner than if the meal happened to be large. This explains the well known positive correlation between meal size and duration of the subsequent interval; that no such correlation is usually found between meal size and duration of the preceding interval indicates that the switch of P_o for feeding to high values is set largely by a critical decrease in satiety (Wiepkema, 1971a).

This example shows that no model of behavior can be at all realistic unless it includes interaction effects, for no reliable prediction of responses pertaining to any one system can be made if only the "specific" motivation of that system is known. Rather, the state of all motivations, and the ways in which they interact, must be taken into account. Here, even more than when we are dealing with a single specific motivation, direct physiological measurement is out of the question. Therefore, the other approach, *i.e.* through behavioral observation, must be exploited to the utmost. Now there are several different behavioral measures of overall motivational state, such as frequency or intensity of elements, or analysis of behavioral sequences (Cane, 1961). Leaving aside the problem of how these measures are related, we will only mention one aspect of frequency measures. P_o values derived from such data are most informative if the observations are restricted to a time interval during which the state of the subject remains essentially constant. Fulfillment of this condition appears to require some caution. If the state is constant over a given interval, "interruptions" of the "dominant" behavior during that time will occur at randomly scattered moments, and be predictable only in terms of relative P_o values. To take feeding behavior as our test case again, we have already seen (p. 489) that positive feedback mechanisms operating during the meal impose a certain degree of patterning on the occurrence of the interruptions. An even more striking case was described in the barbary dove (McFarland, 1970). This bird under conditions where feeding is the dominant behavior, though the thirst motivation is also activated to some extent, when given access (through operant responding) to both food and water, will invariably work for and ingest food first, but will switch to water-directed responding at a rather precisely predictable point in time (about 2 min after the start of feeding). The duration of this interruption is brief, and the bird soon returns to feeding. If drinking is prevented during the interruption, feeding will yet be resumed at the usual moment, even if water is made accessible again at that time. Moreover, the timing of the interruption is independent of the thirst level over a wide range of the latter.

Such examples show that great caution is needed in behavioral measurement of the motivational state of the individual. Even under carefully controlled laboratory conditions, detailed observation reveals subtle, but profound changes over periods that are short in comparison to the time needed to gather data for frequency measurement or sequential analysis. Under natural, more variable conditions, the state of the individual may be even more protean.

To sum up this section, it is clear that a graphic representation of the formal structure of the causation of behavior will be essentially like the diagram of Fig. 5b rather than that of Fig. 5a. Now, if we view the machinery of behavior as one vast, unitary

Fig. 5. Diagrams to illustrate two different concepts of the causal structure of behavior. The repertoire is assumed to consist of only two systems, each containing a single element (A and B, respectively). a: the performance of each behavior is governed exclusively by its own specific motivational (α or β, respectively) and external response-eliciting (a or b respectively) messages. b: this diagram includes a mechanism (I) for interaction between the two behavior systems. Performance of A or B is now governed, in addition to its "specific" signals, by interaction messages (i), the contents of which are influenced by all inputs received by the mechanisms subserving behavior. Note that it becomes more or less arbitrary here, which signals we choose to call "specific" for a given behavior (*cf.* p. 499).

complex rather than a mosaic of discrete specific systems, one trait of the complex deserves emphasis. Many of the specific internal motivational messages probably wax and wane rather gradually and to some extent independently of one another. Yet, even under constant external conditions the individual appears to switch from one behavior system to another in a resolute manner and then to adhere to its choice for some time. If this impression is correct, this must mean that the complex can be "tripped" from one relatively stable state to another. Within such a state fairly wide variation in input appears permissible, but at certain critical changes in total input, a rapid transition to another state takes place under the combined influences of intrasystem positive feedback mechanisms and intersystem interactions. This then, would be the basis on which the phenomenon of behavioral stability rests. Admittedly we have not yet given an operational definition of the concept of stability, but we shall come back to this below (p. 501).

BEHAVIORAL FUNCTIONS ON THE HYPOTHALAMUS

Let us now attempt to apply these views to the problem of the behavioral functions of the hypothalamus. Research in this field has made enormous advances over the past few decades as shown by other contributions in this volume. Yet there is still much confusion as to the behavioral interpretation of the findings. This appears to be due not in the last place to the fact that experimental techniques have evolved more rapidly than a generally accepted theoretical framework for evaluating their results. In consequence each author tends to discuss his data in terms of his own set of concepts. This makes it hard to see what points are certainly established, what others present real controversies, and finally what is mere terminological confusion. A further difficulty is that many experiments are restricted to one or a few behaviors, and carried out with different animal species under special conditions, which vary from author to author, so that the generality of results cannot be assessed. This state of affairs may be illustrated by the following somewhat random anthology from the recent literature, presenting verbatim quotations of typical statements by various authors on the behavioral effects of experimental manipulation of ventromedial or lateral (and posterior) hypothalamus. These quotations will be listed alphabetically because a systematic treatment seems impossible.

As regards the VMH, Campbell *et al.* (1969, p. 186) conclude that this area "may affect eating in some other way than by mediating specific satiety signals. In particular it is conceivable that one of the functions of the VMH, if not the major one, is that of keeping the animal alert while eating, so that any novel or threatening stimulus would instantaneously produce some protective action". In other words, "VMH lesions may make the animal less distractable by novel environmental stimuli".

Grossman (1971, p. 30) concludes that "Stellar's (1954) model of a medial hypothalamic "satiety center" which exerts inhibitory influence on a lateral feeding center may be appropriate", but on the other hand he has repeatedly stated that "some or all of the peculiar effects of ventromedial lesions on appetitive behaviour may be due to a modification of affective processes that produce an exaggerated affective reaction to all sensory inputs" (Grossman, 1966, p. 7). More specifically he has suggested (Grossman, 1972, p. 282) that "a reduction in fear combined with increased sensitivity to pain could account for the classic VMH lesion syndrome as well as the observed increase in shock-induced fighting".

Margules and Stein (1969, p. 475) conclude from the differential effects of application of atropine on free feeding behavior and food-rewarded operant behavior that "cholinergic synapses in the VMH are the hypothalamic link of a brain system that mediates behavioural suppression".

Miczek and Grossman (1972, p. 328) on the other hand find that this treatment results "in a broad spectrum of behavioral effects which are not readily accounted for in terms of disinhibition of punished or satiety-suppressed responding, but may reflect general disruptive effects on behaviour".

Panksepp (1971, p. 393) on re-examination of the role of the VMH in feeding behavior, finds that "the commonly accepted hypothesis of the VMH mediating satiety is conceptually vague and has been often found sterile on close scrutiny". He prefers the hypothesis "that the VMH monitors nutrient depletion and repletion, while the LH mediates postprandial satiety".

Rabin (1972, p. 337), after putting side by side over 100 papers relevant to the possible role of the VMH in food intake, concludes that there is "no anatomical or electrophysiological evidence which can provide unequivocal support for a ventromedial hypothalamic "satiety center" which acts to suppress food intake by virtue of its inhibitory connections to a lateral hypothalamic feeding center".

De Ruiter and Wiepkema (1969) mention that in mice treated with goldthioglucose (GTG), which causes brain lesions in the VMH and elsewhere, "behaviour sequences are broken off due to slight external disturbances to which the intact mouse would pay no heed" (p. 474). They conclude that the GTG syndrome "may be due to combined effects of injury to the mechanisms of several different behaviour systems, and in addition to disturbed interaction between systems" (p. 478 and 479).

Sclafani and Grossman (1971, p. 165) suggest that "the VMH may, in part, be concerned with the mediation of the influence of gustatory stimuli on consummatory behaviour. More specifically it may function to maintain consummatory behaviour in spite of noxious taste sensations, when the organism's energy or fluid reserves are low, and to curtail food and water intake in the presence of positive taste sensations when its needs are met".

As regards lateral (and posterior) hypothalamus, Caggiula (1970) suggests that "(posterior) hypothalamic stimulation may combine with and augment the behaviour-eliciting properties of sexually arousing environment stimuli, rather than "elicit" copulation directly" (p. 404), and considers "the possibility that 'non specific' arousal is channeled under the direction of internal and external environmental factors into specialized sensory-motor systems and that this conversion occurs at the preoptic-hypothalamic level" (p. 412). He further suggests that the hypothalamic systems underlying copulation and feeding are "themselves functionally specialized" (p. 411).

Devor *et al.* (1970, p. 231) support "the idea that (lateral) hypothalamic stimulation can trigger neural substrates which regulate normal feeding and drinking".

Panksepp (1971, p. 327 and 328), in a study of aggression elicited by LH stimulation in rats, obtained data that "superficially are more consistent with a fixed neural hypothesis than a plasticity hypothesis, though there are enough loop-holes to allow either viewpoint to be or not to be supported". Yet "the author feels there is more support for the idea of hypothalamic stimulation activating a non-specific neural system rather than overlapping systems associated with discrete motivated behaviours, because the electrically elicited response was probably not determined by specific functions of the tissue under the electrode but by the personality of the rat".

Roberts (1969, p. 339), reconsidering the specificity–plasticity controversy that has arisen around the behavioral effects of hypothalamic stimulation, concludes: "It appears that the stimulation potentiates the capacity of rather specific proximal cues to elicit specific consummatory responses, but approach responses toward many distal cues of goal objects or toward associated environmental cues must be learned through consummatory reinforcement".

Valenstein (1969) in another thoughtful survey of the specificity *versus* plasticity controversy stresses that "the behaviour produced by brain stimulation is in part the concomitant of the responses that tend to be dominant as a result of the interaction of such factors as environmental conditions, the presence of compelling stimuli, and a particular aroused animal. The term "prepotency" was suggested to summarize these factors that are major determinants of the behaviour elicited by stimulation" (p. 313). Therefore "it would appear more profitable to view the behaviour elicited by hypothalamic stimulation as prepotent responses initiated by the interaction of "general states" and environmental conditions prior to the postulation of motivational states related to biological needs". However, to this the author adds in italics: "*the term "general states" is not meant to imply that it will not be possible to differentiate the effects of stimulation at different hypothalamic regions, but rather that the application of specific terms such as hunger, thirst and sex may not be justified*" (p. 300). In a later paper (Valenstein *et al.*, 1970, p. 29) this view is stated as follows: "hypothalamic stimulation does not create hunger, thirst, or gnawing drives but seems to create conditions which excite the neural substrate underlying well-established response patterns. Discharging this sensitized or excited substrate, is reinforcing and it can provide the motivation to engage in instrumental behaviour which is rewarded by the opportunity to make the response".

In contradiction to this Wise (1969) maintains his earlier statement (Wise, 1968, p. 379) that the effects of LH stimulation are "consistent with the theory that separate fixed neural circuits, functionally isolated from each other by biochemical specificity, mediate eating and drinking in this area of the brain", but adds the important qualification that "not even the initial (presumably dominant) response is produced on the first trial. Rather, it develops with stimulation experience, (giving) the animal an opportunity to learn by trial and error just what acts and what goal objects are appropriate to the drive state (or states) elicited by the stimulation" (Wise, 1969, p. 931).

In spite of the *prima facie* appeal some of these interpretations may have, it appears that for the moment there is not a single point on which a more general theory of behavioral functions of the hypothalamus can be regarded as definitively accepted. This may explain why some recent publications, notably those on anatomical substrates of behavior, are restricted to precise statements on what concrete behavioral changes were noted after interruption of what pathways, without any attempt at generalization (Grossman, 1970, 1971, 1972; Paxinos and Bindra, 1973). Such data are certainly useful. Moreover, in several quarters attempts are under way to broaden the scope of behavioral observations, and to study the effects of hypothalamic manipulation under a variety of conditions (Valenstein, 1969). A general theory may therefore come within reach in the not too distant future.

Indeed one encouraging point already emerges. In many cases stimulation of the hypothalamus, particularly in the lateral, anterior and posterior areas, induces patterns of responsiveness closely similar to behavioral changes arising under the influence of some specific motivation (Miller, 1957, 1961). For instance, in the case of sexual behavior elicited by posterior hypothalamic stimulation in the rat, "the stimulated male did not attempt to mount indiscriminately but instead courted the female by nibbling her ears, nipping the back of the neck, and chasing her around until lordosis occurred. Copulation would then ensue. The behavior was also fully motivated in terms of learning and performing an instrumental response to gain access to the female" (Hoebel, 1969). In line with these facts, ICSS at such a stimulation site proved to be positively correlated with the male hormone level in the subject (Caggiula, 1970).

Moreover, the effects of stimulation parallel those of normal motivational changes also as regards interaction of the stimulated system with other behaviors. We have already seen this above in the case of mutual inhibition of feeding and drinking. Again, in the male rat the stimulus bound sexual behavior just mentioned is inhibited by simultaneous stimulation of a lateral hypothalamic feeding site, and *vice versa* (Caggiula, 1970). A further example, extensively analyzed from the behavioral angle, was found by Koolhaas in our laboratory. He showed that in the male rat hypothalamic stimulation at a particular point (Koolhaas, 1973) will elicit completely normal intraspecific aggressive behavior *versus* another male, if the latter is not dominant over the stimulated male. If the opponent is dominant, the same stimulus will induce fleeing. Finally, if an estrus female is presented, the main effect of the stimulation is a practically complete suppression of the sexual responses the male would normally perform in this situation, whereas only very little overt aggression is now elicited by

References p. 505–507

the stimulus. Apparently stimulation at this site induces a functional state of the brain entirely comparable to that which is normally obtained in "agonistic" behavioral situations.

Caution is required in the interpretation of such findings, because most commonly used stimulation techniques are crude in comparison to the structure of the brain (Roberts, 1969). In terms of the concepts schematized in Fig. 5b, there are several possible explanations. For instance, if the stimulus increases P_o^A, this may be so either because it successfully mimics *a* or α, or because it gives rise in the interaction system (I) to an i pattern favoring A. Most, if not all of these messages presumably are encoded in neural activities too complex for straightforward simulation by electrical stimulation. Therefore, if an integrated behavioral effect results from the stimulus, this may well be largely because the brain itself gives a meaning to the noise we put in. One may, of course, have some misgivings how far this capacity of the brain goes, and whether the stimulation may not disrupt its normal functioning, perhaps especially in the case of the presumably very complex I mechanisms. In many cases, however, behavior under stimulation appears so well integrated that we will for the moment assume that the stimulation in some cases does not disturb normal functions but merely adds information.

Even then the 3 possible interpretations are not easily distinguishable operationally in experiments restricted to behavioral effects of brain manipulation. It is true that ethologists are making efforts (Hinde, 1970) to differentiate operationally the cases where a given behavior is furthered by an increase in its specific motivation from those where its occurrence is due to such interaction processes as, *e.g.* "disinhibition" (McFarland, 1971). Even if these attempts would result in criteria that are generally applicable, this would still not completely solve the physiological problem we are facing here. For, if it were thereby established that an increase in P_o^A is due to artificial manipulation of i (Fig. 5b) rather than a or α, the question remains whether I has been stimulated directly, or via b or β. It will already be clear at this stage that several statements made on page 485 as to the neural mechanism of feeding behavior must be carefully reconsidered; some further complications will follow below.

In some cases theoretical analysis of the expected behavioral effects of manipulation of individual motivational signals may suggest observations that can provide at least a partial answer to questions of this kind, as we have shown elsewhere in the case of the behavioral effects of VMH lesions (de Ruiter and Wiepkema, 1969). A more direct approach, however, will be indispensable. Here we shall have to wait for the further development of electrophysiological techniques suitable for defining the functional state of the behavioral mechanisms of the brain as determined by all the inputs it receives (including possible direct artificial stimulation).

Perhaps the main point in these remarks, and certainly an obvious one at least in retrospect, is that they envisage the behavioral output of the brain as dependent on the sum total of all messages impinging upon it. In this context, the recent controversy over specificity or plasticity of hypothalamic mechanisms of behavior, which looms large in the above sample of the literature, appears almost unnecessary. The concept of specific networks stems from an approach to behavioral physiology that, although of proven heuristic value, is essentially naive. This approach, exemplified by pages 484–491 of the present paper, assumes that each behavior system is controlled by its own set of "specific" messages. A specific network, then, is a neuronal system dealing exclusively with such a set of messages. However, the common phenomenon of interaction makes this concept untenable in its strict form, for it shows that at some stage in the control mechanism other messages must come in. From that stage on, the network is no longer specific, the activity of its neurons being affected by a host of other signals in addition to the messages that on common sense grounds one may continue to consider "specific" for the behavior system in question (*e.g.* osmotic signals for drinking), although they have profound effects on other behaviors as well (*e.g.* feeding). The physiological data now available on descending control of transmission in the afferent systems of the brain (Milner, 1970; Pribram, 1971) suggest that this intermingling of specific and non-specific messages may occur as a rule at many stages in the control systems, beginning at the very entrance of the CNS or even in the sense organs themselves. However that may be, the main point is that adherents of a specific network theory will find themselves compelled sooner or later to admit that the functional state of the entire complex of motivational networks, rather than a single one, determines the behavioral output. This is true of behavior under local artificial brain stimulation, just as much as the normal activities of the undisturbed animal.

Starting point of the "plasticity" hypothesis was the finding that manipulation of the subject's environment may completely change the behavior elicited by artificial stimulation of its brain (Valenstein *et al.*, 1970). For instance, a rat that was a stimulus-bound feeder at first when its LH was stimulated in the presence of food and water, could be turned into a stimulus-bound drinker by prolonged exposure to the stimulus with only water present; when the food was subsequently returned, the brain stimulation elicited both feeding and drinking. On the basis of such examples the view was put forward that local artificial stimulation of the hypothalamus (and certain other brain areas) will elicit any response that happens to be "prepotent" in the subject due to the general conditions of the experiment, *i.e.* the overall motivational state of the subject. On the other hand this so-called plasticity is limited in two respects (see also the above quotation from Valenstein, 1969).

Firstly, as regards neural mechanisms there is some topographic specialization (Roberts, 1969, Caggiula, 1970), *e.g.* in the sense that in a male rat equipped with two electrodes, one in the lateral and one in the posterior hypothalamus, stimulation via the former will often reliably elicit feeding (and inhibit sexual behavior), whereas the other electrode has the opposite effects. Secondly, as regards the overt manifestations, it appears from these studies that there are limits to the range of responses elicitable by artificial stimulation. Of particular interest in this connection is whether this range coincides with the range of the possible effects of a particular natural motivational state. We touch here on the concept of "moods" as used by von Holst and von Saint Paul (1963); or of a "thematic stream" that unifies the responses elicited by local brain stimulation (Valenstein, 1969, p. 298). The latter author was unable to discover any simple theme in his results, yet he points out that the literature contains suggestions

References p. 505–507

that stimulation of some regions "may produce a predictable direction of change in mood", and in a later paper he quotes von Holst and von Saint Paul's statement with implicit approval (Valenstein *et al.*, 1970). In view of the paucity of data on this important point it is worth stressing that in Koolhaas' study mentioned above the stimulus appears to bring about a completely natural "agonistic mood".

In practice, then, there is little difference between the more sophisticated versions of the 2 hypotheses. Much more work will be needed, however, to elaborate them into a useful model. On the behavioral side, the formal rules must be defined which relate total input as represented, *e.g.*, by a vector in a multidimensional space, to behavioral output. On the neural side, there is the problem of what wiring patterns of brain cells underlie the formal rules. We will only discuss one obstacle that must be overcome to answer the latter question.

We have so far assumed that experimental manipulation (*e.g.* artificial stimulation) sets up patterns of activity in the brain that are essentially similar to those that occur under normal operating conditions of the organ. If this condition is not fulfilled, it will become even more difficult to elucidate the wiring pattern of the machinery of behavior. How can we check whether our manipulations do or do not derange the machine itself? We will consider this problem in connection with the functioning of the interaction system (I in Fig. 5b). For brevity, we shall use the term I-damage for functional impairment of this system. It appears, then, that I-damage may have a variety of effects, some of which we shall discuss here. This discussion will focus on the possibility that I-damage causes delays and increased randomness in response selection, as there appear to be concrete examples of this. There is of course the alternative possibility that I-damage may cause the CNS to "lock" into a single functional state, but this case may be very hard to distinguish operationally from that where brain stimulation, instead of interfering with I, mimics a strong positive specific motivational signal. We shall mention 4 different, but related aspects of the former kind of I-damage, giving an example of each.

(1) Changes in relative response frequencies. If each of the functional states among which the brain normally switches back and forth is characterized by high P_o for the elements of one behavioral system, and low P_o for all other systems, one possible manifestation of I-damage is that for one or more states the difference between the high and low values of P_o becomes less marked. If so, behaviors that would normally be preponderant under the conditions chosen will now occur less frequently, whereas responses that would normally be rare will be performed relatively often. This prediction seems to fit, *e.g.*, the effects of atropine inhibition of the VMH, which appears to result in an increase of various low-rate behaviors, and a decrease of high-rate responses (Grossman, 1966; Miczek and Grossman, 1972).

(2) Changes in bout length. We have seen above that mean length of uninterrupted bouts is a good measure of P_o. If the effect of I-damage is to make response selection more random (*i.e.* decrease the P_o differentials) we may expect bouts of dominant behavior to be abnormally short. A clearcut example of this was found by one of us (Wiepkema, 1968) in the sleeping behavior of mice with GTG-induced brain lesions. Sleeping bouts are shortened here to about one third of their normal duration,

whereas total duration of sleep has increased up to 1.4 times the normal value (*i.e.* to about half of total available time). This may mean that in the lesioned mouse the "sleeping state" is switched on longer than normal, but that the switch is less effective in favoring sleep and suppressing other behaviors.

(3) Incomplete sequences of behavior. Particularly in the case of social behavior, fulfillment of a function may require performance of many different elements. The order in which these appear depends on the reactions of the partner. Therefore, the actual course of events may vary widely from case to case, but in broad outline certain more or less fixed sequences may be discerned, such as the sequence approach–courtship–mounting–ejaculation–genital grooming in male sexual behavior in rodents. Obviously the effects on such response sequences of I-damage of the type here considered will be to decrease the probability that a sequence once begun will also be completed. We have commented before (de Ruiter and Wiepkema, 1969) on the incompleteness of sequences of sexual and aggressive behavior in the GTG-treated male mouse, but a better documented example can now be taken from a study by one of us (Veening, in press) on the effects of unilateral stimulation of the VMH in male rats. Fig. 6a presents the results of sequential analysis of the normal behavior of the non-stimulated male in the presence of food and an estrus female. Two sequences can be clearly discerned: a feeding cycle, and a sexual cycle. The non-stimulated male devotes some 60% of its total time to these activities; the greater part of the remainder goes to resting, grooming, and various sundries. Fig. 6b illustrates better than many words that the sequences are badly disrupted by VMH stimulation. Only one aspect of this must be emphasized in the present context: it is particularly the frequency (calculated over the total duration of the experiments, see also below) of mounting, one of the terminal links of the sexual sequence, that is severely reduced (about 15 times), whereas that of approaches to the female is reduced only about 4 times. Clearly the risk that an initiated sexual cycle will break off before completion has increased about 4-fold. It may be added, as a further example of decreased bout length (see point 2) that if a sequence does reach completion, the bout length of *e.g.* genital grooming in the sexual cycle, or of actual eating in the feeding cycle, is shorter than normal.

Before we pass on to the final example, it must be emphasized that the 3 points presented so far prove that there are experimental manipulations that impair the stability of behavior as expressed in the activity measures here discussed. This fact provides an operational basis for the concept of stability.

(4) Prolonged suspension. We have so far assumed that the brain switches from one functional state to another in a rapid, decisive manner, and that the only overt sign of the transition is a sudden shift in P_o values. However, it appears from the rat study just mentioned that the transition may involve a state of suspension with a quite distinct overt manifestation, *viz.* the behavior called scanning. The scanning animal is awake and alert, but sitting or standing quietly, moving its head from side to side in a manner suggesting that it is gathering all sorts of visual, auditory and olfactory information. This behavior is not oriented to any particular stimulus source. The morphology of scanning therefore suggests that it is the expression of a state of

References p. 505–507

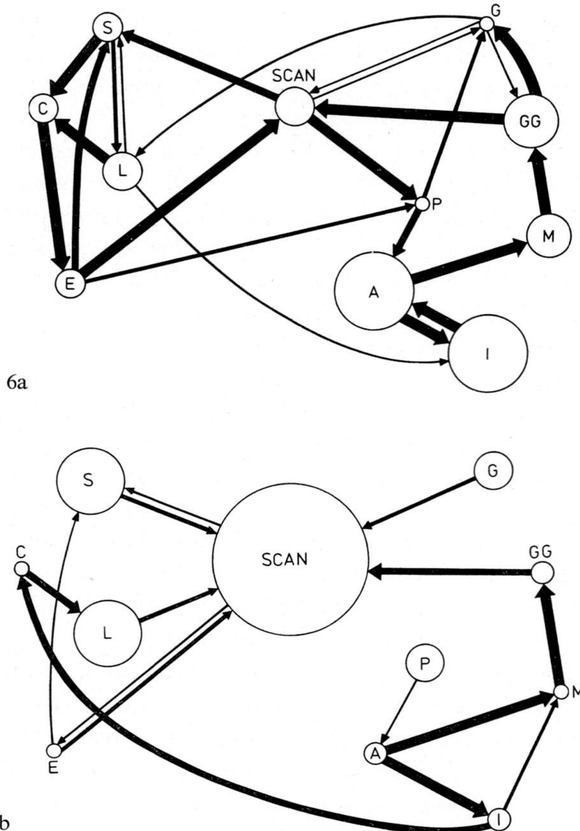

Fig. 6. a: summary of the structure of normal behavior of a male rat in the presence of food and an estrous female. Elements: SCAN, scanning; L, locomotion; S, sniffing; C, carrying and sniffing at food; E, eating; P, attend to partner; A, approaching and following partner; I, investigating partner; M, mounting; GG, grooming genitals; G, grooming. These elements together take up over 75% of total time; of the remaining 25% about half goes to resting, the rest to rarer elements omitted for the sake of clarity. Size of circle for each element indicates total time spent on that element. Arrows indicate preferential sequences in behavior, *i.e.* transitions from one element to another seen more frequently than expected on the assumption that elements are performed in random order. Width of arrow denotes degree of preference (other sequences, performed not more often, or actually less often than expected are omitted from the diagram). Starting from the scanning state, the rat has two clear preferred cycles: a feeding cycle (SCAN → S → C → E → SCAN) and a sexual cycle (SCAN → P → (A ⇄ I) → M → GG → SCAN). b: modified structure of behavior during unilateral electrical stimulation of the VMH (*cf.* Fig. 6a). Note the following changes. Much more time is devoted to scanning. As regards the feeding and sexual cycles, more time is spent on early (appetitive) elements, such as L, S, and P, and much less time on terminal (consummatory) elements (E, M). Some sequences normally preferred (*e.g.* S → C → E) are no longer so under stimulation; on the other hand some new preferential sequences are seen indicating that the normal cycles are easily interrupted (I → C, C → L).

suspension of activities. The results of sequential analysis (Fig. 6a) also indicate that when scanning occurs, this marks the time when the animal must decide what to do next. Sometimes, however, behavior cycles may follow each other without a noticeable pause for scanning. Now, to take again the case of the male rat in the presence of food and an estrus female, it is interesting that scanning, which normally occupies some 4% of total time, is increased by VMH stimulation to as much as 40–50%. This is due partly to an about 7-fold increase in frequency of occurrence of scanning partly also to doubling of the mean duration of scanning bouts. The former effect is explicable to a large extent from the fact that many sequences are prematurely broken off, each time necessitating a renewed decision on the part of the male. It is, however, the increased duration of the individual scanning bouts that is of interest here, as it suggests that the decision process itself is slowed down by the artificial stimulus (either because the I-system is actually temporarily impaired, or because the stimulus presents it with a puzzling "nonsense" input). Finally, in connection with point 3 above, it is worth pointing out that due to the increased amount of time spent in suspension, time actually devoted to sexual behavior cycles is reduced by about 70%. This means that during the time sexual behavior is switched on approaches to the female occur no less often with than without VMH stimulation. In other words, the motivation level appears not to be changed, but only response selection is abnormally slow.

We believe that these examples constitute a strong warning that in experiments on the neural basis of behavior the possibility of disturbance of decision processes, resulting in increased randomness of response selection, must be seriously considered. One may speculate that it makes no great difference whether such disturbance is caused by artificial stimulation or by lesions in the I-system. In this connection it is of interest that McLean (1969) has already commented on the similarity of the behavioral effects of these two kinds of interference when applied to the hypothalamus.

Disturbance of response selection in principle is an effect quite independent of the specific motivational changes that brain stimulation may also induce. Operationally, however, the two effects are not easy to separate. Taking P_o as a measure of motivation, one should realize that increased randomness will also affect this, and it will do so in opposite directions for high-rate and low-rate behaviors. We have no readymade, generally applicable answer to these problems, but the above examples indicate that in a concrete case analysis of bouts, sequences, delays in selection etc., may provide at least some clues.

CONCLUDING REMARKS

For the present, development of a general physiological model of behavior, as defined in the beginning of this paper, may seem too ambitious an aim, but if one therefore restricts oneself to attempts to predict the time pattern of activities subserving a single function (*e.g.* food intake), it turns out that one must nevertheless take into account the entire motivational state of the subject. We encountered this for instance in the experiments on the VMH, from which examples 2–4 in the preceding section were taken, but which originally aimed only at elucidating the satiety function of this area.

In terms of neural substrates, this of course does not disprove that parts of the system controlling responsiveness may be "specific" in the sense that they handle only one kind of message (*e.g.*, on the input side an internal satiety signal or an external

References p. 505–507

food signal, or on the output side a simple motor command such as "mount"). However, it seems that these specific parts form a minor fraction of the control system as a whole, and that cellular activity in most of that system will prove to be influenced by a wide variety of messages, relevant to many different functions.

It seems an inadequate model to consider the control system as a mosaic of specific networks, only one of which is allowed to "speak" at a time. We prefer to view the system as a unitary complex, which can exist in a number of discrete states. Physiological analysis of these states is hardly feasible at present, but they can be precisely described in terms of behavioral input–output relations of the individual. For this purpose the concepts of element of behavior and behavioral system prove useful.

For the further elucidation of the neural mechanisms it is a fact of great importance that manipulation of selected brain sites can induce precisely defined, normal functional states of the control system. That the same effect can be obtained in multiple sites is only to be expected as all important brain functions appear to be entrusted to extensive networks.

However, from the phenomenon of specific effects it cannot be concluded that the site manipulated is part of a specific network in the above sense. This phenomenon proves merely that cells at that site are critically involved in bringing about a particular state of the complex. In principle, only electrophysiological investigation can reveal the range of messages handled by these cells. Yet the fact that in many sites the effect depends partly on what responses are "prepotent" in the individual contains at least a warning against hasty conclusions as regards specificity.

A major obstacle for the further development of physiological models of behavior may lie in the danger that the manipulations necessary for elucidation of the functional properties of the brain derange its subtle mechanisms so much that the patterns we find may bear but little meaningful relation to those underlying the normal functions of the undisturbed brain. Much further work is needed on this question. We feel that it is important, therefore, that our above examples indicate that this point is not altogether beyond the reach of operational verification through ethological observation.

Finally, such different disciplines as functional neuroanatomy (Wall, 1972) and ethology appear to converge to a similar concept of the brain as a vast functional whole that can be programmed for one or another input–output relation, depending on the exigencies of survival in the ceaselessly varying situations with which the natural environment confronts the individual. It seems to us that this convergence holds promise for the eventual development of satisfactory physiological models of behavior.

SUMMARY

In terms of ethological concepts some features of behavior, notably response selection and stability, are outlined that physiological models of behavior will have to account for.

Behavioral stability is attributed partly to graded positive specific motivational feedback messages from performance of responses, and partly to mutual inhibitory interactions between behavior systems.

A picture emerges of the causal mechanism of behavior as one unitary complex that can be switched from one to another relatively stable functional state. These transitions occur when the total input (conceived *e.g.* as a vector in multidimensional space) passes through certain critical states.

Throughout the paper behavioral and physiological studies of feeding behavior are used as typical examples.

No adequate predictive model can be based on presently available data on the physiology of behavior. Combined behavioral and physiological investigations, guided by systems-theoretical analysis, will be needed to achieve a satisfactory model.

As regards the hypothalamus, although no general theory of its behavioral function can yet be constructed, it is argued that the apparently conflicting views on "specificity" or "plasticity" of its neuronal networks are bound to converge on further analysis of the facts. The specific behavioral functions of the hypothalamus probably are subserved by a general network (with specific inputs). Some behavioral outputs of the hypothalamus seen in experiments aiming at revealing its normal functions probably reflect abnormal functional states of its response selecting mechanisms, and might therefore give rise to erroneous conclusions, but careful ethological analysis may provide safeguards on this point.

REFERENCES

ANAND, B. K. (1967) Central chemosensitive mechanisms related to feeding. In *Handbook of Physiology, (Section 6, Vol. I, Ch. 19)*, F. Code (Ed.), American Physiological Society, Washington, D.C., pp. 249–270.

CAGGIULA, A. R. (1970) Analysis of the copulation-reward properties of posterior hypothalamic stimulation in male rats. *J. comp. physiol. Psychol.*, **70**, 399–412.

CAMPBELL, J. F., BINDRA, D., KREBS, H. AND FERENCHAK, R. P. (1969) Responses of single units of the hypothalamic ventromedial nucleus to environmental stimuli. *Physiol. Behav.*, **4**, 183–187.

CANE, V. (1961) Some ways of describing behaviour. In *Current Problems in Animal Behaviour*, W. H. THORPE AND O. L. ZANGWILL (Eds.), Cambridge.

CODE, F. (Ed.) (1967) Control of food and water intake. In *Handbook of Physiology*, *Section 6, Vol. I*, American Physiological Society, Washington, D.C., pp. 1–459.

COLLIER, G. (1969) Body weight loss as a measure of motivation in hunger and thirst. *Ann. N.Y. Acad. Sci.*, **157**, 594–609.

COONS, E. E. AND CRUCE, J. A. F. (1968) Lateral hypothalamus. Food and current intensity in maintaining self-stimulation of hunger centre. *Science*, **159**, 1117–1119.

DEBONS, A. F., KRIMSKY, I. AND FROM, A. (1970) A direct action of insulin on the hypothalamic satiety center. *Amer. J. Physiol.*, **219**, 938–943.

DEVOR, M. G., WISE, R. A., MILGRAM, N. W. AND HOEBEL, B. G. (1970) Physiological control of hypothalamically elicited feeding and drinking. *J. comp. physiol. Psychol.*, **73**, 226–232.

FITZSIMONS, J. T. (1957) Normal drinking in rats. *J. Physiol. (Lond.)*, **138**, 39P.

FITZSIMONS, J. T. AND LE MAGNEN, J. (1969) Eating as a regulatory control of drinking in the rat. *J. comp. physiol. Psychol.*, **67**, 273–283.

FONBERG, F. (1967) The motivational role of the hypothalamus in animal behaviour. *Acta Biol. exp. (Warszaw)*, **27**, 303–318.

GEERTSEMA, S. P. (1973) *Ontwikkeling, analyse en toepassingen van enige modellen der regulatie van voedselopname*, *Thesis* (with an English summary), Groningen.

GROSSMAN, S. P. (1966) The VMH: a center for affective reactions, satiety, or both? *Physiol. Behav.*, **1**, 1–10.

GROSSMAN, S. P. (1967) Neuropharmacology of central mechanisms contributing to control of food and water intake. In *Handbook of Physiology, Section 6, Vol. I*, F. CODE (Ed.), American Physiological Society, Washington, D.C., pp. 287–312.

GROSSMAN, S. P. (1971) Changes in food and water intake associated with an interruption of the anterior or posterior fiber connections of the hypothalamus. *J. comp. physiol. Psychol.*, **75**, 23–31.

GROSSMAN, S. P. (1972) Aggression, avoidance and reaction to novel environments in female rats with ventromedial hypothalamic lesions. *J. comp. physiol. Psychol.*, **78**, 274–283.

HAMILTON, C. L. (1967) Food and temperature. In *Handbook of Physiology, Section 6, Vol. I*, F. CODE (Ed.), American Physiological Society, Washington, D.C., pp. 303–318.

HINDE, R. A. (1970) *Animal Behaviour*. Second edition, McGraw-Hill, New York, 876 pp.

HOEBEL, B. G. (1968) Inhibition and disinhibition of self-stimulation and feeding. *J. comp. physiol. Psychol.*, **66**, 89–100.

HOEBEL, B. G. (1969) Feeding and self-stimulation. *Ann. N.Y. Acad. Sci.*, **157**, 758–778.

HOEBEL, B. G. (1971) Feeding: neural control of intake. *Ann. Rev. Physiol.*, **33**, 533–568.

HOEBEL, B. G. AND THOMPSON, R. D. (1969) Aversion to lateral hypothalamic stimulation caused by intragastric feeding or obesity. *J. comp. physiol. Psychol.*, **68**, 536–543.

HOOFF, J. A. R. A. M. VAN (1971) *A Structural Analysis of the Social Behaviour of a Semi-captive Group of Chimpanzees*, Thesis, Utrecht.

HOLST, E. VON AND SAINT PAUL, V. VON (1963) On functional organisation of drives. *Anim. Behav.*, **11**, 1–20.

KISSILEFF, H. R. (1970) Free feeding in normal and recovered lateral rats monitored by a pellet-detecting eatometer. *Physiol. Behav.*, **5**, 163–173.

KOOLHAAS, J. M. (1974) Intraspecific aggression elicited by electrical stimulation of the lateral hypothalamus in the rat. *Brain Res.*, **66**, 364.

MAGNEN, J. LE ET TALLON, S. (1966) La périodicité spontanée de la prise d'aliments *ad libitum* du rat blanc. *J. Physiol. (Paris)*, **58**, 323–349.

MARGULES, D. L. AND STEIN, L. (1969) Cholinergic synapses of a periventricular punishment system in the medial hypothalamus. *Amer. J. Physiol.*, **217**, 475–480.

MCFARLAND, D. J. (1970) Adjunctive behaviour in feeding and drinking situations. *Rev. Comp. Anim.*, **4**, 64–73.

MCFARLAND, D. J. (1971) *Feedback Mechanisms in Animal Behaviour*. Academic Press, London, 279 pp.

MCLEAN, P. D. (1969) The hypothalamus and emotional behaviour. In *The Hypothalamus*, W. HAYMAKER, E. ANDERSON AND W. J. H. NAUTA (Eds.), Thomas, Springfield, Ill.

MILLER, N. E. (1957) Experiments on motivation. *Science*, **126**, 1271–1278.

MILLER, N. E. (1961) Analytical studies of drive and reward. *Amer. Psychol.*, **16**, 739–754.

MILNER, P. M. (1970) *Physiological Psychology*. Holt, Rinehart and Winston, London, 531 pp.

MICZEK, K. A. AND GROSSMAN, S. P. (1972) Punished and unpunished operant behavior after atropine administration to the VMH of squirrel monkeys. *J. comp. physiol. Psychol.*, **81**, 318–330.

MORGANE, P. J. (1969) Neural regulation of food and water intake. *Ann. N.Y. Acad. Sci.*, **157**, 531–1216.

NICOLAIDIS, S. (1969) Early systemic responses to orogastric stimulation in the regulation of food and water balance, functional and electrophysiological data. *Ann. N.Y. Acad. Sci.*, **157**, 1176–1203.

OOMURA, Y., KIMURA, K., OOYAMA, H., MAENO, H., IKI, M. AND KUNIYOSHI, M. (1964) Reciprocal activities of the ventromedial and lateral hypothalamic areas of cats. *Science*, **143**, 484–485.

PANKSEPP, J. (1971) Aggression elicited by electrical stimulation of the hypothalamus in albino rats. *Physiol. Behav.*, **6**, 321–329.

PANKSEPP, J. (1971) A re-examination of the role of the ventromedial hypothalamus in feeding behavior. *Physiol. Behav.*, **7**, 385–394.

PAXINOS, G. AND BINDRA, D. (1973) Hypothalamic and midbrain neural pathways involved in eating, drinking, irritability, aggression and copulation in rats. *J. comp. physiol. Psychol.*, **82**, 1–14.

PRIBRAM, K. H. (1971) *Languages of the Brain*, Prentice Hall, Englewood Cliffs, N.J.

RABIN, B. M. (1972) Ventromedial hypothalamic control of food intake and satiety: a reappraisal. *Brain Res.*, **43**, 317–342.

REDDINGIUS, J. AND GEERTSEMA, S. P. (1974) Preliminary considerations in the simulation of behaviour. In *Motivational Control Systems Analysis*, D. J. MCFARLAND (Ed.), in press.

ROBERTS, W. W. (1969) Are hypothalamic motivational mechanisms functionally and anatomically specific? *Brain Behav. Evol.*, **2**, 317–342.

RUITER, L. DE (1967) Feeding behaviour of vertebrates in natural environment. In *Handbook of*

Physiology, Section 6, Vol. I, ch. 7, F. CODE (Ed.), American Physiological Society, Washington, D.C., pp. 97–116.
RUITER, L. DE (1968) Feeding behaviour and the hypothalamus. *Proc. IUPS*, **6**, 198–199.
RUITER, L. DE (1971) In *Vergelijkende Dierfysiologie*, S. DIJKGRAAF EN H. J. VONK (Eds.), Oosthoek, Utrecht.
RUITER, L. DE AND WIEPKEMA, P. R. (1969) The goldthioglucose (GTG) syndrome in mice. *Psychiat. Neurol. Neurochir. (Amst.)*, **72**, 455–480.
RUITER, L. DE, WIEPKEMA, P. R. AND REDDINGIUS, J. (1969) Ethological and neurological aspects of the regulation of food intake. *Ann. N.Y. Acad. Sci.*, **157**, 1204–1216.
SCLAFANI, A. AND GROSSMAN, S. P. (1971) Reactivity of hyperphagic and normal rats to quinine and electric shock. *J. comp. physiol. Psychol.*, **74**, 157–166.
STEFFENS, A. B. (1969a) Blood glucose and FFA levels in relation to the meal pattern in the normal rat and the ventromedial hypothalamic lesioned rat. *Physiol. Behav.*, **4**, 215–225.
STEFFENS, A. B. (1969b) The influence of insulin injections and infusions on eating and blood glucose level in the rat. *Physiol. Behav.*, **4**, 823–828.
STEFFENS, A. B. (1970) Plasma insulin content in relation to blood glucose level and meal pattern in the normal and hypothalamic hyperphagic rat. *Physiol. Behav.*, **5**, 147–151.
STRUBBE, J. H. (1974) *A Reappraisal of the Glucostatic Theory*. Thesis, Groningen, in press.
THOMAS, G. (1974) The influences of encountering a food object on subsequent searching behaviour in *Gasterosteus aculeatus* L. *J. Anim. Behav.*, **22** (in press).
TEITELBAUM, P. AND CAMPBELL, B. A. (1958) Ingestion patterns in hyperphagic and normal rats. *J. comp. physiol. Psychol.*, **51**, 135–141.
VALENSTEIN, E. S. (1969) Behavior elicited by hypothalamic stimulation. *Brain Behav. Evol.*, **2**, 295–316.
VALENSTEIN, E. S., COX, V. C. AND KAKOLEWSKI, J. W. (1970) Re-examination of the role of the hypothalamus in motivation. *Psychol. Rev.*, **77**, 16–31.
WALL, P. D. (1972) Somatosensory pathways. *Ann. Rev. Physiol.*, **34**, 315–336.
VEEMING, J. G., *Behavioural Function of the VMH*, Thesis, Groningen, in press.
WIEPKEMA, P. R. (1961) An ethological analysis of the reproductive behaviour of the bitterling (*Rhodeus amarus* Bloch). *Arch. neerl. zool.*, **14**, 103–199.
WIEPKEMA, P. R. (1968) Behaviour changes in CBA mice as a result of one goldthioglucose injection. *Behaviour*, **32**, 179–211.
WIEPKEMA, P. R. (1971a) Behavioural factors in the regulation of food intake. *Proc. Nutr. Soc.*, **30**, 142–149.
WIEPKEMA, P. R. (1971b) Positive feedbacks at work during feeding. *Behaviour*, **39**, 266–273.
WISE, R. A. (1968) Hypothalamic motivational systems: fixed or plastic neural circuits? *Science*, **162**, 377–379.
WISE, R. A. (1969) Plasticity of hypothalamic motivational systems. *Science*, **165**, 929–930.
WYRWICKA, W. (1969) Sensory regulation of food intake. *Physiol. Behav.*, **4**, 853–858.

General Discussion

LEVINE: In view of the statements made during this summer school I should like to warn against the concept of hypothalamic "centers" that has been used during this conference. I consider the hypothalamus a part of a larger, integrated system. The rest of the brain is not only there to keep the hypothalamus warm! I think we lose a lot of insight in the action of the system by looking at isolated centers. More recently the whole concept of feeding centers is being questioned, and even the actual existence of feeding centers is in doubt. This is certainly a thing we need to be concerned about.

ARIËNS KAPPERS: Would it not be better to talk about "a link in a chain", or to use the word "circuit"?

JÜRGENS: Even if one were to speak about a circuit, one must admit that there are critical points, since there are places where, if the circuit is disrupted, the function is totally lost, while at other places it is disturbed only to a minor degree.

DE RUITER: Not even when talking to my students do I use the word "center", because I think it is a bad term. I believe that whatever function is carried out by the central nervous system, it will always be a matter of the entire brain. I don't think therefore that there is any behavioral function that is the exclusive property of the hypothalamus. I expect that when we know more about the electrophysiological activities of neurons during behavior, we will find changes in activity patterns anywhere in the brain when behavioral events are occurring. In this respect the concept of "centers" has gone quite out of fashion.

WAYNER: The hypothalamus has many functions and this fact was also emphasized during the general discussion of the Calgary conference in May 1973 in Alberta, Canada. Therefore I believe we should ask the question: can you think of any function that does *not* have a hypothalamic involvement? Because the hypothalamus is involved in motor functions, both autonomic and somatic, it tends to be involved whenever the animal does something.

In addition, the hypothalamus has multiple afferent inputs which further extends the involvement. As Professor Lammers pointed out, these are polysynaptic connections with the exception of the visual input.

DE RUITER: Am I right in assuming that your view on hypothalamic functioning is largely one of causing an increased activity in motor systems?

WAYNER: Yes, an increase in the excitability of the reflex pathways at the level of the spinal cord.

DE RUITER: But this concerns a rather general increase.

WAYNER: It is the general increase in excitability at that level, which forces one to think about non-specificity.

DE RUITER: On that point I believe Dr. Jürgen's presentation was more balanced than mine, because he, at least, introduced the concept of a general arousal reaction that may occur during stimulation.

JÜRGENS: There is no doubt that a general arousal reaction does occur after stimulation of the lateral hypothalamus — which means that the probability of occurrence is raised for many behavior patterns simultaneously and not only for one single pattern. Nevertheless, this does not exclude that the stimulation facilitates some behavior patterns more than others.

WAYNER: I might be able to explain how the general component occurs. At the level of the spinal motor neuron, the excitability level is high during lateral hypothalamic stimulation. In terms of behavior, this means that the probability that any response in the animal's repertoire will occur to any stimulus is increased. The increased excitability can be demonstrated by the increase in the evoked ventral root reflex discharge.

DE RUITER: But surely not all responses are facilitated to the same degree. To this extent, the effects of lateral hypothalamic stimulation are specific.

DE WIED: We still have to come back to Prof. De Ruiter's presentation about insulin antibodies. If you put antibodies against insulin into a certain brain structure and you obtain an effect, then you should be able to get the opposite effect by adding insulin to that structure.

DE RUITER: Before answering this I should perhaps explain the background of this work, which I mentioned only briefly in my presentation. The story goes back to Jean Mayer's glucostatic hypothesis, which was based originally on rather scanty evidence for glucoreceptors in the ventromedial hypothalamus. He thought that the receptors would be activated by the presence of glucose in the blood. If sufficiently stimulated, the receptors would activate the ventromedial hypothalamic satiety system, which in turn would terminate feeding behavior. Many investigators have tried to verify this by raising blood glucose levels simply by injecting glucose into the blood, but the results of those experiments were quite ambiguous: sometimes there was a satiety effect, sometimes there was not. Mayer then refined his hypothesis. He stated: it is not so much the glucose concentration *per se*, it is rather the availability of glucose to tissue cells. This availability of glucose for the cells depends on two things: presence of glucose, and presence of insulin. The criticism was raised, however, that it is a well-known fact that brain cells do not need insulin in order to take up glucose. Mayer then added to his hypothesis that the glucoreceptors would be an exception to this rule, because they would need insulin in order to take up glucose. Satiety would then occur only if both glucose and insulin are present. There is some evidence that indeed this may be true. If you administer for instance a poisonous glucose-derivative (gold-thioglucose), this damages the ventromedial hypothalamus. The same compound given to an alloxan diabetic animal will not cause such damage. If you administer both insulin and GTG to this animal you will get the gold-thioglucose damage again. This fits in with Mayer's ideas.

ARIËNS KAPPERS: Are the receptor cells located on the ventromedial hypothalamic neurons?

DE RUITER: A more recent hypothesis is that they are glial cells rather than neurons.

WAYNER: Your work is on the mouse, isn't it?

DE RUITER: You can get the same effect, as shown by Smith, with gold-thioglucose in the rat when you administer it directly into the brain.

WAYNER: You reported that, if you destroy the basal medial hypothalamus bilaterally, the animal still eats and responds to insulin. But if you leave the basal medial hypothalamus intact and then destroy the lateral hypothalamus bilaterally, the animal will no longer eat in response to insulin. This provides evidence for the existence of cells in the lateral hypothalamus which are critical for eating. Some recent evidence which is relevant in this context is the elegant work of Prof. Oomura in Japan. He recorded intracellularly from several cells in the lateral hypothalamus of the rat. He was able to deposit glucose extracellularly and results indicate very clearly that when glucose is taken up by the cell, it enhances the operation of the sodium pump. The cell became hyperpolarized, firing rate decreased, but the membrane conductance did not change. As ion permeability did not change, the hyperpolarization was due to the sodium pump. These cells in the lateral hypothalamus seem to be very sensitive to glucose. Latency was long and explained in terms of some electron microscopy data. Oomura stated that a collaborator found cells which had thick coatings of glia in this region of the lateral hypothalamus and that the glucose has to get through the glia somehow before it gets into the cell from which they are recording. The existence of these cells seems to be well established, at least in the lateral hypothalamus. Some of these cells which decrease in discharge frequency in response to glucose increased when free fatty acids were applied, indicating that these cells probably respond under normal conditions to changes in the ratio of at least these two substances. Insulin might be involved in getting the glucose across the membrane.

DE RUITER: That fits in beautifully with Steffen's data on free fatty acids, which go up when the animal becomes hungry, whereas glucose goes down. As regards the VMH lesioned rat responding to insulin, as you put it, this — as indicated by Steffen's results — is not a response to insulin *per se*, but to insulin-induced hypoglycemia. For that matter the same is true in the intact animal, when insulin levels are raised artificially. However, in the VHM lesioned rat under normal conditions, the blood insulin does not show a drop prior to the meal. However, to return to Prof. De Wied's question, Strubbe in our group made some preliminary observations on the effect of insulin injected directly into the VMH, but all I can say about the results is that if there is any effect it is a slight one. It seems better to postpone the interpretation of this till the data are complete.

ISSIDORIDES: I would like to draw your attention to another possibility for cells to respond to glucose. They may have a genetic resemblance to yeast. If you add glucose you suppress oxidative metabolism, and the cell changes from an aerobic to an anaerobic metabolism. Complete suppression of the oxidative metabolism would set off the sodium pump. Possibly there could be such cells in the lateral hypothalamus.

DE WIED: After this discussion I don't understand how an alloxan diabetic animal can regulate its feeding behavior, because there is practically no insulin in these animals.

DE RUITER: I believe that's right, but we never determined the insulin levels. There are, however, also other satiety signals.

HERBERT: If an animal is in what we call a high motivational state, it means that the animal is likely to respond to a particular class of sensory input. This means two things: first of all, it will behave in such a way that other classes of input are reduced, and secondly it will behave in such a way that other afferent inputs will become added to the initial one. This is possibly a mechanism by which behavior is stabilized.

DE WIED: The place which you stimulate, is this the lateral or ventromedial hypothalamus? And is the stimulation unilateral or bilateral?

DE RUITER: We performed unilateral stimulation of the ventromedial hypothalamus, in other words, it was complete nonsense that we offered the brain!

BOHUS: What were the stimulus parameters for the unilateral stimulation?

DE RUITER: It was a mild stimulation of 10–15 μA.

Subject Index

DATE DUE

JUN 1 '8[illegible]			
DEC 12 '8[illegible]			

Integrative hypothalmic activity / edited by D. F. Swaab and J. P. Schade. -- Amsterdam ; New York : Elsevier Scientific Pub. Co., 1974.
xi, 516 p. : ill. ; 26 cm. -- (Progress in brain research ; v. 41)
Ne74-45

Includes bibliographical references and index.

ISBN 0-444-41239-5

1. Hypothalamus--Congresses. I. Swaab, D. F. II. Schadé, J. P. III. Series.